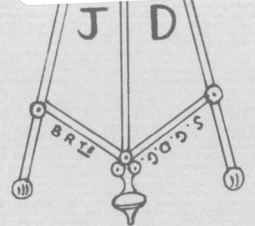

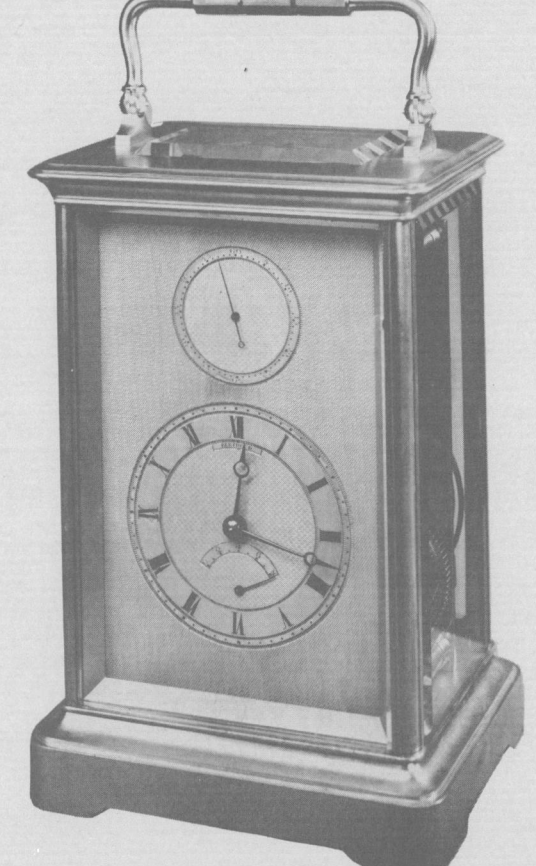

CARRIAGE CLOCKS

Their History

&

Development

By

Charles Allix

Illustrated by

Peter Bonnert

ANTIQUE COLLECTORS' CLUB

© Copyright 1974 Charles Allix and Peter Bonnert
World copyright reserved

Reprinted 1981
Reprinted 1989

ISBN 0 902028 25 1

All rights reserved. No part of this publication may be reproduced, stored in a retrieval system, or transmitted in any form or by any means electronic, mechanical, photocopying, recording or otherwise except for the quotation of brief passages in criticism.

Published for the Antique Collectors' Club
by the Antique Collectors' Club Ltd.

Printed in England by Antique Collectors' Club Ltd.
Church Street, Woodbridge, Suffolk

Contents

		Page
	INTRODUCTION	iii
	PREFACE	iv
	ACKNOWLEDGEMENTS	v

Progenitors of Carriage Clocks

I	**THE FIRST SPRING-DRIVEN CLOCKS**	3

Knight, Death and the Devil. Mainsprings. Fusees and Stackfreeds. Louis XI's Travelling Clock. Drum-Shaped Clocks. Hans Sachs. Table Clocks with Travelling Boxes. Tabernacle Clocks. Coach Watches. Small Travelling Bracket Clocks. Paulet. Mudge. Ferdinand Berthoud. Marine Chronometers. Pendules d'Officier. Capucines. Miscellaneous Small Travelling Clocks. "Sedan" Clocks. Mail-Guards' Watches.

French Carriage Clocks

II	**THE BEGINNING, LATE 18TH AND EARLY 19TH CENTURY**	35

The Breguet Family and Their Successors. Paul Garnier. Exhibitions. Leroy. Other Early Makers. Horlogers du Roi.

III	**LATER MANUFACTURE, 19TH AND 20TH CENTURY**	81

Background. The Strange Organisation of the Manufacture of French Carriage Clocks.

IIII	**SAINT-NICOLAS-D'ALIERMONT, PAST AND PRESENT**	87

Croutte and Papin. Pons. Delépine & Cauchy. Dessiaux, Martin, Sauteur, Villon, Baveux, Guignon, Couaillet, Duverdrey & Bloquel, Bayard, Holingue, Dumas and Others. French Craftsmen in Russia.

V	**PARIS TODAY AND YESTERDAY**	103

Breguet, Daniels, Pitou. Garnier, Bourdin, Oudin-Charpentier, Lefranc, Chartier, Boseet, Berthoud. Jacot, Drocourt, Margaine, Le Roy/Leroy.

VI	**THE CARRIAGE CLOCK INDUSTRY IN THE FRANCHE-COMTE**	127

The Forges and Foundries of the Doubs. The Country of Montbéliard. Japy at Beaucourt and Badevel. L'Epée of Sainte-Suzanne. Carriage Clock Makers in the Jura.

VII	**CASE STYLES. THE NAMES, SHAPES AND SIZES OF STANDARD FRENCH CARRIAGE CLOCKS**	155

The "One-Piece" Case. The "Multi-Piece" Case. Later Case Shapes. Distinctive Decorative Styles. Minor Case Styles. Decorative Panels. Carriage Clock Sizes. Conclusions. Carriage Clock Prices.

VIII	**THE RARER FRENCH CARRIAGE CLOCKS**	191

Alarm Work. Striking Work. Carriage Clock Trains. Calendar Work. Digital Dials. "Flick" Clocks. Unusual Escapements. Tourbillons and Karrusels. Four-Dial and Two-Dial Clocks. Bottom-Wind Clocks. Singing Bird Clocks. Musical Clocks. Sundial Clocks, etc. Year Carriage Clocks.

English Carriage Clocks

IX **ENGLISH WORK AND WORKMEN** — 229
Introduction. Difficulty of Establishing Dates. Carriage Clock Production. Some Celebrated Workmen. The Coles. B.L. Vulliamy. J.R. Arnold. The Dents. Thomas Earnshaw II. Charles Frodsham and Family. Webster. Whitelaw. Barwise. Desbois. Thwaites & Reed. Barraud and Successors. Sir John Bennett. James McCabe and Family. The MacDowall Family. Victor Kullberg. Usher & Cole. Bridgman & Brindle. S. Smith & Son. Jump. C.R. Hinton. Philip Thornton. Col. Quill and J.S. Godman. Nicole, Nielsen. Thomas Mercer. Other Makers. French Fear of English Competition. Huber.

Other Carriage Clocks

X **SWISS CARRIAGE CLOCKS** — 305
The Courvoisier Family and Other Early Makers. Later Clocks sold by Bautte, Moulinié and Henry Capt. The Mathey-Tissot Montres Pendulettes de Voyage of circa 1900. The Stauffer Family. D. Elffroth's Clock.

XI **GERMANY AND THE AUSTRO-HUNGARIAN EMPIRE** — 323
Vienna's Cultural Influence upon Surrounding Territories. Small Austrian and German Travelling Clocks. Viennese Carriage Clocks. Later German Alarms.

XII **AMERICAN CARRIAGE CLOCKS** — 339
Introduction. Waterbury Clock Company. Waterbury Watch Company. Ansonia Clock Company. Jerome, New Haven, Welch and Seth Thomas. Joseph Eastman. The Boston, Chelsea and Vermont Clock Companies. Fasoldt.

XIII **JAPAN, ITALY AND THE ARGENTINE** — 387
Seikosha of Tokyo. Masetti and Maransei, both of Bologna. One Argentine Clock.

APPENDICES
(a) Paul Garnier — Supplementary Information — 395
(b) Paris *Exposition Universelle* 1889:— — 399
   i) *Clockmaking* by T.D. Wright
   ii) *Watch and Clockmaking* in 1889 by J. Tripplin
(c) *The French Carriage Clock after 1900* — A Personal Account by Maurice A. Pitcher — 404
(d) The Japy Factory at Badevel. — 407
(e) *Nail and Cork* and *Rule of Thumb* by D.S. Torrens, 1938. — 409
(f) La Société Des Horlogers. — 413

GLOSSARY OF TERMS — 415

ALPHABETICAL LIST OF NAMES ASSOCIATED WITH FRENCH AND SWISS CARRIAGE CLOCKS TRADE MARKS — 431

BIBLIOGRAPHY — 453

GENERAL INDEX — 467

INDEX OF PLATES — 480

Introduction

All who can recognise a labour of love when they see one will find in the following pages clear evidence of more hard work, devotion to detail and sheer love of the subject than they have come across for a long time. Considering the great volume of horological literature in existence at the present time, it is most surprising that the whole subject of carriage clocks has until now received so little attention. This is the more surprising in view of the quantities of such clocks which have been produced, and the fact that they are owned and keenly appreciated by both specialist collectors and laymen alike. At last here is a book which will provide answers to many questions in the most pleasant and interesting way.

A.G. and A–M.J. Randall.

Preface

Carriage clocks have a charm and fascination all of their own. Beyond question they are by far the most popular and at the same time the most practical type of "antique" clock which has ever been available. Having made such a bold statement it is necessary to define the term "carriage clock" and to explain how it will be used or varied in this book.

By definition "carriage clock" should mean a portable timekeeper designed for use on a journey, and so it does. For most people, however, "carriage clock" means simply a French carriage clock; one of those small balance-controlled pieces housed in a rectangular brass-and-glass case with top carrying handle and with an outer travelling box covered in real or imitation morocco leather. This definition will do very well for a start.

The first chapter is devoted to clocks and watches, not carriage clocks in their developed sense, yet pieces from which they may have been derived. Among such must be numbered not only coach watches, *pendules d'officier, Capucines* and perhaps even the so-called "sedan" clocks, but also very many portable pieces from an altogether earlier era. The only arbitrary restriction which has seemed necessary has been to try to exclude any travelling piece not controlled by a balance and which, therefore, could not tell the time during a journey.

The reasons why French carriage clocks are so immensely popular today are not far to seek. Firstly, although decorative, old and interesting, they are usually entirely practical as timekeepers, going for eight days and affording a reasonable degree of accuracy. Secondly, carriage clocks look completely at home in almost any setting, however modern or period. Finally, carriage clocks offer both aesthetic and antiquarian interest while becoming more valuable with every year that passes.

Strange though it may seem, no book has previously been written on the subject of carriage clocks. The present offering is intended to be at least a start. It does not pretend to be either exhaustive or infallible. By far the greater part of the work has depended upon research done at first hand. In not a few cases it has been necessary to choose between the conflicting recollections of living people, brought up in the nurseries of the carriage clock, and the written records of 19th century writers who were often more journalists than horologists.

In conclusion I should like to say that the task of writing the book has been lightened, and also made far more enjoyable, by the collaboration and advice of Peter Bonnert.

C.R.P.A.

Acknowledgements

Apart from the continual co-operation and forbearance of my wife Julia for the best part of four years, this book could never have been written or illustrated without the unstinting help of friends. The first of these was the late Maurice A. Pitcher. In addition to other help and encouragement he and his son Harvey loaned most of the carriage clocks used to illustrate the Chapter on French case styles and which started the book rolling. In France Jean-Claude Sabrier joyfully gave up many days exploring with us in Saint-Nicolas and Paris, helping endlessly with transport and research, especially in connection with Leroy. In England by far the most significant contribution has been made by Eve Cross. Her unflagging interest and constructive suggestions, not to mention her patience and care in preparing the manuscript, leaves both writer and publisher greatly in her debt. Very special thanks are also due to Anthony Randall and to his wife Anne-Marie who together read through the entire manuscript, providing invaluable criticisms and corrections. George Daniels was good enough to read the parts relating to Breguet, while Guiseppe Brusa and Michael Worsley both checked the Progenitors Chapter. Almost all the illustrations and descriptions of early Austrian clocks are due to the generosity of Dr. Hans von Bertele. Ted Crom kindly read the Chapter on American clocks.

There are, however, a number of other people who have taken so much trouble to provide material of use in producing the book that they must be singled out even at risk of embarrassing them. In France must be mentioned Madame Henry L'Epée and Henry Belmont who provided valuable leads to Japy and L'Epée history. Madame Jeanne Lejoille-Grard did the same in connection with several makers in Saint-Nicolas-d'Aliermont.

In America special thanks are due to Al Odmark, to his wife Gertrude, and to Bryson Moore, Phil Kohl, Walter E. Mutz, Samuel J. Kutner, Herschel B. Burt, Robert G. Spence and J. Hagen Antiques.

Thanks are also due to private and trade owners who supplied most of the photographs used to illustrate the English Chapter. Both the Progenitors Chapter and the American Chapter are illustrated largely with photographs begged, borrowed or commissioned from many different sources. It would also have been difficult to illustrate enough Breguet clocks without the help of Sotheby's and Christie's. The names of various people whose photographs have been used are acknowledged in the captions, but many wish to remain anonymous. Thanks are also due to those who sent illustrations which it has not proved possible to use.

Individuals, firms and institutions who have helped in one way or another include:—

Peter Ashworth.
Seth Atwood.
P.A. Borel.
Mrs. E.W. Bovill.
Marian Brock.
Eric Bruton.
Daniel Patrick Buckney.
J.W. Coe.
René and Claude Couaillet.
Ernest P. Conover Jr.
J. Campbell.
J.E. Coleman.
M. & Madame Alfred Delabarre.
Paul Denis.
Donald De Carle.
Dr. A.B. Dickie.
David Evans (Birmingham).
R.K. Foulkes.
Henry Fried.
Howard M. Fitch.
C.H. Gibbs-Smith.
Dr. E. Gschwind.
John Hawkins.
The Lord Harris.
G.L. Harvey.
B.L. Hurst.
R.W. Husher.
André Imbert.
Albert Japy.
E. Jurmann.
Henri Lengellé ("Tardy").
Harold Malies.

Col. N. McCallum.
Dr. Vaudrey Mercer.
Luis Monreal y Tejada.
Lisa Minoprio.
Robert Millspaugh.
M. Pitou.
Edouard Péquignot.
Col. H. Quill.
Jean-Louis Roehrich.
Gordon Roach.
Col. A. Simoni.
Madame de Tévray.
Charles Terwilliger.
J. Varipati.
Mrs. H. Viner.
T. White.
D.O. Wyatt.
H.C. Young.

The Lord Chamberlain's Office (Geoffrey de Bellaigue).
The British Museum (Beresford Hutchinson).
The National Maritime Museum (Lt. Cdr. H.D. Howse).
The Guildhall Library, London (J.L. Howgago, J.F. Bromley, D.D. Daw, I.F. Maxted).
Christie, Manson and Woods (Simon Bull).
Sotheby & Co. (Michael Webb, Tina Miller).
Sotheby's Belgravia (Philippe Garner).
E. Dent & Co. Ltd. (K.A.G. Butcher).
Alfred Stradling, Cirencester, Ltd. (J.F. Callaghan).
The Clockmakers' Company.
Victoria and Albert Museum.
The French Embassy, London.
The British Horological Institute Library (O.S. Janson).
The French Institute Library, London.
The Science Museum (V.K. Chew).

Fitzwilliam Museum, Cambridge.
The Ashmoleon Museum, Oxford.
Meyrick Neilson of Tetbury Ltd.
Charles Frodsham & Co. Ltd. (F.L. Thirkell).
Garrard & Co. Ltd. (R.E. Smith).
Evans & Evans, Alresford.
Graus Antiques.
Camerer Cuss & Co.
Algernon Asprey Ltd.
Asprey & Co. Ltd.
The Society of Antiquaries of London.
Bibliothéque Nationale, Paris.
Musée Internationale D'Horlogerie, La Chaux-de-Fonds.
Réveils Bayard (Edmond Forest).
L'Ecole D'Horlogerie, La Chaux-de-Fonds (Samuel Guye).
Kugel, Paris.
Caisse Nationale des Monuments Historiques.
Chateau des Monts, Le Locle.
Leroy, Paris (Pierre and Phillipe Leroy).
Mathey-Tissot (Etienne Ch. Mathey).
Société Belfortaine de Mécanographie (M. Peugeot).
Département de La Seine Maritime, Archives Départementales.
Institut Français du Royaume-Uni.
Mairie, Saint-Nicolas-d'Aliermont (Paul Caron).
The Metropolitan Museum of Art, New York.
The Swiss National Museum, Zurich.

It is a pleasure to make these acknowledgements. If there has been left out the name of anyone who should have been mentioned, then please will he or she accept this assurance that the omission has not been intentional.

C.R.P.A.

Progenitors of Carriage Clocks

Chapter I

The First Spring-Driven Clocks

Plate I/1. *ALBRECHT DURER. Engraving entitled* Knight, Death and the Devil. *Date 1513. (The hour-glass symbolises the inexorable flow of Time and is emblematic of Vanity.)*

I
The First Spring-Driven Clocks

Knight, Death and the Devil. Mainsprings. Fusees and Stackfreeds. Louis XI's Travelling-Clock. Drum-Shaped Clocks. Hans Sachs. Table Clocks with Travelling Boxes. Tabernacle Clocks. Coach Watches. Small Travelling Bracket Clocks. Paulet. Mudge. Ferdinand Berthoud. Marine Chronometers. Pendules d'Officier. Capucines. Miscellaneous Small Travelling Clocks. "Sedan" Clocks. Mail-Guards' Watches.

Durer's famous engraving, *Knight, Death and the Devil,* is a good point from which to begin this book. The artist depicts a Christian Knight, as conceived by Erasmus, "passing resolutely through the terrors of life and undismayed by the prospect of death". The date is 1513. A large hour-glass with a dial is prominent in the picture. Durer (1471-1528) was born, lived and worked in Nuremberg in central Germany, one of the first cradles of horology. Plate I/1 reproduces his engraving. The prominent hour-glass is appropriate and also symbolic in the context; but it is certain that Durer by 1513 had seen spring-driven travelling clocks.

The history of early spring-driven clocks is far from being fully documented. New researches and discoveries are constantly being made. At present no one even knows for certain either the date or the country of origin of the first examples, although there are grounds for belief that they may have come from Northern Italy. Unfortunately also, the last word has yet to be written upon the origin of the mainspring, an invention without which no portable clockwork could initially have been made. The earliest reference to a mainspring, about which there could be no argument, is that quoted by Col. Simoni of Bologna, one of the great authorities on early Italian clockwork. Simoni speaks of a letter written by an engineer, Comino da Pontevico which refers in 1482 to "...a ribbon of tempered steel fastened in a brass barrel round which is wound a gut line.... so that it has to pull the fusee". This is a very early reference, not only to a mainspring, but also to a fusee. Baillie in *Watches,* 1929, refers to the preface of a sonnet written by Gaspare Visconti in 1493, which translates "There are made certain small portable clocks which, though with little mechanism, keep going... striking the proper time...". While no early Italian spring-driven clock appears to have survived, there certainly seems to be every reason for believing that examples existed at least during the last twenty years of the 15th century. Italian horology, having influenced the world, apparently was eclipsed gradually after about 1500, Germany and later France thereafter leading Europe. Peter Henlein (1480-1542) working in Nuremberg, and reputed to have been a locksmith, was the earliest known German in the field of spring-clocks. None of his work has survived; but in 1511 J. Cocclaeus mentioned him as "...a young man, makes things which astonish the most learned mathematicians, for he makes out of a small quantity

Plate I/2. *THE JACOB ZECH CLOCK. Date circa 1525. Probably the earliest complete surviving spring-driven clock. (By courtesy of the Society of Antiquaries, London)*

I The First Spring-Driven Clocks

of iron horologia devised with very many wheels, and these horologia, in any position and without any weights, both indicate and strike for forty hours even when they are carried on the breast or in the purse."[1] Plate I/2 shows the universally-illustrated Jacob Zech clock dated 1525. This piece, which was made in Prague, was long regarded as being the earliest spring-driven clock.[2] Plate I/3 shows the movement. Note the central bar-balance and also the fusee and mainspring barrel. The clocks made by Henlein are supposed to have had stackfreeds. The excessively dangerous question of whether stackfreeds were made before fusees is perhaps best left aside here, and more especially in view of the new material mentioned in Footnote 2.

The earliest recorded reference to a travelling-clock as such is almost certainly to that supposed to have been made for King Louis XI of France (1423-1483). The King is said to have carried the clock with him whenever he travelled. Tradition has handed down the name of Jean de Paris as the maker and several writers have quoted the year of its delivery as being 1480.[3] Portal and de Graffigny in *Les Merveilles de l' Horlogerie*, 1888, refer on pages 127-128 to an anecdote recorded by an old French author, Antoine Duverdier. This narrative is said to have been based upon contemporary writings of Louis himself, and it not only makes plain that the clock struck but also that it was small enough to be hidden in the sleeve of a garment.[4] Unless and until the whole account is

Plate I/3. *THE MOVEMENT OF THE JACOB ZECH CLOCK. Note the exceedingly early fusee. (By courtesy of the Society of Antiquaries, London)*

proved to be a fabrication, then it must be accepted as proving that Louis XI owned a clock which was not weight-driven. In any case, there seems little doubt that the closing years of the 15th century, if not the actual year 1480, saw the first appearance of clocks driven by springs, and thus for the first time made portable while going.[5] Plate I/4 depicts a very delight-

---

(1) See Baillie *Clocks and Watches*, 1951, and Britten's *Old Clocks and Watches and Their Makers*, 1956, p.19.

(2) It is illustrated here because it is so fundamental in the context of early spring-driven clocks. Zech's long-held position of priority is, however, constantly being challenged. Not long ago Dr. Zinner found in the Bavarian National Museum at Munich the remains of a clock dated 1509, and having a fusee. The Brussels Manuscript must also be taken into account. It illustrates a manuscript of the *Horloge de Sapience* dating from between 1451 and 1488. It shows, amongst other instruments for time-measurement, a spring-driven table clock with plated movement having a fusee. Evidence has also been advanced by Prager (see Bib.) that Fillippo Brunelleschi (1377-1446), the famous Florentine Renaissance architect, may have devised clocks driven by helical springs (drawn out into tension and used in conjunction with fusees). This new information, if finally proved, might establish the dates of both mainsprings and of fusees as earlier than is at present accepted. The position now is that the Zech clock, which is at the Society of Antiquaries in London, is still the earliest known complete and dated clock having a fusee. Count Lamberti of Rome has a drum-shaped fusee driven clock which is possibly even earlier.

(3) For instance, Ernst Zinner in *Aus der Frühzeit der Räderuhr*, 1954, and Clutton and Daniels in *Watches*, 1965 and 1971.

(4) According to Portal and de Graffigny, the clock stood on the corner of a table in a room where Louis gambled with his courtiers. It was stolen by a Baron who had "lost his shirt". The unsuccessful gambler concealed the clock in his sleeve; but it proceeded to strike so loudly and insistently that the confused Baron was obliged to confess what he had done. The King, although not normally of a kind disposition, was so amused by the incident that he gave the clock to the Baron. Portraits exist of Louis XI beside a table on which stands a small clock. This fact alone would appear to lend some credence to the anecdote of Duverdier, even if (as suggested in 1949 by Gelis in *Horlogerie Ancienne*) the portraits are posthumous. Pietro Aretino reported the same anecdote in 1543. It is not clear, however, whether the King concerned was really Louis XI or Louis XII. The latter reigned at the beginning of the 16th century.

(5) Much research on the subject of early clocks has been made since the war by Col. Simoni of Bologna and by Prof. Morpurgo of Amsterdam, besides by others. The article entitled *Brunelleschi's Clock?* by Frank D. Prager of Abington, Pennsylvania is of outstanding interest in its own right, besides containing references to much important literature on the subject of early spring clocks (see Bib.).

I The First Spring-Driven Clocks

ful 18th century representation of mainspring-making. The methods used in the 15th century would have been no different. It must be emphasised that escapements controlled by balances[6] were in existence by a far earlier date, so that the first of the two inventions necessary to make possible a travelling clock had already been made. Early clocks were, however, very erratic timekeepers. It was not until the advent of the first satisfactory pendulum clock in December 1656 and until the general adoption of balance-springs for balance-controlled pieces from about 1676, that any significant improvement was made. Many readers are no doubt under the very natural impression that the first clocks had pendulums; but nothing could be further from the truth. While undoubtedly Galileo (1564-1642), but *not* as long supposed Leonardo (1452-1519), thought of the use of pendulums in conjunction with clockwork, possibly as a means of achieving automatic counters, it was left to Christaan Huygens (1629-1695) and to Salomon Coster (died 1659), working in the Haague, to develop and construct the first satisfactory pendulum clocks.[7]

The Almanus Manuscript[8] which records the practical side of mediaeval clockwork as noted in Rome by the German Brother Paulus Almanus, during about the years 1475 to 1485, illustrates no less than eight spring-driven clocks; but apart from this and other evidence it is certain that as soon as clocks, anywhere, were made capable of going without weights they became immediately, potentially at least, "travelling clocks". Once this fact is appreciated, and especially in view of what has been said about the still-evolving history of early spring-driven pieces, it becomes apparent that it would be more than unwise to attempt to attribute the invention or evolution of portable clocks to any one clock-making city, country or tradition. Almanus, who according to Zinner was

Plate I/4. MAKING MAINSPRINGS FOR CLOCKS. *This illustration appeared originally in Diderot and D'Alembert's Encyclopédie published between 1751 and 1780, but the method of spring-making would have been no different in the 15th century.*

---

(6) See Beeson, *English Church Clocks, 1280-1850*, 1971, p.12. The writer says that historians are now generally agreed that the verge-and-foliot escapement was invented in the second half of the 13th century, and possibly even before the year A.D. 1283. However, evidence seems to be coming forward that an escapement illustrated by Leonardo is even older, and that Richard of Wallingford in England may have used something similar. A letter in *Antiquarian Horology*, Dec. 1971 (G. Brusa) illustrates Leonardo's escapement and gives some interim comment upon what promises to be a fascinating new chapter of horological history.

(7) See Lee *The First Twelve Years of the English Pendulum Clock ... 1658-1670*, 1969. Drawings dating from the beginning of the 17th century illustrated by Prof. Morpurgo, a great authority on early clockwork, show pendulum clocks; but there is no evidence that they were necessarily ever made. Morpurgo also refers to a small Italian pendulum clock made by Camerini dated 1656, and now displayed in the Science Museum, London. This very interesting piece remains to some extent an enigma, but almost certainly it has been converted from a balance-controlled escapement.

(8) The Almanus Manuscript (see Leopold in Bib.) of circa 1475-85 was "discovered" about thirty years ago in the Augsburg City Library. This MS. contains what has long been supposed to be the earliest known drawing of a fusee. Leonardo's sketches, which also show several fusees, have the approximate date 1490-1500. Now, however, it may be that sketches believed to date from the era of Brunelleschi, i.e. prior to 1446, and illustrating fusees, may prove to be the earliest so far discovered.

I The First Spring-Driven Clocks

of South German origin, almost certainly saw fusees in his own country before ever he journeyed to Rome. The Almanus MS. is supposed to date from 1475-85 (Zinner says 1477-78) and Leonardo's sketches are usually said to have been made between 1490-1500. The fact that Leonardo depicts fusees, even though they would almost certainly have been known before his time, does not prove that Almanus first saw them in Italy. Also very important in this connection is the fact that Leonardo, according to C.H. Gibbs-Smith[9] did not publish his notebooks and sketches during his lifetime. Instead Leonardo's friend Francesco Melzi "sat" on these, without making them public, from 1519 until his own death in 1570. Even after this date, little of Leonardo's scientific work was known to the world until Napoleon looted the bulk of the material from Italy; after which the French scholar J.B. Venturi was able, in 1797, to publish his *Essai sur les Ouvrages Physico-Mathématiques de Leonardo de Vinci*. Whatever the facts, the point of immediate concern is that at least some 16th century spring-driven clocks were provided with fitted travelling-boxes, irrespective of whether they came from Italy, Nuremberg, Prague, Augsburg or from France. The very existence of a travelling-box in itself made a *pièce de voyage* or in other words a "travelling-clock".

A very interesting early drum-shaped table clock, which may be regarded as another milestone in the progression of travelling pieces, is shown in Plate I/5. The clock would have been made in the middle of the 16th century. It is described and illustrated in Sotheby's Catalogue, 19th October 1954 (Webster Sale, Second Portion). Unlike the Zech clock, this example has a gilt metal case innocent of all engraving though with restrained mouldings top and bottom. The movement is made entirely of iron, having a tall early fusee and a large balance with no spring. The upward-facing dial has arabic numerals and is centred by an astrolabe made for the latitude of Toledo, suggesting that either the clock was made in Spain or else was intended for export there. A further feature of particular interest is the original leather case in a beautiful state of preservation, complete with lock and key.

Plate I/6 shows a type of drum-shaped table clock, with detachable alarm, current on the Continent from circa 1540 until perhaps 1600. This particular French example with its original embossed leather case of circa 1590 was in the Webster Collection dispersed

Plate I/5. *GERMAN DRUM-SHAPED TABLE CLOCK. Date circa 1550. (Private collection)*

---

(9) *A History of Flying*, 1953, and *Leonardo da Vinci's Aeronautics*, 1967.

I The First Spring-Driven Clocks

in 1954. The separate alarm attachments were very useful for night use and for travel. These clocks stood about six inches high when their alarms were in place. For the rest of the time they were simply small, attractive tambour clocks about three inches in diameter. The upward-facing dials had single hands and also touch-studs to make it possible to tell the time in the dark. Some early examples had feet. On later clocks ball-and-claw feet are generally found. The alarms, while so conveniently removable when not required, were a most useful, if not essential, adjunct for clocks intended for travel. In the Ilbert Collection in the British Museum is a magnificent piece of the same type, probably also French and with rock-crystal cases for both clock and alarm. It is illustrated by Tait Figs. 29-30 (see Bib.).

By 1568 Hans Sachs (1494-1576) and Jost Amman of Franckfurt-am-Main had published plates and verses dealing with trades.[10] Plate I/7 reproduces the drawing of a clockmaker's shop by Jost Amman. Sachs' verse

Plate I/7. *I MAKE THE CLOCKS FOR TRAVELLERS. By Hans Sachs and Jost Amman, 1568.*

translates, according to Baillie:—
> I make the clocks for travellers,
> Correct and bright, as they should be,
> Of clearest glass and finest sand
> So that they last for many years.
> I make a case and colour it,
> With green and gray and red and blue,
> And from the glasses one can tell,
> The hour and the quarters too.

Although the words refer to hour-glasses, the drawing shows a clockmaker's workroom in full production.

Plate I/8 shows a late 16th century table clock by Roweau of Paris, complete with its travelling case. It is in the Swiss National Museum at Zurich. The maker's punch mark "I. ROWEAU" is stamped boldly on the inner side of the base plate. According to Tardy,

Plate I/6. *EARLY FRENCH TRAVELLING TIMEPIECE WITH DETACHABLE ALARM. Date about 1590. The original leather case is cut with acanthus pattern and with a lion rampant. The alarm is set off by the single (hour) hand of the clock. (By courtesy of Sotheby & Co., London)*

---

(10) Baillie *Clocks and Watches*, 1951, pages 18 and 19.

I The First Spring-Driven Clocks

Plate I/8. *FRENCH TABLE AND TRAVELLING CLOCK. Date circa 1580. Signed "I ROWEAU", Paris. Balance-controlled escapement. Original travelling box and keys. (By courtesy of the Swiss National Museum, Zurich)*

Roweau was a *Maître Horloger* in Paris in 1580. Tardy gave this information to M. Claude Lapaire of the Zurich Museum. The clock has a circular balance, but no balance spring.[11] A fusee of characteristically early shape is incorporated in the going train, as seen in Plate I/9. Note the very important and significant feature of going and striking trains with their arbors planted vertically instead of horizontally, and with the movement arranged in two tiers, one above the other.[12] This is very much a table clock, and yet the beautiful leather carrying-box, with lock and with window to show the dial, proves not only that it was designed to be portable but also to remain in use during journeys. It is not surprising, on reflection, that when spring clocks were both rare and expensive, they were endowed with this ubiquity almost as a matter of course. Even the rather cumbersome European "Tabernacle" clocks, especially the smaller and earlier examples, were given travelling boxes. Very few of these boxes, however, have survived. Plate I/10 shows an admirable example. The clock on the right of the same photograph has been converted to pendulum. "Tabernacle" clocks were made mainly in the last quarter of the 16th century and into the first quarter of the 17th. The smaller examples are usually earlier, while those with many galleries and pinnacles tend to be after 1600. Style depends, to some extent, upon where the clocks were made. The

Plate I/9. *THE MOVEMENT OF THE ROWEAU CLOCK. Notice the two-tier arrangement of the trains and the vertical arbors. (By courtesy of the Swiss National Museum, Zurich)*

---

(11) Balance springs were developed in both Holland and England, appearing in use for the first time in both countries in about 1658. However, another twenty years were to pass before balance springs came into general use.

(12) A clock with very similar mechanical arrangements, but in a hexagonal case with corner pilasters, was in the Webster sale, lot 187. This clock is also French, and the base is stamped "TOVRS". The date is circa 1590. Rare as these clocks are, the British Museum is fortunate in possessing several examples. One of them is signed by "Bartilmewe Newsum" who is supposed to have been a Yorkshireman.

I   The First Spring-Driven Clocks

feet of these clocks are a feature about which no-one would care to be too dogmatic. Easily knocked off or damaged, many feet are undoubtedly replacements. Some clocks may have had no feet at all, while others have buns, claws, lions, mythical animals, and even human heads.

Plates I/11 and I/12 illustrate two other travelling, striking table-clocks which are to be found in the

France, and in England coach watches. They continued to be made until well into the 19th century, and even the great A-L Breguet sold one in 1810. The

Plate I/11. *FLEMISH TABLE AND TRAVELLING CLOCK. Date circa 1590. Signed simply "VALLIN". Note the dial with single hand, visible through the top of the fitted travelling box. (By courtesy of the Musée d'Horlogerie, La Chaux-de-Fonds, Switzerland)*

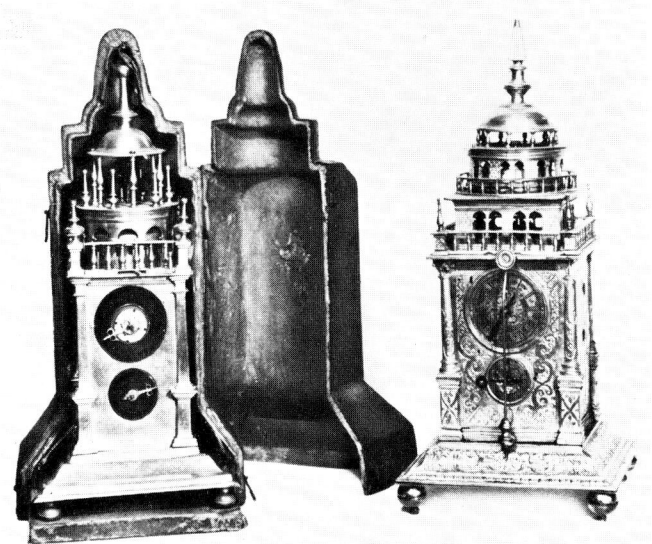

Plate I/10. *"TABERNACLE" CLOCKS. This type was current during the last quarter of the 16th century and into the early years of the 17th. (Private Collection, Milan. Photograph by courtesy of Col. Simoni, Bologna)*

Musée d'Horlogerie at La Chaux-de-Fonds. Both are described aptly by the Museum as "Horloge de table et de voyage avec étui" (travelling table-clock with case). The "square" clock, which would have been made about 1590, bears the name "VALLIN".[13] This maker was born about 1530-35 in Flanders in the town now known as Lille. The "round" clock is signed "Volant à Paris", and the Museum Catalogue says that he was "Master in 1612".

In the 17th century appeared very large travelling watches, often measuring five or six inches across the dials, and with correspondingly thick movements. These pieces were termed *montres de carrosse* in

Plate I/12. *FRENCH TABLE AND TRAVELLING CLOCK. Date end of 16th century. Signed "Volant à Paris". This beautiful piece now lacks its movement. Note hinged lid for seeing the dial on journeys. (By courtesy of the Musée d' Horlogerie, La Chaux-de-Fonds, Switzerland)*

---

(13) A very similar clock from the Schloss Collection was illustrated in the *Horological Journal* of Feb. 1898, page 79, and subsequently by Britten. Some table clocks, especially those made later, were given hexagonal cases.

11

I The First Spring-Driven Clocks

Plate I/13. *FRENCH COACH WATCH. Date c.1620. Note the wonderful engraving and early type of pendant with ring set at right angles to the dial. (Formerly Pérez de Olaguer-Felíu collection. From the book* Relojes Antiguos *by Luis Monreal y Tejada)*

early examples, intended beyond all question for use on the road, almost invariably had alarms. Usually in addition they struck the hours like clocks, or in other words, they were clock-watches. Later not a few were provided with pull-repeats intended for use in the dark. The lovely early coach watch shown in Plate I/13 is French. Its date is circa 1620. It has a number of features characteristic of its origin and period, namely a glorious and complicated engraved dial, split-shouldered bezel, prominent bottom latch, and an early type of bow. It is illustrated by Monreal, whose book is mentioned in the Bibliography. According to Monreal the watch has a single, silver pierced and engraved case, and in addition a leather outer travelling box. The dial engraving, which is attributed to Etienne Delorme, depicts Mathematics, Intelligence, Astrology and Study.[14]

Plates I/14 and I/15 show a particularly fine English coach watch by David Boquet, who died in 1665. It is the property of Her Majesty the Queen and was made about 1650. This watch was given to King William IV by the Countess Howe, and was used as a travelling clock by the King during his journeys. Its earlier history is not known. Note the retention of several features inherited from an earlier period (style of bow, bezel, etc.). Boquet's design, especially as regards the dial has, however, moved forward to reflect changing taste. He continues the early system of placing a single (hour) hand on the alarm ring, giving the alarm setting hand pride of place in the centre of the dial. The time shown by this watch is 3.50 approximately, while the alarm is due to go off a few minutes after 9. The Webster Sale Catalogue for 19th October 1954 illustrated the back of a very similar case, this time belonging to a superb French *montre de carrosse* by Goullons à Paris circa 1640-1660. In this watch the case is silver-gilt, pierced and engraved with sprays of pinks, daffodils, tulips, marguerites, and other garden flowers. The movement has locking plate striking and alarm. Watches much the same were produced by such famous 17th century makers as Edward East, William Knottesford and Tompion in England, and in France by such artists as Jean Baptiste Vallier and Claude Duclair (both in Lyon), G. Gamot of Paris, and by Jacques Taillade and N. Norry at Gisors, all in about the same period.[15] An example of a Tompion coach watch is to be seen in the Metropolitan Museum, New York, and another particularly fine one is at the Fitzwilliam Museum in Cambridge. The Clockmakers' Company Museum at the Guildhall in London has a good example of a Knottesford watch. Two views of

---

(14) See calendar and phase of moon at top of dial. In lower dial note the early system by which the hour pointer is carried on a continuously rotating alarm ring. The alarm setting hand is in the centre of the dial and also rotates continuously. In the photograph the hour pointer and the alarm hand are shown in line; in other words, the alarm is due to go off at twelve o'clock. The right dial sector shows name of month and number of days with the sign of the Zodiac corresponding to the month. The left sector shows weekdays and their symbols.

(15) The Bloch-Pimental collection, sold by auction in Paris at the Hôtel Drouot on 5th May 1961, included choice examples of *montres de carrosse* by the last four makers.

I The First Spring-Driven Clocks

in trying to show minutes on the dial of an erratic timekeeper.[16] From about 1676, however, the introduction of balance-springs transformed the timekeeping properties of balance-controlled pieces out of all recognition almost overnight. The importance of the balance-spring as a "breakthrough" in timekeeping is difficult to exaggerate. Suddenly clocks and watches which previously could not be trusted became capable of keeping time to within perhaps less than five minutes over a period of twenty-four hours. Older existing balance-controlled pieces everywhere had a balance-spring added almost immediately, to the vast improvement of their performances and to the detriment of their originality.[17] Balance-springs were de-

Plate I/14. *ENGLISH COACH WATCH. Date circa 1650. Maker David Boquet who died in 1665. Clock watch striking the hours in passing, and with alarm. (Reproduced by gracious permission of Her Majesty the Queen)*

it are shown in the *Register of Apprentices*, 1931, facing page 172. This Knottesford coach watch of circa 1690 has a calendar dial round the periphery of the bezel, and it is generally a beautiful example of best English late 17th century work. Knottesford was admitted to the Company in 1663 and was Master in 1693.

The two inventions necessary to make possible a portable clock were mentioned earlier in this chapter, namely the balance and the mainspring. Important as these were, they did little to produce a clock capable of keeping time. One hand was more than good enough for nearly two hundred years, and there was no point

Plate I/15. *THE BOQUET WATCH. The beautiful and engraved back of the silver case. (Reproduced by gracious permission of Her Majesty the Queen)*

---

(16) The reader will notice that the dials of one-handed clocks or watches have four, and not five, divisions between each chapter or hour numeral.

(17) Others were converted to pendulum-controlled escapement for precisely the same reason. The pendulum, even more than the balance-spring, represented so fundamental an advance in principle as to render pieces having neither almost completely useless by comparison. There was, of course, no room inside the cases of table or "Tabernacle" clocks for pendulums, so they were often hung in front of the dials.

I The First Spring-Driven Clocks

veloped independently in Holland by Christiaan Huygens (1629-1695), and in England by Robert Hooke (1635-1703) from 1658 if not earlier. In their lifetimes the two men competed with each other as to who had priority of the invention, and historians have competed about it ever since. Readers wishing to study the great controversy will find clues to most of the known facts set down between G.H. Baillie in *Watches* (1929), Clutton and Daniels in *Watches* (1965), and by Alan Lloyd in *Some Outstanding Clocks. . .* (1956). It is not a matter for the present book. Perhaps the recently published Huygens' letters and Hooke's Diary may hold further clues. Baillie's summing up is "Hooke was certainly first to have the conception. Huygens was the first to publish the invention fully to the world, and to have watches made that were successful". It was, of course, no coincidence that pendulums and balance-springs first appeared at almost the same time. The influence exerted by a balance-spring upon a balance, in any position, is strictly analogous to that imposed by gravity upon a vibrating pendulum.

In the 18th and 19th centuries, coach watches tended to become far less practical. Indeed, it cannot be supposed seriously that many of the later pieces were ever really meant to be taken on journeys. A superb English example by William Hughes (circa 1780) is illustrated in Plate I/16. It is nearly 7 inches in diameter, the bezel is set in diamond and ruby pastes, and there is a carillon of six bells played hourly. The back is enamelled with a scene of children and dogs. At the hour, figures pass over a bridge and a waterfall flows. This watch was in the Webster Collection and came from the Summer Palace in Pekin, being taken at the sacking of 1861 from the bedroom of the Emperor of China. These elaborate and decorative coach watches were intended as *tours de force* of the watchmaker's art. Many continental examples,

Plate I/16. *ENGLISH COACH WATCH WITH MUSIC AND AUTOMATA. Date circa 1780. Maker William Hughes, London. Note the beautiful Bilston enamel back showing two children with dogs. (By courtesy of Methuen & Co. From the book* Watches *by G.H. Baillie, 1929)*

I The First Spring-Driven Clocks

especially those made for the Eastern market, had elaborate enamelled cases set with split pearls or semi-precious stones. Others had carillons or music with automaton scenes. Such watches were all very amusing in their way, but they were scarcely intended or fit to be taken seriously as travelling clocks. Another case in point is the superb but also very large and heavy continental silver watch which is illustrated in Plate I/17. This watch strikes the hours and quarters as it

Plate I/18. *MINIATURE ENGLISH TRAVELLING CLOCK. Date circa 1710. Signed "WATSON, LONDON" on the dial, and on the back plate "S. Watson, LONDON". Very rare. Small dial in arch is for setting alarm. Dial at Chapter XI is for regulation. (By courtesy of Sotheby & Co., London).*

Plate I/17. *CONTINENTAL COACH WATCH. Date probably late 18th century or early 19th century. Maker Gautrin, Paris. Hour and quarter striking with alarm. The escapement of this watch is most interesting. It is a cylinder but it is arranged to work in two different planes like a verge. Saunier illustrates a similar arrangement in the* Revue Chronométrique *of August 1872. The watch was sold at Sotheby's on 28th October 1963. (By courtesy of Georges Baptiste, Brussels).*

goes, and repeats hours. The striking work is so complicated that it is constantly giving trouble, even when the watch is kept on a stand and handled with immense care by a knowledgeable owner.[18] The escapement is a cylinder arranged to work in two planes like a verge. The balance is planted in cocks at the bottom of the movement, and was so placed with the clear intention that the watch should always be kept upright (as on a stand).

---

(18) This watch is wound by a left-hand turning tipsy-key. If an ordinary key is used, and the going train winding square is turned in the wrong direction, then the watch is damaged. In addition, it is necessary to remember to wind the striking train first. Meticulous attention to quite complicated winding and hand-setting procedures is essential if the mechanism is not to become deranged.

15

I The First Spring-Driven Clocks

In England at least, coach watches were overlapped in the period just after 1700 by very small travelling bracket clocks usually with alarms and having escapements controlled by balances. These clocks, which are very rare, were mostly made in the form of the full-sized wooden-cased domestic clocks of the day, suitably scaled down. Plate I/18 shows a front view of a beautiful example produced by the important maker Samuel Watson of Coventry and London. It stands only 7¾ inches high, but with its classical proportions and superb case veneered in ebony, the reduction in size is achieved with great distinction and elegance. Plate I/19 shows the movement, in which the same meticulous sense of balance is maintained. There is a pull-wind for the alarm train and also a pull-wound quarter-repeating mechanism. This clock was sold by Sotheby's in London on 15th November 1971. It may well be the clock mentioned by Britten in the sixth edition of *Old Clocks and Watches and Their Makers* on page 855. Before leaving Watson, it is worth looking at the photographs again, if only to decide how he has scaled down a full-sized clock so successfully. Attempts in this direction usually look quite wrong. Even rarer than the English miniature wood-cased clocks are those in metal cases.

Tompion made about three small clocks with composite escapements, offering both pendulum and balance, either one of which could be selected at will. One of these clocks which has a firegilt brass case was described and illustrated by R.K. Foulkes in *Antiquarian Horology,* March 1958. Its date is about 1700. A similar clock, dated 1693, is illustrated by Symonds in *Thomas Tompion, His Life and Work,* 1951 *and* 1969.[19] H. Alan Lloyd in *The Collectors' Dictionary of Clocks,* 1964 and 1969, shows a further example, this time having a wooden case. Both authors show plates of the escapements. It is tempting to believe that these clocks were intended for travelling, although Tompion made one full-size clock of the same type and having a wooden case. The reader will remember a resolve made in the Preface of this book to try to exclude any "travelling" clocks which did not have balances. This point is mentioned again in view of the necessity of now having to omit several delectable miniature Tompion bracket clocks, which, however, have pendulum-controlled escapements; to say nothing of a George Graham lantern clock provided with a fitted travel-box, a Quare 30-hour brass clock with anchor-shaped pendulum, and another Quare, this time a bracket clock but with articulated ball and socket pendant. In the Webster Collection was an important mid-18th century miniature pendulum bracket clock, signed "Peckover, London", and having a gilt-metal case. Later the same clock was in the S.E. Prestige Collection, and it is illustrated on page 37 of Sotheby's Auction Catalogue for

Plate I/19. *MOVEMENT OF THE SAM WATSON CLOCK. This clock has a verge escapement. Note the balance above the movement. (By courtesy of Sotheby & Co., London).*

---

(19) Apparently the only known dated Tompion clock.

I The First Spring-Driven Clocks

29th April, 1968. Other clocks which we should have liked to have illustrated, besides numerous continental pendulum travelling clocks, include the Roman-striking Thomas Mudge clock, mentioned by Thomas Reid in his *Treatise,* and which General Clerk is supposed to have taken with him on campaigns. Mr. R.K. Foulkes has seen a small balance clock by Harris, and there was certainly a 30-hour clock of similar intent made by Daniel Delander shown in the British Clockmakers' Heritage Exhibition in 1952. The case was made of padouk (wood) according to the catalogue, and the glass-covered balance was planted on the back plate of the movement.

The opening years of the 18th century saw in England yet another type of small travelling clock. Plate I/20 illustrates a fine example made by Paulet and preserved in the Victoria and Albert Museum since 1869. Several clocks of this type have survived. A similar Paulet clock is mentioned and illustrated in the 6th edition of Britten *Old Clocks and Watches. . . 1932,* and H. Alan Lloyd shows yet another in his *Collectors' Dictionary of Clocks,* 1964. These pieces, which are decidedly continental in feeling, stand about 9½ inches tall including their extended pendant rings. The cases, either of gilt metal or of silver, have break-arch tops. The whole of the back plates, as well as the spandrel areas of the dials, are overlaid with cast pierced and engraved strapwork in silver. Britten in *Old English Clocks,* 1907, illustrates in Fig. 44 a normal-enough looking English long case clock in marquetry to which he gives the date 1695. The maker is "John Paulet, London".

Britten also illustrates in Fig. 642 of *Old Clocks and Watches. . . 1932,* what is described as the "Carriage Clock of Marie Antoinette", said to have been made by Robert Robin (1742-1809). The piece is still in the Victoria and Albert Museum in the Jones Collection. It is also illustrated by Tardy, together with a similar clock, in *La Pendule Française,* Part II, pp.303 and 304. The Victoria and Albert clock has a leather travelling box bearing a Royal cypher. It may well originally have been a pendulum clock. Today it has a fairly early *pendule Paris* movement, apparently converted to accept a platform escapement. A clock rather similar in appearance was sold at Sotheby's on 24th April, 1972, Lot 62, and carried the name "Lepaute à Paris". According to the Catalogue, the case was mounted with white porcelain panels decorated with gilt Royal monograms and crossed palms. The *pendule Paris* movement had a platform escapement; but a recess in the bottom of the case betrayed that originally it had a pendulum.

Plate I/20. *PAULET, LONDON, TRAVELLING CLOCK. Date circa 1720. This clock has a verge escapement with the balance on the back movement plate, and it strikes hours in passing and repeats hours and quarters. Note the central dial for setting the alarm and the top pointer for the calendar. Although these Paulet travelling clocks are rare, several examples have survived. Some have pierced gilt metal cases and some silver. (By courtesy of the Victoria & Albert Museum, London).*

I The First Spring-Driven Clocks

Plate I/21. *ENGLISH TRAVELLING CLOCK BY THOMAS MUDGE. Date probably prior to 1769. This clock has a very early lever escapement. (By courtesy of the Lord Polwarth)*

Readers will be aware of the importance of Thomas Mudge (1715-1794) both in connection with the development of timekeepers to enable navigators to find longitude at sea, and also as the inventor of the lever escapement which was the forbear of those used universally today in most clocks and watches.[20] The beautiful small travelling clock, belonging to Lord Polwarth, shown in Plate I/21 is very important in its own right. It is signed "THO. MUDGE, LONDON", and it has a lever escapement which, in the opinion of Mr. Richard Good[21] is earlier than that in the world-famous "Queen Charlotte's watch". This watch, with a Mudge lever escapement and a date of 1769 has long been considered the earliest detached lever escapement.[22] Richard Good bases his view upon the fact that the escapement of the Polwarth clock is less developed than that in the "Queen's Watch". The Polwarth clock is described briefly by Col. H. Quill in *Pioneers of Precision Timekeeping,* where some of its provenance is given. A top view of the movement of this extraordinary clock is shown in Plate I/22. It is to be hoped that some day Mr. Good will write a full des-

Plate I/22. *THE MUDGE CLOCK. A top view of the movement showing the balance and part of its compensation and regulating devices. (By courtesy of the Lord Polwarth.)*

---

(20) Almost any general knowledge horological book contains the basic information; although, in fact, this is a complicated subject.
(21) Mr. Good is the author of several articles on Mudge. See Bibliography. He has worked upon both the "Queen's Watch" and the Polwarth clock.
(22) The "Queen's Watch" was long thought (and widely quoted, including by Allix in 1950) to have a date letter on the silver case for 1759. In about 1965 the 1759 date, which is illogical, was questioned; and the date was then established by experts to be 1769. Since that time a letter has come into our hands dated 10th Dec. 1929 and written by the Assay Office at the Goldsmith's Hall in London. It states firmly the opinion of the writer, A.D. Bishop, that the date letter on the watch case is for the year 1769!

I The First Spring-Driven Clocks

cription eventually. A number of very excellent photographs already exist, but have never been published. The late Philip Coole and Mr. Beresford Hutchinson of the British Museum examined the piece in 1969. They discovered that the escapement and balance assembly had been repeatedly modified, almost certainly by Mudge himself as he proceeded in the development of the final form.

In the old Observatory at the National Maritime Museum at Greenwich is preserved a very unusual French *pendule portative* having very much the appearance of a *pendule de voyage*. The clock was made by Ferdinand Berthoud in 1795, and although unsigned by him it is in fact his No. 61. The piece now bears Breguet's name and his No. 11. It is shown in Plate I/23, and a side view of the movement is in Plate I/24. This clock is fully described by Berthoud in his *Histoire de la Mesure du Temps*, 1802 under the heading *Horloges Astronomiques Perfectionnées*. Berthoud says specifically that he made the clock at the beginning of 1795, so there is no question whatever about the date.[23] According to

Plate I/23. *ASTRONOMER'S PORTABLE CLOCK. Made in 1795 by Ferdinand Berthoud. It is his No. 61. This piece also has some connection with A-L Breguet. (By courtesy of National Maritime Museum, Greenwich.)*

Plate I/24. *BERTHOUD'S ASTRONOMER'S CLOCK. A side view of the movement. (By courtesy of the National Maritime Museum, Greenwich.)*

---

(23) "C'est d'après les observations qui précèdent, que j'ai (FERDINAND BERTHOUD) construit l'horloge portative à balancier, No. 61, exécutée au commencement de 1795, et représentée planche XV, fig. I."

19

I The First Spring-Driven Clocks

Berthoud the clock takes the place of an astronomical regulator ("tient lieu d'une horloge astronomique à pendule"). He goes on to speak of the piece in the context of its being moved from one place to another in a carriage. The clock was acquired by the Museum in 1937 from Sir James Caird. Previously it was found by Captain Jauncey, a notable and a knowledgeable collector. The value in 1937 was £37. The clock stands 7½ inches tall with the handle up, and is approximately 4½ inches wide by 3¾ inches deep. The brass and glass case has bun feet and a folding top handle for carrying. There is a front door. The button shown at the bottom left hand side of the clock is a stop device acting directly upon the balance. The balance and its staff are run horizontally on relatively very large "anti-friction" wheels, two at either end. There are also brass cocks above the wheels to prevent any possibility of the balance rising up.[24] The balance is substantially the same as that illustrated by Gould in the *Marine Chronometer,* fig. 25 opposite page 96, except that there are four brass weights. Of these two are fixed to the crossings, and their positions are adjustable, while two are mounted on bi-metal strips and their travels are limited by small shrouds carried upon two of the four arms of the balance. The escapement is a form of pivoted-detent, very like that shown in Gould in fig. 26 opposite page 97. There are, however, detailed differences. The "anti-friction" rollers mentioned above are run in special frames inside the main frames of the clock. End-shake is controlled by two jewelled end-pieces set in cocks. The fusee embodies Harrison's maintaining work. A centre-seconds hand is carried on the extended pivot of the fourth wheel of the train. The top half of the dial is cut away to show the escapement, and there is an amplitude scale behind the balance. Behind this again is a device working through a pinion and rack and used for regulating the clock. The distance between the index pins for the balance spring is adjustable by means of a screw with a squared head. The outer terminal curve of the balance spring is adjustable in several planes and directions by means of what is sometimes termed a "geometric chair".

The clock was made in the early months of 1795. The description of it will be found in Volume One of the *Histoire de la Mesure du Temps,* pp. 220-224, and it is illustrated in Volume Two of the same work

Plate I/25. *FRENCH TRAVELLING CLOCK. Date circa 1780, signed "CH. LE ROY A PARIS". The shape to some extent foreshadows the later* pendules de voyage *of which the successors of Charles Le Roy became prominent manufacturers. This clock was lot 123 in the Sir John Prestige Collection sold at Sotheby's on 28th October 1963. (Private Collection)*

in plate XV at figs. 1 and 2. No satisfactory explanation has ever been offered for why Breguet's name alone should appear on this clock. It is, however, more in the nature of a pre-*pendule de voyage* than of a *pendule de voyage*, and hence its inclusion in this chapter. This is a special clock designed for the use of an astronomer. It is typical of Ferdinand Berthoud's work and thinking in most of its features. Neither in design nor in execution does it in any way resemble the work of Breguet.

Both in England and in France from the middle

---

(24) Vide Colonel Quill on the Grasshopper Escapement in *Antiquarian Horology,* September 1971.

I The First Spring-Driven Clocks

of the 18th century, if not before, the gradual evolution of marine timekeepers (chronometers) could be regarded as a facet in the development of portable pieces designed for journeys. In one sense a box chronometer, suspended in gimbals and with outer carrying box, represents a highly refined, if specialised, form of travelling clock. Readers wishing to study this subject are referred to *The Marine Chronometer* by R.T. Gould, and also Col. Quill's *John Harrison*.

In France from about 1775 had appeared a whole succession of attractive small travelling clocks which we shall call pre-*pendules de voyage* to distinguish them both from the far earlier pieces and from the *pendules de voyage* proper. Plates I/25 to I/34 illustrate clocks belonging to this category and all having been made in very much the same period. These pieces are usually called *pendules d'officier*, but one type was also variously styled *Capucine*, *Foncine* or *lanterne d'écurie*.[25] Let us consider a number of pre-*pendules de voyage* in something approaching their correct date order. Plate I/25 shows a fine small clock, signed "CH. LE ROY A PARIS". The maker was Basile Charles, who is mentioned in Chapter II. Le Roy was in the Palais Royal from circa 1785. Note how very elegant this clock is, despite a basic severity. In a way it foreshadows the *pendule de voyage* shape more than several of the later clocks. Plate I/26 shows

Plate I/26. FRENCH TRAVELLING CLOCK. *Date circa 1780. Signed "J.B. DU TERTRE A PARIS". Another variant of a basic design. Compare with Plate I/25. Note the glass sides. (Formerly Pérez de Olaguer-Felíu Collection. From the book* Relojes Antiguos *by Luis Monreal y Tejada)*

Plate I/27. FRENCH PENDULE D'OFFICIER. *Very typical of the best Parisian work of circa 1780 (Louis XVI). Compare with Plate I/26. (By courtesy of Sotheby & Co., London)*

---

(25) *Pendules d'officier* are supposed to have been taken on campaigns, and no doubt sometimes were. The nickname stable-lantern needs no explanation; but the names *Capucine* and *Foncine* are far more interesting. They may well provide clues to the origin of at least some of the French pre-*pendules de voyage*. More will be said about this matter.

1 The First Spring-Driven Clocks

a clock of much the same kind by J.B. Du Tertre (see List of Names). This clock is important because it is typical of the pre-*pendule de voyage* as sold by another important Parisian maker in almost exactly the same period.

Plate I/27 illustrates a French *pendule d'officier* in its most typical form as made from about 1780 to

Plate I/28. *PENDULE D'OFFICIER. Signed "Cugnier, Leschot". Date circa 1780. (By courtesy of the Musée d' Horlogerie, La Chaux-de-Fonds)*

perhaps 1820. It is characteristically in the style of Louis XVI, and a more happy example of these clocks at their best could scarcely be found. The Swiss and the Austrians also made *pendules d'officier* of very similar appearance. It is probable that all three types

Plate I/29. *FRENCH TRAVELLING CLOCK. Period last years of the 18th century. Maker Hessen, Paris. Height 7¼ins. with handle up. Pull-repeat for hours. Pull-wind for alarm. Regulation dial in arch. Platform cylinder escapement. (By courtesy of Kugel, Rue St. Honoré, Paris.)*

are to some extent related.[26] Plate I/28 shows a *pendule d'officier* bearing a Swiss name.

Plates I/29, 30 and 31 illustrate a very satisfactory example of a French timepiece pre-*pendule de voyage* by a well-known maker. He is André Hessen of Paris.[27] The Hessen clock owes much to its predecessors and yet it is beginning to set the *pendule de voyage* pattern. Plate I/29 shows the front view of the clock,

---

(26) As a matter of interest, see Gounouilhou in Alphabetical List of Names.

(27) This maker was presumably the author of the verge-lever escapement, noted by Baillie, and described and illustrated by Dr. Vaudrey Mercer in *Antiquarian Horology*, March 1964.

I   *The First Spring-Driven Clocks*

Plate I/30. *THE HESSEN CLOCK. Back of movement, showing external repeating work. This clock has a fusee for the going train. Note the increasing use of glass. (By courtesy of Kugel, Paris.)*

Plate I/31. *THE HESSEN CLOCK. Top view showing cylinder platform escapement. The conventional pendule de voyage form is just beginning to emerge. (By courtesy of Kugel, Paris.)*

which is interesting to compare with some of the Central European clocks illustrated in Chapter XI. Plate I/30 shows the back plate of the movement with external repeating work. Plate I/31 gives a very good idea of the escapement platform. It is a cylinder and anticipates future *pendule de voyage* practice. This clock has fusee and chain while *pendules de voyage*, once a standard form was reached, were furnished with going barrels.[28] Hessen was active during the last quarter of the 18th century. He was born at St. Tuna, Dalarna, Sweden in 1745 and died in Paris in 1805.

The type of French pre-*pendule de voyage* most usually called a *Capucine*, but sometimes termed a *Foncine* or *lanterne d'écurie,* is shown in Plate I/32. The movement is shown in Plate I/33. Another clock is seen in Plate I/34 complete in its travelling box.[29] It is very typical and would have been current in about 1800. *Capucines* both over-lapped and eventually replaced the true *pendules d'officier,* and they usually went for either eight or fifteen days at a winding. E. Lebon, writing in 1860 in *Etudes . . . sur l'Horlogerie en Franche-Comté* gives a clue to what may well be

---

(28) French pre-*pendules de voyage* often had fusees and chains for their going trains, but barrels for striking work and for alarms. There exists one real *pendule de voyage* in an early "multi-piece" case having striking work on the back plate. It has fusee and maintaining work on the going side, but going barrel for the *grande sonnerie* striking. It is a Franche-Comté clock. The dial is signed "Jorosay H[gers] Palais Royal".

(29) To prepare the clock for travel the top handle and shank are unscrewed. They are afterwards screwed back on to the bell standard which projects through a hole in the lid of the box.

I The First Spring-Driven Clocks

the origin of the names *Capucine* and *Foncine*. He says that in about the year 1660, according to a family tradition which has not been questioned, the brother doorkeeper of the monastery of the Capucins of Saint-Claude asked the village priest in Morbier whether any

Plate I/32. CAPUCINE CLOCK. *Date circa 1810. This type of clock was otherwise styled* Foncine *or* lanterne d'écurie. *(By courtesy of Mallett of Bourdon House Ltd.)*

Plate I/33. CAPUCINE CLOCK MOVEMENT. *The cylinder escapement is just visible at the top of the back plate. (By courtesy of Mallett of Bourdon House Ltd.)*

1 The First Spring-Driven Clocks

because of their "hooded" appearance. As to *Foncine*, the Mayet family were in the village of that name too, and Lebon also particularly mentions Foncine-le-haut and Foncine-le-bas with other villages, including Bellefontaine and the Rousses, as becoming centres of clockmaking serving Morez. In Chapter V is described and illustrated a *pendule de voyage* made in the 19th century by Michoudet of Foncine-le-bas.

An even more exciting link in the chain is to be found in a *Capucine* clock (Plate I/35) signed "JANVIER Cadet" numbered 62 and to be seen in La

Plate I/34. *CAPUCINE CLOCK IN TRAVELLING BOX. Note the carrying-handle shank in its travelling position above the lid of the box. The actual handle is missing. (By courtesy of Mallett of Bourdon House Ltd.)*

of the ironworkers, already there in some numbers, was capable of repairing the monastery clock. He was told to go to the four brothers Mayet who were locksmiths. According to Lebon, the Mayets pronounced the clock beyond repair, but they subsequently conceived the idea of making iron clocks themselves, and this led to the birth of those clocks of the type called variously *Morbier, Morez, Comtoise,* or *horloge de Comté*. Baillie lists some members of the Mayet family in the Jura from the middle of the 17th century, including *les frères* from 1647-1660; so there may well be some truth in the story. Of course it is always possible that the name *Capucine*, meaning either nun or nasturtium in French, was applied to these clocks

Plate I/35. *JANVIER CADET. A* Capucine *clock with pendulum. This clock is most important as being further evidence to support the contention that* Capucines *originated in the French Jura. (By courtesy of the Musée d'Horlogerie, La Chaux-de-Fonds)*

25

I The First Spring-Driven Clocks

Chaux-de-Fonds Museum. It is known that the famous Antide Janvier's home was at Saint-Claude, while his younger brother *(cadet)* remained there for almost the whole of his life. Further notes on this subject will be found in Chapter VI. Most but not all *Capucines* had alarms and struck hours and halves in passing. Many also were provided with a side lever by which the striking train could be released in order to repeat the last hour. The alarms were wound by pulling cords usually emerging from the tops of the cases. A few clocks struck quarters in addition to hours. Others again were furnished with four bells and eight hammers, playing one of a succession of simple tunes in rotation at the hours. The appearance of *Capucines* was fairly standard. Usually the feet matched the four top finials. The shapes were often those of acorns, tulips or *piques révolutionnaires*.[30] The first *Capucines* sometimes had convex enamel dials without bezels and reaching to the sides of the clock cases. Later examples had glass-less bezels with cast decorations of palm leaves, wreaths, or key-pattern. Plain moon hands were the rule. These were sometimes gilt and sometimes steel. The chapters were Roman rather than Arabic. The earliest *Capucines* had verge escapements with short pendulums, but later many had cylinder escapements with balances. These escapements were set within the movements, the balances lying on their sides inside the back plates. One maker produced a model in which the escapement could be seen working in the dial. The hammer or hammers were inside the bells, well protected by these and by the four corner finials. Most *Capucines* stand about twelve inches high including their stirrup-like top handles. Although the clocks themselves are not particularly rare, it is quite unusual to find one still having its original travelling box.

Swiss clocks having the appearance of the example illustrated in Plate I/36 may be regarded as a type of pre-*pendule de voyage,* even though it is probable that they were overlapped by the advent of the true Swiss carriage clock. Their dates are somewhat uncertain. The example illustrated, which is signed simply "Robert" has a lever platform escapement which no-one, so far as we know, has ever suggested is not original. On the other hand, a piece identical in appearance and signed "Robert & Courvoisier" (which should make it a later clock) has only a verge

Plate I/36. *LA CHAUX-DE-FONDS TRAVELLING CLOCK. Date uncertain. It is not clear which member of the Robert family made this piece. (By courtesy of the Musée d'Horlogerie, La Chaux-de-Fonds.)*

escapement. Clocks of this type offer *grande sonnerie, petite sonnerie* or silence, and may be made to repeat *grande sonnerie* by means of a pull cord. Their *cadratures* are laid out on the back plates. Their going and striking trains have going barrels, while their alarms employ standing barrels. Detailed Courvoisier history, showing the link with Robert, is given in Chapter X. The Courvoisier family name appears upon the finest examples of true Swiss carriage clocks.

Plate I/37 shows an anonymous clock which is described as having Neuchâtel origins. It stands just under seven inches tall with the handle up. It goes for eight days, which sets it sharply apart from those

---

(30) *Capucine* feet, and also finials, are easily broken. Very many are replacements.

1 The First Spring-Driven Clocks

Plate I/37. *NEUCHATEL TRAVELLING CLOCK. Date circa 1820-30. Grande sonnerie striking and repeat. Lever platform escapement visible through "porthole" in the top of the case. (By courtesy of Musée de l'Horlogerie, La Chaux-de-Fonds)*

Plate I/38. *"SEDAN" CLOCK. Date post 1806. Signed "Reid & Auld, Edinburgh, 700". (By courtesy of Meyrick Neilson of Tetbury Ltd.)*

Austrian clocks shown in Chapter XI, but which it nevertheless in some respects resembles. The date of this clock would appear to be circa 1820-1830. Several features are of interest. The striking is *grande sonnerie* on thin Austrian-like wire gongs disposed round the movement. There is a top repeat button. The pin-pallet underslung lever platform escapement is of very good quality and has an overcoiled balance spring. The plain brass balance is above the movement and visible through a circular glassed "porthole" on the top of the drum-shaped case. There is a stop-start device hidden inside the bezel of the front door, besides a regulation lever or index. Note that the case is gilt and decorated all over with engine-turning used in conjunction with a silver dial and moon hands. The bezel has gadrooned edging. This clock would almost certainly have been made at a time when it would have been possible to obtain a true *pendule de voyage*, but it is probably rather earlier than the Swiss carriage clocks in their developed form. It is best described as a very high class travelling clock with alarm.

An English type of portable piece which has been the subject of a good deal of correspondence in *Antiquarian Horology* between December 1970 and March 1972 is the so-called "sedan clock". Plates I/38 and I/39 show a front view and also the movement of the example which first raised the issue. It is signed "Reid & Auld, Edinburgh" and is numbered "700". The engraved silvered dial (more usually these are white enamel) is about four inches in diameter. The movement, which is very like that of a large verge watch, but having rectangular movement plates with radiused tops, goes for thirty hours. Thomas Reid, the author of the famous *Treatise*, was born in Dysart (Fife) in 1746 and died in 1831. He was apprenticed on 9th October 1762 to James Cowan (1744-1781), then of Lawnmarket, Edinburgh. Reid's shop, in about the

I   The First Spring-Driven Clocks

Plate I/39. *MOVEMENT OF REID & AULD "SEDAN" CLOCK. Note the shape of the plates. Goes for 30-hours. (By courtesy of Meyrick Neilson of Tetbury Ltd.)*

Plate I/40. *THE MOVEMENT OF THE BUCHANAN "SEDAN" CLOCK. Date late 18th century. Note the shape of the plates of a very interesting custom-built movement, going for several days. (By courtesy of John Williams, Haslemere)*

year 1794, is shown in a famous picture entitled *The Parliament Close and Public Characters of Edinburgh Fifty Years Since.* In 1806 Reid took into partnership William Auld.[31] The "Sedan" clock, therefore, could not have been made before 1806. It has never been altered in any way. A very similar clock is signed "Archibald Buchanan, Dublin", and has a duration of three days rather than the usual thirty hours.[32] Buchanan, according to Geraldine Fennell (see Bib.), was active from "about 1760". This fact does not in itself make the "sedan" clock of that date, but from the appearance of the movement, and judging by the style of the cock, the piece was probably made before 1800. (Plate I/40)

Very briefly, the two questions which seem to have occupied the writers to *Antiquarian Horology* have been: – firstly, whether "sec'an" clocks were ever used in sedan chairs, or if the term was coined by antique dealers; and secondly, whether the clocks were a type in their own right, or simply made as a convenient way of re-using discarded full-plate watch movements.[33] However, not one shred of conclusive evidence seems to exist to establish that "sedan" clocks were ever used in sedan chairs, much less designed specifically for them. Straus in *Carriages and Coaches*, 1912, says that "chairs" were in decline by the end of the 18th

---

(31) Reference Allix (see Bib.).

(32) The British Museum possesses an eight-day "sedan" clock signed "Thomas Astley, Liverpool, N.180. GOES EIGHT DAYS".

(33) Watch movements were undoubtedly used or re-used at times both in England and in France to form the basis of small portable clocks rather like carriage clocks in appearance. The French type often incorporate a separate alarm train in rectangular plates. The English type occasionally were made to go for eight-days by the addition of a going barrel in external frames. Plate I/41 shows an example. Clocks embodying watch movements are sometimes known as watch-clocks.

28

1 The First Spring-Driven Clocks

Plate I/41. *WATCH-CLOCK MOVEMENT. Made in the 19th century utilising an 18th century watch movement. Note the large barrel added to provide a duration of 8-days. (By courtesy of E. Pitcher & Co.)*

Latin word *sedes* meaning seat, than from any connection with Sedan in France. The French called their sedan chairs *chaises à porteur*.

English "sedan" clocks seem to fall into four main categories. The first type have verge watch movements. These movements are usually 19th century, but sometimes they are earlier. Examples are even found based upon watches sold by such makers as Quare and Tompion. The overwhelming probability is that such pieces assumed their present form only in the first half of the 19th century, being then re-constituted as inexpensive hanging clocks. It is easy now to forget how poor minor clock and watch repairers used to be, and how hard it was for them to scratch a living.[35] There would have been plenty of customers delighted to purchase for a few shillings thirty-hour clocks which would hang on a nail anywhere. If the great Tompion had set his mind to make clocks for sedan chairs, it is surely inconceivable that he would not have produced something better. The second type of "sedans" have watch movements usually with cylinder, duplex, or lever escapements.[36] These movements are too late to have been made for "chairs"; but apart from this they would seem to have been far too expensive in the first place to waste upon a cheap and insignificant type of clock. The third type of "sedan" had the "rectangular-plated" type of movement, which is also found at times in "cottage" clocks.[37] These movements were almost certainly produced with the manufacture of small clocks in mind, but any which we have seen are decidedly 19th century in their origins. When a "sedan" clock is found with such a movement, it is usually manifestly original, looking far less improvised than any of the examples employing watch movements. A few "rectangular-plated" movements, as already noted, go for at least two clear days, and do not have the trains of the normal watches of the period. It is probable that the "rough movements" of the "rectangular-plated" clocks would have been available, as were the ordinary

century, although they lingered on in Scotland until about 1820. Most "sedan" clocks seen by us, or at any rate those giving the impression of having been produced from start to finish as custom-made pieces, were manifestly made well after 1800.[34] The Buchanan clock is perhaps an exception. On the basis of date at least, this clock may have overlapped sedan chairs. It would not be surprising if in Dublin they had persisted well after vanishing from the scene in London. The origin of the term "sedan" chair is obscure; but it seems more likely to have been derived from the

---

(34) Such a piece, a typical "sedan" if ever there was one, was sold at Sotheby's on 12th July 1971, Lot 25. It has a white enamel dial, brass bezel, turned mahogany case 6½ inches in diameter, rectangular movement, and pierced and engraved backcock. The movement backplate is signed "Barnett, London, AD 1839".

(35) This fact also explains why so many fine English verge and duplex watches were "butchered" in the 19th century by being "converted" to bad lever escapements for the sake of a pound or two. Nelthropp in *A Treatise on Watchwork*, 1873, hits the nail on the head on pages 247-249 under the heading *Fallacies of the Trade. Converted.*

(36) An example with a Cooper duplex movement was sold at Christie's on 21st March 1972, lot 2. Cooper, as every watch collector knows, worked at the very end of the 19th century.

(37) The so-called "cottage" clock, very like a "sedan" but often having four ball feet supporting an ebonised wood case with gilt embellishments, is certainly a close relation.

I The First Spring-Driven Clocks

together. A few exceptions, like the Buchanan clock, may prove the rule. One writer to *Antiquarian Horology* mentions "sedan" clocks as being made for "... the fashionable public...". It seems far more probable that, at best, such pieces were always inexpensive, and in the nature of a poor man's general purpose hanging clock of short duration.

Clocks somewhat in the nature of "sedans" were made in France at the end of the 18th century. The most usual sort is perhaps the type with octagonal brass case, enamel or metal dial with gilt ornamental bezel, like the *Capucines* but with glass, and an eight-sided bow for carrying or hanging. Monreal (see Bib.) illustrates an example. (Plate I/42). Another, signed on the dial "LE ROY & FILS, HR DU ROI, Palais

Plate I/42. *FRENCH OCTAGONAL-CASED PORTABLE OR HANGING CLOCK. Date circa 1800. (Formerly Pérez de Olaguer-Felíu Collection. From the book* Relojes Antiguos *by Luis Monreal y Tejada)*

verge and certain other English watch movements, from the material shops in Clerkenwell. The fourth type of "sedan" has an eight-day modern movement. Some examples give the appearance of having been made throughout in the 20th century (remember that even until the last war, the making of hands, enamel dials, wood cases, brasswork, etc. would have presented little difficulty or expense). Other clocks now with modern eight-day movements certainly once had either "rectangular-plated" movements or those intended for watches.

To sum up, "sedan" clocks appear in various forms, but usually only those with "rectangular" plates have cases, dials and movements which have always been

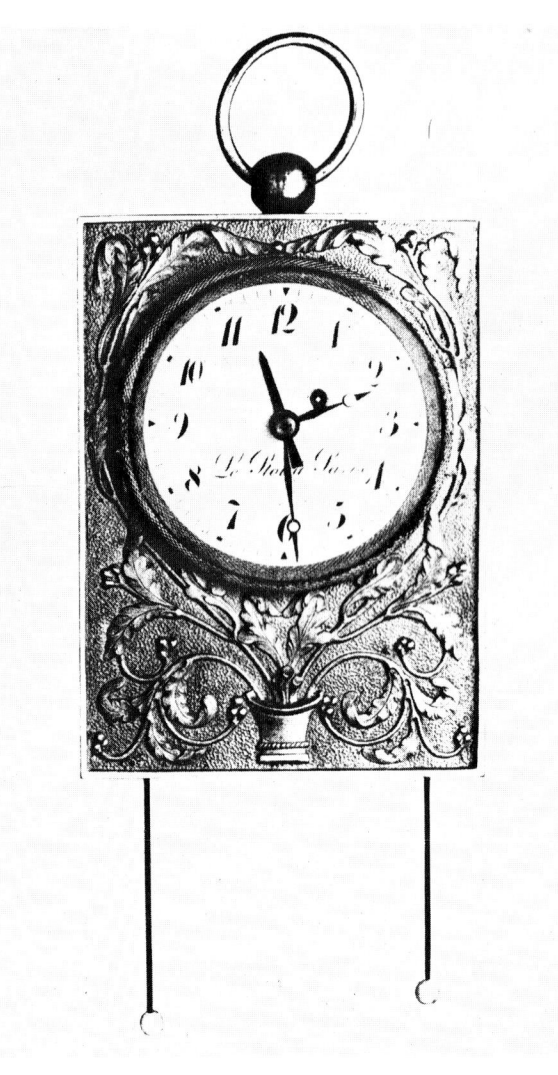

Plate I/43. *FRENCH PORTABLE OR HANGING CLOCK. Date circa 1820. (Formerly Pérez de Olaguer-Felíu Collection. From the book* Relojes Antiguos *by Luis Monreal y Tejada.)*

30

I   *The First Spring-Driven Clocks*

Plate I/44. *FRENCH PORTABLE CLOCK. Another variant of the type of clock which could be hung upon a nail anywhere. (By courtesy of Mme. Ducatez, Paris))*

Plate I/45. *MAIL-GUARD'S WATCH. Date between 1830 and 1837. (Private Collection)*

Royal", has a movement signed "Veyrin à Paris No. 825". Baillie notes Jean-Antoine Veyrin as a *Maître Horloger* from 1773-1792. These octagonal clocks normally had pull-repeats, often with plaited "cows' tails" for the pulls hanging below the cases. The escapements were usually cylinders but some were verges. The Veyrin clock has a fusee with maintaining work. Another French "sedan-like" type of clock was given a rectangular case with top pendant on one of the short sides. (Plate I/43) Usually in these clocks a white enamel dial with arabic numerals was set high in a repoussé front. Monreal illustrates one signed "Le Roi à Paris". Simoni (see Bib.) shows another signed "Le Roy et Fils, Parigi", made for Italy. Both the octagonal and the rectangular clocks were provided with alarms. Another variant is shown in Plate I/44.

Plates I/45 and I/46 show the last type of small portable clock with which we need concern ourselves in this chapter. It is a mail-guard's watch, and there is no question as to its period. The inscription "W.R." ("William Rex") refers to the short reign of King William IV from 1830 to 1837. The same pattern of watch would, however, have been current from about 1825-1845. These years for all practical purposes, saw the rise, heyday and decline of the English stage coach systems. George Littlewort, the maker of the clock, was Free of the Clockmakers' Company from 1822. The mail-guard's watch has a lever escapement and goes for four days. The guard was required at certain stages of the route of his mail coach to hand his locked timepiece to the local postmaster. The time of arrival was then entered on the waybill in a space provided.

The stage coach had been a feature of English travel since the 16th century, but its golden age began in 1825 with the placing of the "Wonder" on the London to Shrewsbury road. This, the first stage coach to be timed for so long a journey in a single day, was the forerunner of many similar coaches which, by their speed, punctuality, comfort and

I The First Spring-Driven Clocks

Plate I/46. *THE MOVEMENT OF THE MAIL-GUARD'S WATCH. A rather similar piece is in the Clockmakers' Company's Museum at the Guildhall in London. (Private Collection)*

smartness, made the English stage coach famous. The irony of fate ordained that the appearance of the "Wonder" should almost synchronise with the invention of the steam locomotive. Some would place the ending of the coaching age as early as 1838 when the London and Birmingham railway was opened to traffic. But, despite the rapidly spreading network of railways, good stage coaches continued to hold their own in the remoter parts of the kingdom up to about 1845.[38] According to Bovill, the stage coach of the 19th century was the child of the mail-cart introduced by the Postmaster General in 1784. From this year Post Office vehicles carrying mail also carried passengers in order that the latter should defray the whole expense.

Post-chaises were something different again. They were usually drawn by one pair of horses, and were driven by a post-boy and not by a stage coachman. These conveyances travelled only short distances between posting houses, but very rapidly. They served mainly the areas away from the coach routes. Many people had to finish their journeys by post-chaise. Others, who disdained public transport, made even very long journeys "travelling post", despite the great extra cost and the many stops and changes involved. Others again used their own travelling carriages or barouches (known as "bounders"), hiring post-horses to draw them from one stage or posting house to the next. It is not difficult to see the need for portable clocks against such a background. A few further remarks on this subject will be found at the beginning of Chapter IX. E.W. Bovill's splendid book, already mentioned, contains much further information upon the road. Post-chaise watches were like those used by the mail-guards, but by late Victorian times the inexpensive Swiss "giant" ("Goliath") watches were loosely called post-chaise watches, and leather holders were provided for them in private vehicles ranging from broughams to dog-carts. Even first class "Sleepers" on the railways provided "watch-holders" until recent times.

Much more could be written about the various progenitors of carriage clocks, but it is hoped that the foregoing will serve to show that they had a long and very interesting pre-history.

---

(38) The last paragraph is quoted directly from E.W. Bovill's *The England of Nimrod and Surtees, 1815-1854,* Chapter XV.

French Carriage Clocks

Chapter II

The Beginning.
Late 18th and Early 19th Century

A.L. BREGUET. (By courtesy of the Clockmakers' Company, London)

II

The Beginning.

Late 18th and Early 19th Century

The Breguet Family and Their Successors. Paul Garnier. Exhibitions. Leroy. Other Early Makers. Horlogers du Roi.

THE BREGUET FAMILY AND THEIR SUCCESSORS

There would seem to be no reasonable doubt that the first French carriage clock was made in Paris early in the 19th century under the auspices of the great Abraham-Louis Breguet (1747-1823) the most inspired of all watchmakers. Breguet has never been credited with this achievement; but then the *pendule de voyage* in its developed form was an evolution and not an invention. It followed naturally and inevitably from a long progression of earlier portable pieces. Amongst those made in Europe may be numbered the *montres de carrosse*, the *Capucines* and the so-called *pendules d'officier*. It would have been natural for Breguet to appreciate that a market existed for a thoroughly practical travelling clock complete with not only striking work but also with alarm and calendar. Others of course had the same idea. With Breguet the metamorphosis was quickly accomplished. Accordingly, the historian finds himself faced abruptly with the fact that suddenly, and certainly by 1810, the first Breguet carriage clocks appeared "fully-fledged".[1] It must be stated immediately, however, that such rare masterpieces are in a class by themselves. Despite the fact that Breguet's first *pendules de voyage* are generically so very early, it is usually neither possible nor fair to compare them with later French carriage clocks by other makers. Always enormously expensive and complicated, often made only to special order, the Breguet clocks evince an innate superiority in conception, design and execution which is difficult to convey in words. Breguet was without question the most versatile horologist in history. It is not surprising that his *pendules de voyage* were both the first and the best ever made in France.

Breguet's son, Louis-Antoine, joined him as a partner in the business in about 1807. After Breguet's death in 1823, his son and then his grandson carried on the business until 1880.[2] In this year the firm passed to Mr. Edward Brown, the English *chef d'atelier*. Brown's son and grandson continued to conduct the business until 1970 when Mr. George Brown retired and sold out to Maison Chaumet of Paris and London. The Breguet firm apparently used the style "Breguet et Fils" from 1807 onwards, although some pieces have various other signatures. It is also necessary to mention the *raison sociale* "Maison Breguet, Neveu et Cie"[3] covering the period circa 1830 to 1860, after which the name reverted to "Breguet et Fils". Mr. George Daniels' new definitive

---

(1) These clocks were almost certainly related to the early Swiss *pendules de voyage* produced at La Chaux-de-Fonds from about the same period. Compare Breguet No. 2020 (Plates II/4 & 5) with the early Swiss clocks (Plates X/1-4). It must be remembered that not only was Breguet of Swiss birth, but also that he returned to that country during the French Revolution.

(2) Breguet's son, Louis-Antoine, died in December 1858. The grandson was François-Louis Breguet.

(3) The name of Breguet's nephew was Savoye. It is not clear exactly how he fitted into the picture.

II The Beginning

book *The Art of Breguet,* to be published in 1974, will solve all the ancient mysteries of the Breguet number sequences, trading names, etc.

represented by Plate II/10. This last style was copied very successfully by the famous Cole brothers in England from 1823 (the year of Breguet's death),

Plate II/1. BREGUET. Clock No. 179. Sold in 1810. Manufacture probably started prior to 1800. This appears to be the earliest unaltered Breguet carriage clock known. (By courtesy of Sotheby & Co.)

Breguet carriage clocks will usually be found to have one of three basic types of case. It would perhaps be as well not to venture any firm opinion as to whether style of case may be used as a guide to date, since on the whole surviving clocks suggest that there are contradictions. Breguet cases may be classed briefly as (1) wooden, Empire style, (2) metal cases, Empire style, and (3) "humpbacked" silver cases. The wooden cases are either very plain or else decorated with ormolu. The metal cases are typified by Plates II/1 and II/2, and the silver ones are well and at a later stage by Jump of London in 1889 and 1901. In still more recent times, as will be mentioned in Chapter IX, clocks of varying sophistication have been made in England having the same style of case. Of Breguet's carriage clock cases, perhaps the wooden ones are the earliest, although almost certainly they were overlapped by the gilt-brass Empire style which has continued ever since.[4] The "humpbacked" style, also in favour to the present time, was perhaps the last pattern to be introduced by Breguet.[5] There seems little doubt that all three case styles were current

---

(4) Some wooden cases almost certainly had ormolu embellishments added to them in the time of the nephew.

(5) The correct name for the "humpbacked" shape is *borne,* meaning milestone.

II The Beginning

The earliest Breguet *pendule de voyage* seen by us is No. 179 sold to S.M. François de Bourbon, King of Naples in 1810. It is shown in Plate II/1. This clock until very recently was thought to have been made in 1804; but research made by Mr. George Daniels from the Books of the firm has since indicated that the clock was probably in the course of manufacture prior to 1800.[6] The clock is a timepiece, but it may be made to repeat hours and quarters.[7] It also has the additional complications of calendar and moon dial. The gilt brass case, in the style of

Plate II/2. *BREGUET. A magnificent and very complicated clock of uncertain history and date, and now bearing the number 780. (By courtesy of The Ashmolean Museum, Oxford)*

by 1813.

Breguet's nephew started a new number series in 1834. However, work already in progress seems to have been completed with the original numbers belonging to the sequence started by A-L Breguet himself. Both earlier and later clocks were, however, often altered by the firm, and serial numbers are apt to be misleading except in cases where full histories of the clocks are known.

On pages 48 to 54 will be found a table showing a few representative carriage clocks and other portable pieces sold by A-L Breguet, his son, nephew and their successors.

Plate II/3. *BREGUET. Clock No. 780. A back view of the movement showing the Tourbillon carriage and also part of the equation and calendar work. Note a movement shape suggesting that it was originally intended to be housed in a "humpbacked" case. (By courtesy of The Ashmolean Museum, Oxford)*

---

(6)   Certificates based on the Breguet workbooks have been issued in the past to authenticate clocks as being the genuine work of the firm. When a certificate combines a late date with a low serial number, it sometimes means that this date is inaccurate. The firm of Breguet, as already noted, often bought back their own work, which was then re-vamped and re-sold.

(7)   The repeating train is wound and set in motion when a long plunger on the top of the case is depressed.

39

II The Beginning

Empire, has corner pilasters, lions'-head attachments for the carrying handle, and top and bottom friezes decorated with stylised foliage in relief. All the evidence points to the fact that No. 179 is the earliest known Breguet *pendule de voyage* in its original state. Another early piece might be a wooden-cased clock sold, according to Tardy, in 1806 to the Prince de Talleyrand. However, Tardy does not give the serial number of the clock, any more than he says whether it has a certificate or not. The piece is fairly simple, offering repeat and alarm. Tardy also illustrates a clock rather like No. 179 and which he says (without mentioning its number) was sold in 1811.

Plate II/4. BREGUET. Clock No. 2020. Mahogany case. Echappement naturel. *Note the startling resemblance to the Cugnier, Leschot clock in Plates X/3 and 4. (By courtesy of Sotheby & Co.)*

The Ashmolean Museum in Oxford possesses a superb Breguet Tourbillon carriage clock having many complications. Plate II/2 shows a front view, while Plate II/3 illustrates the back of the movement. The clock is numbered 780, but it is doubtful whether this serial number is the original one. There is also no evidence as to the date of manufacture. It is unlikely that work on this clock was started earlier than about 1810–1812, while the case gives every appearance of belonging to the period circa 1820. The shape of the movement suggests that it was intended to be placed

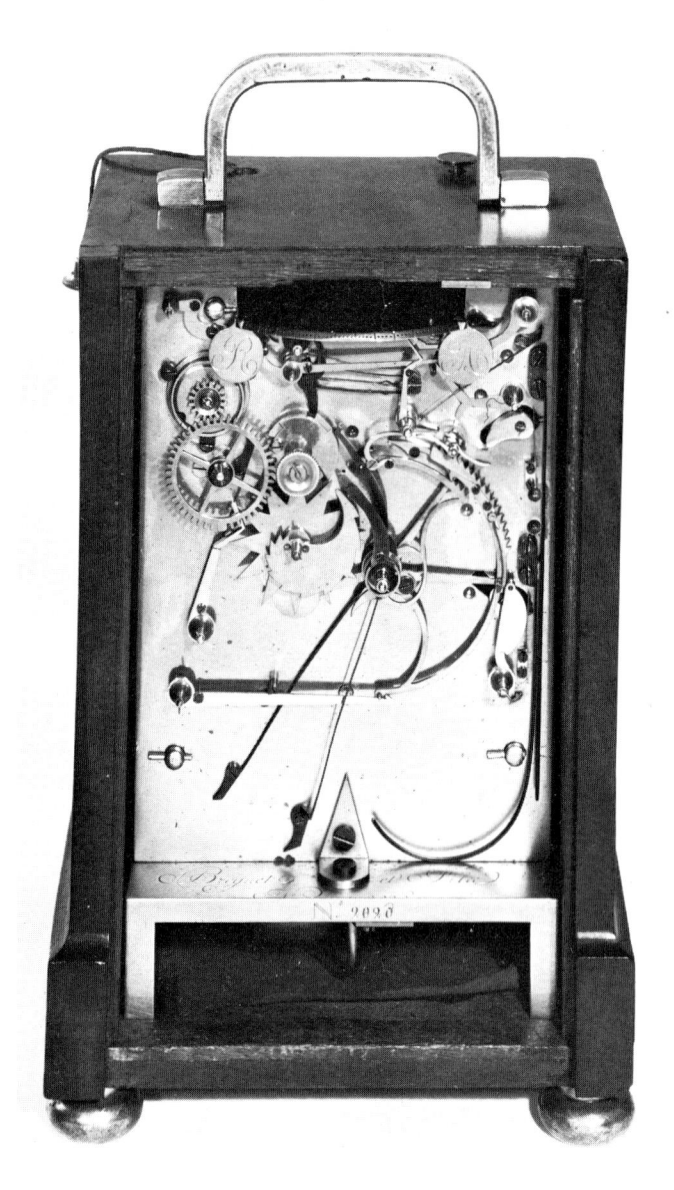

Plate II/5. BREGUET. Clock No. 2020. *Rear view of case and movement. (By courtesy of Sotheby & Co.)*

40

II The Beginning

Plate II/6. BREGUET. *Clock No. 2767. Walnut case with ormolu mounts. Special type of Robin escapement. (By courtesy of Cecil Clutton. Photograph by Christie, Manson & Woods)*

Plate II/7. BREGUET. *Clock No. 2848. Date 1815. Case veneered in thuyawood (or possibly walnut). Robin escapement with impulse on exit pallet. (By courtesy of Sotheby & Co.)*

in a "humpbacked" case, and indeed the whole clock represents something of a mystery.

At Sotheby's on 19th June 1972, Lot 194, was sold a very important Breguet *pendule de voyage* No. 2020. It is shown in Plate II/4. While this clock is almost identical in appearance to the one mentioned by Tardy as having been sold to Prince de Talleyrand in 1806, it is probable that the date of No. 2020 is nearer 1815. One of the most interesting features of the clock is the *échappement naturel*. This escapement, developed from that attributed to Robin and intended to combine the best qualities of both chronometer and lever, was probably first devised by Breguet as early as 1789. Both going and striking trains of No. 2020 are driven by a common mainspring. The back of the movement is shown in Plate II/5.

Plate II/6 illustrates a wooden-cased clock No. 2767. It is an excellent example of a decorative walnut case embellished with ormolu mounts.

The next Breguet carriage clock which may be dated with any degree of certainty is No. 2793 of 1813 from the Salomons collection. Clocks have also appeared in recent years having dates of origin 1815, 1819, 1826, 1828, 1834 and 1835. Jeanneret in *Biographie Neuchâteloise,* 1863, Volume 1, page

41

II The Beginning

also a repeater. There is a pull-wound alarm sounding on a bell in the base. The escapement is of special interest, being a form of Robin, but giving an additional impulse on the exit pallet. The number of the clock is 2848 and its date, according to the Certificate, is 1815.

On 13th July 1821 Breguet et Fils sent to Lord Spencer No. 3050. The *blanc* of this clock was obtained from Pons, and the necessary work was carried out by Jacob, Louis Vuital, Couët and Kessels.

At this stage must be mentioned Breguet No. 3629, a "humpbacked" silver-cased clock sold in 1822 and which may be seen in the Ilbert Room of the British Museum. This beautiful clock has perpetual calendar, showing day, date, month and year, besides the phases of the moon. It is also provided with half-quarter

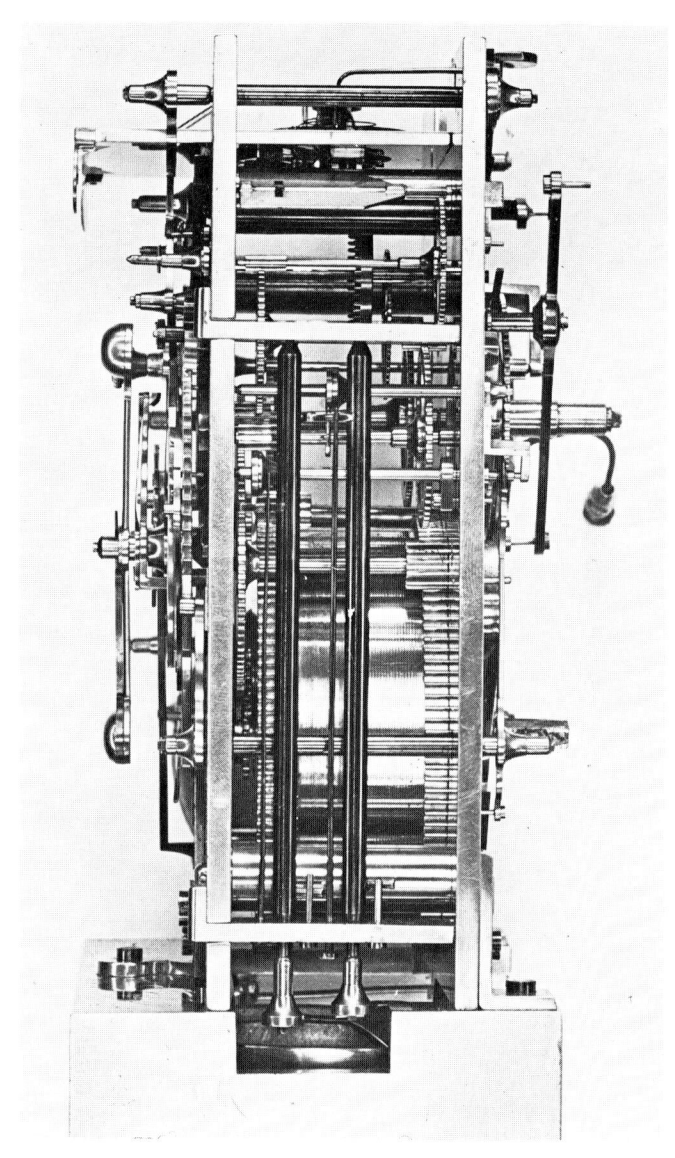

Plate II/8. BREGUET. No. 2848. Side view of movement. (By courtesy of Sotheby & Co.)

108, mentions Breguet's carriage clocks shown at the Paris Exhibition of 1819. He says that Breguet exhibited ". . . several carriage clocks, with repeating, alarm, age of the moon, and complete calendar" (". . . plusieurs pendules de voyage à répétition, réveil, mouvement de la lune et quantième complet . . ."). The calendar is likely to have been of the type aptly called *à rouleaux*.

Plate II/7 illustrates a very fine clock in a plain wooden case, being either thuya-wood or possibly walnut, veneered on a mahogany carcass. Plates II/8 and II/9 show two views of the movement. This clock strikes *grande* and *petite sonnerie* on bells, and is

Plate II/9. BREGUET. Clock No. 2848. View under the dial. (By courtesy of Sotheby & Co.)

II The Beginning

Plate II/10. *BREGUET. Clock No. 3629. Sold 1822. Silver "humpbacked" case. (By courtesy of The British Museum)*

Plate II/11. *BREGUET. Clock No. 3629. Rear view of movement. (By courtesy of the British Museum)*

pull-repeating, all achieved upon a single gong, and with a pull-wound alarm sounding upon a double gong by means of a double-ended hammer. See Plates II/10 and II/11. In this clock the lever of the straight-line escapement is banked on the 'scape pinion, but the brass wheel with twenty broad-tipped teeth is neither slit nor drilled to retain oil. There is a safety roller and guard-pin. This escapement was in all probability made without draw; but it has since been clumsily modified, leaving a situation where there is effectively no draw whatever upon the exit pallet and far too much on the entry. The escapement lies on its side on the back plate. The two-arm compensated balance is provided with both *parachute* and overcoiled balance spring. Apart from the escapement, the movement is unusual in having fusee and chain with Harrison-type maintaining. The mainspring is wound from the front right-hand side via a pair of wolf-teeth wheels, one of brass and one of steel. The hands are set from the front left-hand square which is disengaged, except when pushed in by the key.

In the 1834 Paris Exhibition, Maison Breguet, Neveu et Cie exhibited two clocks, of which one was described as "Petite Pendule de voyage et de cabinet sur les principes des chronomètres à équation, marchant huit jours, à grande et petite sonnerie et répétition à volonté; boîte d'argent massif, à pilastres et ornements ciselés; le tout surdoré; glaces sur toutes les faces, permettant de voir tout l'ouvrage; les boutons servant à ouvrir les deux petites portes de devant et derrière, sont en brillants. Cadran d'argent indiquant l'heure et les minutes concentriques par deux aiguilles d'or, les secondes courantes, quantième et phases de lune, réveil, développement de ressorts; quantième

II The Beginning

des jours de semaine, date du mois, noms des mois et millésime, avec petit thermomètre métallique donnant les degrés de température, suivant Réaumur. Cette belle pièce, qui est tout ce que nous avons fait de plus extraordinaire, de plus riche et de plus parfait en ce genre, est la propriété de M. le Comte A de Demidoff." The number of this piece is not mentioned.

The second clock, no less interesting, had *grande* and *petite* sonnerie striking, alarm, repeat, calendar and moon. It was housed in a gilt-brass case with pilasters, being glassed front, back and sides in order to show off the work. This piece was specifically designed to go in any position "construite pour supporter toutes les positions sans se déranger;".[8] The travelling box was designed to remain open in order to allow a traveller both to see the dial and to make use of the repeat etc. The advertising material emphasises that the *pendule de voyage et de cabinet* was equally useful when standing on a mantelpiece or on a piece of furniture. It was a man's clock, and *cabinet* in this context should be translated as study rather than as drawing room.

In the Ilbert Student's Room at the British Museum is now kept Breguet No. 4685. This well-known clock, formerly in the Sir John Prestige Collection, was sold in 1835 to Madame Baudin. It has a semi-Empire case which bears a striking resemblance to the Swiss pieces

Plate II/12. BREGUET. *Clock No. 4685. Date 1835. (By courtesy of The British Museum)*

Plate II/13. BREGUET. *Clock No. 4685. View of platform showing* parachute *and Breguet's late type of lever escapement, closely resembling Swiss work. (By courtesy of The British Museum)*

---

(8) All the information concerning the two clocks exhibited in 1834 is taken from a *notice advertissement* issued at the time by Maison Breguet, Neveu et Cie.

II The Beginning

made in La Chaux-de-Fonds at almost exactly the same period. This fact is perhaps scarcely surprising in view of Breguet's Swiss background and contacts. The clock is a timepiece, repeating hours and quarters both on the same bell. The speed at which the repeating train runs is controlled by an escapement (the invention of Julien Le Roy) instead of by a fly. There is also a pull-wound alarm. The straight-line lever escapement has no draw, and is banked on the 'scape pinion. The escapement has a "modern" double roller, *parachute* and compensation balance with overcoiled spring. See Plates II/12 and II/13.

In the *Exposition* of 1844, Maison Breguet, Neveu et Cie showed a number of carriage clocks. The first one was described as *Une petite Pendule de voyage et de boudoir*. According to Breguet, Neveu's own advertising literature, it went for eight days, had a chronometer escapement, *grande* and *petite sonnerie* striking, alarm, repeating at will, and a special calendar. The ornate case, very much in the taste of a Parisian lady of fashion, had pilasters and was embellished all over with filigree work and semi-precious stones. Even the knobs on two doors were set with brilliants. This clock was specially made for Prince Demidoff. No serial number was mentioned. Another clock was described as *Petite pendule de voyage et de cabinet*. This also went for eight days, had *grande* and *petite sonnerie* striking, and could be made to repeat hours and quarters at will. In addition, the clock was provided with a button, or more probably a plunger, for minute repeating. Other clocks shown by the firm were rather more ordinary. They were described as *plusieurs petites pendules de voyage*, and the descriptive literature of the Exhibition says that they struck hours and halves in passing, repeated hours and quarters (the mechanism for the latter almost certainly being wound by the depression of a plunger). The clocks all went for eight days and had chronometer escapements.

In Plate II/14 is shown the last "humpbacked"

Plate II/14. BREGUET. *A clock numbered 759 completed in 1931 for Ettore Bugatti (1882-1947), the Italian-born designer of the world-famous French car. (By courtesy of George Daniels)*

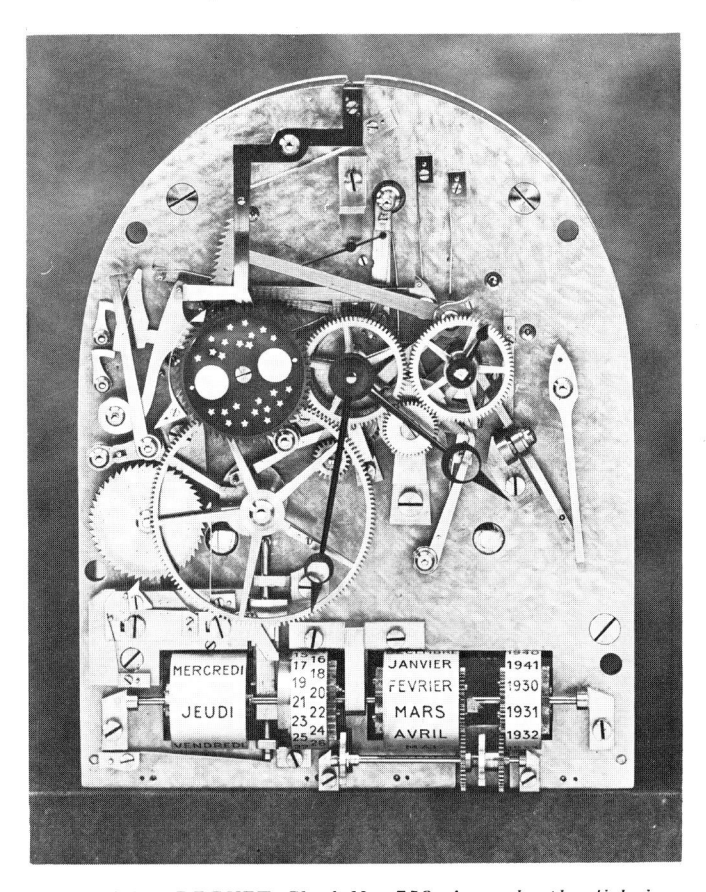

Plate II/15. BREGUET. *Clock No. 759. An under-the-dial view of the movement of the clock shown in Plate II/14. (By courtesy of George Daniels)*

45

II The Beginning

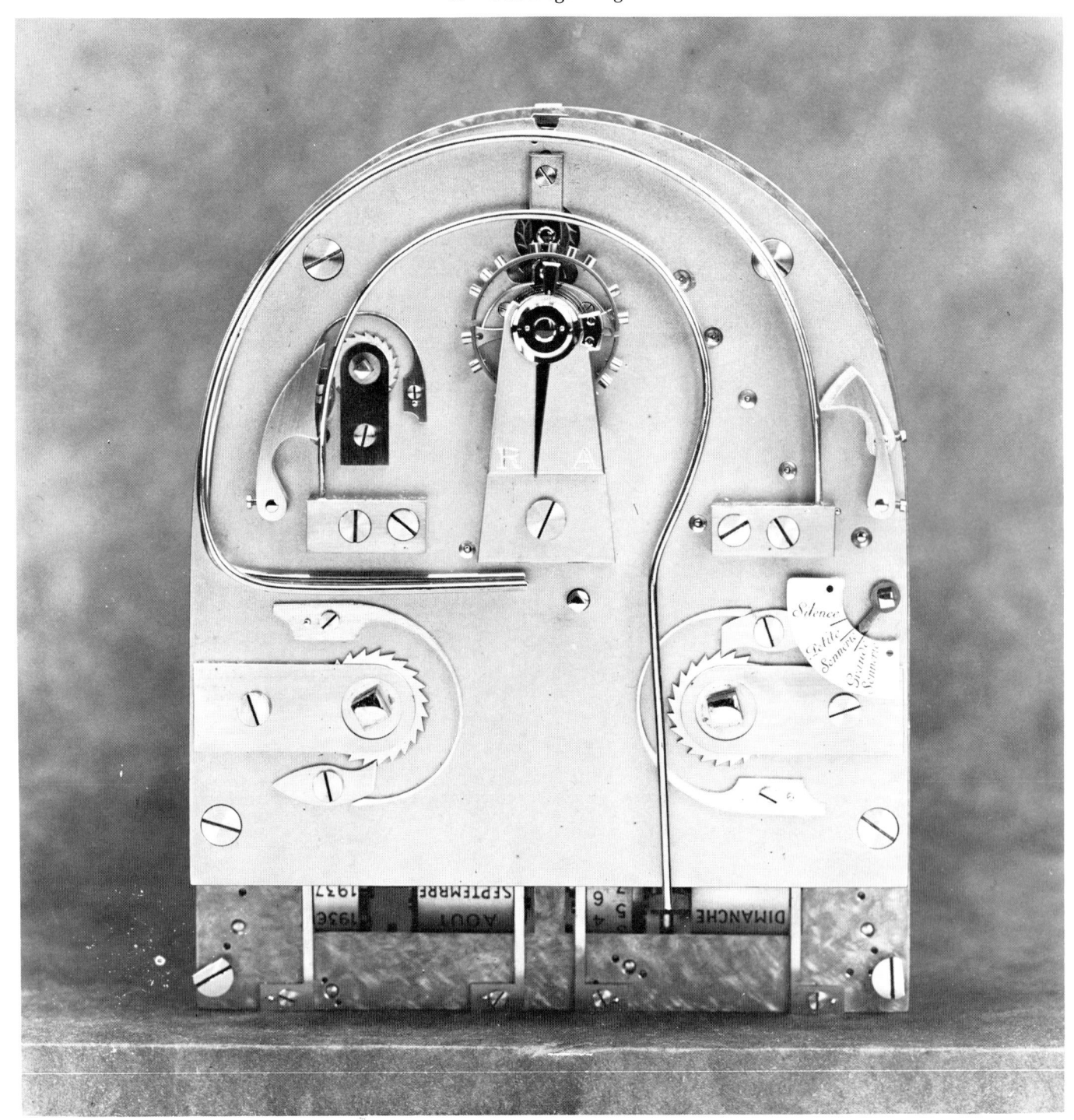

Plate II/16. *BREGUET. Clock No. 759. The back movement plate of the clock shown in Plate II/14. (By courtesy of George Daniels)*

clock ever sold by the Breguet firm. It was begun in 1928 and was delivered to M. Ettore Bugatti on 30th May 1931 for the price of Fr. 60,000. The clock, numbered 759, is by way of being a copy of an earlier piece No. 2940, sold in 1818 to the Comte de Pourtalès. Externally the two pieces are exceedingly similar. The movement of No. 759, however, is very largely Swiss in its origins, the "rough" pinions being made by Samuel Lecoultre of Le Sentier, while the escapement work was produced in La Côte-aux-Fées

II The Beginning

Plate II/17. *BREGUET. A clock completed in 1970 by M. Pitou for the House of Breguet. Note a new case in traditional Breguet Empire style used to house a top quality standard carriage clock* roulant *(rough movement) made perhaps 70 or more years ago but only recently finished and given its escapement. (By courtesy of M. Pitou, Paris)*

Plate II/18. *BREGUET. A rear view of the repeating carriage clock shown in Plate II/17 and completed by M. Pitou in 1970. Note that the alarm setting is done from the back of the clock so as to leave an uncluttered dial. (By courtesy of M. Pitou, Paris)*

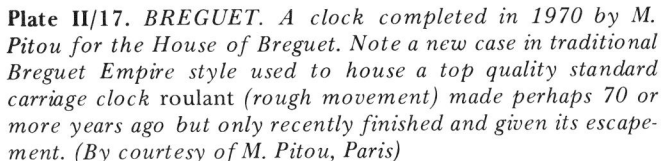

by the Piaget brothers. On the other hand, much work was done in Paris by M. Gastellier, *chef d'atelier* of the Maison Breguet. Plates II/15 and II/16 show two views of the movement from which its modern nature is apparent.

Plates II/17 and 18 show a clock completed in 1970 by M. Pitou who was the last finisher of carriage clocks in Paris.[9] Pitou used top quality standard *blancs roulants* which he had had in stock for upwards of forty years, many of them emanating from Jacot. For the last thirty years of his business life M. Pitou worked almost exclusively for the House of Breguet, finishing *pendules de voyage* which were given gilt metal cases in the traditional Breguet Empire style. Further notes about M. Pitou will be found in Chapter V.

The superb Breguet clocks of the type known as *pendules Sympathiques*[10] were sometimes made in *pendule de voyage* form. A good example of this type was made circa 1822 by Breguet's pupil Raby. It was

---

(9) Another Pitou-type clock, striking *grande sonnerie* was sold at Sotheby's on 16th October, 1972, Lot 193.

(10) A *pendule Sympathique*, one of the most inspired and exciting of A-L Breguet's inventions, winds a special companion watch every day, and also sets it to time. (A simplified version, made in the time of Breguet, Neveu, did no more than set the watch to time.)

II The Beginning

exhibited at the Musée Galliera in 1923 (Item 213), together with its watch No. 722 (Item 129).

It cannot be said that the standard French *pendules de voyage* developed from those by Breguet, although his work helped to establish their general design. His earliest examples were so perfect, and also of such refinement and complication, as to leave no room for subsequent improvement. Manufacture was more or less confined to the period 1810-1900, this embracing not only Breguet's own lifetime but also those years when the firm remained in the Breguet and Brown families, continuing to sell work of the highest order.[11] At a still later date, clocks signed "Breguet et Fils" tended to become rather more ordinary. However, they retained to a late date many of those complicated features for which Breguet was famous. Two Breguet specialities, namely the Empire and "humpbacked" styles of case, have not only been copied ever since, but have persisted to the present time.

SOME REPRESENTATIVE "BREGUET" CARRIAGE CLOCKS AND OTHER PORTABLE PIECES

SERIAL NO. DATE	CASE STYLE ESCAPEMENT HEIGHT	STRIKE REPEAT ALARM	GENERAL REMARKS:—	SOLD TO:— LATER OWNERS:—	WHERE ILLUSTRATED
No. 179 Date: Manufacture probably started prior to 1800.	Gilt-brass Empire case. Pilasters. Lions' heads hold handle. Engine-turned dial. Lever escapt. Height: 14.7cm.	Timepiece ¼ repeater on bell. Alarm on bell at back.	Possibly the earliest unaltered Breguet carriage clock known. Calendar for day, date, month and year. Age & phase of moon. This clock was long thought to have been made in 1804, but research by Mr. G. Daniels has proved otherwise. Certificate No. 2889.	S.M. François de Bourbon (Roi de Naples). Sold in 1810. 1) S.E. Prestige 2) Private Collection	This book: 1 showing both case and movement (Plate II/1). Also illustrated in Sotheby's *Catalogue*, 29.4.68, Lot 5
No. Date: 1806	Wooden case with plain dial and Arabic numerals.	Repeat. Alarm.	Appearance almost identical to clock No. 2020.	Sold in 1806 to Prince de Talleyrand.	Tardy, *La Pendule Française*, Part II, p. 404.
No. 757 Date: 1810	*Pendule Sympathique* Tourbillon.		Gold balance spring. Two barrels. Exhibited Musée Galliera, 1923 (Item 79).	Sold by Moreau to the House of Russia, 31st Dec. 1810 with watch No. 528 for Fr. 8,000.	Not apparently illustrated anywhere at present.

---

(11) Tripplin in *Watch and Clock Making in 1889* records that M. Brown of Breguet exhibited ". . . also a quarter carriage clock, with tourbillon escapement, both hand-made from the beginning to the end at his place, attracted attention by magnificent and unequalled workmanship".

II The Beginning

SERIAL NO. DATE	CASE STYLE ESCAPEMENT HEIGHT	STRIKE REPEAT ALARM	GENERAL REMARKS:—	SOLD TO:— LATER OWNERS:—	WHERE ILLUSTRATED
No. Now 780 Date: Open to question.	Gilt-brass Empire case. Pilasters. Engine-turned dial. Chronometer escapement on tourbillon carriage. Height: 18cm. approx.	*Grande* and *petite sonnerie* and repeat. Alarm.	Calendar for day, date, month and year. Seconds dial. Dial for age and phase of moon. Thermometer. Equation of time. Up-and-down dial. The radiused tops of the movement plates suggest that it was originally destined for a "humpbacked" case. The whole clock presents something of an enigma, and its history is not apparently known.	Not known. 1) Mallet Collection. 2) Ashmolean Museum, Oxford.	This book: 1 of clock (Plate II/2) 1 of back of movement (Plate II/3) Tardy illustrates a clock similar in appearance.
No. 1641 Date: 1810	Silver case. The piece described as *montre de carrosse*. Enamel dial. Turkish numerals.	Striking as it goes, with pull-repeat.	Exhibited Musée Galliera 1923 (Item 226).	Sold to Count Galowskin, *suivant compte de vente de Moreau*, 1st May 1810 for Fr. 1,800. 1) Grande-Duchesse Vladimir of Russia	Not apparently illustrated anywhere at present.
No. 2020 Date: c1815	Mahogany case with gilt bezel. *Echappement naturel.* Height: 18.5cm.	*Grande* and *petite sonnerie.* Semi-circular racks. Repeat. Alarm.	Signed "Breguet" on the dial and "Breguet et Fils" together with No. "2020" on the base of the movement.		This book: 1 of clock (Plate II/4) 1 of movement (Plate II/5) Also illustrated in Sotheby's *Catalogue*, 19.6.72. Lot 194.
No. 2516 Date: 1811	Case described as *Boîte bronze doré.* *Echappement naturel.*	Quarter striking.	Calendar and moon. Exhibited Musée Galliera, 1923 (Item 48).	Sent to St. Petersbourg for Baron de Blome, 29th Jan. 1811, for Fr. 2,000.	Apparently not illustrated anywhere at present.
No. 2767 Date:	Walnut case with ormolu mounts. White enamel dial. Robin escapt. with an additional impulse on exit pallet. Height: 18.5cm.	*Grande* and *petite sonnerie.* ¼ repeat. Alarm.		Not known 1) A lady. 2) Cecil Clutton.	This book: 1 of clock in case (Plate II/6) Also illustrated in Christie's *Catalogue*, 12.7.67. Lot 157.

II The Beginning

SERIAL NO. DATE	CASE STYLE ESCAPEMENT HEIGHT	STRIKE REPEAT ALARM	GENERAL REMARKS:—	SOLD TO:— LATER OWNERS:—	WHERE ILLUSTRATED
Number erased. Date: About 1812	Gilt-brass Empire case. Engine-turned dial. Lever escapement. Height: 16cm.	*Grande sonnerie* only. Repeat.	Calendar for day, month, year. Moon.	Not known. 1) Sir David Salomons. 2) Tel Aviv.	Salomons' *Breguet* (see Bib.) 1 of clock in case (P.313, French edition, P.211, English edition).
No. 2793 Date: 1813	Silver "humpback" case. Described as *pendule de voyage en argent*. Lever escapement. Height: 15.5cm.	Pull ¼ repeat on gong. Pull-wind alarm.	Calendar *à rouleaux* for day, date, month, year. Salomons says originally going for only 3 days, but altered in 1920 to go 8 days. Exhibited Musée Galliera, 1923 (Item 163). Certificate dated Aug. 1920.	Grande-Duchesse de Toscane. Sold 26th Aug. 1813 for Fr. 4,000. 1) Salomons. 2) Tel Aviv.	Salomons' *Breguet*: 2 of movement 1 of clock in case. Galliera *Catalogue*: 1 of clock in case.
No. 2848 Date: 1815	Mahogany case, veneered in thuyawood (or walnut) Robin escapement with an additional impulse on exit pallet. Height: 22cm.	*Grande* and *petite sonnerie* on bells. Repeat. Semi-circular racks. Pull-wind alarm on bell in base.	Signed "Breguet et Fils". Certificate No. 3032	General Ramsey 1) Major Searight 2) Private Collection	This book: 1 of clock in case (Plate II/7) 2 of movement (Plates II/8 and II/9) Also illustrated in Sotheby's *Catalogue* 9.3.64. Lot 179.
No. 2940 Date: 1818	Silver "humpback" case. Described as *pendule de carrosse*.	Striking. Alarm.	Calendar. Moon. Exhibited Musée Galliera, 1923 (Item 102) See also No. 759 (1931)	Sold to the Comte de Pourtalès, Seigneur de Gorgier on 10th Feb. 1818 for Fr. 4,800.	Tardy *La Pendule Française*, Part II, p.459.
No. 3050 Date: 1821	Special escapt. (type not clear) with some form of helical spring.		The movement *blanc* of the clock was supplied by Pons, while Jacob, Couët and Kessels at least had a hand in its manufacture.	Sent to Lord Spencer on 13th July 1821 for Fr. 2,400.	Not apparently illustrated anywhere at present. (Documentation: Col. H. Quill)
No. 3135 Date: 1819	Gilt-brass Empire case. Engine-turned dial. Lever escapt. Height: 14.5cm.	*Grande sonnerie* with semi-circular racks. Repeat. Pull-wind alarm on bell in base.	Calendar for day, date and month. Moon. One barrel drives both going and striking trains. Certificate dated Aug. 1920	Duc de Fernand Nunez. Sold 13th Nov. 1819 for Fr. 5,000 1) Salomons 2) Tel Aviv	Salomons' *Breguet*: 3 of movement 1 of clock in case (pp. 309-312, French edition. pp. 216-219, Eng. edition).

II The Beginning

SERIAL NO. DATE	CASE STYLE ESCAPEMENT HEIGHT	STRIKE REPEAT ALARM	GENERAL REMARKS:—	SOLD TO:— LATER OWNERS:—	WHERE ILLUSTRATED
No. 3620 Date: 1822	Silver case. Described as *pendule portique*. Lever escapt.	Striking as it goes.	Exhibited Musée Galliera, 1923 (Item 104)	Sold to Général de Yermoloff, 20th Nov. 1822 for Fr. 2,700. Sale never completed. Then sold to Pozzo di Borgo, 9th Dec. 1822 for Fr. 2,700.	Not apparently illustrated anywhere at present.
No. 3629 Date: Sold 1822	Silver "hump-back" case. Engine-turned dial Lever escapement.	Pull ½-¼ repeat on one gong. (Repeat train employs an escapt., instead of a fly). Alarm on two gongs, struck with double-headed hammer.	Calendar for day, date, month, year. Seconds dial. Lever banks on 'scape pinion. Triangular-shaped ruby pin. Guard pin probably gold. Divided lift. Brass 'scape wheel (not pierced or slit) with wide tips to teeth. Escapt. has been "bodged". Now has some draw on entry pallet, none on exit. Probably designed to have no draw. This clock has fusee and chain (and maintaining work) instead of going barrel.	Not known 1) Sir J. Prestige 2) British Museum	This book: 1 of clock in case. (Plate II/10) 1 of back of movement. (Plate II/11)
No number quoted. Date: c.1822	*Grande Pendule Sympathique*. Gilt-brass Empire case. Chronometer escapt. Height: 19.5cm.	Striking on gong without repeat. Alarm.	Calendar for day, date and month. Up-and-down dial. Remontoire. Made by Breguet's pupil Louis Raby in 1822. Re-winds and sets to time a special watch, provided that the error, fast or slow, does not exceed seven minutes. Exhibited Musée Galliera, 1923 (Item 213) with watch No. 722 (Item 129) which belongs to it.	Sold in 1822 for a reputed price of Fr. 25,000.	Salomons' *Breguet* 1 of clock in case 1 of back of clock (pp. 303-304, French ed. pp. 220-221, Eng. ed.)
No. 3749 Date: 1828	Silver "hump-back" case. Lever escapement.	½-¼ repeat. Alarm.	Calendar for day, date, month and year. Age and phase of moon. Equation of time. Certificate No. 3022.	Sir Charles Cockerill for Fr. 5,750. 1) R.H. Muir 2) Private Collection Portugal	Sotheby's *Catalogue* 28.10.63, Lot 94 (incl. movt.)
No. 3816 Date: 1826	Gilt-brass Empire case. Pilasters. Engine-turned dial. Lever escapt. Height: 14.5cm.	¼ repeat. Alarm.	Certificate No. 3100	Lord Gower 1) A lady 2) Bell	Sotheby's *Catalogue* 1.11.65. Lot 134

II The Beginning

SERIAL NO. DATE	CASE STYLE ESCAPEMENT HEIGHT	STRIKE REPEAT ALARM	GENERAL REMARKS:—	SOLD TO:— LATER OWNERS:—	WHERE ILLUSTRATED
No. 4464 Date: 1832	Illustrated in Galliera Cat. and described as *pendule à boîte d'argent carrée*. Beyond question a *pendule de voyage* in what is today usually called an Empire case.	*Grande sonnerie* and repeat. Alarm.	Calendar for day, date, month and year. Age and phase of moon. Equation of time. Up-and-down. Exhibited Musée Galliera, 1923 (Item 97)	Sold to Count A. de Demidoff, 4th Dec. 1832 for Fr. 12,940.	Musée Galliera *Catalogue,* 1923 (Appearance of clock very similar to that of No. 780 above)
No. 4663 Date: 1832	Silver-gilt Empire case. Engine-turned dial in silver. Rosette top frieze. Tapered-corner pillars. Lever escapement. Height: 16.5 cm.	Timepiece with ½-¼ repeating. Alarm.	On dial "Breguet et Fils" above chapter VI. Offset seconds at chapter XII. Calendar for day of week and date of month. Misprint in catalogue at Zurich showed number as 4643. Certificate No. 3234.	1) Purchaser in Zurich 2) Private Collection, Essex.	Illustrated in Sotheby's *Catalogue* 27.4.70, Lot 87 and Galerie Neumarkt *Catalogue* 30.4.71. Lot 157. 1 of clock in case.
No. 20 Date: 1833	*Pendule Sympathique*		Exhibited Musée Galliera, 1923 (Item 64). Described as *pendule Sympathique nouvelle construction*.	Sold to King Louis-Philippe, 23 Aug. 1834 for Fr. 600	Not apparently illustrated anywhere at present.
No. Date: 1834	*Petite pendule de voyage et de cabinet* in very elaborate case of silver. Chronometer escapt.	*Grande* and *petite sonnerie*. Repeat. Alarm.	Calendar. Moon. Up-and-down dials. Thermometer. Equation of time. The case was glassed front back and sides in order to show the movement.	Count A. de Demidoff	Mentioned in 1834 *Exposition* but without illustration.
No. Date: 1834	*Petite pendule de voyage et de cabinet.* Gilt-brass case glassed to show movement. Chronometer escapt.	*Grande* and *petite sonnerie*. Repeat. Alarm.	Calendar. Moon. Intended for use both at home and on journeys. Will go in any position. Travelling box gives access to both dial and repeat button.		Mentioned in 1834 *Exposition* literature but without illustration.

II The Beginning

SERIAL NO. DATE	CASE STYLE ESCAPEMENT HEIGHT	STRIKE REPEAT ALARM	GENERAL REMARKS:—	SOLD TO:— LATER OWNERS:—	WHERE ILLUSTRATED
No. 4685 Date: 1835	Gilt-brass semi-Empire case. Engine-turned dial and case top. Straight line lever escapt. Parachute. Double-roller.	Timepiece repeating hours and quarters on a single bell. Repeat train has escapt. instead of a fly. Alarm with pull-wind.	Seconds dial. Calendar for day, month, date. Escapt. has no draw. Lever banks on 'scape pinion. Impulse pin sapphire of very shallow 'D' section. Safety dart in shape of arrow. This escapt. is in form of a top platform. Contrate pinion is recessed into the front plate in order to plant it far enough in front of clock to leave room for the very long lever. Front wind.	Madame Baudin 1) Sir John Prestige 2) British Museum	This book: 1 of clock in case (Plate II/12) 1 of escapement (Plate II/13)
No. 5100 Date: 1834	Ordinary plain brass-and-glass case. Duplex escapt. Height: 22cm.	*Grande* and *petite sonnerie* on gongs. Repeat. Semi-circular racks. Alarm on bell in base.	Signed "Breguet Neveu & Compagnie No. 5100". Both going and striking trains driven by a single barrel. Centre alarm-set hand silvered to avoid confusion with minute hand. Front wind. Certificate No. 3059.	1) Not known 2) Barclay	Sotheby's *Catalogue* 27.7.64. Lot 246
No. Date: 1844	*Petite pendule de voyage et de boudoir.* Special ornate case with filigree work and set with semi-precious stones. Chronometer escapt.	*Grande* and *petite sonnerie.* Repeat. Alarm.	Complicated calendar.	Prince Demidoff	Mentioned in Breguet's *Notice advertissement,* 1844. (see Bib.)
No. 262 Date: 1844	*Petite pendule de voyage et de cabinet.* Chronometer escapt.	*Grande* and *petite sonnerie.* Repeat. Extra plunger for minute repeating.			Mentioned in Breguet's *Notice advertissement,* 1844. (see Bib.)
No. Date: 1844	Several other *petites pendules de voyage.* Chronometer escapt.	*à sonnerie ordinaire* (presumably striking hours and halves in passing).			Mentioned in Breguet's *Notice advertissement,* 1844. (see Bib.)

II The Beginning

SERIAL NO. DATE	CASE STYLE ESCAPEMENT HEIGHT	STRIKE REPEAT ALARM	GENERAL REMARKS:—	SOLD TO:— LATER OWNERS:—	WHERE ILLUSTRATED
Not Numbered Date:	*Petite pendule de voyage.* Lever escapt. with *spiral cylindrique.*	Strike and ¼ repeat.	Calendar. Exhibited Musée Galliera, 1923 (Item 212)		Not apparently illustrated anywhere at present.
No. Date: 1889	Escapt. with tourbillon carriage.	Quarter striking.	Exhibited in Paris *Exposition* of 1889.		Not apparently illustrated anywhere at present. (Documentation: Tripplin, *Watch and Clockmaking in 1889*)
No. 759 Date: 1931	Silver "hump-back" case. Described as *pendule de carrosse.* Engine-turned dial. Lever escapt.	*Grande* and *petite sonnerie* on gongs. Repeat. Alarm.	Manufacture began in 1928. This clock has a perpetual calendar *à rouleaux* for day, date month and year. The design was based upon an earlier clock No. 2940 (which see).	Made for Ettore Bugatti, 30th May 1931 for Fr. 60,000.	This book:— 1 of clock in case (Plate II/14) 2 of movement (Plates II/15 & 16) Documentation: undated description on Breguet writing paper, given to Mr. George Daniels)
No. Date: 1970	Gilt-brass Empire case. Lever escapt.	Hour and half hour strike and repeat. Alarm.	Finished by M. Pitou. (See Chapters II and V)		This book:— 1 of clock in case (Plate II/17) 1 of back (Plate II/18)

PAUL GARNIER

There is no doubt whatever that the first production or semi-mass-produced carriage clocks, entirely standard and satisfactory, were made in Paris from 1830 by Paul Garnier (1801-1869). This splendid achievement has been quite forgotten for more than a hundred years, and is unlikely ever to have been known in England anyway. In France, Garnier's contemporaries were well aware of the facts, which they have recorded faithfully here and there in the now rare chronometric treatises and periodicals published during the second half of the 19th century. To avoid over-statement, it seems best to confine the evidence given in these pages to one brief translated quotation of proof, offering the original text in Appendix (a) together with further important corroborative evidence and with biographical notes. The crucial statement, with reference to the first Paris *Exposition*

II The Beginning

Universelle is to be found in Volume I of the *Revue Chronométrique,* December 1855, page 79. It translates to read: "We waited to see M. Paul Garnier receive his award, and even if his exhibits do not appear impressive at first sight, it is none the less true that M. Garnier is the creator of the Parisian carriage clock industry." Garnier was able to "create" the Parisian carriage clock industry through his introduction of a simple basic design used in conjunction with the escapement for which he had obtained a *Brevet* in 1830.[12] It must, however, be emphasised that Garnier was by no means the developer of the French carriage clock. That honour almost certainly belongs to Breguet thirty two years earlier. It would seem that few other makers offered carriage clocks during the intervening period. In this connection the literature of the 1827 Paris Exhibition[13] is of particular interest as perhaps showing when a change began to take place. On the one hand in 1827 the well-known and inventive Blondeau merited attention for what was probably no more than an extremely complicated coach watch *(montre de voyage).* On the other hand, Lepaute presented a *petite pendule portative,* but there is no evidence whether or not it was a true carriage clock. It was described as having *grande sonnerie* striking, repeat, alarm and calendar; and it is interesting to note that it had what could only have been a platform escapement, probably a lever (". . . échappement libre, trous en pierres fines, montée sur un plateau qui permet de l'élever séparément . . ."). Paul Garnier in the same Exhibition received a Silver Medal for various excellent horological items, but he certainly did not display carriage clocks.[14] However, on the occasion of the 1834 Exhibition[15] Garnier showed a very comprehensive selection of carriage clocks indeed. Some of these were newly developed, being shown for the first time, while others were clearly stated to have been his standard pieces previously in current production. Garnier received a Silver Medal for his various exhibits, while the Report said plainly that he had sold more carriage clocks in the preceding two years than had previously been made in the entire history of horology. It was expressly stated that, because of the low price of the machine-made escapements of Garnier's clocks, they had become a most viable proposition for sale in England. Those new items shown for the first time by Garnier in 1834 included a carriage clock having repeat, alarm and calendar, in addition to an unspecified, but newly-developed escapement, which Garnier proposed to apply to chronometers. Another rather similar carriage clock was described as having "échappement à ressorts." Both escapements, clearly considered to be of a precision nature, were obviously quite different from the "bread-and-butter" frictional-rest invention of 1830 still to be used for many years. The wording of the report is, as so often, imprecise; so it is impossible to be sure of even the general nature of the new escapements. It is, however, overwhelmingly likely that both were nothing more than variants of already known types. Under the heading of *L'Horlogerie Ordinaire,* the report said that Garnier showed a variety of small carriage clocks with repeat and alarm; some striking quarters, some hours and half hours and with an escapement of his own invention. This last was undoubtedly the two-plane frictional-rest escapement of 1830 but now used in rather more ambitious clocks. Before leaving the Report of the 1834 Exhibition it is worth saying that among many other items exhibited by Garnier on that occasion was a watch made with the two-plane escapement, said to have offered the same advantages in watches as it did in clocks.

The Catalogue of the 1839 *Exposition des Produits de L'Industrie Française*[16] said that Garnier obtained yet another Silver Medal in that year for carriage clocks, for a chronometer and various other items. Once again the citation said that he had by then made

---

(12) Garnier, writing to the *Tribune Chronométrique,* said that he first made this escapement in 1829. The wording of the *Brevet,* together with an English translation and also comment on Garnier's letter, appear in Appendix (a).

(13) *Compte Rendu des Produits de L'Industrie Française à L'Exposition de 1827,* pp. 96-105 (Bib. Nat. No. V.38337).

(14) Since these words were written in the early 1970s, it has come to light that Garnier did in fact show carriage clocks in the 1827 Exhibition. Afterwards, he went out of his way to forget all about them (and to mislead historians!). Discussion of this subject, together with much additional Garnier and other information, is planned for publication in 1988 in a separate *Supplement* to *Carriage Clocks.*

(15) *Notice des Produits de L'Industrie Française à L'Exposition de 1834,* pp. 206-207 (Bib. Nat. No. V.38339). Also exhibiting carriage clocks in the same exhibition were Raingo, Deshays, Blondeau and possibly others.

(16) *Rapport du Jury Central,* pp. 224-248 (Bib. Nat. No. V.38342) *Quatrième Commission, Première section: Horlogerie.*

II The Beginning

Plate II/19. *PAUL GARNIER. Clock No. 799. Date circa 1834. This was the first type of carriage clock produced by Garnier. It stands 5¾ins. tall with the handle up and is 3¾ins. in width. Note the "one-piece" case with front glass sliding upwards to give access to the typical Garnier dial with "watered-silk" engine-turning and with painted numerals. A larger version of the "Series I" was also made. No. 1096 is an example. It stands 6¾ins. tall with handle up and it is 4½ins. wide. No. 1127 is the same size as No. 799.*

a very large number of *petites pendules portatives ou de voyage* of his own design,[17] and that nine years of production had done nothing except to show how good they were. In 1839 at least six other firms exhibited carriage clock innovations, while the preamble to the Commission's Report went out of its way to say that the efforts made by distinguished artists in order to perfect de-luxe horology, and particularly small portable clocks, had given birth to a new branch of the industry of prime commercial importance to Paris.

Garnier's unique place in the history of the carriage clock rests upon the cleverness of his beautifully straightforward initial design, used in conjunction with a simple and inexpensive frictional-rest escapement, introduced at a time when travelling clocks were scarce, complicated and costly. In its own field, Paul Garnier's clock was a "breakthrough" of considerable importance, and overnight he made the *pendule de voyage* and the *pendule portative* no more expensive than the ordinary mantel clock of the day. Garnier's early *pendules de voyage* are as plain as Breguet's are elaborate. Plate II/19 shows a typical example of one of Garnier's first production pieces made from circa 1830. These clocks strike hours and half hours, but they do not repeat. While they may look very straightforward and unassuming, they are in fact beautifully made and also thoughtfully planned. Those functional "one-piece" cases are accurately finished, with well-fitting glasses even slotted into the base castings in order to help exclude dust. The movements are particularly good.[18] Each barrel is provided with a stop-work to allow only a few turns of mainspring to be used, and also to permit some variation in initial set-up. The rectangular plates are separated by four completely unornamented pillars. The trains are unusually light and nicely proportioned, the pinions being made of good steel and having really fine (small diameter) pivots even by French standards. There are many other small touches of refinement. For instance, Garnier chose to use rack striking work at a time when most French clockmakers were content with locking-plates. Not only this, but in the early days he took the trouble to support the gathering-pallet arbor by means of a cock beyond the pallet itself. Today, after perhaps as much as one hundred and forty years of almost continuous use, these "Series I" Garnier travelling clocks remain as practical as they are attractive. See Appendix (a) in which lighthearted tests made in recent times are compared with results observed by Dubois in trying a similar clock circa 1850-52.

The above first generation Garnier carriage clocks will be found to have several very peculiar features

---

(17) The term "small portable clocks" in this connection would have included balance-controlled domestic clocks as well as *pendules de voyage* as such.

(18) Very likely the *blancs-roulants* were specially made by Pons of Saint-Nicolas-d'Aliermont. See Chapter IIII.

II The Beginning

associated with their date and origins. Let us take a look at the typical example No. 799 illustrated in Plate II/19. For a start it is housed in a "one-piece" case which includes the extraordinary feature of having a block of wood set in its hollow base casting, and covered with coloured paper.[19] The next point worth noticing is that the clock is wound from the front and has sturdily-made steel hands designed to be set to time with the fingers. The case has no doors. Instead the front glass slides upwards vertically complete with the front top case rail in order to give access to the dial. The dial is made of brass, engine-turned and silvered to a soft finish. Note the scalloped radial pattern within the chapter ring and the vertical ribbon-like (moiré) background used elsewhere on the dial.[20] The chapters are simply painted in black as is the signature on a *cartouche* below. The signature also appears on the top left side of the back movement plate, together with the inscription "Paul Garnier H. DU ROI A PARIS",[21] and under the bell the punched inscription "P.G. Breveté" and the number 799.[22] By far the most unusual and also interesting feature of the clock is to be found in its escapement. This is a development of a two-plane frictional-rest design of the type supposed to have been invented by De Baufre in the 18th century, and which has been "re-invented" at intervals ever since.[23] The Garnier escapement, of which a close-up view is shown in Plate II/20, is built into the clock instead of being carried upon a separate platform. Also because a change of plane is inherent in the design of the es-

Plate II/20. *PAUL GARNIER. The two-plane escapement of a typical early Garnier carriage clock. Garnier devised this escapement in 1829 and patented it in 1830. Note the twin steel 'scape wheels turning in the direction away from the balance and acting upon a single "disc", probably made of sapphire.*

---

(19) The use of these blocks, which are often found in ordinary early 19th century French mantel clocks, seems to be confined in carriage clocks to those sold by Garnier. The purpose was probably to provide inexpensively not only a safe and solid anchorage for the movement, but also a continuous underside to a hollow-based metal or wooden case, thereby making it far less likely to be toppled accidentally off the edge of a table or mantelpiece. The blocks may also have been intended to prevent the metal bases from scratching polished surfaces.

(20) Some early front-winding Garnier *pendules de voyage* will be found to have ringed winding holes. When they do not, it is possible that the holes in the dial have been enlarged to admit inferior modern winding keys of larger outside diameters than the originals. On the other hand, plain holes could well have been a manufacturing economy.

(21) It has not so far proved possible to discover when Garnier gained the right to use the title *Horloger du Roi;* but the fact remains that most, if not all, of his early carriage clocks are so signed.

(22) Back plate signatures show small variations in both lettering and wording, probably the work of different engravers. For instance clock No. 881 has the engraved inscription "PAUL GARNIER HR DU ROI PARIS", but No. 1127 is inscribed "PAUL GARNIER HER DU ROI PARIS". Both are different from the clock already mentioned. It is likely that the production of "Series I" clocks spanned at least the years 1830-1840, although other models were introduced in the meanwhile. That Garnier eventually allowed some of these clocks to be sold under other names is apparent from one numbered 1085 which has the name "Silvani & Cie, PARIS" deeply engraved at the top left-hand corner of the back plate. The acknowledgement "P.G. Breveté" is inscribed nearby.

(23) A rather similar escapement, also dating from the 18th century, was that of Sully. He employed one 'scape wheel and two pallets or "discs", while De Baufre used two 'scape wheels and one "disc". Another variant made in much the same period by Enderlin was very similar to that of De Baufre. See Chapter VIII where a Sully-type escapement used in a clock produced in about 1867 by J. Samuel is described and illustrated.

II   The Beginning

Plate II/21. *PAUL GARNIER. Clock No. 799. The back of the movement. Notice the signature "Paul Garner H. DU ROI A PARIS" and the punched inscription "P.G. Breveté". The number is concealed by the bell.*

Plate II/22. *PAUL GARNIER. The travelling box of the clock No. 799 illustrated in Plates II/19 and 20. Note the red morocco leather covering and the backward-opening top with brass carrying handle and secured by side latches and turnbuckles.*

capement itself, the expense and difficulties of a contrate wheel in the train are avoided. The plain gold balance and the very shape of the balance cock set into and screwed to the back plate, are highly characteristic of Garnier alone. The long blued steel index is also typical. While these early Garnier portable pieces, despite their peculiarities, both foreshadow and set the trend for later conventional standard carriage clocks, they retain much in common with the early to mid 19th century *pendules de Paris* to which they are so closely related.[24] Plate II/21 shows the back movement plate of one of these clocks, while Plate

II/22 shows the type of travelling box sold with Garnier's first standard carriage clocks.

Plates II/23, 24 and 25 are important as showing precisely what is meant by the term *pendule portative* as opposed to the more specific *pendule de voyage*. This Garnier *pendule portative* is numbered 1117 and it was probably made close to the years 1838-40. It is wound from the front and it stands nine inches tall with the handle up. It strikes hours and half hours on a bell by means of a locking plate and, above all, its round-plated movement bears Paul Garnier's name in conjunction with H^{ER} DU ROI" and "P.G. Breveté" inscribed close to the two-plane escapement. A rococo case standing upon a cast brass tray completes the picture of a clock intended to be home-based upon the tray but used about the house in between times. It is emphatically not a *pendule de voyage*. The going train is provided with a peculiar stopwork allowing five turns of the mainspring to be used. Compare this clock with the far less expensive Garnier/Leroy & fils *pendule portative et de voyage* mentioned later on in this chapter. At least one radically different Paul Garnier *pendule portative* exists having a cast brass rococo case. It is numbered 895. This clock which strikes *grande sonnerie* on bells has rectangular move-

---

(24) The name perhaps most commonly associated with the ordinary *pendule de Paris* is that of Henry Marc. Henry Marc clocks are often found in plain wooden cases, relieved by inlay and stringing and having a top "window" and sometimes a carrying handle. Such pieces have well made movements with round plates, locking-plate striking on a bell and anchor escapement with a very light pendulum suspended by a silk thread. Readers wishing to study in detail the design and construction of ordinary French pendulum clocks (*pendules de Paris*) will find detailed descriptions in Moinet's *Nouveau Traité . . . 1853*.

II The Beginning

case, down to the last detail of the glossing and finishing of the movement. It is a timepiece having plunger-wound quarter repeating work, alarm and simple calendar. It is wound and set from the front, the holes in the engraved dial being provided with gold escutcheons to prevent marking caused by careless use of the key. These escutcheons, the planning of the dial as a whole, not to mention the layout of the plunger-wound quarter repeating work used in the movement, suggest some affinity with certain clocks by Breguet (compare for instance with the clock shown in Plate II/1.) This watch-type repeating work employs the same principles of chain and roller used by Breguet in those of his clocks which are basically timepieces. The elegant lever platform is unusual in having the balance cock planted sideways. The right-angled escapement has a steel wheel, divided lift,

Plate II/23. *PAUL GARNIER. PENDULE PORTATIVE. Clock No. 1117. Date circa 1838–40. This is a true* pendule portative *in the exact meaning of the term as distinct from being a* pendule de voyage. *The clock is not provided with an alarm, and it was certainly not intended to be taken on journeys. (By courtesy of E.B. Gent).*

ment plates and employs the justly-famous two-plane escapement. The date of the piece is perhaps circa 1835-40.

Plates II/26, 27 and 28 show both case and movement of a most attractive and exceptionally well-finished Garnier *pendule de voyage*. It was made almost certainly for the Paris Exhibition of 1839; but it is so utterly dissimilar both from the standard Garnier clock already described and from his later ordinary productions, that it must stand to some extent alone and should not be regarded as typical.[25] It has already been emphasised that Garnier's clocks are well-made and finished, and so they are; but this particular piece reflects the most expensive work possible from its elaborate gilded and engraved

Plate II/24. *PAUL GARNIER. PENDULE PORTATIVE. A rear view of clock No. 1117. Note the separate cast base upon which the four feet of the clock are located by steady-pins. The side glasses are bevelled. (By courtesy of E.B. Gent).*

---

(25) "Exhibition work" was always one thing, and everyday work another!

II The Beginning

Plate II/25. *PAUL GARNIER. PENDULE PORTATIVE. Note the use of a pendule de Paris movement in conjunction with Garnier's escapement together with his signature "PAUL GARNIER HER DU ROI" and the inscription "P.G. Breveté". This clock has locking plate striking. (By courtesy of E.B. Gent).*

single roller action and compensated-balance with flat spring. Unlike most Garnier pieces, this special carriage clock is not numbered. While at first it may appear a little surprising to find a "multi-piece" case used at this period, it is reasonable to suppose that Garnier's latest and best models would have been more up-to-date in all respects than his *horlogerie ordinaire* in production at the same period.[26] In Chapter VII (Case Styles) it is explained that the "multi-piece" cases both overlapped and eventually superseded the original "one-piece" style. It seems certain that the Garnier "one-piece" clock shown in Plate II/30, and which will be discussed directly, is later in date than the exhibition "multi-piece". Plate II/29 reproduces illustrations of four Garnier clocks, two having "one-piece" cases and two "multi-piece". The four clocks were presumably available at much the same period. Neither the source nor the date of the illustration is known. The plate, however, must have been published

before 1848 because between the beginning of the 2nd Republic and throughout the 2nd Empire until 1870 the use of the title "Horloger du Roi" would have been out of the question. It is very interesting to compare the clocks of which that in Fig. V (Plate II/29) would seem to be a "Series I" but with the addition of repeating work. The clock in Fig. VII is

Plate II/26. *PAUL GARNIER. Almost certainly an Exhibition piece shown in Paris in 1839.*

much the same, but with calendar and alarm instead of repeat. Both these clocks have "one-piece" cases and almost certainly their designs are older than those shown in Figs. IV and VI. Between the two clocks are found the "multi-piece" case and also the handle of the Exhibition clock. Also featured is a pillared design, besides a *recherché* calendar *à rouleaux*.

A further stage in Garnier's carriage clock development, but still using his original escapement, is repre-

---

(26) Incidentally Breguet's metal Empire cases were never anything else but "multi-piece".

II The Beginning

Plate II/27. *PAUL GARNIER. The movement of the clock shown in Plate II/26.*

Plate II/28. *PAUL GARNIER. The back movement plate of the Exhibition clock shown in Plate II/26. Compare with Breguet clock No. 179 shown in Plate II/1 on page 38.*

sented in Plate VII/4 by a clock made close to the middle of the 19th century. The example shown was made for the famous London firm of Dent (who never had a shop in Paris)[27] but it is a typical Garnier piece, and examples enough exist bearing Garnier's name. Note the use of the now popular "multi-piece" case in a standard clock and how the designer has used canted corners, white enamel dial, trefoil hands, and generally has reduced the area of glass. The fashion portrayed by this example is beginning to look rather more modern, while maintaining some links with tradition.

Plates II/30 and II/31 bring us to another standard carriage clock in the Garnier progression. It is numbered 2563 and bears the legend "PAUL GARNIER H^{ER} DE LA MARINE A PARIS". This clock will be seen to be housed in a late type "one-piece" case having a shuttered back door instead of the earlier system of the upward-sliding glasses. The movement of this clock, apart from the fact that the winding squares have now been transferred to the back, is little different from that of the clock shown in Plate

---

(27) In this connection it is worth recording that the French newspaper *La Patrie* accused Dent at the time of the 1851 Exhibition in Hyde Park of representing Garnier *pendules de voyage* as his own. This slur is largely refuted in *La Tribune Chronométrique*, page 283. One of Dent's brief, and also apparently frustrating, visits to Paris was reported in the *Revue Chronométrique* of October 1877, page 336, under the heading "Monsieur Dent à Paris".

II The Beginning

II/19. No. 2563 has Garnier's escapement and strikes hours and half hours on a bell and repeats hours. It is also furnished with an alarm. Note the trefoil hands replacing the earlier favoured Breguet moon style,

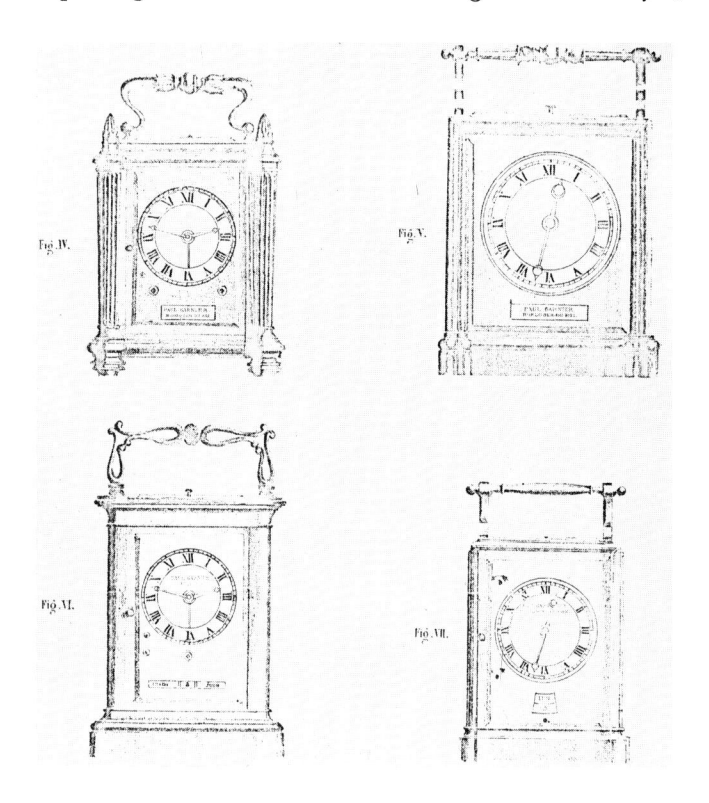

Plate II/29. *PAUL GARNIER. This illustration, from an unknown source but probably dating from before 1848, shows four Garnier carriage clocks all presumably available at the same period. Note that two clocks have "one-piece" cases and two "multi-piece" cases, and that both front and rear winding are used.*

while the white enamel dial set in an engraved brass surround has superseded the silvered dials of the two earlier pieces. The date of this clock, to judge from the signature, is post 1848, and probably "Horloger de la Marine" implies service to the Second Empire which began in 1852.

A final clock in terms of date, at least so far as the movement is concerned, is shown in Plate II/32. This example bears the unexpected signature "PAUL GARNIER H^{ER} DU ROI PARIS" and is numbered 2982. This clock has little externally to distinguish it from any late 19th century standard French carriage clock. The escapement is a normal lever platform. The movement is housed in an early *Corniche* case of unusual proportions, 7 inches tall by 4½ inches wide by 3½ inches deep. The clock strikes hours and halves, and repeats hours on a bell which is also used for the alarm. However, the great interest of this piece lies in the question of date which it poses. With such a case and dial it would not be reasonable to believe that the clock could possibly have been made before 1848; and yet it bears the signature "H^{ER} DU ROI" which, as will be explained at the end of this chapter, was absolutely unthinkable in France between 1848 and 1870. No-one would have dreamed of using such a title, simply because it was far too dangerous. This fact must leave the date when the clock was sold as being post 1870. There seems little doubt that after this period some old firms (L. Leroy & Cie among them) on occasions at least made use of their old pre-revolutionary titles.

Plate II/30. *PAUL GARNIER. Clock No. 2563. Date post 1848. A late "one-piece" case used in a standard clock. Note trefoil hands and also dial reflecting a change in fashion. (By courtesy of Charles Terwilliger, New York).*

II The Beginning

escapement with a steel disc, in addition to an alarm mechanism, and a stopwork on the going barrel only. The inlaid wooden case has a flush-fitting top carrying handle, while a rectangular well in the base shows that, as made, it was intended to house a pendulum movement. Hand-written operating instructions are gummed to a solid tuck-in back door. A quaint leather-covered

Plate II/31. *PAUL GARNIER. A rear view of the clock No. 2563 shown in Plate II/30. Note the shuttered winding holes in back door. (By courtesy of Charles Terwilliger).*

Plates II/33 and 34 bring us to Garnier No. 3077 probably made about 1870 or even later. It stands 7 inches tall and the pillared case is very heavy. The movement is wound from behind and offers hour and half hour striking, repeat and alarm. It is signed "PAUL GARNIER H^{ER} DE LA MARINE" and it is furnished with a lever escapement. The white enamel dial has blue chapters.

The reader may now be interested to consider a wooden-cased travelling clock of a very peculiar and old-fashioned disposition. It is illustrated in Plate II/35. The round-plated movement, which is numbered 315 and which never had a pendulum, has engraved on the back plate "Leroy & Fils H^{er} du Roi à Paris". Above this inscription is the punching "P.G. Breveté". While basically a typical *mouvement de Paris*, this example is different in having Garnier's

Plate II/32. *PAUL GARNIER. Clock No. 2982. Note the late-type broad* Corniche *case, white enamel dial and trefoil hands found in conjunction with the contradictory inscription "H^{er} Du Roi". (Private Collection).*

travelling box completes a *pendule portative et de voyage* which is certainly something of an enigma. In theory at least, assuming the serial number 315 is Paul Garnier's, this clock should have been made circa 1830, and it is certainly most logical to believe that the first applications of the two-plane escapement would have been to round-plated movements. However, while it is very tempting to conclude that this unsophisticated travelling clock may represent some form of Garnier *pendule de voyage* prototype, the

II The Beginning

Plate II/33. *PAUL GARNIER. Clock No. 3077. Note the signature "PAUL GARNIER H^{ER} DE LA MARINE PARIS". This clock, which has a lever escapement, was probably made after 1870. (By courtesy of E.B. Gent).*

Plate II/34. *PAUL GARNIER. Clock No. 3077. A rear view of the movement of the clock shown in Plate II/33. Note that the number of the clock, which is hidden by the bell, is repeated on the back door of the case. (By courtesy of E.B. Gent).*

signature "Leroy & Fils"[28] establishes that, whenever the piece was made, it could scarcely have been sold before 1839. By this time, as the reader will know, Paul Garnier had not only sold a very large number of *pendules de voyage* but also had a variety of standard types in full production. In fact, Garnier's escapement is found in a number of widely differing small domestic clocks, most of them having round-plated movements but apparently being made, or at least finished and sold, at quite different periods.[29] One early example (Plate II/36) is mounted for exhibition purposes on a satinwood base under a glass dome and has mainsprings bearing the scratched inscription "Janvier 1831". The movement is numbered "14" under the bell along with the inscription "Pons MEDAILLE D'ARGENT 1823". A similar clock is signed "Raingo Frères". Most of these domestic clocks could be classed as *pendules portatives,* but they are emphatically not *pendules de voyage*.

Plate II/35. *PAUL GARNIER. PENDULE PORTATIVE ET DE VOYAGE. This wooden-cased clock, bearing the name of Leroy & Fils, was almost certainly sold after 1839 at a time when conventional* pendules de voyage *were in full production. Note the rather quaint leather-covered travelling box and also the alarm-setting hand. The round-plated movement, which has Garnier's escapement and which bears his punched inscription "P.G. Breveté", is virtually identical with that of the metal-cased Garnier pendule portative No. 1117 shown in Plates II/23, 24 and 25.*

---

(28) This signature suggests that the clock was sold by the Galerie Valois firm (see Chapter V).
(29) It would appear that not all of them were made by Garnier, or under licence from Garnier. See Appendix (a), footnote (2).

64

II The Beginning

Plate II/36. PENDULE PORTATIVE. *A clock included to show that the term* pendule portative *did not necessarily imply anything approaching a travelling clock. (By courtesy of John Kendall. Photograph by F.W. Mancktelow, Sevenoaks).*

Despite the very important contribution made by Paul Garnier to horology in general, and to carriage clocks in particular, he also had many interests and achievements in other spheres. As early as 1829 he presented to the Académie des Sciences a medical instrument in connection with the circulation of the blood. This was followed by all manner of counters[30] and other instruments in connection with industry, railways and steam boats, while by 1845 Garnier was associated with the first *télégraphie électrique*. In 1847 he presented to the Institut de la Société d'Encouragement the first "master and slave-clock" system ever made in France.

The son of the distinguished Garnier, also Paul, carried on the business after the death of his father, receiving awards in 1878 and 1889. Paul Garnier II formed a notable collection of early watches which in 1916 was given to the Louvre. An illustrated catalogue was published in 1917.

EXHIBITIONS

As already made plain, the parts played by both Breguet and Paul Garnier in the history of the carriage clock are not in doubt. These two men, in their totally different ways, made such significant contributions that evidence is not hard to find. In the case of other early makers, there is considerable difficulty in establishing the facts. Today, almost the only way of discovering the names and dates of at least some of a number of firms offering carriage clocks during the first half of the 19th century, lies in perusal of catalogues and reports relating to the various Exhibitions staged in Paris. These *Expositions* were of two main kinds. In the beginning, they were simply *Expositions des Produits de l'Industrie Nationale;* but in time they became *Expositions Universelles,* in all that the term implies.

The National *Expositions* developed from being mere *fêtes*. They were first held at the end of the 18th century each September from 1798 to mark the anniversary of the establishment of the First Republic. They became full-scale displays of French industry and agriculture. They took place at various sites, but always somewhere in Paris. A brief history is included in a preamble to the catalogue of the *Exposition* of 1834, under the heading *Historique des Précédentes Expositions* (Bib. Nat., Paris, No. V. 38339, pages III to VII).

Exhibitions of the National type were held in 1798, 1801, 1802, 1806, 1819, 1823, 1827, 1834, 1839, 1844 and 1849.[31] The *Revue Chronométrique,* in its first issue of June 1855, says that only three horologists, of whom Breguet was one, exhibited anything at all in 1798. No carriage clocks were shown.[32] The article goes on to say that in subsequent *Expositions* more and more horology appeared, with F. Berthoud, Breguet, and Antide Janvier receiving Gold Medals in 1802. The Bibliothèque Nationale in Paris does not have readily accessible any catalogues of National Exhibitions before 1827, although Monsieur Jean-Claude Sabrier of Acquigny found some from 1819. To judge from these catalogues few *pendules des voyage* were shown in Paris between 1819 and 1827. It seems fairly unlikely that such clocks were shown before 1819, in which year Breguet et Fils showed, as already noted, several complicated examples. From

---

(30) Garnier used his two-plane escapement even in his counters, signing himself "Paul Garnier Ing^{eur} M^{cien}".

(31) M. de Champagny, who will be mentioned in Chapter IIII as being responsible for sending Pons to re-organise the horological industry in Saint-Nicolas-d'Aliermont in 1806, was in the same year, as *Ministre de l'Intérieur*, the enterprising organiser of the very successful *Exposition*, staged at a time when French industry as a whole had reached a low ebb and badly needed revitalising.

(32) Couturier showed a clock beating decimal seconds; Breguet a new free constant-force escapement in a *pendule Sympathique*, and also a *chronomètre musical;* Lemaire a clock playing on flutes (an organ clock?) and a *boîte à carillon*.

II The Beginning

EXPOSITION DE 1867,

Plan de l'Exposition de l'horlogerie française

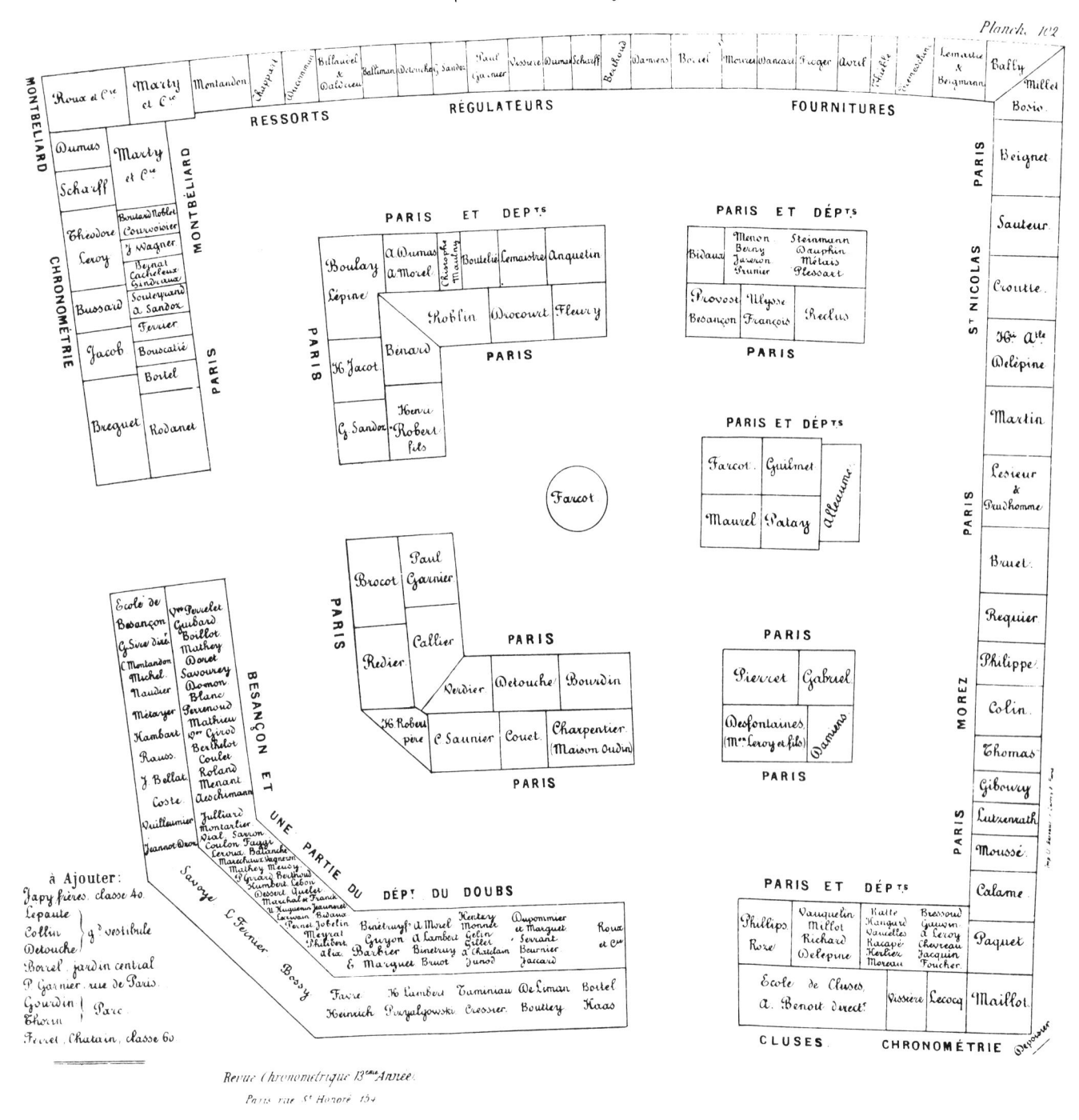

1819 there are two kinds of documents. The first are the *Catalogues Officiels*. These are not a great help since they are just lists of the firms exhibiting. Admittedly the lists supply both the names of the firms and their places in the *Expositions;* but no information whatever is given about their products. On the other hand there are four other types of document which are far more helpful. These are respectively the *Catalogue Explicatif et Raisonné des Produits les Plus Remarquables admis à l'Exposition, Le Compte Rendu des Produits de l'Industrie Française à l'Exposition,* the *Rapport du Jury d'Admission,* and the *Rapport du Jury Central sur les Produits de l'Exposition,* which are variously available. Unfortunately, none of these mentions, much less describes, everything which was exhibited in any year. The fact is that usually nothing was included unless it was considered to be either important or new. As a result, once "ordinary" carriage clocks had come to be taken for granted then neither they nor their makers were necessarily mentioned at all. In theory at least, for a clock to figure in a catalogue it had to embody either some innovation of design, or else to show a technical advance, such as a new form of repeating mechanism.

Early makers found in *Exposition* documents include Bruneau, Raingo, Breguet, Le Roy (Leroy), Garnier, Blondeau, Campbell, Deshays and Berolla, all showing carriage clocks; and Pons, Cailly, Delépine and Douillon of Saint-Nicolas-d'Aliermont exhibiting clock movements which would inevitably have included some for *pendules de voyage*.[33] It has not so far proved possible to trace documents and catalogues relating to all the pre-1827 Exhibitions. There is ample scope for others in the future to go into the matter far more fully. As already noticed, the French *Expositions Nationales* all took place in Paris. Some were held in the Louvre, for instance that of 1819. The reporter on Horology at the 1839 *Exposition* (M. Mathieu, *Président,* working in conjunction with MM. Pouillet, Savart, Savary and the Baron Séguier) says in his introductory note "Already the efforts made by distinguished artists towards the perfection of de-luxe pieces, and particularly of small carriage clocks, has brought into being a new branch of industry of great commercial importance to Paris."

In 1855 the French staged their first *Exposition Universelle.* This was a very vast and grand affair indeed, showing products of many descriptions from countries all over the world. Horologically speaking, the 1855 *Universelle* sought to redress the poor showing which the French felt that their own people had made in London at the Great Exhibition of 1851 at which many had not even troubled to exhibit. (*Tribune Chronométrique,* p.256, and *Revue Chronométrique,* June 1855, p.5). A still more successful *Exposition Universelle* was held in 1867. It was followed by others, the size and scope always increasing, held in 1878, 1889 and 1900. These latter events are documented in the *Revue Chronométrique,* published from 1855. The *Tribune Chronométrique,* which had a short life from 1850, covers fairly fully the French *Exposition* of 1849, and the London Great Exhibition of 1851.[34] The Universal *Expositions* became of such importance that the *Revue* of 1867 republished its report as a separate volume. This book contains some 328 pages, including an index and many folding plates. There are also plans showing the countries involved, besides a layout of the French Horological Section indicating the Stands, names of Exhibitors, and roughly what they showed. The Sections are grouped under the titles *Paris et Départements, Besançon et une Partie du Dép. du Doubs, Montbéliard, St. Nicolas* and *Morez.*

The Paris *Exposition Universelle* of 1889 attracted attention hitherto without precedent. Parts of the descriptions written by T.D. Wright and by J. Tripplin, the English reporters, are quoted in Appendix (b). As will be noted more than once in this book the year 1889 saw the heyday of the French carriage clock when both its popularity and its production were at a peak.

Latterly, there were set up, or at least proposed, for Orléans, Besançon (1860), Nantes (1861), Dijon, Marseilles, Bordeaux, Metz, etc. *Expositions Universelles Permanentes.*[35] There is no doubt that the London Great Exhibition of 1851, which was the first-ever International Exhibition, spurred on the French to try to surpass it in their own *Universelle* of 1855. The London Exhibition of 1862 was horologically speaking such a successful English reply that the *Revue Chronométrique* devoted space to little else for many issues on end. On that occasion at least fifty French firms exhibited in London, and plenty of carriage clocks were shown.

---

(33) Breguet No. 3050, for instance, used a Pons *blanc.*
(34) Leroy et Fils of Galerie Montpensier 13 & 15 Palais Royal showed a minute-repeating carriage clock in London in 1851. (*Tribune Chron.* page 256).
(35) Vide *Revue Chronométrique* of December 1861, pages 72 to 74 (continued in later issues).

II The Beginning

LEROY

An important point which must be emphasised is that neither of two firms, the one at first spelling their name Le Roy but latterly Leroy, and the other normally but not always spelling the name as one word, had any connection with Julien and Pierre Le Roy of chronometer fame.[36]

Plate II/37. THEODORE LEROY. *Date circa 1827. A "giant" carriage clock with* grande sonnerie *striking. (By courtesy of P.H.C. Mathieson).*

The top award given to *Exposants Horlogers* at Paris Exhibitions was the *Décoration de la Légion d'honneur*. Next came the *Grand Médaille d'honneur (en Or)*, then *Médailles d'honneur (en Or)*, *Médailles de 1re Classe (Argent)*, *Médailles de 2e Classe (Bronze)*, and finally *Mentions honorables*.

Other important Exhibitions held between 1850 and 1900 included Besançon in 1860, Vienna 1873, Sydney N.S.W. 1879, Melbourne 1881, and Amsterdam in 1883.

Plate II/38. THEODORE LEROY. *The movement of the "giant" clock shown in Plate II/37. Note the early design. (By courtesy of P.H.C. Mathieson).*

---

(36) The eminent Pierre died without issue. None of his three brothers (Jean Baptiste, Julien David, or Charles) produced any sons; so the line died out completely. Antide Janvier was explicit in *Manuel Chronométrique* (1815 edition, p.246, 1821 edition, p.213) when he wrote "Julien Le Roy n'a plus d'héritier de son nom, exerçant l'art de l'Horlogerie". The same fact was reiterated in 1937 (see *Tribunal* in Bibliography) when it was stated in the *Jugement* "Pierre Le Roy est décédé sans enfant, ayant légué ses biens à ses trois frères dont aucun n'a été horloger et n'a laissé de descendance masculine".

II The Beginning

The name Leroy appears on an early *grande sonnerie* carriage clock made by Theodore Leroy circa 1827, and which is shown in Plate II/37. In fact, this is a "giant" clock, standing 8¾ inches high. Its movement, as shown in Plate II/38, will be seen to be un-carriage clock-like in its conception. This early Leroy *pendule de voyage* still pursues practices inherited from the traditions of an earlier era. The plain frames and pillars, the clickwork mounted on the front plate, the large mainspring barrels and heavy wheelwork, are all reminiscent of house clocks of the Louis XVI period. This style of work is well demonstrated in fig. 34 of Winthrop Edey's *French Clocks*, published in 1967.[37]

The "giant" Leroy carriage clock is signed simply "LEROY A PARIS". It has an early "multi-piece" case and circular white enamel dial set in an engraved gilt-brass surround. Other than Breguet's first *pendules de voyage,* this Leroy piece is about as early an example of a French carriage clock as one is likely to find. It would have been made before 1827, because in this year Theodore Leroy obtained for himself and for his new partner Auguste Pierre Lepaute the *Brevet des Horlogers du Roi.*[38] It is surely inconceivable that Theodore would have omitted to record his recent and hard-won honour upon a clock which was a fine example of a then almost new innovation.

There were two firms both in the Palais Royal in Paris during much of the 19th century whose clocks bear the name of Le Roy or Leroy. This fact is far from being generally known, much less understood. The situation is further complicated because even their Paris and London contemporaries habitually confused them.[39] Even further confusion subsequently arose when the dials of clocks were restored; the name then being spelt either way quite indiscriminately. In fact, the situation became so serious that eventually twenty-five years of litigation resulted between the two firms. The matter was only resolved in 1960 resulting in the closing of the firm originated (but subsequently sold) by Theodore Leroy. Further notes on the two houses, whose addresses during the greater part of their careers were Galerie Montpensier,

Plate II/39. *RAINGO FRERES. A fairly early carriage clock. Note the moiré dial. It is probable that the hands are a later addition. (By courtesy of The Antique Collectors' Club, Woodbridge, Suffolk).*

13 and 15 Palais Royal, and Galerie Valois, 114 and 115 Palais Royal respectively, are given in Chapter V.[40]

OTHER EARLY MAKERS

It is one thing to notice the names of makers of early carriage clocks exhibited in the Paris Exhibitions, but quite another to find examples of their work. Carriage clocks by Raingo (Plate II/39), Campbell (Plate II/40) and Berolla, although rare,

---

(37) Another clock signed "Auguste Paris", having both movement and case bearing strong resemblance to the Leroy "giant" is shown in Plates VIII/36-38. It is described in Chapter VIII.
(38) All the correspondence exists in the Conservatoire National des Arts et Métiers in Paris.
(39) This fact is repeatedly demonstrated in the literature relating to the Paris Exhibitions where, because of inconsistencies in addresses and in the spelling of names, it is now difficult to decide which firm was exhibiting.
(40) This book is indebted to MM. Pierre and Philippe Leroy of Leroy, 4 Faubourg Saint-Honoré, Paris for access to many family documents, as well as to M. J-C Sabrier, who photo-copied these documents.

II The Beginning

Plate II/40. *CAMPBELL. A clever composite photograph showing three views of a fairly early carriage clock. (By courtesy of Charles Terwilliger, New York).*

are far from being unknown; but the same cannot be said for examples of the work of Bruneau, Blondeau and Deshays, who are also on record as having exhibited carriage clocks prior to 1840. The work of Deshays in particular might be of importance because he is on record as having produced by 1839 what was probably some type of jig for drilling holes in clock movement plates.

Plate II/41 shows an early carriage clock in a "one-piece" case and having a duplex escapement.[41] This is a very high class clock, and it is signed on the dial "LEPINE A PARIS". The main firm of Lépine, which

---

(41) Carriage clocks with duplex escapements are nearly always fairly early in date. An exception is a late 19th century clock made by Japy of Beaucourt. This piece has a double-duplex or "Chinese-duplex" escapement which is illustrated in Plate VIII/24. The reason for the re-introduction of an escapement by then long considered obsolete is that this clever variant enables a centre-seconds hand to "show seconds" from a balance beating quarter-seconds. In pendulum clocks, escapements of the type classed as *coup-perdu* enable seconds to be shown from a half-seconds pendulum. The name "Chinese-duplex" is derived from the fact that this escapement was used extensively in elaborately decorated centre-seconds watches made in the 19th century by the Swiss for export to the Orient.

II The Beginning

Plate II/41. *LEPINE. Date circa 1840. Carriage clock with duplex escapement. Note the early "one-piece" case with sliding back glass. (Private Collection. Photograph by P.H. Kohl).*

Plate II/42. *LEPINE. A top view of the movement and duplex escapement platform of the Lépine clock shown in Plate II/41. (Private Collection. Photograph by P.H. Kohl).*

was started in the 18th century by Jean, continued in the same name under various ownerships until about 1916. The carriage clock shown was, however, made close to the year 1840 and it is fully representative of the first generation of post-1830 *pendules de voyage*. We have purposely shown a rear view of this clock because everything is done from the back, access to the winding and setting squares being obtained by removing the glass vertically. The dial of this clock is of silvered brass having a sunk centre and painted-on Roman chapters. The movement is of unusually high quality and was certainly made with little regard to cost, there being much expensive hidden work such as pinion heads hollowed and glossed. The duplex escapement, Plate II/42, is very well made and was produced complete with its platform by some special-

ist concern. It has a plain gilt brass balance, flat balance spring, ruby roller and a ruby impulse pallet set chronometer-style in a large roller. The balance spring stud is arranged to be movable radially in order to help set the spring "in circle". This clock, while it may appear superficially rather similar to the first pieces sold by Paul Garnier, was in fact very much more expensive to produce in the first instance.

Bolviller was another of those makers associated with the fundamental period of early carriage clocks. His clock No. 2 shown in Plate VII/3 is very typical of the earliest model which will be encountered bearing his name. Its appearance and beautifully dignified early "multi-piece" case and fine white enamel dial is fully in accord with the more expensive carriage clocks of the period close to 1840. This clock, standing seven inches high, strikes hours and half hours by means of a rack and snail and has an alarm. It also repeats hours. The high quality is emphasised by the calendar and seconds dial. The latter feature shows the maker's confidence in the timekeeping abilities of the finished clock. Note the delicate Breguet-style "moon" hands, far removed from the somewhat coarse and heavy pattern favoured by Garnier in his first standard clocks. This Bolviller piece deserves a more than cursory description. The case has a large area of glass on all sides. It is hand-engraved and gilded all over, including the solid back panel which has shutters

II The Beginning

Plate II/43. *BOLVILLER. Clock No. 2. Date close to 1840. The club-toothed lever escapement of the beautiful* pendule de voyage *shown in Plate VII/3. Very similar escapement platforms were also used by Jules, by Auguste and by Beguin at about the same period. (Private Collection).*

Plate II/44. *BOLVILLER. A view under the dial of clock No. 2 shown in Plates VII/3 and II/43. Note an hour rack cranked in order to avoid the seconds pivot. (Private Collection).*

Plate II/45. *BOLVILLER. Clock No. 67. A less characteristic piece, having chronometer escapement and incorporating a musical box which at each hour plays one or another of two tunes alternately. (By courtesy of Charles Terwilliger, New York).*

covering the winding and setting holes. The club-tooth straight-line lever escapement (Plate II/43) is in the form of a platform with a circular sub-frame. It is planted deeply into the movement of the clock in order to leave room inside the case to accommodate the height of the sub-frame and balance bridge. The brass and steel balance is "cut". This means that it was intended to compensate for variations in temperature and is not just an ornament. The brass "blocks" and timing screws show the influence of Earnshaw, the great English chronometer maker. The balance has a helical spring, and is provided with a regulation index pointing to a scale on the back of the platform. Under the dial, as will be seen in Plate II/44, the hour rack is cranked to avoid the seconds pivot. Once again, this clock by Bolviller was expensive in its day. It drops

II The Beginning

Plate II/46. *AUGUSTE. Clock No. 357, date circa 1840. The platform escapement of this piece is very similar to those found in both Bolviller No. 2 and Jules Nos. 844 and 847. Note a very early engraved "multi-piece" case, looking like a "one-piece". (Private Collection).*

Plate II/47. *JULES. A typical Jules platform escapement as found in clock No. 844. Compare with that of Bolviller No. 2 shown in Plate II/43. (Private Collection. Photograph by P.H. Kohl).*

Plate II/48. *JULES. The movement of a typical Jules clock numbered 844. Note the locking plate striking on a bell, the radiused corners to the movement plates and a platform escapement deeply sunk into the frames. (Private Collection. Photograph by P.H. Kohl).*

no stitches and gives no indication of being other than the best of its kind and generation.

A very different and rather later clock signed "BOLVILLER A PARIS" is seen in Plate II/45. The gilt brass case of this clock is cast in deep relief with a bird motif and stands 8¼ inches high with the handle up. The movement strikes hours and half hours on a bell, also offering repeat and alarm. There is a chronometer escapement. In the base of the clock, which accounts for its extra height, is a musical box playing two tunes, one every twenty-five minutes after the hour. The clock is numbered 67.

Other early makers contemporary with Bolviller were Auguste, Jules and Beguin. All of these sold clocks with strong resemblances to each other. Plate II/46 shows Auguste No. 357. It is very similar in

73

II   *The Beginning*

Plate II/49. *JULES. Clock No. 847. The early Jules' clocks are sometimes found in early "one-piece" cases, either plain or engraved, and sometimes in ornate "multi-piece" cases of which the above is an example.*

Plate II/50. *JULES. Clock No. 133. Date circa 1840. This clock has rack striking. It strikes hours and half hours and repeats hours. Note the engraved "one-piece" case. The highly unusual platform escapement of this clock is shown in Plate II/51. It is not like the usual Jules/Bolviller/Auguste platforms. (Private Collection).*

many respects to the rack-striking Bolviller No. 2. Other clocks in much the same idiom, but having locking-plate striking, are Jules No. 844 (Plates II/47 and 48) and Jules No. 847 (Plate II/49). Both these clocks have sub-frame, divided-lift escapements and also "multi-piece" cases. A somewhat different clock is Jules No. 133 (Plates II/50 and 51). It has locking-plate striking and a "one-piece" case, but its chief interest lies in its very unusual escapement. This is an "underslung" pointed-tooth lever, about as different from the club-toothed sub-frame platforms as it is possible to imagine. The top surface of the platform, which would otherwise be distinctly devoid of interest, is decorated with straight-line engine-turning. We suppose the period of all those clocks mentioned in this paragraph to be circa 1840. At least some of them may be more closely related than has been hinted. The same basic movement could so easily have been employed to accommodate either rack or locking plate striking. It would only have been necessary to plant the gathering pallet wheel in a different place, if rack striking was required, in order to achieve the correct positions for the trains and *cadratures*. Who is to say that at least some of these basically similar early clocks, having sub-frame escapements and with low serial numbers, are not in fact all part of the same manufacturing sequence, despite the different names which appear upon them?[42]

---

(42) An entirely different Auguste clock, which does not have a serial number at all, and which in many ways resembles the "giant" Leroy clock shown in Plates II/37 and 38, is described and illustrated in Chapter VIII.

II The Beginning

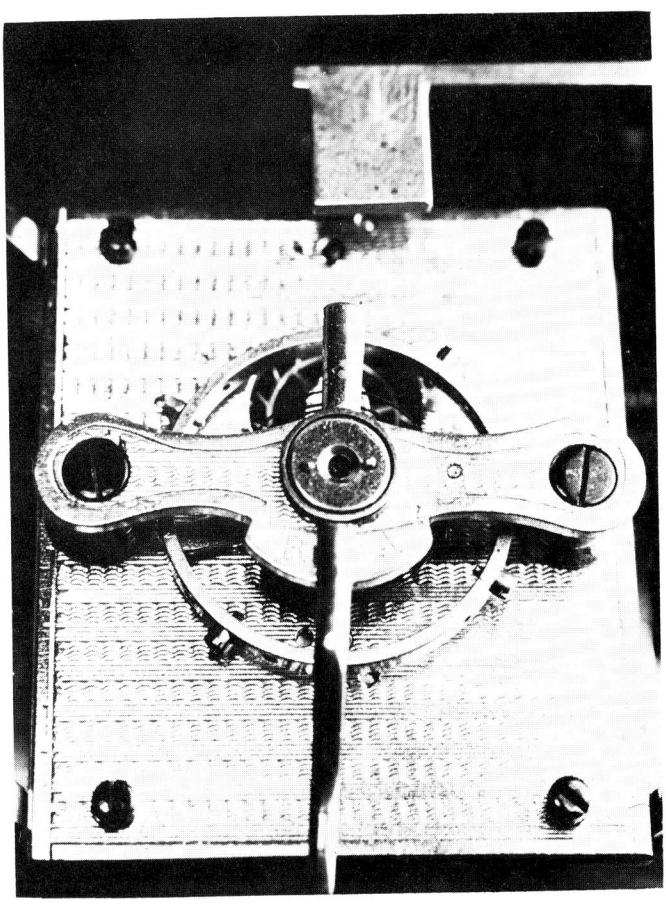

Plate II/51. *JULES. The unusual platform escapement of clock No. 133 shown in Plate II/50. Note the "underslung" pointed-tooth lever escapement and a platform decorated by straight-line engine-turning. (Private Collection).*

Plate II/52. *EARLY MASS-PRODUCED CYLINDER PLATFORM ESCAPEMENT. This platform was probably made after 1850 somewhere "in the Doubs".*

Plate II/52 is interesting because it shows the escapement platform of a clock made circa 1850 and which has an early "one-piece" case, trefoil hands, early enamel dial, but no maker's name. The comparatively crude cylinder platform escapement must be an early attempt to market a mass-produced component for a portable clock, and there is little doubt that it would have been made in the Franche-Comté, and probably somewhere "in the Doubs". The platform is very unlike the ones made in large quantities at a later date. Despite a manifestly cheap-as-possible construction, the cylinder of this escapement remains unworn and still performs well after perhaps as long as some 120 years.

A clock signed on the dial "D.C. RAIT PARIS" is shown in Plate II/53. It is worth noticing because, while it is a comparatively early clock, almost certainly made before 1865, its "multi-piece" case seems to foreshadow the standard *Corniche* which was to follow. Plate II/54 shows a rear view of the movement which has a platform lever escapement and a peculiar form of locking plate striking in which the warning piece and the lifter are planted on the back plate.

The final illustrations in this chapter, Plates II/55, 56 and 57 show an unusual small carriage clock in a "multi-piece" case and signed "PINCHON A PARIS". This piece, which has repeat and alarm, has both its going and its striking trains driven from the opposite ends of a common mainspring. The movement is numbered 3. While the platform escapement of this clock is in the English manner, it is worth saying that many Franche-Comté escapements were deliberately finished in the English style using single

75

II The Beginning

Plate II/53. *D.C. RAIT. A striking carriage clock in early "multi-piece" case, possibly foreshadowing the standard Corniche. (Private Collection).*

Plate II/54. *D.C. RAIT. The movement of the clock illustrated in Plate II/53 showing the peculiar locking plate striking in which the warning piece and lifter are planted on the back plate. (Private Collection).*

rollers, plain steel balances and pointed tooth 'scape wheels.

Among the early carriage clocks, on the whole the first examples made tended to have absolutely plain parallel-sided pillars. These gave way gradually to ornamentation which at first took the form of one trumpet-end and one turned rib. Later pillars were tapered and ribbed at either end.

HORLOGERS DU ROI

Since the title *Horloger du Roi,* which carried great prestige, appears upon a number of French carriage clocks, some explanation is desirable concerning its implications. Unfortunately these are far more difficult to understand today than might be supposed, while also there is no doubt that the system changed from one reign to another.

It would appear that only four French-speaking authors of consequence, writing at different periods, have attempted to shed any light upon the subject of *Horlogers du Roi*. While the various accounts seem at times to contradict each other (possibly because of differences in terminology) the one most useful and believable is that of the earliest writer of them all.

The first writer was François Béliard writing in *Réflexions* . . . published in 1767. He himself was an *H. du Roi* at the time. The second writer seems to have been S. Franklin, who gave in *La Mesure du Temps,* 1888, a list of the *H. du Roi* from the reign of Louis XII to that of Louis XV. The third author was Alfred Beillard writing his *Recherches* . . . published in 1895. Beillard, while drawing heavily upon the work of Béliard, also transcribed Franklin's list.

76

II The Beginning

Finally, Gelis in *L'Horlogerie Ancienne* published in 1949, brought up the subject yet once again. While Gelis doubtless studied the earlier authors, it would appear from what he says that he had access to other sources of information.

Béliard begins with the great advantage of having lived in the period about which he writes,[43] namely the second half of the 18th century, besides himself being an *H. du Roi*. He says that in 1767 there were at the same time eight legitimate *Horlogers du Roi*. Of these four were *par charge*. Two others were *par brevet*. Yet two others were *reçu en survivance*. Of the eight, four or even five did not maintain establishments in Paris. The eminent Pierre Le Roy did not use his title at all, while in 1767 at least a dozen Paris makers were unlawfully calling themselves *Horlogers du Roi*.

The exact implications of the various types of *Horloger du Roi* are nowhere made clear, even by Béliard. It would seem, however, that those *par charge* were expected to follow the Court everywhere. They paid for the privilege, even though as officers of the Court they were necessarily obliged to neglect their businesses. Their duties seem to have entailed both the supplying and the winding of clocks

Plate II/55. PINCHON, PARIS. *A fairly early carriage clock having both going and striking train driven from a common barrel. (Private Collection. Photograph by P.H. Kohl).*

Plate II/56. PINCHON. *The movement of the clock shown in Plate II/55. (Private Collection. Photograph by P.H. Kohl).*

---

(43) For instance, he brings out the interesting fact that from 1750 onwards, and at an increasing rate, new types of imported Swiss watches were undermining in France the old-fashioned native production of superior quality. For roughly twenty years, however, both the old and the new remained on the market side by side. By 1767 only one third of what was sold was entirely French, and this also applied to clocks, which unscrupulous merchants bought as cheaply as possible wherever they could be obtained.

77

II The Beginning

Plate II/57. *PINCHON. A further view of the movement of the clock shown in Plate II/55. Note the English style of escapement. (Private Collection. Photograph by P.H. Kohl).*

and watches for the Court, besides keeping them in order.[44] The status of *Horloger du Roi par brevet* was apparently confined to such clock and watchmakers as held their posts by direct Royal Appointment. This was given only on merit for actual work. Men in this category were really producing clocks or watches for the King, and usually they lived in the *Galeries du Louvre*, and received payment for their services. Pierre Le Roy and Lepaute were in this class.

They only served for three months at a time. For a *Horloger du Roi* to be *reçu en survivance* meant that he was the trained successor of a *Horloger du Roi par brevet*. In 1767 Béliard was one of these, while the other was Le Faucheur fils.

Various authors, for example Beillard, Gelis and Winthrop Edey in 1967, mention *Horlogers suivant la Cour*, but it is not clear whether this was another term for *Horloger du Roi par charge* or if it implied a different office altogether. Another title not satisfactorily explained is *horlogers valets de chambre*. It would seem that these men were personal servants of the *Grands Seigneurs*.

According to Gelis, the *Corporation des Horlogers* occasionally bent its rigid rules in order to help the widows of clockmakers by allowing them to continue with suitable helpers the businesses of their deceased husbands. One such lady in the 17th century was Elisabet Coupe, *Horlogeusse du Roi*.

The system upon which *Horlogers du Roi* were based always depended upon patronage. After the Revolution[45] and during the reigns of Louis XVIII, Charles X and Louis-Philippe it is likely that the whole conception of the title underwent radical changes. Certainly during the 2nd Republic lasting from 1848 to 1852 no-one in their senses would have dreamed of using the title *Horloger du Roi*, which would have been very dangerous for them. During the 2nd Empire between 1852 and 1870 the use of such a title would not have been allowed by Napoleon III, who would have thrown in jail anyone who had anything to do with the king. After 1870 and under the 3rd Republic, the use of *Horloger du Roi* could just have been possible again because by that time there was no longer the possibility of any restoration of the monarchy.

Returning briefly to the subject of the absolute taboo on the very mention of the word "King" in France during and after the Revolution, an unsigned typescript,[46] found after this book had been sent to the Publisher, throws some further light on the matter. The writer quotes Planchon as saying that any object, decoration or inscription which could con-

---

(44) Béliard implies that people were buying the *charge*, but that the members of the Court were buying their clocks and watches from other sources.

(45) M. Sabrier of Paris has lately found a letter from Lamy Gouge, one of the last two Horlogers to Louis XVI. He wrote to Prince Talleyrand (1754-1838) after the death of Louis XVI and at the beginning of the 1st Empire asking to be allowed to become *Horloger de l'Empereur*.

(46) *French Clock Making during the 1789 Revolution*, presumed to have been written by A. Mongruel of Paris. See Bib.

II The Beginning

ceivably be regarded as royal or royalist in intent had to be destroyed or defaced under penalty of confiscation. According to Planchon, clocks with fleur-de-lys hands came into this offending category, and many such hands were modified by the removal of their "ears".[47] The typescript also quotes Planchon as recounting that an artist called "Citoyen Serrier" living in Rue des Poitevins advertised that he could remove the word "King" from enamel clock dials without damaging them and could substitute "People" or "Nation" instead. Many dials, however, had the word "Roi" crudely scratched out. Any clock found today bearing traces of such mutilations is certainly likely to have been in existence during the Reign of Terror.

---

(47) Such hands are illustrated in a bound album entitled *L'Aiguille de Pendule son Evolution Décorative* and which shows Mongruel's collection.

Chapter III

Later Manufacture, 19th and 20th Century

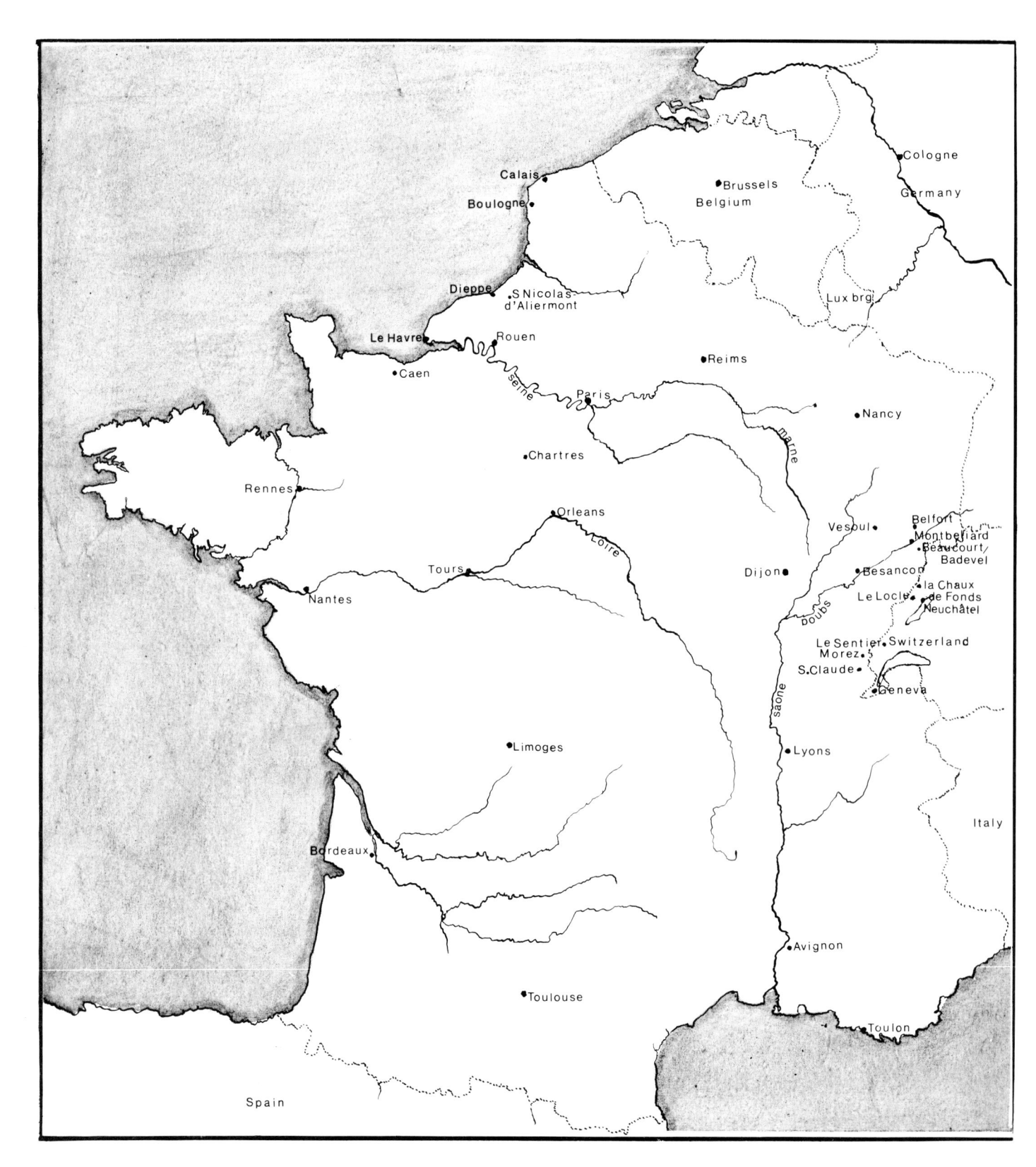

FRANCE AND SWITZERLAND. The relative positions of Paris, Saint Nicolas-d'Aliermont and the Jura.

III

Later Manufacture, 19th/20th Century

Background. The Strange Organisation of the Manufacture of French Carriage Clocks.

Although it might have been possible to buy a French carriage clock in London as long ago as 1830,[1] the peak of manufacture was not reached for almost another sixty years in late Victorian times. England soon became the principal market, absorbing progressively a high percentage of the entire production. For all practical purposes, the era of the French standard carriage clock may be regarded as one hundred years, from approximately 1830-1930, although clocks were made both before and after these dates. There is no doubt that *pendules de voyage* at their best represent the highest quality ever attained in mass-produced clocks.

The forty years following 1815 and the Battle of Waterloo saw a period of social change without precedent in England. The twenty years of war with France were over, the Industrial Revolution was fully under way, and man had suddenly, in the words of G.M. Trevelyan, "... acquired formidable tools for re-fashioning his life before he had given the least thought to the question of what sort of life it would be well for him to fashion". The rich became richer, and moreover were joined by an ever-increasing number of parvenus all with money to spend. During the same period, from perhaps 1825 to 1845, vast ramifications to the stage coach systems rose almost overnight,[2] and as suddenly fell to be ousted by the railways. All was "progress", and the consumer market for manufactured goods, although in fact still restricted to a limited proportion of the total population, was suddenly very great. The illustrated Catalogue of the Great Exhibition in London of 1851 gives a clear picture of what was required and offered. Such a period could scarcely have been more conducive to the potential sale in England of carriage clocks. All the same, almost four decades were still to pass before French industry and British importers became fully geared to the situation. The heyday of production was not in fact reached until 1889, the year of the Paris *Exposition Universelle*. That a very large output had been achieved by this date is apparent not only from the Exhibition Catalogue, but also from various contemporary reports. For example, the *Revue Chronométrique* speaking of carriage clocks in the Exhibition refers to "... the craze for these elegant objects", and then goes on to describe them as "... an indispensable part of the luggage of any self-respecting traveller". T.D. Wright, who was the clockmaker selected by the Mansion House Committee to visit the Exhibition, also makes it plain that carriage clock

---

(1) That is to say not more than fifteen years after Wellington's victory; or, put another way, when the Crimean War had yet to be fought in full-dress uniform, the regiments marching into battle behind their bands.

(2) E.W. Bovill in *The England of Nimrod and Surtees 1815-1854*, published in 1959, says on p. 137 that at the peak period seven to eight hundred coach horses were stabled at Barnet alone.

III Later Manufacture

output had by then about reached its height, although he was surprised at the cottage-industry manner in which manufacture was organised.[3] A good idea of the scene is also given by J. Tripplin in *Watch and Clock Making in 1889*...[4] D. Rousaille, writing of the same Exhibition on behalf of the clockmakers of Lyon, says what translates as "In spite of their small size, carriage clocks occupy a respectable position in Class 26 of the Exhibition. This industry, which is essentially French, is being expanded considerably. France supplies these clocks to the rest of the world, and as output increases so artists are encouraged to perform marvels of good taste in designing stylish and graceful cases, which double and quadruple the value of this kind of clock, which leaves nothing to be desired, either in its execution or in its often very complicated functions". Rousaille continues with the important evidence that "The carriage clock. . . sells freely in England, but seldom in the centre of France . . .". However, having briefly and intentionally run ahead of history, it is now necessary to return the attention of the reader to the years close to 1850 and to those clocks immediately succeeding the examples by early makers described in the last chapter.

The London 1851 Exhibition Catalogue shows that a number of French firms exhibited carriage clocks there. It is quite clear that these were by then considered to be nothing new. No special importance was attached to them, and not even one was illustrated. The *Revue Chronométrique* in December 1855, p.79, refers to *pendules de voyage* as a matter of course, leaving no room for doubt that any reader might not know what was meant.[5]

The French carriage clock industry in 1889, as noted by T.D. Wright, was indeed very strangely organised. Wright had expected to see in Paris large factories making carriage clocks throughout, but nothing could have been further from the truth. Paris was supposed and represented as being the centre of French clock manufacture, but very few *pendules de voyage*, or for that matter clocks of any kind, were latterly made there in their entirety. Instead, two small country areas, at opposite ends of France, supplied the capital with semi-completed clocks to be "finished", "escaped" and cased in such style and to such standard as the "maker" chose. The "rough" movements were called *blancs-roulants*, or simply *roulants*, and more will be said about them as this book progresses. The first of the two rival sources of supply was the small town of Saint-Nicolas-d'Aliermont in the Seine-Maritime near Dieppe, where there were many comparatively modest manufactures. The second source was in the Jura region of the Franche-Comté, near Montbéliard and close to the border of France with Switzerland. Here the main producer was the large Japy factory in the village of Badevel. Both the Saint-Nicolas and the Franche-Comté *roulants* (other than for timepieces) were supplied to the trade with the external striking and repeating work, called the *cadrature*,[6] all but completed and also correctly "planted" on the front movement plate. Latterly both Duverdrey & Bloquel and Couaillet Frères of Saint-Nicolas-d'Aliermont, and their competitor Japy Frères, near Montbéliard, sold completely finished clocks. Others may have done the same. Japy then made two qualities of *blancs*, one for Paris and one for their own use. On the whole, Saint-Nicolas *blancs* were of a better quality and looked less machine-made than those produced in the Jura.

The platform escapements for the carriage clocks (*porte-échappements*) and all the detail parts (*assortiments*) that were required to make up platform assemblies (such as cylinder or lever 'scape wheels, pallets, levers and rollers) came from manufacturers scattered round Montbéliard, Morteau and along the French-Swiss frontier. A few such factories were certainly in Switzerland. According to the Catalogue of the 1889 Paris Exhibition, C. Gelin and Coulon &

---

(3) It was very much the same as he was used to seeing at home in Clerkenwell or Prescot. It was inconceivable to Wright that the large numbers of well-made and elaborately cased *pendules de voyage* exhibited were not the product of large and highly mechanised establishments.

(4) Relevant quotations from the actual writings of T.D. Wright and J. Tripplin are included in Appendix (b).

(5) The amazing fact is that while carriage clocks were a distinct rarity before 1830 (when Paul Garnier followed by others began to produce them in mass), by 1855 Redier alone was on record as selling between 35,000 and 40,000 travelling pieces, which included alarms, to the English market. *(Visite à L'Exposition Universelle de Paris 1855)*.

(6) That part of the striking work associated with the *cadran*, or dial. The same parts of the striking work are still correctly termed *cadrature* even when planted on the back plate of a clock.

III Later Manufacture

Molitor were two of a number of rival French makers of platform escapements. The firm of L'Epée at Sainte-Suzanne, which today still produces many *porte-échappements,* was then also engaged in making music boxes. Mainsprings came from many sources of which the large concern of Montandon Frères or the smaller Ducommun were typical.[7] Cutters *(fraises)* for the wheels and pinions of the *blancs-roulants* were available from the famous firm of Louis Carpano at Cluses, although many factories undoubtedly made their own. Small detail parts such as hands were "bought-out" from home workers. There were several specialist dial makers in Paris, and a casemaker, A. Kremer, was one of many working at the period. On the other hand, Japy and some of the Saint-Nicolas-d'Aliermont makers produced their own cases and dials, while others were made elsewhere both in Paris and in the provinces. The Paris "makers" faced the task of "finishing" the *blancs-roulants* and of completing the clocks with cases, dials, hands, etc. of their own choosing. Appendix (b) will help to complete the picture in the reader's mind.

The manufacturing practices outlined in the previous paragraphs were more or less standard procedure with French clocks once the industry became established. To say that a Paris *pendule de voyage* was produced from a *blanc-roulant* is in no way to criticise it. The *roulants* were made to a high standard and could be finished to any quality required. In the case of the top establishments these standards were very high indeed.

The *Revue Chronométrique* for July 1878 contains a very interesting article[8] in which, amongst other data, *roulant* production for small clocks in France for the year 1867 is compared with that of 1878. First is covered the Franche-Comté with specific reference to the factories of Japy Frères at Beaucourt and Badevel, those of Roux et Cie and Marty et Cie at Montbéliard, and that of Louis Japy at Berne-Seloncourt. Next is covered the production of clocks at Saint-Nicolas-d'Aliermont in the Seine-Maritime. Lastly the Parisian clock industry is analysed. Break-down figures are not given for carriage clocks, so that it is not easy to compare the output of *roulants* in the Franche-Comté with that of the Seine-Maritime. The *Revue,* however, says that by 1857 or earlier out of a total of 170,000 *roulants* of all types, produced by between 2,000 and 2,500 workers near Montbéliard, 140,000 were bought by Paris alone, while the remaining 30,000 were sent unfinished to be completed and cased in establishments situated in the Doubs and Jura.[9] The writer says that for several years prior to 1867 changes had taken place in the industry, and that whereas formerly the majority of small clock movements were made at Saint-Nicolas-d'Aliermont, by 1867 most of the ordinary clocks were produced in the establishments of the Doubs. He goes on to say that the production of carriage clock *roulants* of late had been greatly increased in both places and states that the production of clocks at competitive prices was achieved largely by the use of machine tools. By 1867, according to the *Revue,* the annual production of small clock *roulants* had risen to 200,000 pieces, while 2,500 workers were employed. This figure, however, did not include some 30,000 music boxes mostly made by L'Epée of Sainte-Suzanne, any more than it did a further 100,000 miscellaneous items of clockwork which included platform escapements. This last point is easily missed because Saunier, writing in 1878, did not re-quote his article of 1867 in its entirety. The total production of medium-sized clockwork made in the Départements of the Doubs and Haut-Rhin in the year 1867, was said to be worth 3,500,000 francs. By 1878 output had further increased. Of clock movements of all types a total of 400,000 were produced. In the same period about 18,000 platform escapements " ... destinées aux pièces de voyage" were made near Montbéliard. Not all of these escapements would have been used in Franche-Comté clocks. Many of the platforms would have found their way to Paris and Saint-Nicolas.

---

(7) According to M. Pitou of Paris, Jacot obtained some of their mainsprings from Sweden. So did Couaillet, according to Claude Couaillet. According to Tripplin, p.123, some French and Swiss springmakers obtained their steel from Sheffield, selling each year some £20,000's worth of finished mainsprings back to England.

(8) *Fabrique d'Ebauches de Pendules* (usually Vol. X, beginning p.103). Saunier had previously published a more detailed version of the first part of this article in the *Revue* between July 1867 and December 1868 as part of his report upon the 1867 Paris Exhibition *(later re-published in book form).*

(9) This information was collected by M. Monnin-Japy and given to Dr. Muston. (See Chapter VI). Carriage clock *roulants* would have accounted for only a fraction of this enormous production of clockwork.

III Later Manufacture

At the other end of France, Saint-Nicolas-d'Aliermont in 1867 had 2,500 inhabitants of whom less than 1,000 worked at horology.[10] Those who did so are said to have been engaged "... above all in carriage clocks, alarms and electrical apparatus". According to the *Revue*, enquiries made locally showed that 144,000 clock movements of one kind and another were made in the year, mostly for Paris but some for London.[11] The article concludes by saying that the horologists of Saint-Nicolas-d'Aliermont had previously tended to work as cottage industries like the Swiss; but lately they had begun to change and to imitate their rivals in the French Jura, necessitating the building of large factories with foremen and managers. No actual production figures are given for 1878, but the *Revue* says that turnover had risen from 1,100,000 francs to 1,500,000 francs in about eleven years.

With reference to Paris, the *Revue* makes it plain that the work done there, so far as *pendules de voyage* were concerned, was confined almost entirely to the finishing of *roulants* obtained from either the Jura or Saint-Nicolas, and to the production of such component parts of clocks as mainsprings, hands, dials, etc., and particularly of cases. Writing in 1867, Saunier estimated that there were 3,700 workers in the horological industry of Paris; but that the figure was misleading because so many were working upon telegraphic apparatus. In 1878 there were 2,000 *patrons* and 6,000 workers employed by the Parisian clock industry. He says that 250,000 clocks and more than 300,000 alarms, pieces of telegraphic equipment, and other clockwork items were handled during the course of a year, and that in the main all the finishing of electrical parts was done by clockmakers. By 1878 the Parisian horological industry, in its broadest sense, showed a turnover in the region of 23,000,000 francs.

However hard one tries, it is not really possible to determine with any degree of accuracy the relative outputs of carriage clocks of Saint-Nicolas and the Doubs. While it seems clear that the latter produced a vastly larger turnover of clockwork in general, it is not altogether unlikely that the majority of all French carriage clocks ever made saw their beginnings in Saint-Nicolas. The figures given in the *Revue Chronométrique* are apt to be misleading, since they do not always refer to the production of the same items. Moreover, the industry at the time was undergoing a process of change. On the other hand the *Revue* figures are interesting if only because they give a general and rough idea of the state of French clock manufacture between perhaps 1855 and 1878.

---

(10) The report of the Paris Exhibition of 1819 says that in about 1719, and until the time of Pons, horology occupied about three hundred people in Saint-Nicolas.
(11) The reader will realise that, when completed, most carriage clocks were sold to England. The movements sent to London may explain the anomaly of the French carriage clock sometimes found in an original English case.

Chapter IIII

Saint-Nicolas-d'Aliermont, Past and Present

Plate IIII/1. *THE ARMS OF THE TOWN OF SAINT-NICOLAS-D'ALIERMONT. The crosier denotes the fact that the Archbishops of Rouen were Seigneurs de l'Aliermont from 1197 to 1789. The other symbols represent the progress of horology at different periods thriving alongside a rural economy. (By courtesy of the Mairie de Saint-Nicolas-d'Aliermont).*

IIII
Saint-Nicolas-d'Aliermont, Past and Present

Croutte and Papin. Pons. Delépine & Cauchy. Dessiaux, Martin, Sauteur, Villon, Baveux, Guignon, Couaillet, Duverdrey & Bloquel, Bayard, Holingue, Dumas and Others. French Craftsmen in Russia.

Sancte, sancte Nicolas,
Tute patronus noster es:
Laus et Deo gloria;
Tu pro nobis exora.

(Patronal hymn, SS. Mary & Nicolas College, Lancing, Sussex.)

Saint-Nicolas-d'Aliermont is near Dieppe, almost on the boundaries between Normandie and Picardie. The town possesses a civic document dated 1197, and has a horological history going back to at least the 16th century. To this day, the fifteen or so thriving factories there include Réveils Bayard, a firm exporting alarm clocks all over the world, besides still making a few carriage clocks for the English market. The Arms of the town (Plate IIII/1) feature a crosier to underline the fact that the Archbishops of Rouen were Seigneurs de l'Aliermont from 1197 to 1789. There are other emblems also. The sand-glass symbolises horology, the true and proper industry of the place. The pin-pallet 'scape wheel represents more modern developments. The Count's coronet commemorates "Le Comté de l'Aliermont, possession des archevêques de Rouen". The sheaves of corn represent both the local agriculture and the unspoiled rural nature of the district.

The early origins of the horological industry in Saint-Nicolas-d'Aliermont are obscure and uncertain, although it is known that long ago the town boasted skilled metal workers who were called *"maignens"* or *chaudronniers*. In 1572 the Saint-Barthélemy massacre of the Huguenots by the order of Charles IX and of his mother, Catherine de Medici, lasted for thirty days over the whole of France. More than thirty thousand Protestants are said to have been slaughtered. This religious persecution removed in a month a high proportion of all French craftsmen. In 1598 the Edict of Nantes, issued by Henry IV of France, secured freedom of religious practices to Protestants and re-opened all official appointments to them. This tolerance was unfortunately short-lived. In 1685 the Revocation of the Edict of Nantes by Louis XIV was disastrous for France, causing some two to three hundred thousand people, amongst them many of her worthiest subjects, including craftsmen, to go into voluntary exile.[1] A further set-back occurred in 1694 when an English squadron under Admiral the Lord Berkeley bombarded the port and town of Dieppe (a name said to be derived from "deep"), burning most of it down and forcing the unfortunate inhabitants to seek safety inside the country.[2] Some of those who were clockmakers settled in Saint-Nicolas-d'Aliermont, but on

---

(1) Very many Huguenot families of craftsmen and horologists migrated to the Swiss Jura and to Outre-Rhin. Others came to England.
(2) This was an episode in the so-called War of the English Succession, with James II, helped by Louis XIV, trying to take back his throne from Dutch William. Dieppe was largely built of wood, thus burning easily. The area was noted for forests. Many ships were built at Dieppe, which was famous also for good navigators — men such as Abraham Duquesne, Jehan Ango, Vera Zane, Cousin, etc.

the whole the ones who had remained after 1665 were said not to have been equal to those who left. A number of the carvers of ivory for which Dieppe was very famous also came to Saint-Nicolas, as did a number of lace-makers. Such towns as Dieppe, Rouen and Beauvais, so seriously depleted of craftsmen in 1572 and 1685, had long been nurseries of instrument making and horology in France. For the next hundred years Saint-Nicolas-d'Aliermont saw hard times, both clockmaking and ivory-carving being "en grande récession". There were many crises, and frequently the municipality debated what could be done to provide work. A particularly bad year was 1791, during which some tradesmen were able only to find ten days work.

After the Revolution, and particularly in the Empire period of success and after the Battle of Austerlitz in 1805, matters gradually improved. Local tradition has handed down the names of two men from Dieppe, Croutte and Papin, as being the best horologists in Saint-Nicolas during the 18th century. The year 1806 saw the turning point in the fortunes of the town. At that time M. Savoie Rollin, *Préfet* of the Seine-Inférieure, addressed a long memoire to M. de Champagny, Minister of the Interior, setting out the precarious state of the horological industry in Saint-Nicolas, and asking that someone should be sent who could instruct the clock makers there in more up-to-date methods. The Minister immediately sent a young master horologist called Honoré Pons (otherwise known as Pons-de-Paul). This clever man, the son of a musical instrument manufacturer in Grenoble, had learned his horology in Paris, and in particular it was said that he was very conversant with clock movements made there by the famous firm of Lepaute.[3]

What Pons in effect did, after a careful investigation into the 18th century methods used in Saint-Nicolas, was to form the clockmakers into a sort of Guild called Fabrique d'Horlogerie de Saint-Nicolas-d'Aliermont, directed by himself and designed to co-ordinate their efforts and business interests. He introduced the use of machinery, some of it of his own devising, and generally he brought an industrial revolution to Saint-Nicolas. Within ten years of his arrival he had drastically reduced the cost of the production of clock movements.[4] The *Tribune Chronométrique,* page 91, confirms some of this history and adds that from 1830 the products manufactured under the auspices of Pons were exceptionally well made as regards toothforms, etc., and that Pons was eventually succeeded by MM. Delépine & Cauchy. It is also interesting to note that as early as 1819 the Fabrique d'Horlogerie obtained at the Paris Exhibition two Silver Medals for "rough" movements of clocks *(blancs-roulants).* One of these medals was given to the town in general as an encouragement to a by then rapidly expanding industry, and the other to Pons personally.[5] *Blancs-roulants,* literally translated, means running movements "in the grey", and this is exactly what they were. Examples of carriage clock *blancs-roulants* found in Saint-Nicolas and in Paris, now in the author's possession, consist of frames (i.e. plates and pillars) left straight from the file, but with the barrels in place and with wheels which have been cut and crossed and also mounted on their pinions, planted with the depths correct and with the holes and pivots finished to size. This pre-finishing embraced also the *cadrature* (that which went under the *cadran* or dial) which in carriage clocks meant the whole of the external striking and repeating work (if any) and probably also

---

(3) *Paris Normandie,* 25th-27th September 1932. Article entitled *A Saint-Nicolas-d'Aliermont; Les Confidences d'un Reveille-matin.*

(4) Pons is said to have improved first the factory of Croutte, which made *roulants*. According to the *Bulletin de la Société d' Encouragement* of 1809, Vol. 8, pp. 326-327, eight machines designed by Pons, mainly in connection with wheel and pinion cutting, increased production enormously. When Pons died he left money in the care of some sort of Committee. Annually every workman in Saint-Nicolas received a small cash gift in memory of Pons. This was fine at first, but eventually there were too many workers. In the end the arrangement ceased. The last time the husband of Madame Lejoille received the present, she said that "it was just enough to buy a glass of wine". (Madame Lejoille-Grard will be mentioned again shortly.)

(5) *Rapport du Jury Central* of the Paris Exhibition of 1819. The 1823 Exhibition *Rapport* said that Pons exhibited portable clock movements *(mouvements pour pendules portatives).* It was stated that the output of clock movements of all kinds in the town was approximately five or six thousand a year. Pons received a Gold Medal in 1834, and finally he was awarded the Légion d'Honneur in recognition of his outstanding services to French horology. See additional notes in List of Names.

under-the-dial complications such as calendars when they were included.[6]

Once the clock *blancs-roulants* industry in Saint-Nicolas-d'Aliermont became established it was immediately very successful and expanded rapidly. Marine chronometers were the most refined clockwork ever made there and latterly they were produced in large numbers. In addition, carriage clocks and most if not all types of ordinary domestic 19th century French clocks and alarms were made. The house clocks included standard eight and fifteen day pendulum clocks cased in any of a number of ways. Other specialities included later, if not at first, the so-called *régulateurs*. These last, not to be confused with astronomical regulators, were really little more than mantel clocks. Latterly they were often provided with Brocot-type escapements and suspensions and also with mercury pendulums. *Régulateurs* usually had those cases now often described as "four-glass" and having glass panels front, back and sides. In the end, alarms, novelty alarms[7], barographs, gas lamp time-clocks, bulk-head clocks and "flap-jack" fold-flat travelling clocks were produced in quantity.

When the demand for *pendules de voyage* was first established, the Saint-Nicolas factories immediately produced the *blancs-roulants* which were supplied in bulk to be finished by Paris firms (some already famous) under whose names they were to be sold. Latterly, and especially by the beginning of the 20th century, only a few firms sent their clocks to Paris to be finished. The Saint-Nicolas records have not survived the two World Wars; so there seems to be no accurate, much less easy, way of ascertaining how many firms were occupied in carriage clock making at any given period. The issue is further complicated by the fact that a number of Paris "makers" used Saint-Nicolas addresses, without in fact maintaining factories there. On the basis of reading and also researches made in Saint-Nicolas and in Paris in 1970, it seems doubtful whether more than a few firms out

Plate IIII/2. *REVEIL FANTAISIE. Alarm clocks accounted for a very large part of the horological production in Saint-Nicolas-d'Aliermont during the 19th century. The example shown was made by E. Dessiaux circa 1880-1885. (By courtesy of M.E. Forest, Le Président Directeur Général, Réveils Bayard)*

of a total of perhaps fifteen in the town were ever engaged at any one time in making either carriage clock parts, their *blancs-roulants* or finished carriage clocks. Jacot and Drocourt of Paris used Saint-Nicolas as part of their addresses; but while there is some evidence

---

(6) In contrast, French and for that matter English 18th/19th century "rough" movements of *watches* were far less finished. The train "depths" were not "run", and the unfinished and unpivoted pinions, with wheels mounted, floated in clearance holes. There were some exceptions. For instance, "rough" verge watch movements, circa 1670-75, in the Clockmakers' Company Museum in London are almost finished, except for their unpierced cocks; and an advertisement in *The Watch and Clock Maker* of October, 1883 inserted by P.W. Roberts and Co. of 249 Park Road, Liverpool, described them as "Manufacturers of Watch Movements with Pitched Depths in a more Finished State than ordinary Movements. A great Saving in Time by Using Them".

(7) See Plate IIII/2 taken from the Catalogue of E. Dessiaux of Saint-Nicolas, circa 1880-1885. This firm specialised in *Réveils Fantaisie* as well as *pièces de voyage*. The model illustrated was called *Rouget de l'Isle* (after the composer of the Marseillaise). Old Dessiaux alarms are still to be found. Villon and Dessiaux together took *Brevet* 97,946 of 4th February 1873 "improvements and additions to ordinary alarms".

IIII Saint-Nicolas-d'Aliermont

that the famous Drocourt maintained a small establishment there while himself living in Paris, we have not so far traced any factory in Saint-Nicolas as having belonged to the even more famous Jacot.[8]

In about 1847, when Pons wished to retire, he was only too happy to hand over his interests and responsibilities to one Boromé Delépine who was already among the principal manufacturers in Saint-Nicolas. According to the *Tribune*, p.92, Delépine received a Silver Medal in 1844 for the fine collection of clock movements which he exhibited at that time. In 1849 the Jury of the Paris Exhibition were so satisfied with the way in which MM. Boromé Delépine et Cauchy were carrying on the traditions of Pons that they awarded them a further Silver Medal.[9] In 1859 Emile Martin was said to have been the largest manufacturer of clock movements including those of carriage clocks. Martin's work was again praised at the Paris *Exposition* of 1867. On the same occasion, Sauteur received a similar recommendation. In 1867 also, A. Villon started in business, and by 1878 Baveux were already famous for high-class *roulants* and clocks. This last firm was to continue successfully well into the present century. In 1889 Guignon was showing cheap carriage clocks in Paris, and by 1892 Armand Couaillet had started in the carriage clock business with his brothers. Of these old firms the successors of both Villon and of Couaillet are still active today, so that with their help it is possible to give histories of both.

In 1867 Albert Villon was responsible for the setting up of a large factory almost opposite the Château.[10] The Villon concern made complete clocks, of which *pendules de voyage* formed a large part and the business expanded steadily for the next twenty years. In 1887 Paul Duverdrey joined Villon as Director, and in 1910 on the death of Villon Duverdrey was joined by Joseph Bloquel. The firm continued under the name of Duverdrey & Bloquel.[11] In 1922 Robert Duverdrey joined Bloquel and the factory continued to produce an enormous range of finished carriage clocks. These were distinguished by a "Lion" trademark and they are often seen today. The round clock shown in Plates IIII/3 and 4 is typical of one of their cheaper timepieces. A catalogue of circa 1910 illustrates some two hundred and twenty different models. Plate IIII/5 shows a view of the works at the time. The carriage clocks offered included timepieces, timepieces with alarm or centre-seconds, hour and half hour strikers and simple repeaters. *Grande* and *petite sonnerie* and other complicated clocks were not offered; neither did the Duverdrey and Bloquel products pretend to be equal to the top

Plate IIII/3. *DUVERDREY & BLOQUEL. A late but unusual clock having a circular section case and employing an* Obis *movement. D. & B. specialised in inexpensive clocks which they offered in all manner of unusual and usually ornate cases. (By courtesy of Temple Brooks)*

---

(8) On the contrary, there is some evidence that Jacot obtained *roulants* from Baveux in Saint-Nicolas. See Chapter V.
(9) Delépine continued to use Pons' name on his work. (*Revue*, Oct. 1859. Report of Rouen Exhibition).
(10) See Tripplin's notes upon Villon's factory in Appendix b (ii).
(11) Duverdrey & Bloquel's English agents were Landenberger and Company, 91 Aldersgate Street, E.C.

IIII Saint-Nicolas-d'Aliermont

Plate IIII/4. *DUVERDREY & BLOQUEL. The back of the clock shown in Plate IIII/3. Note the lion trademark (sometimes turned the other way) and also the characteristic hand-setting arrow. (By courtesy of Temple Brooks)*

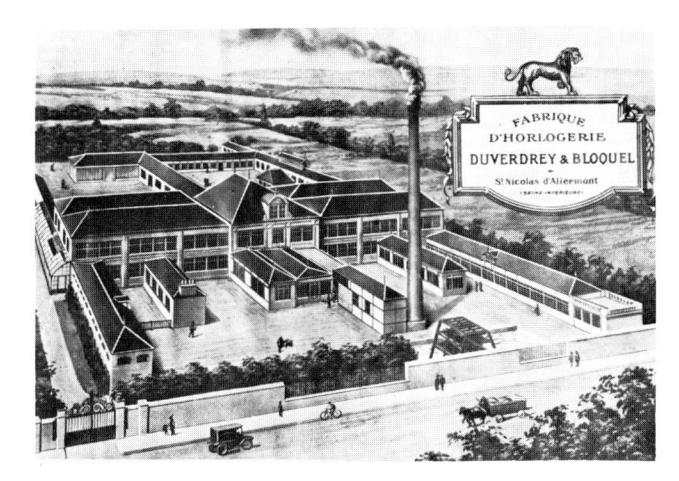

Plate IIII/5. *DUVERDREY & BLOQUEL. A view of the factory as it existed before the 1914-18 war, reproduced from their Catalogue of circa 1910. (By courtesy of Réveils Bayard)*

"Paris" quality. They were, however, soundly constructed, thoroughly practical and relatively inexpensive. Today the modern Bayard clock factory occupies the same site, still using some of the old buildings. While concerned principally with alarms, Bayard even now make a modest timepiece carriage clock for the English market. This current Bayard clock maintains tradition by being signed on the back door and on the backplate of the movement "DUVERDREY & BLOQUEL, FRANCE". The clock has the barrel in a sub-frame, and employs a lever platform lying on its side to obviate the expense and also difficulties associated with a contrate wheel and to allow the use of a back plate less tall than the traditional practice. Bayard are active in seeking further rationalisations for future models. We have seen one prototype incorporating several interesting escapement innovations. M. Forest, who married a daughter of M. Hennion (below) is Président Director Général of Réveils Bayard, and the firm is exceedingly active, even selling to Japan. At the time of our visit, we were shown round both the old and the new factories. The latter is equipped with the most up to date automatic machinery (Plate IIII/6) for cutting wheels and pinions, as well as some old machines dating from pre-1914, which the firm keep for use in an emergency such as unexpected large orders for their cheapest clock. M. Forest was not only more than kind and helpful, but he was also a mine of information on the history of his firm. In the office are plates that read:— Albert Villon, Fondateur des Usines 1867-1887, Paul Duverdrey, Director 1887-1910, Joseph Bloquel, Director 1910-1922, Robert Duverdrey, Director 1922-1947, Raphaël Hennion, 1947-1960. The above

Plate IIII/6. *SAINT-NICOLAS-D'ALIERMONT. The part of the production shop producing alarms as seen in 1970. (By courtesy of Réveils Bayard).*

dates relate to positions held. It is clear from Tripplin, page 85 (quoted Appendix b(ii)) that Villon did not vanish from the scene in 1887. At first the firm was run like a "co-operative", and all the work given out to families to be done at home. These outworkers were paid on a piece-work basis and were expected to buy everything connected with the jobs from the parent factory, even to the oil for their lamps. M. Villon had his own brass band, the musicians being his workers. In about 1900 "modern" tools were introduced and the factory proper dates from this period. A great transformation began in 1914, when the factory made fuses, particularly for ".75" shells. In the earliest days the wheels and pinions were cut in private houses, improved methods being found from time to time. All workers made one item only.[12] In 1957, Bayard took over a wheel and pinion cutting concern founded in 1939 and housed in a fine factory built in 1940. To-day Bayard uses there the latest Swiss Asco automatic pinion cutting machines made in Neuchâtel. These work upon the principle that the steel rod from which the pinions are cut stands still while the machine moves round it. Other modern "automatics" in use are from Tornos and from Petermann of Moutier.

Another firm well known in Saint-Nicolas-d'Aliermont and still in business there to-day is that of Etablissements Couaillet, Mauranne & Quesnel. The present partnership, while no longer concerned with the manufacture of clocks of any type, is in the process of building a new factory in Rue de Milan behind the main street. Their business is general light engineering, producing component parts to special order. The history of the firm is that the original Couaillet enterprise, Etablissements Couaillet Frères[13] was started in 1892 by Armand Couaillet with his two brothers, Ernest and Henri. Plate IIII/7 shows a view inside what was probably their first factory. Ernest was *chef de fabrication,* Henri looked after the business side and Armand travelled on the road. The first factory burned down. It was soon rebuilt, but was again destroyed by fire a few years afterwards. Couaillet rebuilt, but at a later date, some time after 1913, the brothers bought the large factory and house of Delépine-Barrois. This house, today No. 14 Rue Edouard-Cannevel[14] is illustrated in Plates IIII/8 and IIII/9. Henri was the one who actually lived there. Couaillet Frères became one of the largest clock-

Plate IIII/7. *SAINT-NICOLAS-D'ALIERMONT. COUAILLET FRERES. M. Ernest Couaillet, probably in the first factory, surrounded by his workers. Taken from an old undated postcard. A note on the back says that there were 200 workers. (By courtesy of Madame Jeanne Lejoille-Grard, Saint-Nicolas-d'Aliermont)*

---

(12) This was the practice also in Prescot, London and Coventry. It accounts both for the sheer quality of the old work and also for the fact (which seems so inexplicable when the reasons are not known) why skilled horological workers usually had little if any general knowledge. As in Prescot, large collections of horological tools have been sold from time to time in Saint-Nicolas, where about ten years ago a considerable quantity was dispersed. There used to be a Museum, but it was burned down in about 1935.

(13) Couaillet's London agent was Ernest S. Pitcher, father of Maurice.

(14) Up until 1946 there were no street names in Saint-Nicolas.

IIII Saint-Nicolas-d'Aliermont

Plate IIII/8. *SAINT-NICOLAS-D'ALIERMONT. COUAILLET FRERES. The entrance to the main factory in what is now 14 Rue Edouard-Cannevel. M. Henri Couaillet lived in this house which at an earlier date was occupied by Delépine-Barrois. The house on the left of the picture is the home of Monsieur and Madame Delabarre.*

makers in Saint-Nicolas. They concentrated upon producing carriage clock *blancs-roulants,* which were supplied to the industry as a whole to be finished and marketed under other names. That they also made and sold completed carriage clocks is apparent from their catalogues. A Couaillet *Obis*, in a case termed in their Catalogue *Classic,* will be found in the Case Styles Chapter in Plate VII/6 on page 162, while Plates IIII/10 and IIII/11 show two views of a Couaillet hour and half-hour strike and repeat clock.

M. Ernest Couaillet formed and conducted his own brass band which competed with the one already in existence and run by Albert Villon. M. Alfred Delabarre, the oldest surviving member of the carriage clock industry of Saint-Nicolas, worked for Couaillet from about 1912 to 1930. At this time, the firm ran into money difficulties. Apparently, one of the brothers was "too experimental and too inventive". This disaster ruined the Etablissements Couaillet Frères as such. It ended the grand days when the main firm made, either completely or as *blancs-roulants,* a high percentage of all French carriage clocks and also boasted impressive headquarters on the Rue Cannevel,

besides other factories and outworkers all over the town. Plate IIII/12 shows the front door of the office of the main factory, the "frosted" pane depicting an actual clock which once hung on the wall there.[15] In the hey-day of the firm, Couaillet made their own carriage clock cases *(boîtes)* and all other parts, except for the mainsprings, which came from Sweden. The

Plate IIII/9. *SAINT-NICOLAS-D'ALIERMONT. COUAILLET FRERES. Another view of No. 14 Rue Edouard-Cannevel showing the main factory building discreetly hidden behind the house.*

platform escapements and parts for them came either from Switzerland or from the firm of L'Epée at Sainte-Suzanne close to the Swiss border.

After the breakdown, the three brothers went their separate ways. Armand was bought out by and became involved with a short-lived clockmaking firm called Exacta, which was financed from Switzerland. This concern lasted from 1930-1935 and is said to have made carriage clocks, although none are shown in their Catalogue of 1933. They certainly made drum platform timepieces, *mouvements de Paris,*[16] and probably the so-called 400-day clocks. Armand reputedly eventually went to Paris, taking some of his workmen. It is not clear what happened afterwards. Henri Couaillet faded from the scene, but his son Henri set up as Couaillet Henri Fils, which he ran from January

---

(15) This could not have been a Couaillet clock because the "clock-door" appears in an old postcard showing the factory still in the ownership of Delépine-Barrois.

(16) Some of the older people living in Saint-Nicolas believe that their town made all the *blancs* and *roulants* for *pendules de Paris* during most of the 19th century. This seems not to have been the case, for Sire, writing in 1870, says plainly that most of them were made round Montbéliard, and we know that such clocks were produced in quantity by Japy, Marti, Roux, etc.   See for instance Appendix (d).

Plate IIII/10. *COUAILLET. A clock striking hours and half hours and repeating hours, as made by Couaillet close to the end of the 19th century. Note the beaded decoration of a case undoubtedly made in competition with Duverdrey & Bloquel who offered a wide range of case style variants at this period.*

Plate IIII/11. *COUAILLET. The back of the clock shown in Plate IIII/10. Note the characteristic hand-setting arrow.*

1930 to October 1959 with M. Delabarre as *chef d'atelier* at 6, Rue Robert Lefranc. They made carriage clocks, and also glass-sided mantel clocks going two weeks and known as *régulateurs quinze jours*. Madame Delabarre worked upon the assembly of these clocks. Ernest Couaillet retired in 1930, but his three sons René, Jean and André made parts of bulkhead clocks until about 1955, by which date René was the only survivor. In 1955 Claude Couaillet bought an interest in the firm of Henri Couaillet Fils, which is now run by himself, his father René and two partners, Mauranne and Quesnel. Their new factory is in Rue de Milan. The Couaillet factories employed over eight hundred people in making munitions during World War I. M. Claude Couaillet showed us a wheel cutting engine unlike any we had ever previously seen and designed for repetition work (Plate IIII/13). It is mostly made of steel and appears to belong to the second half of the 19th century. It is self-indexing and employs a cutter-frame guided by a vertical slide. Number determination is by interchangeable index plates, pre-divided on another engine. Turning the side-handle reciprocates the cutter-frame vertically and at the same time rotates the wheel blank one division of the index per stroke of the cutter-frame. A fly cutter was used.

Another clockmaking family well known in Saint-Nicolas during the second half of the 19th century, and whose name was spelt in any number of different

IIII   Saint-Nicolas-d'Aliermont

worked with Martin. Both were important members of the Saint-Nicolas carriage clock industry. The other Holingue, Alexandre, evidently specialised in striking work. He received medals at Brussels in 1910 and Turin in 1911. Some of his clocks were small, but one specifically mentioned was "une horloge à poids une vraie normande, construite par M. Holingue, lui donne une heure indéréglable, et sonne les quarts". This presumably would have been one of the peculiar traditional-style local longcase clocks of the type known in the French antique trade as *Saint-Nicolas*. While these clocks visually appear somewhat like those made in Morez, the two types are in fact entirely different.

Firms directly connected with *pendules de voyage* and which undoubtedly did operate from the town, although not necessarily at the same time or in this order, included the Delépine family, Martin, Sauteur, Villon, Douillon, Baveux, Guignon, Holingue, Couaillet

Plate IIII/12. COUAILLET FRERES. *The door of the front office of the main factory. The clock panel dates from the days of Delépine-Barrois. It represents* a pendule de Paris *and epitomises one side of the industry.*

ways, was that of Holingue. Of the various members of the family, M. Albert Holingue, who will be mentioned in Chapter V, is on record as having had a connection with Drocourt and also with E. Martin, while the *Paris Normandie* of 1932 mentions particularly Monsieur and Madame Alexandre Holingue, whose establishment was set up in 1869 and who were in business for a total of 63 years. Albert Holingue was almost certainly the manufacturer of whom it was said in the *Revue Chronométrique* of 1859 on the occasion of the Rouen Exhibition that, without having either the workshop or the tools of Martin or Croutte, yet he managed to compete with them. It would seem from Tardy's new *Dictionnaire* that Holingue later

Plate IIII/13. COUAILLET FRERES. *A 19th century wheel engine in which, by turning a single handle, the blank is indexed automatically while the cutter frame is reciprocated. (By courtesy of René Couaillet)*

and Bayard. As already noted, it does not seem likely that there were ever operating simultaneously more than about fifteen "factories" in connection with clocks or of horological mechanisms of any kind. There may well have been less. The *Revue Chronométrique* makes quite plain that less than 1,000 people were employed in horology in 1867. However, in addition to this number, many undoubtedly worked at home supplying concerns employing perhaps eight people, of whom only the name of one would have appeared on the main factory books. Throughout most of the 19th century, and despite the factories and innovations in production introduced by Pons, the whole business of clockmaking was still organised very much on a cottage-industry basis, with much of the work being done in private homes. The same conditions were of course found in the French Jura, in Switzerland, and in the Black Forest,[17] while similar procedures were followed in England at Prescot, London, Liverpool, Coventry and Ashbourne. Factories there were; but most of the real work, especially in the early days, was done in the homes of outworkers who made just one small part. Some of the houses in Saint-Nicolas-d'Aliermont, where clock parts were made, still betray the fact by having their windows larger than usual (Plate IIII/14). Others have quite substantial workrooms built on to them, much in the manner of Prescot or Clerkenwell. Yet others again are obviously small factories, with runs of continuous windows much in evidence. Finally, there are the really large concerns, such as Couaillet and Bayard, usually later in date and appearing to have been built only as factories.

Saint-Nicolas-d'Aliermont, although long dignified by the municipal appurtenances of a town, is really far more like a large village, typical of many in Seine-Maritime. The town consists essentially of one long straight street, which only recent fringe development has really altered at all. Many of the houses are old. Some stand directly on the road, but not a few are of considerable size, hidden discreetly behind high walls and well-tended trees and hedges. At one end of the village stands the small château (Plate IIII/15) set well back in a miniature park and visible through a wrought iron fence. Many of the local buildings are of a farm-

Plate IIII/14. *COTTAGE INDUSTRY, SAINT-NICOLAS-D'ALIERMONT. M. Pigny worked here about 27 years ago. Note the typical* vitrage d'horloger. *(Documentation by Madame Delabarre)*

like character, with open country directly behind, and with cattle, chickens and sheep much in evidence. In autumn, piles of cider apples are to be seen in yards and orchards. The visitor in the 19th century would have found much the same. The comparatively large horological industry of the past was apparently never very obtrusive, being largely tucked away in private houses. This was the fact which surprised T.D. Wright, who expected to find grander arrangements in an industry from which the output

Plate IIII/15. *SAINT-NICOLAS-D'ALIERMONT. Château de Thévray. (By courtesy of Madame Thévray).*

---

(17) Incidentally, the *Revue Chronométrique* says that many so-called "Black Forest" clocks were made in France. (C. Saunier, *Exposition Universelle en 1867, Horlogerie Française, p.29, 1867*).

was so impressive (see Appendix b(i)). The plain fact is that at the best of times the number of men engaged in Saint-Nicolas in connection with small clocks never much exceeded fifteen hundred, and was probably less. The products of the small domestic manufacturies were fed to several parent firms who maintained decently sited factories and were responsible for the co-ordination of the industry as a whole.

The main street has still a few surprises to offer behind the quiet façades. For instance, No. 22, in the part now called Route Dieppe, is a comparatively large house, probably built in the late 18th century and set well back in a railed garden flanked by two tall Wellingtonia trees (Plate IIII/16). We were at once made hopeful by the presence of a large clock with the inscription "O. DUMAS" below it and set high in the centre of the front wall. While the present occupant knew of no horological history in connection with the place, it transpired that this house was at one time the home of Onésime Dumas and at a later date was occupied by Emile Delépine. Onésime Dumas was the famous chronometer maker. He is mentioned repeatedly in this connection in the *Revue Chronométrique*. The issue of August 1879, p.326 (Vol. X) says that Dumas took over Gannery's chronometer interests, installing himself in Saint-Nicolas and becoming the most important maker in France. Dumas was apparently a nephew of Motel and certainly wrote his obituary, which appeared in the *Revue* of 1859, pp.79-82 (Vol. III). O. Dumas seems to have been quite separate from A. Dumas of Paris,

whose name appears on a number of *pendules de voyage*. That the latter must also have maintained workshops in Saint-Nicolas seems implicit in the fact that to this day a street is named after him. The workshops belonging to No. 22, Route Dieppe, were reached by a footpath on the right hand side. Here Delépine reputedly installed a transit instrument for checking chronometer rates. Madame Lejoille-Grard's father worked for Emile Delépine. This remarkable lady, who is a splendid and also explicit correspondent, even remembers sitting on M. Delépine's knee at tea before 1900 in the house with the clock. Rue Cannevel No. 74, (Plate III/17) is a ruinous house; yet it is still perhaps as manifestly a building formerly devoted to

Plate IIII/17. *SAINT-NICOLAS-D'ALIERMONT. An old factory No. 74 Rue Cannevel. No horology has been carried on here during the past 72 years. (Documentation M. Delabarre, 1971)*

Plate IIII/16. *SAINT-NICOLAS-D'ALIERMONT. Onésime Dumas' old house, later occupied by Emile Delépine, another famous chronometer maker. (Documentation: Madame Lejoille-Grard, Monsieur and Madame Delabarre, 1971)*

horology as any still to be seen in the town. A small part of the ground floor seems to have provided living accommodation. It has been renovated since we first saw it a year or two ago. The remainder, including all the top floor, was once dedicated to some clockmaking activity now unknown. It is disappointing that for the time being there seems to be no clue as to what was made here. Note the mill-like opening in the roof to admit raw materials too large to be carried upstairs. The windows are all top-hinged and prop open at the bottom. Monsieur Delabarre says that there has been no horology here since he first came to the town seventy years ago, but that for a while the house was a junk-shop. A few yards further down the road, on the opposite side, lies a path dignified by the name Chemin E. Delépine. Other evocative street names include Rue des Horlogers, Chemin A. Dumas, Chemin H. Pons, besides more modern names such as Rue Raphaël-Hennion and Rue Robert-Duverdrey. Across the road again from Chemin E. Delépine, and on the same side as the Delépine house, lies a complex of old and curious buildings. They are of many shapes and sizes, with much quaint roofing above the now-familiar continuous windows. Our photograph, Plate IIII/18 shows one of the buildings. The name of the family once working here was Mauranne. Both

Plate IIII/18. *SAINT-NICOLAS-D'ALIERMONT. Old disused factory behind the main street.*

Mesdames Delabarre and Grard remember them well. Mme. Delabarre says that horological work was still going on when she used to go there about sixty years ago to play with their children. At least one activity of Mauranne was the finishing and fitting of platform escapements to carriage clock *roulants* for Couaillet. This alley and yard is a place only for the intrepid horological explorer, who is liable to be bitten by a *chien méchant* from behind every fence! One adventurer was shadowed by a brown dog of diminutive size and of unparalleled ferocity. Immediately next door, Rue E. Cannevel 16, is the home of M. and Mme. Delabarre, the oldest horological couple in Saint-Nicolas (Plate IIII/19). M. Alfred Delabarre, now aged eighty-one, worked for the various firms of Couaillet in one way or another for forty-seven years from 1912 to 1959. Madame Delabarre, now aged seventy-one, worked for Couaillet from 1920 to 1959, assembling mantel clocks. They have two Couaillet carriage clocks in the house and were able to lend the Couaillet catalogue for 1914. M. Delabarre's father before him also worked making carriage clocks both for Couaillet

Plate IIII/19. *MONSIEUR & MADAME ALFRED DELABARRE. Monsieur Delabarre worked for Couaillet from 1912 to 1959. Madame Delabarre worked for Couaillet from 1920 to 1959. They are the oldest horological couple in Saint-Nicolas.*

and for Duverdrey. As already noted, Couaillet had factories scattered all over the town. One stood where the infant school now is, and the main later one, previously mentioned, still stands next door to the home of M. and Mme. Delabarre. Before Armand Couaillet bought this factory it was the establishment of Delépine-Barrois. M. Denis (who will be mentioned again) has a post-card showing the house in this ownership as late at 1913. Madame Lejoille-Grard has sent another, this time in colour, showing a tall chimney stack which no longer exists. Apart from the Château, the Delépine-Barrois/Couaillet house is perhaps the most dignified in the town. Here for once we behaved badly and trespassed into the fastness of a well-screened drive. It was fortunate that we did so, for unseen behind the house lies a large range of specially constructed factory buildings, empty but complete and untouched to this day (Plate IIII/9). Here in the hey-day of carriage clock manufacture must have been employed a fair proportion of the eight-hundred souls in the direct employment of the Couaillet enterprise.

About seventy-two years ago, the Russians came to Saint-Nicolas to recruit clockmakers, and all the best were invited to go to Moscow.[18] Two families who were friends of the Delabarres spent several years working in Russia. One lady, Madame Jeanne Lejoille-Grard, who is still living in Saint-Nicolas, is the last survivor of the expedition. She writes most interestingly, so much so that her account should be placed on record. A letter dated 23rd June 1971 translates to read "This is exactly what happened in connection with Russia. During an exhibition in Moscow, inaugurated by the Emperor, His Excellence noticed a very attractive *mouvement* invented by a Russian named Alexandre Bellanovsky. As a result, the Empress asked him to come to the Winter Palace at St. Petersbourg in order to repair clocks there which no one could mend. He went; and when he was in the private apartments of the Empress she questioned him. He explained that he would like to improve his knowledge but was hampered by being poor. The Empress immediately offered her sponsorship, and asked the Russian Ambassador in Paris to tell her where she could send her protégé to further his studies in horology. The Ambassador replied that the best horological centre in France was at St. Nicolas-d'Aliermont and at the house of Monsieur Emile Delépine. Mr. Bellanovsky then came here, and it was my father, M. Bénédict Grard, who looked after him. When Mr. Bellanovsky returned to his country he was named Head of the Horological classes at the St. Petersburg School of Arts and Crafts. One day subsequently, Mr. Bellanovsky came here to ask my father if he would like to go to Russia to fill the vacant post of Professeur of the top class. There was already another Frenchman, M. Baron, who taught the second class. In due course Mr. Bellanovsky asked my father if he knew of a good horologist at Saint-Nicolas who might perhaps accept the appointment of teaching the third class. My father recommended M. Emile Dumanchel who joined us later". In another letter dated 16th July 1971, Madame Lejoille explains how her father eventually agreed to go to Russia on his own terms and only with a contract. Passports and paper-work generally caused much difficulty. The Grard family, having sent their furniture ahead to Russia, were held up in Paris, calling each day at the Russian Embassy. During this period they visited the Paris *Exposition* of 1900. In the end, after a delay of three weeks, they went to England where Madame Lejoille's aunt lived in London. Here, M. Grard was found work by his brother-in-law and proceeded to set up a new home. No sooner was the family re-settled and happy than they were traced by the Russian Embassy in Paris. It was said that their papers had been mis-filed and that all was now in order. By this time, M. Grard did not wish to go at all. However, he had signed a contract, and a few weeks later he travelled to St. Petersbourg with his wife and daughter. A beautiful house awaited them there. Madame Lejoille's letter loses something in translation. It is hoped that the foregoing will have conveyed to the reader something of the charm and dash with which she tells her story. For instance, of the lost papers she says, "Mais l'Ambassade Russe de Paris fit des recherches et après nous avoir retrouvés, envoya ses excuses, ils avaient retrouvés les fameux papiers indispensables qu'un employé distrait avait mis dans le dossier d'un inconnu, et ils les avaient depuis longtemps, et on le pria de partir au plus tôt. . ." Emile Delépine died during the time that the Grards were in Russia.

When the small Château at Saint-Nicolas-d'Aliermont was first mentioned earlier in this chapter, we never sought or expected to see inside the place. In fact, on a return visit we found ourselves the guests of Madame Thévray, who as châtelaine of the house and owner of an engineering works not far away is well versed in the history of the town. Here, thanks to her kindness, we were able to verify many small details. The Château which dates from the 17th century is a fine example of a Normandie period country house. It has never been spoiled by modernisation and has a pleasant air of gentle decay.

It is interesting to compare Saint-Nicolas-d'Alier-

---

(18) M. Delabarre's father, who had worked in his time upon carriage clocks for Duverdrey and for Couaillet (including *grande-sonnerie* for the latter) and for Guignon, was one of those who was asked. M. Alfred Delabarre was seven years old at the time, but he remembers clearly.

IIII Saint-Nicolas-d'Aliermont

mont with Prescot in Lancashire, because in point of history they could be called "twin-towns". Prescot was the small workroom source of the "rough movements" of many English watches from quite early times until their manufacture ceased for all practical purposes in about 1914[19] and of marine chronometers until after the 1939-45 war. The town was also a centre of manufacture of horological tools. What is sad is that Prescot, unlike Saint-Nicolas-d'Aliermont, is today scarcely a manufacturing town.

Denis Frères (Société d'Usinage des Métaux) are an engineering firm still active in Saint-Nicolas and having a horological past. The founder in 1874 was M. Gustave Denis, the father of Ernest and Georges Denis and the grandfather of the present partners, Georges and Paul. Before 1914 a speciality of the firm was small clocks with cases in the shape of churches and with an angelus chime playing daily at 6 p.m. The Denis grandfather was in business as a wheel and pinion cutter, supplying these parts for use in carriage and other clock *roulants* made by others in the town. The last complete clocks made by Denis were bulkhead clocks, produced from 1945-1955 for Auricoste in Paris. The new title, S.U.M., was taken in 1959.

Other firms still busy in Saint-Nicolas and having horological backgrounds include Lemaignen, Lechevallier & Mercier, Enregistreurs Lambert and Ateliers Vaucanson. On the whole, central-heating, plumbing fittings, electrical and electronic components, counters, etc. today form the hard-core of the industries of Saint-Nicolas, rather than anything specifically horological.

---

(19) Latterly some Swiss *ébauches* were used in England, as is evident when centre wheels are seen above barrels.

Chapter V

Paris

Today and Yesterday

BOURDIN. See Plate V/7.

V

Paris, Today and Yesterday

Breguet, Daniels, Pitou. Garnier, Bourdin, Oudin-Charpentier, Lefranc, Chartier, Boseet, Berthoud. Jacot, Drocourt, Margaine, Le Roy/Leroy.

By far the best known clock and watch shop in Paris for at least one hundred and fifty years was that of the Maison Breguet, latterly at No. 28 in the beautiful Place Vendôme. However, the year 1970 saw the retirement of the proprietor M. George Brown and the transfer of the horological side of the business, including the world-famous manufacturing books of the firm, to Messrs. Chaumet of Place Vendôme 12 and Bond Street, London. Thus ended an era stretching back, if somewhat tenuously, to the time of A-L Breguet himself. In their last years, Messrs. Breguet had a brief resurgence of sorts due to the outstanding skill of their agent in London, Mr. George Daniels. Until after 1960 there were still lying about in Paris a number of unfinished, complicated pieces which Mr. Daniels proceeded to complete in the style and to the standard originally planned for them. A unique achievement. It would scarcely be an exaggeration to say that the "cloak" of Breguet fell after almost 150 years upon Daniels, so closely does the work of the latter resemble the original. On a lesser scale, M. Pitou (who will be introduced soon) supplied "Breguet" for more than thirty years in the present century with basically standard carriage clocks which he finished from *blancs-roulants* obtained from Jacot about fifty years earlier. M. Pitou's clocks are fine-looking pieces, still with the "Breguet" Empire appearance and fully in the idiom of the smartest and most fashionable parts of the capital of France.

Happily, there are still some parts of Paris which in the past have almost never seen the casual tourist. In consequence they have not become artificial and phoney like the Montmartre and Pigalle. Such an area is the old manufacturing district in the now somewhat overplayed Marais. This jumble of streets, built on ground which in ancient times was a marsh, became in the 17th and 18th centuries the place where the high-society of France built their large town-houses, or *hôtels particuliers*.[1] From the end of the 18th century and during the 19th much further building took place, hiding most of the beautiful earlier houses. Many of these still exist. Some of the best are now restored to their former glory by the French Government besides being once again made visible by the removal of later, unimportant buildings. The Musée Carnavalet is a typical *hôtel*, now filled with period furnitures. What happened to create the apparent contradiction of the Marais was this. After the Revolution, when there were no more rich people left alive in Paris, their fine houses and possessions were taken over by numberless small industries, mainly connected with horology, jewellery, spectacles, art-work and cheap clothing. This type of occupancy has kept the old houses standing until today, but only just. Now a second metamorphosis is taking place. The small workrooms are fast disappearing and the restored buildings are being let at high rents. In one sense much is being preserved, but at the same time a great deal of 19th century history will soon be lost. The same is happening at home in London, where scores of small clock, watch and similar businesses were once concealed by the façades of the quiet streets and squares of Clerkenwell. Today whole streets are being demolished at an alarming rate; but here, alas, the motive is to make way for new buildings and not to reveal old. Old industries apart, the two areas, Clerkenwell and the Marais, could scarcely be more dissimilar; but it is interesting to compare them. The basic date of much of surviving old London is early 19th century. Until after the Second War,

---

(1) This term means a grand house occupied by one family, as opposed to a house with *appartements*.

V Paris, Today and Yesterday

Clerkenwell was typical. In the "Well", at one time, nearly every dwelling had behind it workrooms housed in long, narrow buildings with continuous windows. These miniature factories were built on to the backs of ordinary, small, brick, terrace houses each with two stories, attic and basement. The manufactures were well insulated from the streets. The workrooms were usually approachable only through the front doors (opened by remotely-controlled string-and-pulley devices) and then via much-barricaded "front-room" offices. The typical Paris workroom, by contrast, was most often to be found upstairs in tall 17th, 18th or 19th century buildings overlooking narrow streets or courts. These hidden houses, by no means all of them once *hotels particuliers,* are still standing almost unchanged in large numbers. No guesswork is needed to see them as they were a hundred or more years ago. Many façades are typically those of the old pre-Revolution Paris. The stone-built houses are narrow, crumbling and stand five or six stories high. Most are divided into *appartements,* one or more flats being found on floors reached by a common staircase. The windows have slatted shutters and like the buildings they are tall and narrow. In some of these houses and flats there were once clock factories, set cheek by jowl with purely domestic premises. To the horologist the blue and white street name plates on the walls to-day often seem strangely familiar, because they are the addresses associated with carriage clocks.[2] Apart from these plates, little if any trace of the old trades now remains. Plate V/1 shows one notable exception, almost a miracle of survival into the present age. It is the workroom of M. Pitou above the Rue Debelleyme. It is approached from 85 Rue Turenne through an archway leading to a courtyard which is a real part of history. The yard is unevenly paved and affords access to houses crowded round it. The peeling stuccoed walls are pierced by dark entrances. The doorway leading to M. Pitou's rooms, with its disused well set in the left pillar of the entrance, seems as old as Paris itself. The worn, curving stone stairway has gnarled and rusted iron banisters and rail.

Plate V/1. *PARIS, 1971. M. Pitou's workroom above the Rue Debelleyme.*

---

(2) To name but a few best-known addresses:

Brunelot	10 Rue Oberkampf
Corpet (Marcel)	84 Rue Amelot
Detouche & Houdin	228 and 230 Rue Saint-Martin
Drocourt	28 Rue Debelleyme
Erbeau (L.)	100 Boulevard Sébastopol
Garnier (Paul)	25 Rue Taitbout
Hour (Charles)	7 Rue Sainte Anastase
Jacot	31 Rue de Montmorency
Joseph (Charles)	114 Rue Amelot
Margaine	22 Rue Beranger
Maurice	75 Rue Charlot
Moser	15 Boulevard de Temple
Requier (Charles)	5 Rue Debelleyme

V Paris, Today and Yesterday

Plate V/2. PARIS, M. PITOU. *The clock which he is holding was "finished" by himself in 1970 from a Jacot* blanc-roulant. *(By courtesy of M. Pitou)*

M. Pitou (Plate V/2) was eighty years old in 1970. He was then still finishing the *blancs-roulants* which he bought fifty years ago when he acquired the remaining stock and materials of the famous firm of Jacot. At one time Pitou employed eight workers. He was apprenticed to Chartier in Rue du Pont-aux-Choux, but later worked for the firm of Jacot until it closed in about 1920. Jacot's workrooms in Paris were upstairs in a building similar to that which houses M. Pitou. He says that during the years in which he worked for Jacot no more than ten people were employed there. According to M. Pitou, Jacot obtained his best *blancs-roulants (grande sonnerie* for instance) from Baveux in Saint-Nicolas; but he also bought more ordinary *roulants* from Japy. We noticed in Pitou's rooms a number of rough movements which from our recent researches were recognisable as having been made by Japy at Beaucourt.[3] As already stated, the Baveux roughs were of a superior quality, and they were thus no doubt favoured by Jacot for his best clocks. M. Pitou believes that Baveux never finished their own *blancs-roulants*. He also confirms a fact already discovered; namely that Saint-Nicolas makers as a whole did not finish their carriage clocks, with the notable exceptions of Couaillet and Duverdrey and Bloquel. A box of punches in one corner of M. Pitou's workroom revealed some interesting history. There would seem to be no doubt that Jacot finished clocks for A.H. Rodanet, Paul Garnier, L. Leroy & Cie, P.E. Dubois, Ch. Pougeois and G. & B., at least, and that he punched the movement plates accordingly.[4]

Plate V/3 shows a selection of *blancs-roulants* photo-

Plate V/3. BLANCS-ROULANTS. *The* roulants *from left to right are for hour and half-hour strike and repeat; timepiece alarm; timepiece; unusual small grande sonnerie. The movement in the foreground is a timepiece alarm. Note that all hammer shanks were originally straight. (By courtesy of M. Pitou)*

---

(3) These *roulants* were presumably acquired by M. Pitou along with Jacot stock.
(4) Naturally, there are also punches for "Réveil", "Hours and Quarters", "Alarm", "MADE IN PARIS", "MADE IN FRANCE", "PARIS", etc., as well as "Slow" and "Fast", and those "makers" once so beloved of auctioneers, "Advance & Retard". We only wish we could truthfully say that Monsieur Breveté and Madame Aiguilles were also represented!! According to M. Pitou, Jacot supplied some clocks direct to the Army & Navy Stores in London.

V Paris, Today and Yesterday

halves in passing, and it is also a repeater. The alarm-setting dial is cleverly hidden from view inside the back door so as not to spoil the frontal appearance. M. Pitou's work is somehow very reassuring in the present age of shoddiness. He proves once again that here and there traditional craftsmanship is far from dead. Plates V/5 and V/6 show another of M. Pitou's clocks. According to M. Pitou, there is still one dial-maker working in Paris, while the last French gong-maker was Fournery, said to have been a hard business man. Apparently a few English gongs (already set in their "blocks") were imported at one time from a maker called Drury. Amant (possibly spelt differently) was a bell maker favoured for Paris carriage clocks at an earlier date.

Plate V/4. *NEW CARRIAGE CLOCK IN ENGRAVED GORGE CASE. A fine clock with hour and half-hour strike, repeating hours and with alarm, finished by M. Pitou of Paris in 1970. (By courtesy of M. Pitou)*

graphed at Rue Turenne. These are the "raw materials" from which M. Pitou has worked since 1920. They include the semi-completed movements of timepieces, timepieces with alarm, hour-and-half-hour strikers with repeat, and even *grande sonneries* with alarm. Plate V/4 illustrates a clock striking hours and halves with both repeat and alarm. This piece, with its engraved *gorge* case and engine-turned dial, could perfectly well have been made, and indeed should have been made, almost exactly 100 years ago; but in fact it was finished in 1971 many years after French carriage clock manufacture is supposed to have ceased! Plates II/17 and II/18 illustrate another "brand-new" *pendule de voyage*, this time made in the style of Breguet. Note the traditional Empire case, still timelessly elegant. This clock strikes hours and

Plate V/5. *A FINE GRANDE SONNERIE MOVEMENT. Another example of one of M. Pitou's clocks completed in 1970. (By courtesy of M. Pitou)*

The Garnier business continued long after the period at which it was left in Chapter II. The first Paul Garnier, as already noted, died in 1869. His son,

108

V Paris, Today and Yesterday

Plate V/6. *A TOP QUALITY ESCAPEMENT PLATFORM. The escapement, in all probability Swiss, used by M. Pitou in the* grande sonnerie *clock shown in Plate V/5. (By courtesy of M. Pitou)*

also Paul, died in 1916, having previously given his watch and clock collection to the Louvre and having carried on the horological business inherited from his father. He in turn was succeeded by his nephew M. Blot Garnier, born in 1871. Blot Garnier in due course became President of the *Chambre Syndicale d'Horlogerie*. According to Tardy, Blot died in 1938; although not before the business had been split and partially sold off.

A.E. Bourdin was a mid-19th century Paris maker whose best work must be taken very seriously indeed. Tardy quotes various addresses for him at different periods, while the *Tribune Chronométrique* shows that in 1850 he was working from 18 Rue de la Paix. The *Tribune* is full of praise for Bourdin, saying that the 1849 *Exposition* was the third in which his work had featured. He received a Bronze Medal in 1844, a Silver Medal in 1855, and a Bronze Medal in 1867 for pieces which included carriage clocks. The *Tribune* (page 82), speaking of his 1849 showing, says that his *pendules de voyage* are not only very well finished and presented in all their details, but are also excellent timekeepers ("... excellents chronomètres marchant avec la dernière exactitude...") and this despite the many complications which they embody. The *Revue Chronométrique* is equally complimentary, referring to Bourdin constantly, and saying in September 1855, page 46, that both his shop window in the Rue de la Paix, and his display in the Palais d'Industrie are a wonderland of beauty. Various fine and complicated clocks of one kind and another are mentioned. The *Rapport du Jury* of the *Exposition Universelle* of 1855 translated says: "*Bourdin*. In this rich exhibition one notices a small portable clock with *grande sonnerie* (10,000 francs) with spring-detent escapement, and a compensation balance with helical balance spring. The dial shows the hour, the minute, the second, the day of the week, the date of the month, the phases of the moon, and even by means of a bimetal strip, the temperature." A similar clock was sold at Sotheby's on 19th June 1972, Lot 195. It is illustrated both front and back in Plates V/7 and V/8. In addition to the going train, there are two separate striking trains, and yet another for the alarm. There is also an up-and-down dial. The piece is signed "BOURDIN HGER BTE A PARIS". The chronometer escapement is of the Earnshaw type. There is a compensation balance having helical spring complete with terminal curves. The beautiful gilt-metal case is set with bloodstone panels, and has an ornate carrying handle with two infants flanking an enamelled coat of arms. The nature of the *cadrature* (external striking work) and also its disposition upon the back plate of the movement, where the signature is repeated, is very reminiscent of the best La-Chaux-de-Fonds clocks mentioned in Chapter X. This *pendule de voyage* is about as fine an example of a complicated travelling piece as it is possible to imagine, even when compared with the Breguet masterpieces, not to mention the best Swiss and English pieces. It would not be fair to compare this specially made Bourdin clock with the excellent everyday clocks sold by Jacot, Drocourt, Margaine and other well known makers. Bourdin also sold more modest clocks. One is signed "BOURDIN

V Paris, Today and Yesterday

Plate V/7. BOURDIN. *This clock may well be the one shown at the Paris Exhibition of 1855 receiving wide acclaim. Its overall quality has hardly ever been equalled, much less excelled. The beautiful case is inset with bloodstone panels. The very complicated movement has four trains and is controlled by a chronometer escapement. (By courtesy of Sotheby & Co.)*

Plate V/8. BOURDIN. *The back of the exceedingly fine carriage clock shown in Plate V/7. One cannot help wondering whether the movement may not owe something to Switzerland. (By courtesy of Sotheby & Co.)*

H<u>ER</u> DU ROI", so it was probably made before 1848. It is very good quality but with its engraved case and *grande sonnerie* movement it is not remarkable in any other way.

Other makers, both those contemporary with Bourdin and some at a later date, produced special clocks on special occasions for their best customers or for Exhibitions. Oudin-Charpentier was one; and there may well have been others whose work has yet to be re-discovered. Charpentier took over from Oudin

110

V Paris, Today and Yesterday

fils, the son of Charles Oudin, the pupil of Breguet. Oudin-Charpentier produced a special book (see Bibliography) to accompany his display at the 1862 Exhibition in London. Here Oudin-Charpentier described himself as "... principal clockmaker to their Majesties the Queen and King of Spain and to the Imperial Navy" (sic). He showed "... a new system to facilitate the regulating of carriage clocks ..." besides examples with chased silver mounts, *grande sonnerie* and calendar, with crystal dial and polished steel motion work visible. Another clock, this time a chronometer, was made for the King of Spain.

Three further names which must be mentioned in connection with distinguished carriage clocks made in the second half of the 19th century were Lefranc, Chartier and Boseet. A particularly fine clock by Lefranc is illustrated in Plates VIII/17, VIII/18, V/9 and V/10. These photographs show respectively the

Plate V/10. LEFRANC. *The platform escapement of the clock shown in Plate V/9 and also in Plates VIII/17 and 18. (By courtesy of E. Pitcher & Co.)*

case, under-the-dial, the rear of movement and escapement of this clock. Plate V/11 shows a very fine clock No. 13393 by Edouard Chartier, while Plates VIII/14 and 15 illustrate a special clock by Boseet having thermometer, barometer, centre seconds and fly-back calendar.

Ferdinand Berthoud, whose many achievements, and also quarrels with Pierre Le Roy have occupied the attention of horological historians ever since, has already been mentioned briefly in Chapter I in connection with an astronomical pre-*pendule de voyage* which he made. He died on 20th June 1807 and his nephew and pupil Louis, who succeeded him, died in 1813. The sons of Louis, who used the style "Berthoud Frères" from this time, were Louis-Simon-Henri (always called Louis) and Charles Auguste (often known as just Charles). Their Paris address was 103 Rue Richelieu in 1819; but they seem to have lived and worked at Argenteuil, concentrating upon marine

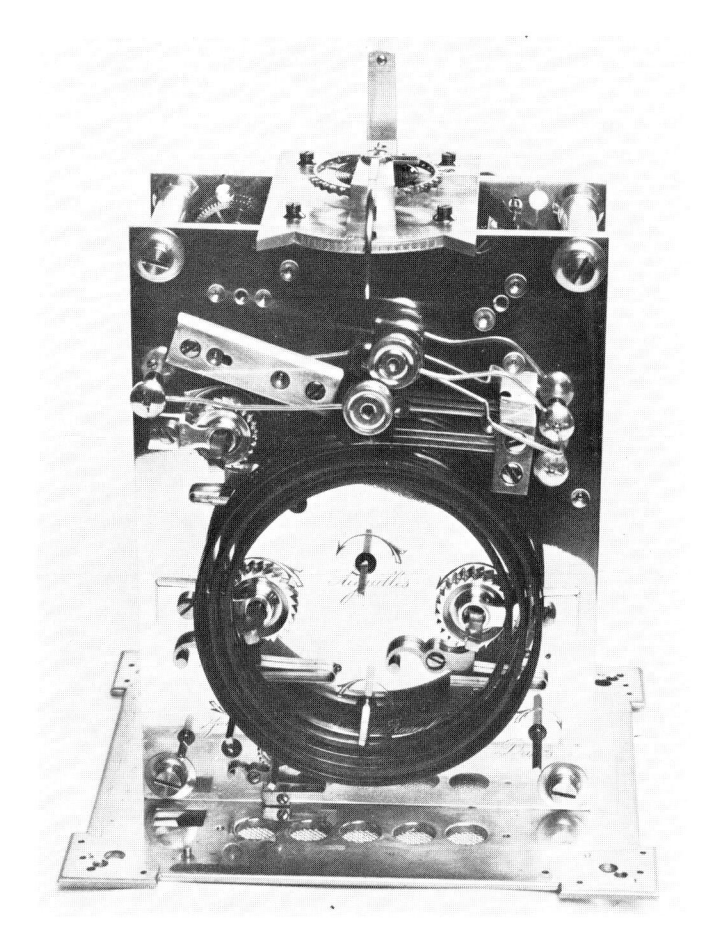

Plate V/9. LEFRANC. *The rear of the movement of the perpetual calendar,* grande sonnerie, *minute-repeating clock with alarm and moon, also illustrated in Plates VIII/17 and 18 and V/10. (By courtesy of E. Pitcher & Co.)*

111

V *Paris, Today and Yesterday*

this clock has a plain and restrained case, set off by an engine-turned dial, large seconds circle, moon hands and up-and-down dial. The chronometer escapement (Plate V/13) maintains the Berthoud tradition by having a pivoted-detent, brought into line with the fashions and style of finishing of the best French/Swiss practices of a fairly late period, circa 1860. Note the fine overcoiled balance spring, the volute detent return-spring (compare with the straight springs used for the same purpose in the earlier

Plate V/11. *CHARTIER. An outstanding case. The number of the clock is 13393. (By courtesy of Meyrick Neilson of Tetbury Ltd.)*

chronometers and precision work. They were the contemporaries and equals of Motel and of Vissière. According to Tardy, Auguste-Louis Berthoud, a nephew of Charles Auguste, was born in 1828 and also became an horologist. A marine chronometer exists, signed "Ate. Louis Berthoud" and numbered "47", which may have been sold by him, although Tardy's *Dictionnaire* shows other Berthouds in business in the second half of the 19th century. The fine *pendule de voyage* shown in Plate V/12 is signed simply "BERTHOUD"; so it is a matter for speculation which member of the family made or sold the piece. The style of the movement, escapement, case and handle, however, suggest a date towards the end of the century. As will be seen from the illustration

Plate V/12. *BERTHOUD. A very fine chronometer carriage clock with pivoted-detent escapement and with up-and-down indicator. Note a fairly traditional dial with large off-set seconds circle used in conjunction with a Corniche case. (By courtesy of Asprey & Co. Ltd.)*

112

V Paris, Today and Yesterday

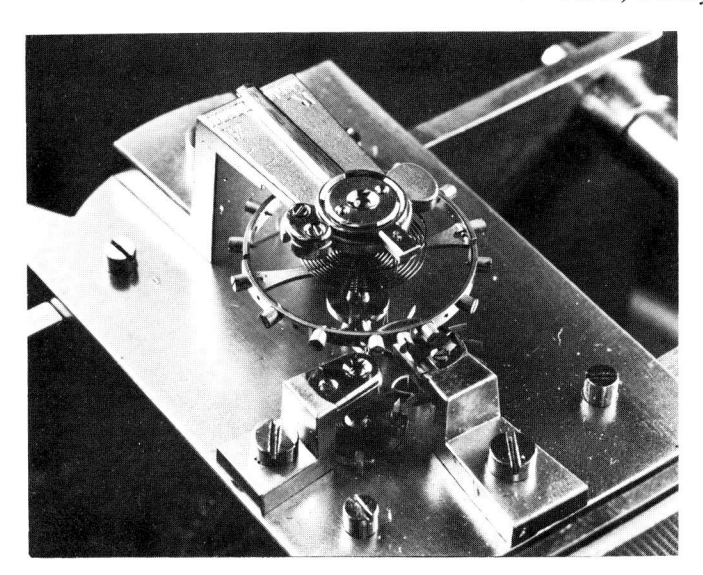

Plate V/13. *BERTHOUD. The beautiful pivoted-detent escapement of the chronometer carriage clock shown in Plate V/12. (By courtesy of Asprey & Co. Ltd.)*

La Chaux-de-Fonds pieces in Chapter X), the flat 'scape wheel, and the style of the cocks, jewelling, index, etc. A beautiful clock.

Other than the superb productions of the Breguet firm, and with the exception of those few really outstanding pieces produced on occasion by such makers such as Bourdin, Boseet, Chartier, Lefranc, etc., then Jacot was probably the best of the four leading Paris producers of carriage clocks towards the end of the 19th century. The four firms were Jacot, Drocourt, Margaine and L. Leroy & Cie. Any difference in excellence between their best clocks is almost more a matter of opinion than of fact. All four firms sold clocks of really high quality. On the whole, Jacot tended to make superb and fairly complicated clocks of restrained elegance. Drocourt and Margaine, on the other hand, often favoured florid designs with much use of engraved cases and of decorative panels. L. Leroy & Cie regularly used both plain and decorative cases, while mechanically their clocks show wide variation in both design and origin.[5]

There were two Henri Jacots in the carriage clock business. The first was Henri senior, the uncle. The second was Henri junior, his nephew and successor. Henri Jacot senior died in 1868. Saunier wrote a short obituary which appeared on page 263 of the *Revue* of August in that year. This translates to read:- "HENRI JACOT. This excellent horologist died on the 31st July last. The whole of Paris horology is indebted to him. It was to him that we owe the development of the splendid carriage clock industry which today provides the livelihood of a large number of families. Beneath an unassuming nature were hidden the qualities of the artist prepared to forgo even material gains for the success of his work. The greater part of his profits went into the invention and perfecting of tools. A cruel mistake at the 1867 Exhibition saddened the end of this hard-working career. The error has since been put right, and his friends have had the consolation of not seeing the old and valiant worker die under the shadow of the profound disillusionment which he had experienced." The "cruel" error was that Henri Jacot had not been awarded a medal at the Exhibition, but only an "Honourable Mention". Commenting upon the awards in general, Saunier was very critical of the way in which the jury had carried out its task, and of the rushed conditions in which it had been compelled to work. His opinion appears to have been vindicated by the fact that after the names of medal-winners had been published, it was discovered that the list included the name of one firm who had not even exhibited! Saunier was particularly scandalised by what had happened to Jacot, and he referred to this event more than once in the *Revue Chronométrique*. For instance in Sept. 1867, p.59 he says "... to think that one of the most outstanding personalities in the Paris industry, the creator of the contemporary style in carriage clocks... in a word, our eminent manufacturer, M. Henri Jacot — to think that he, a medal winner at several exhibitions, should receive no more than an *Honourable Mention!*. We are told that a mistake was made, and that he had been awarded a Silver Medal. Very well! but we are also told that the error, recognised and confirmed by two of the honourable members of the jury, M. Breguet and Mr. Frodsham, can never be put right." The obituary, however, makes it clear that the error was officially rectified after all.

Describing Henri Jacot's display at the Exhibition, Saunier wrote in the *Revue* of December 1867, "The carriage clock industry has become extremely important in Paris. This style of clock is exported throughout the world, and nowhere is it better produced than here. The Exhibition has certainly shown

---

(5) Some came from Japy, some from Couaillet (already finished) and some from Jacot (Baveux). The best clocks were always finished in Paris, some of them latterly by M. Pitou.

us some beautiful examples of the style of the English section, but these were exceptional pieces, produced at great expense and not the result of the regular manufacturing. Moreover we did not observe among the English a single example of a *grande sonnerie*. If foreigners are indebted to us for this article, we owe it in the main to M. Henri Jacot, and he has remained the master of this speciality. In contrast to almost all his colleagues who rely on the large factories in St. Nicolas, Beaucourt, and Montbéliard for their basic materials, M. Henri Jacot makes everything himself, using mechanical aids which he has invented or improved. The reason for the excellence of his production is also to be found in the cleverly designed range of his models, and in the ingenious machinery by means of which he produces both his cases and his movement parts in his workshops at St. Nicolas and Paris."

Jacot clocks will usually be found to have two identifying marks. The first is the Jacot trade mark of a parrot on a perch between the letters "H" and "J". The trademark is usually struck on the lower left-side of the back plate, deeply and sharply defined and quite unmistakable; but sometimes it is positioned near the hand-set square. The second mark is an oval *poinçon* varying in form but listing, depending upon when the clock was made, the medals obtained by Jacot at various Exhibitions. This dated mark is found either on the inside of the back plate, hidden by a barrel, or else it is behind the falseplate of the dial. A typical inscription would read as follows:-

"MEDAILLES EX^{PONS} UNIVER^{LES}
H. JACOT
PARIS
B^{ze} 1855, 1862 A^t 1867, 1878, 1889."

The same basic inscription is also found without reference to the 1889 medal, dating a clock to the period 1878-1889; and without the 1867, 1878 and 1889 medals, meaning that the clock must have been produced during the period 1862-1867. Known Jacot serial numbers range from 387 to over 19,000. No. 387 is a very fine *petite sonnerie* clock with alarm and made for the famous Paris firm of Lepaute. No. 746, a *grande sonnerie* in *gorge* case made between 1862 and 1867, was sold by Le Roy & Fils. Serial numbers 907 and 2806 are known to still exist and both have *cariatides* cases. A typical late Jacot *caryatid* clock is shown in Plate VII/24. Jacot serial No. 7968 falls within the period 1878-1889, while the 1889 models appear with serial numbers above 10,000. If it were only possible to collect enough serial numbers of Jacot carriage clocks and to compare them with the Exhibition medals mentioned on the same pieces, it would soon be possible to form a very accurate idea of the exact date of any Jacot clock from its number alone. This task is something for the future.

That Henri Jacot the nephew successfully maintained the enviable reputation established by his uncle is evident both from contemporary writings and from examination of later examples of clocks sold by the firm. We have never seen a signed Jacot carriage clock which was not beautifully made and finished. As already observed, perhaps the most characteristic Jacot pieces, admired and collected by many people, have *gorge* cases with plain white enamel dials. Plate VII/8 illustrates a good example of such a dial as found in clock No. 825 made between 1862 and 1867, and of which the escapement platform is shown in Plate V/14. Later Jacot dials used with *gorge* cases

Plate V/14. *JACOT. The platform escapement used in the Jacot* grande sonnerie *clock No. 825 shown in Plate VII/8. This type of right-angled lever escapement with "butterfly" endpiece and shaped tail, sometimes with pointed-tooth wheel and sometimes club-tooth, was used extensively by both Jacot and Drocourt. (By courtesy of E. Pitcher & Co.)*

were distinguished by bolder and blacker-looking chapters, often accompanied by five-minute marks in Arabic numerals. For all their lack of ostentation, these clocks with their restrained distinctive designs manage to look at once both superior and expensive. Even the plain Jacot *gorge* timepieces evince no signs of having been made down to a price. They show the same detail refinement as the more complicated

V Paris, Today and Yesterday

certainly difficult to decide in the face of reports which do not agree. Several writers such as G. Sire in 1870 (p.69) and Saunier in *Revue* December 1867 (Vol. VI, p.138) both dealing with the *Exposition Universelle* of 1867, go out of their way to say that Jacot, unlike his contemporaries, makes his clocks, ". . . boîtes et mouvements. . ." in his own workshops with the help of mechanical aids peculiar to himself. Saunier also says plainly that Jacot has *ateliers* both in Saint-Nicolas-d'Aliermont and in Paris. However, it is necessary to take into consideration the information given by M. Pitou. As the reader will know, Pitou was born in about 1890. He is still very much alive, and what is more he not only worked for Jacot but also took over the remaining stock of *roulants* etc., circa 1920. According to Pitou, the Jacot concern only finally ceased making carriage clocks. Jacot's *caryatid* clocks, although they are much rarer than his *gorges* are in fact every bit as characteristic. The example already mentioned as appearing in Plate VII/24 certainly has an engraved dial plate which we have only seen on Jacot clocks. Two other clocks, which are discernibly of Jacot's production on account of their dial surrounds, are shown in Plates V/15 and V/16. A further, but very different style of Jacot, is to be found in the minute repeater No. 3707 illustrated in Plate VIII/8. The escapement of this last clock is shown in Plate V/17. It is interesting to compare this lever platform with the one already shown.

The question of how and where Jacot clocks were made during the last quarter of the 19th century is

Plate V/15. *JACOT. A grande sonnerie clock No. 7467 having a dial apparently peculiar to Jacot. Other Jacot clocks appear in Plates VII/8, VII/24 and VIII/8. (Private Collection).*

Plate V/16. *JACOT. A late clock in an unusual case with pillars having knurled bands. (By courtesy of E. Pitcher & Co.)*

115

V Paris, Today and Yesterday

therefore felt able to claim that he had workshops in Saint-Nicolas-d'Aliermont.

Readers studying Appendix (b) will notice that neither T.D. Wright nor Tripplin so much as mentions Jacot (this would be the nephew) in their reports on the occasion of the Paris Exhibition of 1889 when the business was at its peak. Such an extraordinary omission might at first seem hard to understand, especially in view of the fact that Jacot obtained a Gold Medal for carriage clocks on that very occasion. However, when a French friend was asked (after it was noticed that Jacot did not appear to be a member of the *Groupe Syndical de l'Horlogerie*) he said at once that Jacot was probably not introduced to the repor-

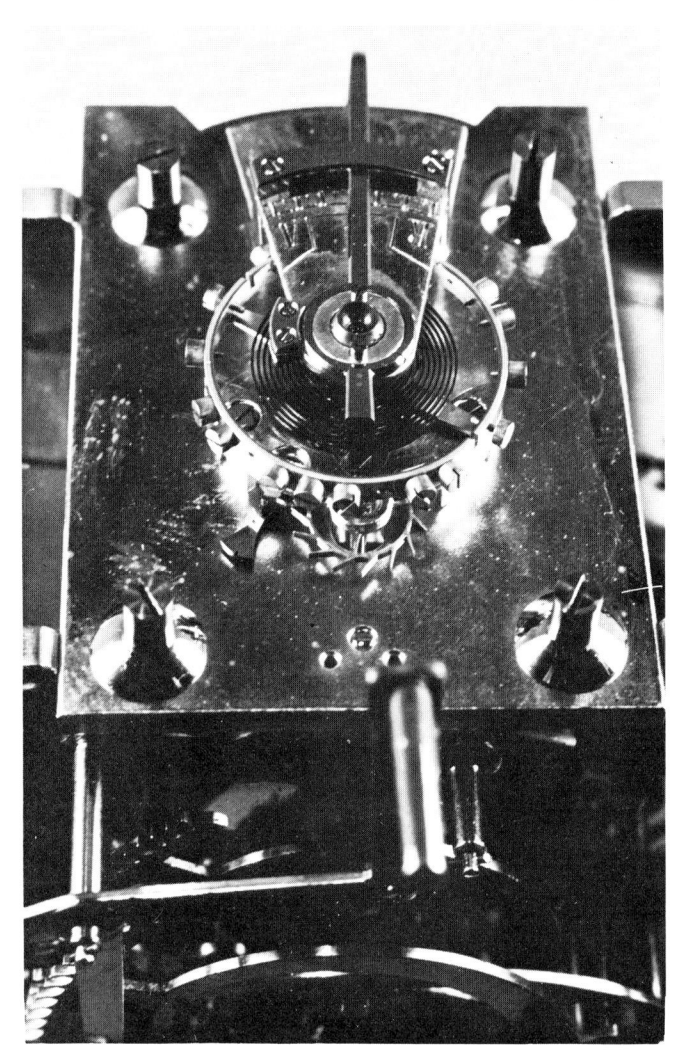

Plate V/17. *JACOT. The very high quality platform escapement used in the Jacot minute repeating carriage clock No. 3707. Other views of this clock will be found in Plates VIII/8 and VIII/9. (Private Collection)*

clocks at about that time. He remembers that their Paris workrooms were upstairs and that only about ten people were employed to finish *roulants* obtained from Baveux at Saint-Nicolas-d'Aliermont. Then there are our own recent findings of Japy (Montbéliard) *blancs* in Pitou's stock. Of course it is always possible that Jacot may latterly have obtained carriage clock *roulants* from Japy; but this solution is not confirmed by the Japy workbook for 1907 mentioned in Appendix (d). It seems impossible at present either to be sure that Jacot maintained a factory at Saint-Nicolas or that he did not. The truth is probably that he obtained most of his *blancs roulants* from there, most likely from Baveux, and

Plate V/18. *DROCOURT. A* grande sonnerie *clock with centre seconds, simple calendar, moon and alarm. (By courtesy of Asprey & Co. Ltd.)*

116

V Paris, Today and Yesterday

Italian work called *pietra dura*. Some Drocourt clocks were even sold complete with matching stands or plinths, while his cases received high praise in the report of the Universal Exhibition held at Besançon in 1860. Saunier, in describing Drocourt's display at the 1867 Paris Exhibition says "M. Drocourt has devoted himself to specialising in carriage clocks. He presented a large number of them well arranged, creating a most favourable impression at first sight, and on closer investigation revealing numerous improvements. His clockmaking, without reaching the level of the late lamented Henri Jacot, the master of the style, holds a very respected position in our industry and the distinction that he has achieved is fairly won". (*Revue*, August 1868, p.257).

It would not be right for the reader to imagine that Drocourt's clocks were always ornate, any more than Jacot's were always plain, although there is a certain amount of justification for both these generalisations. Plate V/18 shows a very complicated Drocourt clock housed in a plain *gorge* case, while Plates V/19, V/20 and V/21 show three views of a specially-ordered Drocourt for which a *gorge* case was considered entirely adequate even by a member of the Upper House. Drocourt carriage clocks are also illustrated in Plates VII/14, VII/33, VII/42, VII/C1 and VII/C13, besides in those plates associated with this chapter.

Drocourt is mentioned by contemporary authors as having had workshops in Saint-Nicolas-d'Aliermont, but it has not so far proved possible to find any evidence of such a factory. On the other hand, Dr. Dickie quotes Mme. Vve. G. Gamard of Saint-Nicolas-d'Aliermont (who died in about

Plate V/19. DROCOURT. *A* grande sonnerie *clock, No. 28436, with alarm and made to special order circa 1887. T. Martin & Co. moved from 225 Regent Street to 151 Regent St. circa 1893. They succeeded Henry Capt at this address. (Private Collection)*

ters, Wright and Tripplin, by the *Président* when he conducted them round, and hence did not find a place in their notebooks. The *Président* at the time was Rodanet.

Drocourt carriage clocks are always thought of and mentioned almost in the same breath as those of Jacot. Often they are very similar, favouring *gorge* cases and Roman chapters on white enamel dials of extraordinary clarity and beauty. Drocourt, however, also specialised very much in ornate and distinctive cases, of which the one shown in Plate VII/C1 is a very good example. Plate VII/C13 gives a good idea of another Drocourt decorative clock, in this instance having inlaid side panels derived from the type of

Plate V/20. DROCOURT. *The top glass of a* grande sonnerie *clock No. 28436 shown in Plate V/19. (Private Collection)*

117

V Paris, Today and Yesterday

Plate V/21. *DROCOURT. The extremely high quality escapement used in the* grande sonnerie *clock No. 28436 shown in Plates V/19 and V/20. This escapement is almost certainly Swiss. (Private Collection)*

1969) as imparting the following information:- "Towards the end of 1875 or early 1876 M. Drocourt bought a small clock factory in Saint-Nicolas from a M. Albert Hollinge.[6] From Paris M. Drocourt transferred M. Auguste le Chevallier who was his Manager at that time and installed him as Manager of the new factory. Following this he wished to expand further and he bought another small factory from M. Dumas, a maker of marine chronometers.[7] There were other factories available but none was equipped to the standards required by M. Chevallier who specialised in *grandes sonneries,* quarter repeaters and other high quality clocks of meticulous workmanship. The other factories in the main produced only alarms. The factory in Saint-Nicolas ceased to produce clocks of this type when M. Drocourt died in 1900".

Margaine was a name synonymous with beautiful carriage clocks towards the end of the 19th century. The firm's best known address was 22 Rue Beranger; but there has also been quoted Rue de Bondy 54. The trade mark was a beehive in the form of a clock flanked by the initials "A.M." Margaine received a Silver Medal for *pendules de voyage* at the Paris Exhibition of 1878. Of his clocks shown in 1889 Tripplin wrote ". . . let us look at the exhibition of M. Margaine, which is more that of a decorator, so elegant and tasteful are all his works, displaying the greatest refinement. As a matter of course all the movements are good, but the excellence is noticed in the decorations; here every taste is studied, rich, plain, highly decorated; all remind you that it is only Paris that can create such things. Margaine, in the way of clocks, is on a par with Champion, the watch engraver". Rousaille (pp.34-35) also describes Margaine's display at the 1889 Exhibition. His offering included a number of carriage clocks with very complex movements. Particularly impressive was a finely engraved *caryatid* model with *grande sonnerie* striking and giving day, date and phases of the moon. "Margaine's trademark", writes Rousaille "has become one of the best in this industry. He is the first, we are certain, to apply precision mechanisms to this style of clock. The numerous exhibitions at which M. Margaine has presented his products and the high prizes which he has always obtained, are sufficient indication that he has always been in the forefront of improvements, (even) where he has not introduced these improvements himself". Rousaille concludes by referring to Margaine, Jacot and Drocourt as the three leading carriage clock manufacturers. He says elsewhere of Margaine "This maker like M. Soldano uses movements from the Franche-Comté area or Saint-Nicolas and cases and finishes them in Paris.[8] By the care apparent in their manufacture, his knowledge of the craft and his personal work, he has created in a few short years an important establishment whose products are much appreciated. M. Margaine is certainly a man of progress having proved this by the manufacture of accessories, his gongs being second to none, and also by his efforts to introduce into carriage clocks the regulating mechanism of larger clocks. M. Tripplin noted that his movements were good but had a special word to say about the cases, every taste being studied, plain or highly decorated". Plates VII/9, VII/19 and VII/C8 show three utterly dissimilar clocks sold by Margaine. Each piece is of really high quality. An undated Margaine Catalogue and Price List, probably issued about 1875 but cer-

---

(6)   It is not clear what relation this Hollinge was to the one mentioned in Chapter III.

(7)   More recent research suggests that this statement may not be quite correct. People still living in Saint-Nicolas, Mme. Lejoille-Grard for one, are adamant that Dumas's house was sold to Emile Delépine. Mme. Lejoille used to sit on the knee of Delépine in this very house and says that the factory was behind it. Perhaps Drocourt bought a factory from the other Dumas.

(8)   In fact a misprint leaves Rousaille as saying ". . . the Franche-Comté area *of* St. Nicolas. . .", but this, of course, is nonsense as printed. Rousaille's remarks about Margaine's priority in complicated work are perhaps somewhat exaggerated.

V Paris, Today and Yesterday

tainly not earlier than 1869[9], contains about thirty-two photographs. A more thoroughbred-looking collection of *pendules de voyage* is hard to imagine. While most of the cases are ornate, the effects are achieved without resort to either over-embellishment or vulgarity. Margaine clocks will be found usually to have either spade or moon hands, the emphasis tending to be upon lightness. Handles, as remarked elsewhere, show great virtuosity in design; but they carry off their unusual shapes without ever once succeeding in looking wrong. Conspicuously absent from the Margaine's Catalogue and Price List is any direct reference to porcelain panels and dials. Probably these were only supplied to special order. They could certainly be used without difficulty in conjunction with standard movements and cases. Porcelain panels are not even mentioned in the catalogue among the "Extras", which offer dial variants such as "Dial engine-turned gilt", or "Dial cut, with coloured back". In fact, the only direct reference to special side panels is "Two sides peinced-coloured". A note adds that the prices are made ". . . for clocks with ordinary enamel dials . . . delivered, examined and timed, in morocco cases". The price list appears as follows:—

LIST OF PRICES OF CARRIAGE CLOCKS
TRADE MARK — With lever escapements, striking on gongs — TRADE MARK

N°⁸ of Movements	0	1	2	3	4	5	6	7	8		9	
DESCRIPTION OF MOVEMENTS	Simple Movement	Movement with Alarm	Simple Striking	Striking with Alarm	Simple Repeater	Repeater and Alarm	Quarter Repeater	Quarter Repeater and Alarm	CHIMING Simple		CHIMING with Alarm	
N°⁸ of the Photographs	1ˢᵗ Q'ᵗʸ	1ˢᵗ Q'ᵗʸ	1ˢᵗ Q'ᵗʸ	1ˢᵗ Q'ᵗʸ	1ˢᵗ Q'ᵗʸ	1ˢᵗ Q'ᵗʸ	1ˢᵗ Q'ᵗʸ	1ˢᵗ Q'ᵗʸ	Superior Q'ᵗʸ	1ˢᵗ Q'ᵗʸ	Superior Q'ᵗʸ	1ˢᵗ Q'ᵗʸ
11	58/– (2ⁿᵈ quality 50/–)	70/–	78/– (62/–)	90/–	90/– (72/–)	102/–	130/–	142.–	194/–	174/–	210/–	190.–
12	66/–	78/–	86/–	98/–	98/–	106/–	138/–	150/–	202/–	182/–	218/–	198/–
13-14	70/–	82/–	90/–	102/–	102/–	110/–	142/–	154/–	206/–	186/–	222/–	202/–
15	74/–	86/–	94/–	106/–	106/–	114/–	146/–	158/–	210/–	190/–	226/–	206/–
16-17-18	78/–	90/–	98/–	110/–	110/–	118/–	150/–	162/–	214/–	194/–	230/–	210/–
19-20-21	82/–	94/–	102/–	114/–	114/–	122/–	154/–	166/–	218/–	198/–	234/–	214/–
22-23	86/–	98/–	106/–	118/–	118/–	126/–	158/–	170/–	222/–	202/–	238/–	218/–
24-25	90/–	102/–	110/–	122/–	122/–	130/–	162/–	174/–	226/–	206/–	242/–	222/–
26-27-28-29-30-31-32	94/–	106/–	114/–	126/–	126/–	134/–	166/–	178/–	230/–	210/–	246/–	226/–
33-34-35-36	98/–	110/–	118/–	130/–	130/–	138/–	170/–	182/–	234/–	214/–	250/–	230/–
37-38	106/–	118/–	126/–	138/–	138/–	146/–	178/–	190/–	242/–	222/–	258/–	238/–
39	122/–	134/–	142/–	154/–	154/–	162/–	194/–	206/–	258/–	238/–	274/–	254/–
40-41	130/–	142/–	150/–	162/–	162/–	170/–	202/–	214/–	266/–	246/–	282/–	262/–
42	162/–	174/–	182/–	194/–	194/–	202/–	234/–	246/–	298/–	278/–	314/–	294/–
Cornice N° 0	40/–	50/–	”	”	”	”	”	”	”	”	”	”

It will be seen that a *Corniche* No. 0 appears at the bottom of the table. This was a miniature clock offered as a timepiece or as a timepiece with alarm. Margaine took out a *Brevet* for a *mignonnette* alarm mechanism in 1869. The inclusion of this clock helps to date the catalogue.[10]

According to the *Revue Chronométrique* of July 1880, a M. Emile Dubois working with Margaine in Paris was awarded a prize for the beautifully executed finishing of a *petite sonnerie* carriage clock movement. That Margaine was still in business in 1914 is made plain from the Japy workbooks mentioned in Appen-

---

(9) The Catalogue offers a type of decoration for which Margaine took out a *Brevet* with Dorius in 1869 and in which background colour is shown through piercing.

(10) Further notes on the Margaine catalogue will be found in the Case Styles Chapter VII.

V Paris, Today and Yesterday

dix (d). It is probable that latterly Margaine depended largely if not wholly upon this source of supply for movements. Certainly this contention might be supported by the existence of some late Margaine carriage clocks which do not seem to be up to the usual standard. The surviving Badevel workbooks of 1907 show Margaine as the main customer of Japy for carriage clock work at that time, while the last surviving Saint-Nicolas-d'Aliermont makers appear in the end to have concentrated upon completed *pendules de voyage*. Unbound copies of the *Revue Chronométrique* for 1912 show Margaine as advertising each month: "MARGAINE. FABRIQUE SPECIALE DE PENDULES DE VOYAGE" from 54 Rue Bondy.[11] About the only other firm then offering carriage clocks was Charles Hour of 7 Rue Sainte-Anastase.

In Chapter II was mentioned the fact that during much of the 19th century there were two horological firms both in the Palais Royal in Paris, the one spelling their name at first Le Roy and latterly Leroy, and the other spelling the name as one word. The reader will remember that a legal tussle eventually ensued. It culminated, after some twenty-five years, in the closing down of the firm "Leroy et Fils" in the Avenue Opéra, while the other firm "L. Leroy & Cie" of 4 Faubourg Saint-Honoré have since continued in business. They are still at the same address, but in the past year or two their trading name or *raison sociale* has become "Leroy". In view of these very confusing facts, it is now necessary to give short histories of the two firms.

As noted above, the firm of Leroy of 4 Faubourg Saint-Honoré is still very much in existence, although no longer engaged in making *pendules de voyage*. This House was established at 60 Galerie de Pierre, Palais Royal soon after 1785[12] by Basile Charles Le Roy[13]. The 1785 date is quite important for in that year the Duc d'Orléans opened the Palais Royal gardens to both public and trade alike, and many shops were set up under the arcaded walks (see Plate V/22). According to a *notice advertissement* put out by L. Leroy & Cie at the time of the St. Louis International

Plate V/22. PARIS. THE PALAIS ROYAL. *This group of buildings was originally built in 1633 for Richelieu and named Palais-Cardinal. In 1643 the buildings were bequeathed to Louis XIV, passing in 1661 to the Princes of the House of Orleans. In 1785 the Duc d'Orléans (Philippe Égalité) opened the Palais Royal for trade. Many horologists set up their shops in the Galleries.*

---

(11) In the same issue L. Leroy & Cie were advertising only marine and pocket chronometers, watches, astronomical clocks, decimal watches and second-hand marine chronometers.

(12) Various documents in the Conservatoire National des Arts et Métiers in Paris suggest that this date may have been 1786 or even 1788. To settle the matter conclusively would require much further research. Similarly there exist conflicting pedigrees for the ancestors of Basile Charles Le Roy.

(13) A fine mantel semi-skeleton clock, No. 68, belonging to this period, was sold at Sotheby's on 16th October 1972, Lot 35A. This piece is signed "Le Roy à Paris" on twin *cartouches* below the dial. The dial itself bears the signature "Le Roy H^{GER} du Roi", while on the backplate of the movement appear the following inscriptions engraved on three separate lines: "B.C. LE ROY", "H^{gr} Palais-Royal", and "No. 68 à Paris", The significance of this clock is that it proves conclusively that Basile Charles used the title *Horloger du Roi*. The second point of interest is that the piece must have been made before September 1792 when the monarchy of France was abolished. By the time of the Bourbon restoration of 1814, when Louis XVIII "regained his native land", the firm of Basile Charles Le Roy had already been "Le Roy & Fils" since 1804. According to Tardy's *Dictionnaire*, published since this chapter was prepared in 1970, Basile Charles eventually even worked for Napoleon.

V Paris, Today and Yesterday

Exhibition of 1904, Basile Charles Le Roy passed through very trying circumstances under the *Terreur*, being obliged to alter his name and use the anagram "Elyor". He was associated during this period with the address 88 Rue de L'Egalité. After the Revolution, Basile Charles moved to Galerie Montpensier, 13 and 15 Palais Royal, an address at which the firm was to remain for almost one hundred years. In 1828 Basile Charles Le Roy died, and his son Charles-Louis (often called Louis-Charles) directed the firm and continued to use the name "Le Roy & Fils". In 1839 "M. Leroy, à Paris, Palais Royal 13 et 15" (note the wrong spelling of Le Roy) received a Silver Medal for watches and travelling clocks. The most important piece shown by the firm was a *grande sonnerie* carriage clock with repeat and an escapement from which was shown dead seconds. On 30th June 1845 Charles-Louis Le Roy sold out to his employee Casimir Halley Desfontaines on condition that he kept the name "Le Roy & Fils".[14] Plate V/23 shows a fine semi-perpetual calendar clock sold by the firm about 1860. It is signed "LE ROY ET FILS H^{RS}, PALAIS ROYAL G^{IE} MONTPENSIER 13 & 15, PARIS". More is said of this clock in Chapter VIII. On 25th December 1883 M. George Halley Desfontaines succeeded his father, keeping the trading name as before. He died in 1888. In 1889 his brother and heir, Jules Halley Desfontaines, took as a partner Louis Leroy who had been employed in the firm since 1879. Louis Leroy was the son of Théodore-Marie Leroy,[15] but apparently he was no relation to Louis-Charles Le Roy above. The firm then changed its trading name and became "Ancienne Maison Le Roy & Fils, L. Leroy & Cie Successeurs".[16] In 1895 Léon Leroy, the brother of Louis, entered the

Plate V/23. *LE ROY & FILS. A fine clock with semi-perpetual calendar and* grande sonnerie, *sold circa 1860 during the time of Casimir Halley Desfontaines. (Private Collection)*

---

(14) Despite this fact, the firm exhibited a minute-repeating carriage clock in 1851 in London under the style "Desfontaines, Maison Leroy & Son, 13 & 15 Galerie Montpensier, Palais National, Paris".

(15) Théodore-Marie Leroy, a most distinguished horologist, was born in 1827, and died in 1899. He was the son of Marie Balthazar Leroy, Horloger à Argenteuil. Théodore-Marie exhibited carriage clocks in 1867, and in 1897 was named as a member of the *Comité d'admission* responsible for French horology in connection with the *Exposition Universelle* of 1900. An oration delivered at his funeral by A-H Rodanet was published in the *Revue Chronométrique* of October 1899, pp.347, 348, and lists Théodore-Marie's many accomplishments and contributions to horology. He never established a large firm, preferring to devote the greater part of his life and energy to precision work mostly in the form of marine chronometers and regulators. According to his obituary, Théodore-Marie was trained by Vissière, worked for the firm of Breguet, and was associated with Philipps in the development of terminal curves for balance springs. Although Théodore-Marie is largely outside the scope of this book, he was very much the most important of the 19th century Leroys.

(16) A fine *grande sonnerie* clock in an aluminium *gorge* case, with back plate numbered 15166 and 29711, exists bearing this style. It also carries the address Bond St. London, and on one side of the back movement plate the inscription "15166, L. Leroy & Cie, 13 & 15 Palais Royal, PARIS".

121

V Paris, Today and Yesterday

PARIS". In 1934 a limited company was formed with Louis and Léon as directors. They still used the style "L. Leroy & Cie" (sometimes engraved "L. Le Roy & Cie"). The firm settled at their present Faubourg Saint-Honoré address in 1938. Léon continued as sole director from the time of his brother's death in 1935 until his own death in 1961. Léon's two sons Pierre and Philippe then took over the business which continues to flourish under their directorship. The present style of the firm is simply "Leroy". The House also had a branch in London at 296 Regent Street in 1854, firm.[17] The House of Leroy remained in the Palais Royal until 1899, moving to the Boulevard de la Madeleine 7 in time for the Paris Exhibition of 1900.[18] Plate V/24 shows a fine example of a *pendule de voyage* sold by Leroy at the turn of the century. It is signed "L. LEROY & CIE, 7 Bᴅ DE LA MADELEINE,

Plate V/24. L. LEROY & CIE. Clock No. 25564 made circa 1900 and having an Empire case, centre seconds, calendar, moon, alarm and repeat. The escapement is provided with a helical balance spring. Note the quaint inscription below the dial: "Ton glas est un ami qu'attendent mes oreilles". (By courtesy of Asprey & Co. Ltd.)

Plate V/25. LE ROY & FILS, BOTTOM-WIND. This clock, which was sold from 57 New Bond St., London, is numbered 9487. Its date is between 1885-1889. Note an elaborate doucine *case with fluted columns. (Private Collection)*

---

(17) He became a partner in 1914.
(18) A Swiss minute-repeating *montre pendulette de voyage*, No. 18430, presumably dates from about this period. It is signed on the dial "L. LEROY & CIE PARIS", on the front of the metal case "LE ROY", and engraved on the case back "Le Roy Horloger du Roi, 7 Bd. de la Madeleine Paris 18430". (Sold at Sotheby's 19th March 1973). Note the resumption of the *Horloger du Roi* title.

V Paris, Today and Yesterday

if not earlier. They also acquired 211 Regent Street sometime prior to 1866, and by 1875 they were occupying both 211 and 213 Regent Street. By 1885 (according to the London Trade Directories) the firm had moved to 57 Bond Street. Plate V/25 shows a high quality clock made between 1885-1889 signed "LE ROY & FILS, 57 New Bond Street, Made in France" It has a *doucine* case combined with fluted columns having semi-Corinthian capitals. The clock strikes hours and halves in passing, repeats at will, and is provided with alarm work. In addition this clock embodies the so-called bottom-winding, or keyless mechanism, which appears to be peculiar to those Le Roy & Fils clocks sold before 1900, but which the surviving members of the family do not even remember as being in production. See Chapter VIII for further details. The London house remained in Bond Street until 1940. The address in 1941 is given as 18-20 Regent Street. In 1947 they moved back to 57 New Bond Street. The last entry appears in the 1952 directory.

The other house of Leroy, Palais Royal, Galerie Valois 114 & 115, mentioned in Chapter II, owes its origin to a M. Théodore Leroy. There seems no evidence that this Théodore was related to Théodore-Marie Leroy mentioned earlier in this chapter. Théodore Leroy, a clockmaker in the Rue Saint-Martin, moved to the Palais Royal in 1813. In 1827 he entered into partnership with Auguste-Pierre Lepaute, who had recently left his uncle Jean-Jacques Lepaute. On the 22nd June 1827, Théodore Leroy petitioned the King and obtained the *Brevet d'Horlogers du Roi* for himself and for Auguste-Pierre Lepaute. From 1839 Théodore Leroy used the trading name of "Leroy et Fils" (often engraved thus:— "LeRoy & Fils") sometimes adding *Horloger du Roi*. In 1843 he sold out to M. Fraigneau, who continued to use the name of "Leroy et Fils".[19] Despite successive different ownerships such as Schaeffer in 1871, Clericetti in 1883, and Thomas Garnier in 1924, the firm continued to use the title of "Leroy et Fils". They had moved during M. Clericetti's ownership from the Galerie Valois to the Avenue Opéra.

Apart from the "giant" Leroy clock illustrated in Chapter II, neither of us has yet seen another clock

Plate V/26. PORTABLE ALARM. *In our opinion clocks of this type were almost certainly made by some entrepreneur rather than by any of the major names often to be found on them. (By courtesy of Garrard & Co. Ltd.)*

from this firm both signed and addressed conclusively. Plates V/26 and V/27 illustrate, however, a portable alarm with painted signature "LEROY & FILS H. DU ROY, PALAIS ROYAL" (sic). There is no evidence when this piece was produced, but it has something in common with the "sedan" clocks mentioned in Chapter I. Almost certainly it was made by a "handyman" using an obsolete anonymous verge watch movement. To further confuse the issue, the reader will be interested to learn that an almost identical clock was

---

(19) The *Revue Chronométrique* of September 1855 on page 46 is very critical of the bad taste shown in the display of M. Fraigneau ". . . who, in his shop, calls himself by the name of Leroy". Aspersions are cast at the ancestry of a chronometer partly hidden by gilt metal pots of flowers and foliage. The reporter says sarcastically that if M. Fraigneau is to receive a medal, then it will have to be from the Society of Horticulture!

V Paris, Today and Yesterday

Plate V/27. PORTABLE ALARM. *These small clocks, which will be found bearing a number of names including "Houdin", "Leroy & Fils", and "Le Roy & Fils" are based upon thirty-hour verge watch movements used in conjunction with separate alarm trains. The engraved gilt brass cases stand about 4¾ inches tall with handles up. (By courtesy of Garrard & Co. Ltd.)*

sold at Sotheby's on 19th June 1972, Lot 40. It has the engraved signature "LE ROY & FILS, Hrs du Roi, Palais Royal Gie. de Valois 114 & 115" (sic)!! As if this was not enough, a gold watch sold at Christie's on 21st March 1972, Lot 61, was signed "LeRoy, Av. Palais Royal, No. 13 et 14 à Paris, No. 2075".[20] Need more be said about the confusion in names and addresses which has always existed?

The two-firm position continued for over one hundred years. Then, in 1936, Léon Leroy decided to invoke an old French law, dating from the period of the Revolution, in which it was stipulated that any trader must work under his own name. The other firm, as the reader will know, had used only the name of their distinguished predecessor; although latterly during the litigation period there is evidence that they probably used the title "Thomas Garnier successeur de Leroy & Fils". The case was only finally resolved in favour of M. Léon Leroy just before his death in 1960. The Thomas Garnier firm then closed down.[21]

Returning briefly to the subject of Jacot, Drocourt and Margaine, a few words should perhaps be said concerning those *good* anonymous clocks (usually in *gorge* cases) which are occasionally encountered and which both owners and sellers are tempted to attribute to one or another of the above makers. It does not, however, really seem very likely that top class established firms would in the ordinary way have released their best clocks to be sold without either the proud advertisement of the maker's name or else a trademark, or both. It is possible that in the early days, and while establishing themselves, the three firms may have been content to supply unsigned clocks to the trade. This theory is perhaps supported by the existence of some Jacot/Drocourt-like clocks having very low serial numbers. Such a clock might be a Size *Zéro* half-hour *sonnerie* No. 82 and housed in a *gorge* case. This piece, which appears to carry no name or trademark anywhere, does indeed give every appearance of being an early Jacot clock. On the other hand Jacot clock No. 746 exists, and that is certainly not anonymous. It is a *grande sonnerie* in a *gorge* case and it is signed on the dial "LE ROY & FILS". Upon first examination it appears to have an unsigned movement, so effectively is the Jacot ellipse giving details of the medals of 1855 and 1862 concealed behind the large strike barrel. A Jacot clock No. 1591 bears Klaftenberger's name and address not only on the dial but also on the outside of the back plate. It is only when one looks behind the barrels that Jacot's name, with references to the medals received in 1855 and 1862, becomes apparent. When Drocourt supplied

---

(20) "Av. Palais Royal" never existed. The watch, in our opinion, is almost certainly a Swiss fake.

(21) Many family and business pedigrees, besides documents relating to the litigation between the two firms, were made available for analysis in 1971 through the generosity of MM. Pierre and Philippe Leroy, heads of the present firm of Leroy, 4 Faubourg Saint-Honoré. Better still, at Christie's on 26th June 1973 appeared some thirty Le Roy/LeRoy/Leroy/T. Garnier watches. They were variously signed and addressed, and confirmed these conclusions reached in this chapter.

V Paris, Today and Yesterday

clocks to be sold by Charles Frodsham, then the inscription "DROCOURT PARIS" nevertheless appeared inside the movement hidden by one of the barrels and where it would only be seen by those whose business it was. One such clock bears the signature "CHA^s FRODSHAM & CO, 19521 PARIS" on the backplate, while the number "F21" appears beside Drocourt's concealed signature.

No doubt it is only a matter of time before fakers will begin to engrave the names of the "big three" upon anonymous clocks in order to enhance their values. When they do so, they will need to be very careful indeed. It will not be good enough merely to slap on the chosen name and somehow to transform the hand-setting arrows, etc. to resemble the distinctive types most often used respectively by J., D. and M.[22] Far more than this will be required in order to make the pose even half convincing. It is most difficult to express in words those subtle combinations of normally-related features which go to make up a Jacot clock; or for that matter one by Drocourt or Margaine (although they may well be fairly conclusive). All sorts of small details are significant when considered collectively; but their relevance lies almost entirely in the combinations in which they are found. Taken alone they mean less than nothing. It would be easy to list here those combinations of features which examination of a good many clocks has shown to be fairly consistent with the signed work of this or that maker; but on the whole it would seem that such generalisations, for the time being at least, are more likely to be dangerous and misleading than in the interests of anyone. Provenance is one thing, wishful thinking is quite another.

There are, of course, a few very high quality clocks which in no way resemble any of the models normally associated with Jacot, Drocourt and Margaine and which bear neither names nor trademarks. The correct attribution of such clocks seems to be impossible at the present time.

---

(22) Three distinct types of arrows are found so consistently in conjunction with the "Parrot", the "Clock" and the "Beehive" respectively as to leave no reasonable doubt that arrows have some significance. Couaillet and Duverdrey & Bloquel certainly tended to use distinctive arrows. Different arrows again are found on various anonymous clocks. It is possible that arrows may eventually provide positive clues as to the origins of clocks.

Chapter VI

The Carriage Clock Industry in the Franche-Comté

LE LION DE BELFORT. Bartholdi's magnificent monument to the defenders of Belfort during the Franco-Prussian War of 1870-71. The lion is 11m. high and 22m. long.

VI

The Carriage Clock Industry in the Franche-Comté

The Forges and Foundries of the Doubs. The Country of Montbéliard. Japy at Beaucourt and Badevel. L'Epée of Sainte-Suzanne. Carriage Clock Makers in the Jura.

Look at a map of Eastern France, where the border with Switzerland close to Germany weaves to and fro to such an extent that it is difficult to see quickly where one country ends and another begins. This is part of the Franche-Comté, a fascinating and unique region comprising the *Départements* of Doubs, Jura and Haute-Saône, together with Le Territoire de Belfort. The river Doubs, winding through its deep gorges, marks for many miles the frontier between France and Switzerland. Here it is a fairly shallow but swiftly-flowing stream with a great industrial past. Today it runs quietly below cliffs and woods, now in one direction, now in another. In the 18th and 19th centuries, however, the Doubs was very different. Rivers were then almost the only sources of power; and throughout the long journey made by the Doubs through the Franche-Comté between Morteau and Besançon it turned mill after mill. Most of these "mills", many of them multi-wheeled and very powerful, were iron foundries or forges. It is to them that France owes much of that wealth of ornamental ironwork which is everywhere so much in evidence. Many of the very place names along the Doubs are evocative, "Les Forges" and "Le Fourneau" being typical. Indeed, at one time the sounds of hammers which echoed back and forth among the hills and mountains were so obtrusive that one writer said that it were as though Vulcan himself had there "... set up his toils".

At a later date, firearms and the like were also made in the *usines* of the Doubs, in time for the Revolution and for Napoleon's wars. Later still, resources were turned over to the production of parts for clocks and watches. Sire, writing in 1870[1], says that the *horlogers* of the cantons of Saint-Hippolyte, Russey and Maîche had equipped themselves specifically for the manufacture of component parts of escapements (apparently cylinder and lever) and that the excellent products of the various Doubs factories were sent to Switzerland, England and Germany. Sire also quotes the then well-known Redier as saying, with reference to the 1867-1870 period, that without France the Swiss would have been hard put to meet the orders which they received from abroad. Instead, they were able to obtain, if not completed articles, at least partly-made pieces such as *ébauches,* detail parts, escapements, etc; but that is now all long past. Today most of the Doubs factories are either half-buried in the undergrowth or else are turned over to other purposes. On the whole, the coming of "steam"

---

(1) *L'Horlogerie à l'Exposition Universelle de 1867,* pages 115-116. Sire quotes Redier's brochure *Les Récompenses de la Classe 23.* (Horology)

VI Franche-Comté

spelled their ruin. It also had the effect of moving industry to where there was coal. Readers with a taste for industrial archaeology will find no difficulty in discovering the remains of the old factories of the Doubs. Before a mill the river is dammed completely across. Often there were factories on both banks. Sometimes a small château, the home of the original owner, still stands nearby. One firm happily still very active is that of MM. Joseph Jeambrum, Fabrique d'Ebauches, Rue Saint-Hippolyte 26, at Maîche. The region of the Doubs called the Territoire de Belfort (embracing Le Pays de Montbéliard) is even more interesting. We shall deal with it presently.

In the French Jura, until very recent times and before large snow ploughs were available everywhere in Europe, outlying homesteads and even whole villages were sometimes "snowed up" for many weeks on end. It was then literally impossible to leave the houses, except on skis from upstairs windows. The people were mostly farmers. In winter the animals were kept indoors, while the families occupied themselves in various small industries, such as making boxes for Camembert cheeses, making smoking pipes, diamond cutting, making spectacles, and above all in horology. If someone died during the height of the winter, their relations could not bury them. They were put out on the roof until the thaw came!! This practice persisted until as late as just before the Second World War in a place called Bois-d'Amont near Morez. It is close to the Swiss border in the very horological area near Le Sentier and Le Brassus.[2]

The Haute-Saône borders very closely upon Besançon, Belfort and Montbéliard, yet the *Département* was never noted for horology.[3] Textiles and timber used to be the chief industries there during the 19th century, along with some metal refining in Vesoul and other places. Apart from being a part of the Franche-Comté, Haute-Saône really has nothing to do with our present history.

The Franche-Comté as a whole had a very turbulent and complicated past. It only finally became a permanent part of France in 1678,[4] and even then more in name than fact until well into the 18th century. The area was originally Celtic, but throughout its entire existence it has been time after time invaded or annexed. The last occasion (other than in the Second World War) was just one hundred years ago during the Franco-Prussian War of 1870-71. Probably the first recorded invasion came with the Romans in B.C.52. Their beautiful amphitheatre remains to this day at Mandeure beside the river Doubs. The Roman town nearby had 25,000 inhabitants, and the theatre was one of the largest in Europe. Today there is a good small museum containing many important domestic items. Not a few of them were found by the grandfather of the husband of the present Madame Henry L'Epée, who will be mentioned later. After the Romans left, the Comté for a time existed as several separate kingdoms; but it was not long in becoming more or less a part of the Holy Roman Empire. From the end of the 13th century to the end of the 15th, the fortunes of the area were directly linked with those of France. At the end of the 15th century the Comté passed to the house of Austria (Hapsburg) being assigned to the Spanish branch after the abdication of Charles V. The influence of this last event persisted well in to what could be called the second French period, taking the form of loyalty to Spain rather than to France.

The way from the horological towns of Switzerland, particularly Le Locle and La Chaux-de-Fonds, to Le Pays de Montbéliard lies across the Doubs and by the pass through Fournet, Charquemont and Maîche to Sainte-Hippolyte. Here again is found the river, although now somewhat changed in character. The road then follows the Doubs first through gorges and then by a plain right to the country of Montbéliard. There Belfort is the largest town, and Montbéliard near by is the centre of Le Pays de Montbéliard, a "country" with a history and character uniquely its own. Montbéliard itself is a fairly large and not particularly attractive industrial city, whose economy is dependent upon the Peugeot motor car factory; but the dispirited traveller soon finds that there is also a fine fortress set upon a rock, several museums, and an old centre dominated by a friendly small

---

(2) The wonderful light reflected from the snow in Switzerland and in neighbouring parts of France contributed greatly to the success of horology in these regions. Many old houses and farms are still found having very large windows *(vitrages d'horlogers)*.

(3) This fact is confirmed by the *Revue Chronométrique* of 1899, p.274. In the Paris Exhibition that year 267 exhibitors had origins as follows: Paris 75, Besançon 70, Morteau et Villers-le-Lac 34, Pays de Montbéliard 19, Haute-Savoie 18, Morez 9, Saint-Nicolas-d'Aliermont 5, pays divers 37. (Saint-Nicolas clock producing appears already to have been well in decline).

(4) Under Louis XIV (La Paix de Nimègue). Capital Besançon.

VI Franche-Comté

Plate VI/1. *LE PAYS DE MONTBELIARD.* *The invitation card of a recent Exhibition of the old clockwork industries in the Montbéliard area and featuring the work of L'Epée, Japy, Marti and others. The three views represent Montbéliard Castle, the River Allan and "la grotte de Sainte-Suzanne". Similar cards will be found inside the lids of old L'Epée musical boxes listing the airs. This card well captures the atmosphere of the district. (By courtesy of Madame Henry L'Epée and Le Conseil Culturel de la Maison des Arts et Loisirs de Sochaux)*

square and by a fountain. In summer the stone walls, ancient roofs, Morez turret clock, pollarded chestnut trees, formal flower beds and gay window boxes, make it easy to feel that here at least a visitor would have seen little different towards the close of the 19th century. A few kilometres away to the southeast lie the adjoining villages of Beaucourt (old name Bocourt, or "village of the woods") and Badevel. They were once the homes of the watch and clock factories of Japy, by far the largest concern of the type anywhere in France. To the south-west of Montbéliard, scarcely out of the town and on the river Allan (Allaine), is a village called Sainte-Suzanne. Here, only yards from the river and near to the "Rhône au Rhin" canal (which connects the Rhine to the Saône) is the thriving platform escapement factory of Frédéric L'Epée & Cie. The two old firms, Japy and L'Epée will form the main subjects of this chapter; but before writing of them it is necessary to say a few words more about the country from which they sprang. The horological traditions and practices of the Pays de Montbéliard developed independently of those of Besançon (the capital of the Franche-Comté), Morteau or Morez, each of which tended to specialise separately in different types of work. Plate VI/1 reproduces the delightful invitation card of a recent Exhibition in connection with the old industries of Le Pays de Montbéliard.

If the Franche-Comté as a whole was different from the rest of France, so in turn is Le Pays de Montbéliard noticeably dissimilar from the other parts of the Comté. People born and bred near Montbéliard are very much aware of this difference, besides being most proud of it. They refer to Le Pays, *tout court*, and mean by it nothing but their own small district. They are proud that the fortress of Belfort held out successfully in the Franco-Prussian War under the command of Col. Denfert-Rochereau.[5] They are proud of their long-standing Protestantism in a predominantly Catholic area. Not least, the people of Le Pays are proud of the fact that, while in a very real sense their own local history has been more closely connected with Germany than with that of France, yet they have kept a peculiarly, if different, French identity of their own. French has always been the language spoken and the one used in official documents, even in times of occupation. Madame L'Epée has explained that the Belfort area has been in trouble since the dawn of history because of its peculiar geographical situation. It lies "on the straight line between the Mediterranean

---

(5)   This event is commemorated by Bartholdi's magnificent monument *Le Lion de Belfort* in Belfort itself, and also by a smaller lion in Place Denfert-Rochereau in Paris.

VI Franche-Comté

and the North", and it is also "impossible to make an invasion from the East between Belfort (here) and Strasbourg". Madame L'Epée further says that Le Pays has always been "... a rich country, able to live alone.... There was coal, a little iron and the fields were good". There were also many bourgeois families who were very important in the community. A great number of these people, other than those in the professions such as doctors and lawyers who tended to live round the fortress at Belfort, were cornmillers. In the 19th century nearly all the old watermills became factories of some kind, many being turned over to the textile industry. Some families, such as Japy and L'Epée took up the production of watches, clocks or music boxes. Others again, and Japy was the chief of these, specialised in the production of hardware. Towards the end of the century the Peugeot family branched out into the then new industry of motor cars. Nearly all of the old bourgeois families in Le Pays are related. For example, the Japys, the L'Epées, Monnins, and Peugeots, to name but four, are all cousins many times over. One tradition which has always been particularly strong in the area is that of good work-relations. Until the French government attended to what are today called public services, these were traditionally supplied free by the employers. Schools and hospitals were the two most conspicuous examples, but in fact this patrimony extended to most aspects of the life of the country. Japy and Peugeot were always notably conscientious in this respect, and the L'Epée family built a hospital as a contribution to the War of 1870-71. It would be easy to continue in this vein, but perhaps enough has been said to explain why Le Pays is to some extent special, and that no doubt the reasons stem largely from its geographical situation. Incidentally, the rolling, wooded countryside is like nowhere else in France, without being particularly reminiscent of neighbouring parts of Germany or Switzerland. Forestry is taken seriously and has long been linked with the local economy. The building style of Le Pays is individual also. While wars, invasions and revolutions have destroyed many of the most important large old buildings, nothing appears to have disturbed at all the domestic architecture of the villages. This is something just as precious, and so much of it is left as to make abundantly clear that the area evolved its own characteristic type of village-house, part-farm, part-cottage and part-barn. These peculiar homesteads in no way resemble those to be seen even a few miles away in any direction. The tall, arched-top barn doors, the oldest ones with radial fan-pattern tops, are as characteristic of the area as anything which we have seen

Plate VI/2. *LE PAYS DE MONTBELIARD. A typical village house in the Montbéliard area.*

Plate VI/3. *LE PAYS DE MONTBELIARD. A church tower with clocks at two different levels.*

VI Franche-Comté

(Plate VI/2). The church with two clocks at different levels, in Plate VI/3, also conveys, but in a different way, the local feeling. In summer the sun is hot, far more so than in England, and throughout these "poor" villages there are crocks of bright flowers, predominantly geraniums, overflowing in every yard and upon every doorstep and windowsill. In the afternoons and evenings grasshoppers two inches long fill the air with a synchronised, soaring sound not easily forgotten. These apparently trivial matters have been emphasised here in the hope that they will help to set the scene for what will be said later of the old Japy factories, now mostly in ruins. Dwellers in crowded countries, where every inch of space is at a premium, will find it almost inconceivable that there should exist in Eastern France derelict factories, apparently abandoned and ownerless, occupying sites of many acres in extent.

It is not easy to write about Etablissements Japy, a firm reputedly having a capital of over £2,000,000 in 1888,[6] if only because clocks, and carriage clocks in particular, were among the least of the activities and productions of this remarkable 19th century complex of manufactures. Yet clocks and parts of clocks were made by Japy in vast quantities, this side of their business alone being far greater than that ever achieved by any rival concern. What else did Japy make? The list is endless, but it included at one time and another watches and chains, locks, ironmongery, garden furniture and cooking utensils of every conceivable type, children's toys, electrical equipment, oil lamps, petrol engines, photographic equipment, phonographs, taximeters, lamp-lighting clocks, typewriters, revolving mirrors to catch skylarks, roundabouts, spinning and weaving machinery, pumps and adding machines.[7] Even the first Peugeot horseless-carriage was driven by an engine made by Japy!! Eventually in Le Pays there were Japy horological factories at Beaucourt and Badevel, and others producing hardware, etc. including those at Voujeaucourt, Laroche, Bas-les-Fonds, Fésches, La Féschotte, Etupes, l'Isle-sur-le-Doubs, Berne and Seloncourt.[8] Workers were drawn from all the neighbouring villages. Japy also had a factory at Besançon from 1920 to about 1933. They were connected with "Jaz" alarms. Two other factories, not horological, were sited in the North of France. One of them was at Anzin, and Japy even had a factory in Roumania. At first most of these factories were based upon watermills, the old wheels providing the necessary power. In due course, first steam and then electricity were used. Almost all the Japy ventures were immediately successful, and perhaps in the end it was the sheer vastness of the concern which partly led to its demise. It is somehow difficult not to feel that a firm which founded its reputation upon Lepine calibre *ébauches* for watches, and with machine-made *blancs-roulants* for clocks, might profitably have steered clear of enamelled stewpans and casseroles, wood screws, cast-iron fountains, etc.; but this is only a personal opinion, voiced in all humility. There is, of course, no doubt whatever that Peugeot[9] early in the 20th century attracted many workers away from Japy to the new motor industry. However, since the origins of Japy go back to the 18th century, it is necessary now to ask the reader to turn his mind to his period.

Frédéric Japy has two main biographers. The first is Dr. Muston in 1882; and he is a useful, but perhaps occasionally unreliable, source of information for up to this period.[10] The second is Ernest Vinter, already mentioned as writing in 1944. His history is invaluable for after 1882, while earlier he is far less detailed than Muston. In writing this book heed has been given to both of these invaluable chroniclers, and also to the writings of a number of others. At the same time we have made our own researches (albeit about one hun-

---

(6) Tripplin *Watch and Clock Making* in 1889, p.82. London 1890.

(7) Vinter *Histoire des Etablissements Japy Frères, depuis leur création jusqu'à nos jours, 1777-1943*. Beaucourt, 1944, 220 pages.

(8) Vinter op. cit. Incidentally Fésches (Feches) and La Féschotte (Lafechotte) are correctly spelled in any of a number of ways.

(9) Caroline Japy married Constant Peugeot. The *usine* Peugeot-Japy at Audincourt once made weaving machinery. (Authority: Madame L'Epée and M. Albert Japy).

(10) Muston (M. le Docteur) *Histoire d'un Village*. Montbéliard, 1882. This is a history of Le Pays de Montbéliard and of the Japy family and business. Muston should always be right; but since he married into the family he may well also be unduly partisan, and his account of F. Japy's period of apprenticeship is certainly at variance with the versions given by various other writers. Muston married first Marie Japy, great grand-daughter of Frédéric Japy. Secondly he married Marie Borneque, daughter of Eugènie Japy and grand-daughter of Pierre Japy. No wonder he was so interested in the family. He lived at Montbéliard.

VI Franche-Comté

dred years too late) and have drawn our own conclusions in certain matters. In others, especially where the written histories of authors of integrity and repute are at variance, then mention has been made of the various conflicting versions.

Georges Frédéric Japy, the founder of the horological industries at Beaucourt and Badevel, was born on 22nd May, 1749 at Beaucourt. To this day the place is little more than a large and pleasant village, despite its industrial past and present. The district is delightful. It is typical of the Jura foothills of the Alps a few kilometres from Montbéliard. The countryside is lush, green, hilly and well-wooded. It takes an effort to imagine how such a setting ever became the home of vast manufacturing enterprises. Frédéric's father, Jacques Japy, was a local wheelwright and blacksmith specialising in lock-making and in the repair of agricultural tools. The family had been well known in the area since at least the 15th century. The father had a good deal of local influence, having been appointed in 1760 by the Prince of Montbéliard-Wurtenberg as Mayor of Beaucourt. Jacques Japy ruled in a village with a population of two hundred and twenty-five. The young Frédéric Japy, who was a second son, was schooled first at Beaucourt. Then, because he did well he was sent away from home to Montbéliard until the age of seventeen. Montbéliard at the time boasted particularly good schools provided by the Montbéliard-Wurtenberg family, and Jacques Japy readily found other parents willing to accommodate his son. After this, Frédéric worked for his father in the forge. He proved an apt pupil, and this period of his life happened to coincide with the sudden upsurge of watch-manufacture in Switzerland in the Jura mountains and in the cantons of Geneva and Neuchâtel. Everyone locally was aware of the success of Daniel Jean-Richard. No doubt the advent of this new and promising industry explains why Jacques Japy decided to send Frédéric to serve an apprenticeship at Le Locle. It is interesting to discover that various reputable historians do not agree to whom Frédéric Japy was apprenticed.[11] What is more, between them they give two completely conflicting descriptions of the apprenticeship itself. According to Muston in *Histoire d'un Village*, Frédéric Japy was apprenticed to Perret and completed the full three years with great credit. He was considered one of the best workers in Le Locle and his master predicted a great future in store for him. Muston says that Frédéric was given a traditional Swiss send-off on completion of his time, and that he returned to Beaucourt in 1772. Chapuis, on the other hand, says in *Grands Artisans*:— "F. JAPY, born in 1749, started his apprenticeship with the famous clockmaker Abram-Louis Perrelet, along with two of his brothers. But at the end of two years, considering that they had learned their trade sufficiently, all three of them left their master, notwithstanding the fact that their term of contract had not expired. Perrelet took Japy's father to court, when such costs were awarded against him that he had to sell parcels of land to meet them."[12]

In view of the conflicting reports it seemed worth while to try to settle the matter for once and for all. Accordingly, at the suggestion of Henry Belmont of Besançon, research was made at Neuchâtel in 1972 by P.A. Borel, with the result that it is now possible to translate an extract from the Register of Isaac Vuagneux, Notary of Le Locle, Volume VI, 1770-1776. This extract concerns a letter of apprenticeship and translated it reads:—

"Honest Jean-Jacques son of Isaac Perrelet, master horologer of Locle, which lies in the sovereignty of Neuchâtel and Valengin in Switzerland, being requested to accord to the honest men PIERRE-ABRAM and Frédrich (FREDERIC) Japy, brothers, sons of Jacques Japy, Mayor of Beaucourt, seignieury of Blamont (Montbéliard) a certificate of term and of the manner in which they served as apprentices in the profession of horologer. The said Perrelet could not refuse. In this respect he declares that the said two Japy brothers who undertook to serve for three years as apprentices remained about 22 months, during which time they worked faithfully and diligently, without giving any cause for complaint as to

---

(11) A Japy *Plaquette* of 1949 opts for Perrelet. Chapuis says Abram-Louis Perrelet in *Grands Artisans* and so does F.A.M. Jeanneret in *Etrennes Neuchâteloises* of 1862. On the other hand Muston, Beillard and Reverchon all say clearly Perret. Vinter is on his own with the typing error of "Perreley".

(12) Chapuis goes on to say "Frédéric Japy next worked for J.J. Jeanneret-Gris, who sold him several of his inventions. A little later, F. Japy made use of the new methods devised by J.J. Jeanneret-Gris as applied to the manufacture of *blancs-roulants* for mantel clocks". Whatever the facts, it seems very likely that Frédéric was at least in touch with J.J. Jeanneret-Gris, who made his reputation as a designer and maker of automatic wheel and pinion-cutting engines mainly in connection with the manufacture of *blancs-roulants* "destinés aux pendules de cheminées."

VI Franche-Comté

their conduct, the two brothers having then found it opportune to leave their apprenticeship of which there was still 14 months to go, to establish themselves at home. They have by agreement with the said master paid damages which they were obliged to do by their agreement, and are thus clear and irreproachable in all respects concerning this apprenticeship. This is the testimony of truth which the said master Perrelet accorded them and which he required the undersigned notary to draw up in writing this form having ratified it by touch of hand, in the presence of David Favre-Bulle, *sautier*[13] of the said Locle and of Jonas-Pierre Petitpierre, of Couvet, citizen of Neuchâtel, master mounter of *boetes*,[14] living at Locle, requested witnesses. Who have, with the said Perrelet, signed before notary on Monday 24th December in the year 1770. Signed: Isaac Vuagneux."

This Notaried Act must surely be acceptable as the final word, both on the subject of the duration of Japy's apprenticeship, and also with whom it was served. It would also appear that Frédéric Japy was apprenticed together with one, and not two, of his brothers, and furthermore that he was apprenticed to Jean-Jacques Perrelet and not Abram-Louis Perrelet. There were many J.J. Perrelet's during the 17th, 18th and 19th centuries according to M. Borel.

Whether F. Japy returned home in 1770 or in 1772, it is certain that, once back, he wasted no time in setting up a work-room in which he began to make "rough" watch movements *(ébauches)*, an art which he had learned at Le Locle. He took on the most promising young men he was able to find in the village as apprentices. Once a month, according to the historians, Frédéric walked to La Chaux-de-Fonds in order to sell his *ébauches*. We have already twice mentioned this particular journey, a distance by road of sixty-nine kilometres. At the time of Japy's first enterprise it was sometimes possible to take a coach from Montbéliard as far as Saint-Hippolyte; but more often the journey had to be undertaken on foot. It involved about fourteen hours of walking, and was customarily accomplished in one day when the weather permitted.[15]

On 16th February 1773 at the age of twenty-four young Japy married Catherine Marguerite Amstutz, daughter of a Swiss *Anabaptiste* then living at Dampierre-les-Bois. This family came from the Bernese Oberland. There were many such farming in the Franche-Comté, where their agricultural skill was much admired. Shortly after the marriage the Amstutz family were moved to work a farm called *Grange-Madame* near Montbéliard, and Japy's young wife persuaded him to move his work-room there. This arrangement lasted, according to Muston and the 1949 *Plaquette,* from 1774 to 1777, and the first children were born to the couple during this period. Of these, five sons grew up to join the family business one after another from about 1806. They were Frédéric-Guillaume, the eldest, who was always called "Fritz" (1774-1834), Louis-Frédéric (1777-1852), Jean-Pierre (1785-1863), Charles (1792-1821), and "Fido". The brothers became in due course the "Japy Frères" known to all horologists.[16] Of Frédéric Japy's surviving sons the eldest, as already noted, was Frédéric-Guillaume ("dit Fritz"), but the fifth was another "Fritz" (variously called "Fide", "Fido", "Frido" or "Fidot"). The three eldest sons, Fritz, Louis and Pierre looked after factories at Beaucourt and at Feschotte-du-Haut, d'Etupes, Bas-des-Fonds, Berne and Seloncourt, while Charles and Fidot were jointly responsible for Badevel.

Soon after 1777 Frédéric Japy decided that the time had come for him to put into effect some ambitious projects which had been in his mind for several years. During his stay in Le Locle, Japy had realised that wholesale watch production and sale was not

---

(13) A *sautier* is a kind of *garde champêtre*, and in small villages implies something between a policeman and an attorney.

(14) Boetes equals boîtes equals cases.

(15) The historian Sahler, writing in 1835 *(Notes sur Montbéliard* and *Ma Vie en Deux Mots,* see Bib.) mentions this very subject, giving the time and distance quoted. The same author says that on Sunday, 25th December, 1791 (which must have been Christmas Day because the calendar was only changed for the period 1793-1806), a messenger employed collectively by the Montbéliard watchmakers for journeys to La Chaux-de-Fonds was prevented from travelling by a representative of the Bureau de Delle (to which canton Montbéliard belonged) who told him that from now on he would not be allowed to pass without paying dues in both directions. An appeal to the French Government was organised the following day ... (179 – Arch. Nat. K 2209).

(16) Vinter's history states that the trading name "Japy Frères" was formalised or "renewed" in 1837, but that from 1854 the brothers traded as "Japy Frères et Cie". In 1928 the firm became "Société Anonyme des Etablissements Japy Frères". In 1955 "Société Japy Frères" formed a branch making typewriters called "Société de Mecanographie Japy". In 1967 this firm became "Société Belfortaine de Méchanographie" ("S.B.M."). In 1969 "S.B.M.", who are very active and prosperous today with a large modern factory close to the old Japy village site at Beaucourt, became a branch of the "Société Suisse Hèrmes-Paillard" (Bolex).

135

beyond his capabilities, and this despite the head start already achieved by Daniel Jean-Richard. At the time virtually all parts of watches were made more or less by hand. Much time was wasted, according to Japy's thinking, in the initial roughing-out with files of the pieces that went to make up *ébauches;* but to produce comparable parts by any other method at competitive prices *(finies à bon marché)* required the help of machines which were not generally available. Japy considered that there were three main problems to be overcome. The first was to develop a whole new sequence of machine-tools. The second was to position them to form a production line, and the third was to train operators in their use. Given these conditions he was convinced that he could produce good *ébauches,* while undercutting substantially the price of those partly made by hand. Accordingly he decided to set up once more at Beaucourt, and acquired a rocky site high up above the main village. Today the new S.B.M. typewriter factory stands on much of this old area, while nearby are many disused Japy buildings, some of them burnt out and derelict, but others in good condition. Muston says that Frédéric entrusted the construction to a contractor named Brand. There is no doubt that at least some of the machinery and methods developed by Jeanneret-Gris were incorporated in the new factory. According to Muston, J.J. supplied certain new tools at this period, and the watch *ébauches* were in full production by 1779.

Japy was soon in the position of being able to make his own machine-tools. From 1786 he had the help of Jean-Henry Dodillet of Courtelary. Sahler cites a contract dated 1st April 1785, which he says he actually saw in the hand of Dodillet's descendants. In this agreement Dodillet bound himself from 1st May the following year, and for three years afterwards, "to faithfully serve" the named Frédéric Japy in the capacity of ". . . a founder and forger, or whatever other work for the said factory and which will be used for the profit, advancement and good-will of Japy . . .". Japy in turn promised to pay the sum of 900 *livres* over the three years. The first year Dodillet was to receive 200 *livres*, in the second 300, and in the third 400. He was also to be provided with his food and his clothes were to be laundered. Both parties signed duplicate copies.

Some of Japy's early machinery, which he probably designed himself, is listed in the *Bulletin de la Société d'Encouragement,* and again by Jacquet and Chapuis in *The Swiss Watch.* A *Brevet* may be found dated 17th March 1799 covering five years in the name of Frédéric Japy (Beaucourt), and entitled *Machines diverses servant à exécuter les principales pièces d'une montre en fabrique.* An abridgement of this *Brevet* was published in a supplement to the *Revue Chronométrique* in 1873.[17] Japy's machines were in connection with full-plate verge watches; but it is clear that the same principles were soon extended for making clocks, and for the manufacture of the Lepine-calibre watches in which he later specialised. The tools mentioned in the *Brevet* were:—

1) A circular saw for cutting brass sheet into strips.
2) A lathe for making plates, fusees, barrels, cocks, slides and racks. (This must have been a versatile machine with good milling capacity.)
3) A machine for cutting the teeth of wheels (in other words a wheel-engine. No doubt it was capable of cutting a stack of blanks at one time).
4) An automatic lathe for turning both round and ornamentally-shaped pillars.
5) A press tool for blanking out watch balances. (According to Jacquet and Chapuis *The Swiss Watch* page 178, in their very interesting comments upon the early tooling of horology, Japy's dies would have been used in conjunction with a fly-press, and the balances would have required a good deal of subsequent finishing. The "two-pillar" block system did not come into use until 1854).
6) Another press tool for making train wheels. (Presumably crossed blanks, ready for the teeth to be cut).
7) A vertical drill.
8) A machine for riveting pillars to movement plates.
9) A screw-head slitting tool.
10) A draw-bench, said to have been used in connection with the manufacture of potence slides.

That these "engines", presses, dies, etc. were very much in advance for their time is evident from the amount of notice which they received. Their great advantage, other than speed, was that only semi-skilled labour was required to operate them. Steam was introduced at Beaucourt by an early date. The site is high up and no water power could possibly have been avail-

---

(17) The *Brevet* abridgements are usually found bound with Vol. VIII after the Index and plates for the years 1874 and 1875.

able. Old prints of the factory, which are examples of artistic licence taken to the last stage of exaggeration, depict vast ranges of buildings peppered with many tall smoking chimneys. We reproduce one of the less imaginative of these drawings in Plate VI/4.

In 1804, Frédéric Japy was decorated by Napoleon I with the *Légion d'Honneur,* one of the first to be given. According to Reverchon *(Petite Histoire...)* by the year XIII, i.e. 1805, Japy was already employing 300 people. He sold *ébauches* of watches at prices of between 28.50 francs and 59 francs per dozen. In 1805, again according to Reverchon, Japy registered further *Brevets* for tools (Messidor 19, l'An XIII)[18]. By 1806 the construction of a factory was started at Beaucourt for making wood screws. Another was started in the water mill at La Féschotte (Usine de Rondelot) between 1800 and 1803, producing square-headed screws, bolts and hinges. Japy soon added to these other articles of hardware made both in cast-iron and in non-ferrous materials.

Plate VI/4. *BEAUCOURT. Japy's main factory. A somewhat exaggerated view reproduced from an old print dating from about the middle of the 19th century. (By courtesy of S.B.M. Beaucourt)*

From soon after 1800 Frédéric Japy had begun to believe that, since he was able to sell watch *ébauches* to the watch manufacturers of La Chaux-de-Fonds, there was no reason why he should not make the *blancs-roulants* of the types of clocks most in favour in Paris to be sold to "manufacturers" who were still making them largely by hand.[19] Accordingly, in about 1809 he bought the corn-mill at Badevel where he set up a new factory only for making clocks. According to M. Péquignot, who will be mentioned shortly, this mill was bought from his family. The machine tools already designed by Japy and made by Dodillet served as a basis for the new installations, so that the equipping of the factory proceeded quickly. It was in full production by 1810. Staff were recruited from the surrounding district, and the whole venture was an immediate success.[20] The village of Badevel lies at the foot of the hills of Beaucourt. The two places are connected by several roads, one of which winds up through an escarpment thickly covered by extensive

---

(18) Much early Japy machinery survived until 1942. We know because our friend Monsieur Henry Belmont, President of Yema International of Besançon, was told directly by Monsieur Linard, who was in charge of the last horological activities at Beaucourt. The old equipment was stored on the upper floors at Beaucourt, and probably also at Badevel. It was "lost" during the last war, no doubt in much the same way as the railings in the London Squares. The official report of the Paris *Exposition* of 1855 actually discusses in detail, together with production costs, the machines and methods then in use by Japy, comparing them with the systems in use at Saint-Nicolas-d'Aliermont. M. Belmont also discovered F. Japy's link with Dodillet.

(19) Evidently sometimes it is not a mistake to "carry coals to Newcastle".

(20) According to Reverchon, the American author C.T. Higginbotham published in 1912 an article on this subject in *The National Jeweler and Optician*.

VI Franche-Comté

beech woods. On the opposite side of the narrow valley is the source of the small river Féchotte, and right in front of this at the eastern end of the crooked main street stands Japy's mill. The Féchotte is a tributary of the Doubs, joining the river Allaine (or Allan) at Fesches-les-Prés after a course of only seven kilometres. The Féchotte rises dramatically and suddenly from what used to be a very large and copious spring below a bluff. In the early 19th century there was sufficient fall and volume of water from the source to turn a large mill with two wheels. Today the spring is nearly dry; but this is due only to the municipal water undertakings from 1897 in which M. Eugène Borneque-Japy played no small part. Plates VI/5, VI/6 and VI/7

Plate VI/5. *BADEVEL, 1971. Frédéric Japy's mill at the source of the Féchotte. This clock factory was in full production by 1810 and it is where the Japy carriage clocks were made.*

show the mill as it is today, a pathetic near-ruin in a highly dangerous condition. No words will express adequately the risks inseparable from entering this building. A Cerberus at the main gate could not provide too strong a warning! Neither of us had ever before in the whole of our lives, including service in the armed forces, stood upstairs in any structure which could be felt quaking beneath our feet so that we felt impelled to leave instantly. Even an outside iron staircase which we ascended is ready to collapse at any moment. The Japy mill-factory is most interesting. It is fascinating to notice, as is still possible, which buildings Frédéric Japy found, and what alterations and additions he made. The original building was two storeys tall. Some kind of mill had probably stood on this site from early times but the one which Japy took over seems to have dated from the early

Plate VI/6. *BADEVEL, 1971. Japy's clock factory. The main entrance of the old cornmill as enlarged by F. Japy 1809-1810. The wheels were on either side of the bow.*

18th century, and to have had a mansard roof. The "bow" main doorway is roughly in the centre of the old building. From behind on either side flow the twin leats or lades bringing water to turn what were probably latterly either breast or Poncelet wheels. The spring is only a few yards away. It rises in what was once a wood of beech, birch and nut trees, but which is now redolent of decay and of elder, the symbol of rural neglect. The spring water is blueish in colour, almost as though glacial in origin. The local people say that it "... comes from Switzerland in the Jura mountains". Whatever the source of supply, there was never any need or provision made for a millpond to store water. Two hefty penstocks were simply

Plate VI/7. *BADEVEL, 1971. Another part of the Japy clock factory.*

VI Franche-Comté

set across the main stream to dam it. It is abundantly clear that in happier days the mill was able to use but a fraction of the water normally available. The wheels apparently drove shafting on the ground floors only, machine-shops being confined to this level mainly on account of weight. There are signs that eventually there may have been at least one reaction water turbine. The overall dilapidation is such that it is difficult to judge. Certainly one wheel-pit is fed by a large pipe of round section. Whatever the facts, the used water was never allowed to become "tail-bound". The tail races are enormous in relation to the leats, and the spent water is conducted away underground into the main village. The water course runs up the central aisle between the two facing main blocks of factory buildings, its path being marked by *oubliettes* along the way. Behind the main building and close to the source of the Féchotte stands the remains of a sizeable power-station with a tall chimney. Here a steam-driven plant is said to have been working until 1910. An old enamelled iron notice reads "FORCE MOTRICE DE L'USINE DE BADEVEL VAPEUR 1845, ELECTRICITE AVRIL 1909". Electricity was obviously also used for lighting from quite an early date, for the various buildings are strung like harps with primitive wiring.

It has been observed that the original mill was not tall. Frédéric Japy, or perhaps later his sons, soon changed all that, increasing the height of some buildings by up to three storeys. The ground floors provided "services", stores and machine shops. There still exist at this level the remains of huge boiler houses for heating. On the upper floors are ranges of work rooms, of which the view shown in Plate VI/8 is very typical. All the windows of the old mill had to be enlarged enormously. In order to preserve the maximum possible strength and "box formation" to the building, besides to allow for the weight of the extra floors where they were added, massive cast iron window-framings, not only very deep but also braced with substantial webs, were added to stiffen every opening on the lower levels. Plate VI/9, which illustrates the very room where the carriage clocks were assembled, shows the construction of these re-inforced windows.[21] The Badevel mill reputedly closed as a clock factory in 1933; but it is clear that it was in decline, if not in actual decay, by a far earlier date.

The present precarious condition of the principal structures, even allowing for forty years of total neglect, is in no way consistent with the time elapsed. Many floors and stairways are on the verge of falling down altogether. The evidently once-fine engine house, presumably built or restored in 1845, is a worse ruin than the Roman theatre of 52 B.C. a few miles away! We were told in the village that the main building on the left side looking into the main gate was devoted to the manufacture of *pendules de voyage* and *régulateurs,* while the one on the right at the end nearest to the gate made *réveils Américains.* At first the meaning of "American Alarms" was not clear,

Plate VI/8. *BADEVEL, 1971. Upstairs in Japy's clock factory.*

Plate VI/9. *BADEVEL, 1971. The room in which carriage clocks were assembled. Note the reinforced windows to help support storeys added above.*

---

(21) Note the pillars or staunchions helping to support the floors above. The *chinoiserie* dates from World War II.

VI Franche-Comté

even after having heard it repeated several times. It was only later on a visit to Monsieur Péquignot (see page 144) that the significance became clear. M. Péquignot has in his collection the small one-day timepiece clock shown in Plates VI/10 and 11. This model made by Japy Frères at Badevel, and said to have been in production from about 1880, was called the *type Américain*. It apparently was made both as a timepiece and as a timepiece with alarm. Almost certainly the alarm version was the *réveil Américain* of which many thousands a month were still being made in January 1907 (see Appendix (d)). It is most significant that the Japy clock illustrated is almost indistinguishable from the American Ansonia *Pert* (Plate XII/46) as shown in the 1914 Catalogue. Alarm mechanism apart, the one is an almost exact replica of the other. It would be most interesting to know whether Japy *réveils Américains* were ever marketed in America, and also to what extent American carriage clocks were exported to Europe.[22]

Plate VI/11. *JAPY FRERES. The back of the clock shown in Plate VI/10. The pin-pallet lever escapement is integral with the movement as in American clocks. (By courtesy of E. Péquignot, Badevel)*

Plate VI/10. *JAPY FRERES. An American-style model. The case is made of thin sheet brass and is nickel-plated. Compare with the Ansonia* Pert *(Plates XII/46-48). (By courtesy of E. Péquignot, Badevel)*

The latest date of construction apparent on any part of the Badevel factory is 1893. One building so dated is obviously far newer than any of the main parts, except for an annexe where in recent times there has been a foundry and also a pressure diecasting plant making carburettor components, probably for sale to Peugeot. A clue to the probable date of a beginning decline in manufacture of *pendules de voyage* at Badevel may be given by Vinter on p. 122 of the official history. Here he says "In 1921 at Badevel, cradle of clockmaking, in this factory where

---

(22) Beillard, writing in 1895 (*Recherches*, page 155), says that but for her own powerful horological industry France would have been swamped with inferior American clocks which, although selling for the same price as Japy's, were far from being of equal value. (Beillard here almost certainly means not only wooden-cased shelf clocks, but probably also American carriage clocks and copies of French mantel clocks like those shown in the Waterbury and Ansonia Catalogues).

VI Franche-Comté

have been made such wonders of mechanism, such as carriage clocks, they erected on the left, opposite the old workplace, a large building in which they installed on the ground floor the workrooms for automatic lathes and machining generally for the production of the detail parts necessary for manufacturing alarms". Vinter continues "On the floor above, in rectangular galleries, they assemble the alarms". The building mentioned appears to be that formerly used for *pendules de voyage* and *régulateurs*.

It is now necessary to steer the reader away from the Japy Badevel factory, with its three-trumpet steam or Klaxon horn, crumbling walls, and with a large fir tree growing from the top of one of the tallest buildings. Today the only sound in the place is the regular thump of the small hydraulic ram powered by the Féchotte and locked away in one of the old wheel-

Plate VI/12. *BADEVEL, 1971. Ex-Japy accommodation and* fontaine. *This accommodation and also the clothes-washing facilities were very advanced for their time.*

pits. It is not clear to whom this potable water is still supplied, but all over Badevel are gravity-fed *fontaines*, watering and washing places donated by the Japy family. Near the factory gate, as seen in Plate VI/12, still stands a sort of half-square of houses all of which once were attached to the factory. In the heyday of the business, the facilities offered by Japy Frères to many of their workers included accommodation and *fontaines*. As already noted, Badevel was run jointly by Charles and Fidot Japy. Opposite the factory houses in some allotments still stands a rather touching monument to Jean Charles Japy (1792-1821). This must have been the Charles above, for Muston says (*Histoire d'un Village*, Vol. II, p.21) that "Charles Japy, le fils cadet de l'illustre fondateur de Beaucourt

est mort jeune, sans postérité". The memorial (Plate VI/13) with its draped urn and depicting in carved stone hourglass, clock, square, dividers, etc., is set against a background of agriculture. The inscription is very strangely worded, reading as follows:— "TOUT PRÈS DU MONUMENT, QUÉRIGEA SA TENDRESSE, SOUS CES JEUNES PEUPLIERS, L'OIN DE L'OEIL DES MÉCHANTS, REPOSE, SEUL, LE CORPS DU MEILLEUR DES ENFANTS. S'IL N'ATTEIGNIT POINT LA VIEILLESSE, IL VÉCUT BIEN ASSEZ, POUR SAVOIR QU'ICI BAS L'ONNE VOIT QUE TROMPEURS, QUE MASQUE À CHAQUE PAS. JEAN CHARLE JAPY NÉ LE 17 JANVIER 1792 DÉCÉDÉ LE 25 MAI 1821". (sic) A free translation of this might read:— "Quite near to the monument, which was erected by the loving one, beneath these young poplar trees, far from the eye of the wicked ones, lies, alone, the body of the best of children. If he did not reach old age, he lived long enough to understand that down on earth one encounters only deceivers who cheat at every step. Jean Charle Japy born 17th January 1792, died 25th May 1821". Perhaps not surprisingly, Muston seems nowhere to make any reference to this monument; but almost certainly it was erected in 1821 by Fidot to the memory of his brother Charles. Muston says (Vol. II, p.77) that after the death of Frédéric (i.e. from 1812)

Plate VI/13. *BADEVEL, 1971. Memorial to Jean-Charles Japy (1792-1821).*

141

VI Franche-Comté

Plate VI/14. *BEAUCOURT, 1971. A bust in memory of Frédéric Japy, commissioned by his family.*

Charles and Fidot were the sole proprietors of "la fabrique de mouvements de pendules de Badevel", and that when Charles Japy died in 1822 (here Muston means 1821) M. Fidot Japy continued the fabrication up to 1828. It is not easy today to understand the exact nuances of the inscription on the Badevel monument, and so far no one asked in the district has been able to explain their significance.

Frédéric Japy, as already noted, had died in 1812. His outstanding endeavours, carried on by his sons, gave work to many people who previously had depended upon the soil for their livings. It is remarkable that Frédéric in particular was able to accomplish so much at such a critical period of history, which included the time of Napoleon's disastrous campaign in Russia involving 800,000 men. According to Muston, Beillard and Vinter, Japy was always greatly liked by those who worked for him, and for whom he did so much in the way of medicine, schooling, crèches, rest rooms, old age pension schemes, libraries, and even bakeries. At Beaucourt F. Japy and his wife presided each day in dining rooms provided for the workers. He is quoted as saying, "Je veux que mes ouvriers ne forment avec moi et les miens qu'une seule et même famille". Plate VI/14 shows a bust of Frédéric in a lobby of a restaurant which was at one time a Japy family home. A similar memorial stands in the garden of another old Japy château, now the canteen of S.B.M. A dozen or more large houses in the immediate vicinity once belonged to members of the family of the founder. Undoubtedly the good relations which the Japy family always maintained with their employees must have played no small part in the success of their businesses throughout times which were, from both national and political points of view, exceedingly troubled. The first workplace on the Beaucourt hill top was burned down in 1815 during an "invasion" by Austrian troops; but the Japy brothers rebuilt it immediately afterwards to incorporate a large new machine-shop. The buildings, according to Muston, were on both sides of Frédéric Japy's original house. Half were devoted to work, and half to rest rooms and dining rooms. Despite what Vinter says in his official history about this reconstruction, the now disused factories seen in Plate VI/15 are based upon far older buildings. They are in far better condition than anything at Badevel, and must have been kept in use and properly maintained until much later. Incised in the stone above one window on the left side of the street, looking down the hill, is the date of the eleventh year of the Republic, "AN XI", or 1803. Above another window on the same side is the inscription "J.F. 1816". The remains of what was perhaps Frédéric's house still also stand. Beside it is a burned-out building which may have been the rest room and

Plate VI/15. *BEAUCOURT, 1971. Old Japy factory buildings, parts dating from as early as 1803. (By courtesy of S.B.M. Beaucourt)*

dining room. Muston says that after 1828 the Badevel factory went through a very critical phase just at the worst possible moment for the three Japy brothers at Beaucourt.[23] Some time later, Ingénu Japy, son of Frédéric-Guillaume and grandson of the founder, was put in charge of Badevel. Ingénu is said to have learned his business in Switzerland with Sandoz and in Paris with Breguet. He is credited with the development of further automatic machines for both cutting and polishing pinions. We are told that under his charge Badevel soon became "la première fabrique de France, en son genre, et un modèle au point de vue de l'ordre et de la méthode". This development all took place in times still very troubled after the *Restaurations* when France still had many internal difficulties. It is best to ignore here the frequent "local" troubles about this time in Le Pays, as also the inevitable involvement of the Japy family; but unrest was very considerable and the worst since the First Republic. The peak of these local *Revolutions* was reached at the time of the uprising of 1830, after which and until 1848, when there was another, there ensued years of comparative peace and prosperity.[24] The peaceful times were when the Japy family organised their medical services, etc. for the people round Beaucourt. In one of these latter periods the Japy brothers decided to form a new outlet for their products and opened a store and accounts department in Paris at 108, Rue du Temple. Monsieur Monnin, a relation of Fritz Japy, was in charge. Muston records that the Depot was particularly useful for selling the clocks from Badevel and for export to Spain.[25]

The production of hardware by Japy Frères greatly exceeded in bulk, if not in quantity, their enormous output of clocks and watches. It has already been noted that the Rondelot factory was set up between 1800 and 1803 making coach-bolts, hinges, etc., and that production of wood screws was started at Beaucourt in 1806. The next expansion was the setting up of the La Féschotte factory at La Casserie between 1824 and 1826, and another at Gros-Pré from 1888. An enamelled notice at Gros-Pré reads "CETTE USINE CONSTRUITE EN 1889-90. A été mis en marche En Février 1891. JAPY FRERES & CIE" (sic). The lower Féchotte factories are close to Dampierre-les-Bois and Fesches-le-Châtel. They are ideally sited for business, each with private railway sidings. At Gros-Pré a huge modern plant (apparently not running to anything like full capacity) and a large shattered old watermill stand side by side. Enamelled pots and pans were perhaps the first and last items made in the La Féschotte factories.[26] Between times the list of articles produced there was long indeed. Many a *bidet* or a *douche* from Gros-Pré is destined soon to become "une antiquité"! The La Féschotte factories are on the river Féchotte not far below the Badevel mill. In 1845 the Japy Frères bought the forges of MM. Bouchot Frères on l'Isle-sur-Doubs to further increase the output of coach bolts, wood screws, etc. A relation by marriage, Louis Meiner (who had married Rose Japy, daughter of Ingénu) was put in charge; but to be fair he proved most capable. Five hundred people were employed in this factory alone and the whole of the Japy screw and bolt manufacture was centralised there.

It is interesting that the family historian, Muston, nowhere mentions Japy in connection with *pendules de voyage*. Such clocks certainly exist made by Japy Frères close to 1850, while by 1862 the brothers are on record as taking out a *Brevet* for a carriage clock with lever escapement and alarm. By 1878 Japy was patenting a carriage clock with hour and quarter repeating, besides another of a special cheap design. It is a curious fact that even Vinter, in writing about Japy at a far later date, makes almost no mention of *pendules de voyage*. He does so at last on pages 89-90, saying simply "This extremely interesting manufacture which was carried on at the factory of Badevel has for

---

(23) According to Galliot the Badevel factory changed hands several times between 1826 and 1832. Muston does not say this in so many words, but certainly he is inclined to hedge with great skill on any Japy history which is not pleasant. Galliot says that in 1820 there were 80 workers at Badevel of whom half were children under 14. In 1826 there were 400. In 1820 the factory delivered 4,800 pendulum clock movements. In 1826 the output was 12,000 clock movements.

(24) From 1848 to 1852 came the Second Republic with all its attendant toils. In 1870 the Franco-Prussian War resulted in the occupation of Beaucourt; but somehow MM. Japy kept their factories running normally, ". . . sans aucune interruption. . .". Vinter says that the Japys emerged creditably from the terrible year 1870-71, which saw the beginning of the Third Republic.

(25) See also Japy, Marty, Roux in List of Names.

(26) Japy's publicity brochures circa 1900 show grossly exaggerated views of the various factories. We know most of them, and so are in the position of being quite sure. In one print in our possession Beaucourt, Lafeschotte, Larouche, etc. are drawn looking like the Ford Motor Works at Dagenham, while Badevel is larger than the Chelsea Barracks which it much resembles!

VI Franche-Comté

various reasons been discontinued, and in 1935 the last Japy *pendules* and *pendulettes de voyage* were made." On pages 185 to 187 Vinter lists "Horology. The Development in Chronological Order of the Different Japy Factories at Beaucourt and Badevel". This table shows that alarms were made from 1825, and that from 1865 there was an "Expansion of horology. Creation of several calibres of watches and clocks. Beginning of manufacture of bronze cases for clocks and alarms". In 1882 and 1890 are mentioned ". . . cases of clocks and alarms of all sorts and styles". It is necessary to wait until 1902 for references to the manufacture of carriage clocks, having alarm, repeating, calendar work and *grande sonnerie*. Two sizes appear to have been made at this period. During visits to Badevel in 1971 we twice had the pleasure of visiting Monsieur Edouard Péquignot who lives at 57, Rue des Usines in the village. His portrait appears in Plate VI/16. He is a splendidly fit and entertaining young gentleman of 94. What is more he worked for Japy from 1890-1940, including on *pendules de voyage,* of which he has two examples. M. Péquignot believes that Japy only ever sold complete carriage clocks. However, with the greatest respect, this information is utterly at variance with what is recorded by such 19th century writers as Sire (1870), T.D. Wright (1889) and Tripplin (1890), not to mention our own discoveries in Saint-Nicolas-d'Aliermont, in Paris and in Le Pays de Montbéliard. In each of these three places we have actually handled Japy carriage clock *blancs-roulants,* including some without punchmarks or inscriptions of any kind, but still packed in their original paper wraps, these bearing the Japy four-star trademark and with printed labels reading "JAPY FRERES & CIE, A Beaucourt et à Paris, Rue du Château-d'Eau 7, 3- VOYAGES, REPETITION SIMPLE No 3, DEPOT A PARIS, Japy, Marti, Roux, Mougin & Cie, 75, Rue de Turenne, 75". The combination of firms mentioned on the labels was not in existence before 1863, so the wrapped *blancs* cannot have been earlier. At present there is no more specific evidence than this as to when Japy Frères first marketed *blancs-roulants* as such. It is apparent from observation made at first hand how Japy's *roulants* may be distinguished from those made in Saint-Nicolas-d'Aliermont. However, enough evidence has perhaps been advanced to satisfy the reader that Japy did make and supply *blancs-roulants* of carriage clocks to the trade as a whole, and what is more in quantity. Japy of course also made finished carriage clocks. Tripplin makes it quite clear that carriage timepieces were among Japy exhibits at Paris in 1889. According to Vinter on page 186, the models made by Japy in 1902 included calendars, sunrise-and-sunset, phases of the moon, repeaters, *grande sonnerie* and miniatures. He is also specific in stating on page 90 that all these clocks were made in Badevel and that production ceased in 1935. On the other hand, on page 215 Vinter says "On fabrique actuellement à Beaucourt en horlogerie, des pendules et réveils dans tous les genres". This observation, in view of the state of the Badevel factory, seems to suggest that some of the last travelling clocks might really have come from Beaucourt.

Plate VI/17 gives an excellent idea of the appearance and presentation of one of the earlier, if not indeed

Plate VI/16. *BADEVEL, 1971. M. Edouard Péquignot. He worked for Japy Frères from 1890-1940 and he is one of the last links with the old carriage clock manufacture side of the business.*

VI *Franche-Comté*

by Japy and sold by Henri Marc. This Japy/Marc clock shows several interesting features. The first is a centre-seconds movement embracing a subsidiary seconds train planted externally on the back plate in exactly the same manner as a later Japy clock which will be mentioned in connection with Plate VI/19. In fact many Japy clocks, of various periods and qualities, are provided with centre-seconds work. The second feature of interest in the Japy/Marc clock is the provision of a spring-detent chronometer platform escapement complete with helical balance spring, without terminal curves, uncut balance (!), bridge cock and long backward-pointing index. It would be inter-

Plate VI/17. *JAPY FRERES. A carriage clock made circa 1850. Both the "one-piece" case and the motif of the engraving are typical of Japy's early productions. Many Japy carriage clocks will be found to have centre-seconds. (Private Collection)*

earliest, types of carriage clock sold by Japy Frères circa 1850. This clock, in its "one-piece" engraved case bearing Lange's punchmark, is not only very typical of early Japy *pendules de voyage* as a whole but it is also in many ways reminiscent of certain even earlier clocks, also often found in Lange cases, and variously bearing the names of Auguste, Bolviller, Jules and even of Moulinie. What connection, if any, there may be between these clocks is not known. The Japy clock in question was sold by Henri Marc of Paris, whose name appears in an ellipse at the bottom centre of the back plate above a roundel reading "JAPY FRERES, MEDAILLES D'OR 1823, 27, 34, 39, 44, 49".[27] So the clock was clearly made

Plate VI/18. *JAPY/RECLUS. Clock No. 28460. This clock, which bears both Japy's and Reclus' marks, was presumably produced by Japy and finished by Reclus. The escapement of this clock is shown in Plate VIII/24. (Private Collection)*

---

(27) These awards were for watch and clock work in general.

145

VI *Franche-Comté*

Plate VI/19. *JAPY/RECLUS. The back plate of the centre-seconds clock shown in Plate VI/18. Note the Reclus and Japy marks above and below the alarm setting square. Note also the external subsidiary seconds train concealed by the bell (removed for the purposes of photography). The Japy clock shown in Plate VI/17 has the same centre-seconds arrangement. (Private Collection.)*

Plate VI/20. *JAPY FRERES. An inexpensive timepiece carriage clock with centre-seconds and alarm of a design introduced post 1867. Size 6.7/8 inches tall with handle up. (Private Collection. Photograph by P.H. Kohl)*

esting to know how much better the timekeeping properties realisable from this platform would have been than those obtainable from the lever escapements made about the same period. One clock of which two views are shown in Plates VI/18 and 19 bears the sunburst trademark "V.R." (Victor Reclus) cheek by jowl with the inscription "JAPY FRERES & CIE EXPOSITION 1888 GRANDE MED D' HONNEUR" [28] We have seen a number of Japy clocks sold by others, for example one signed "O. Berger 6234 Paris" and bearing the inscription "MEDAILLES D'OR, JAPY FRERES, 1823, 27, 34, 39, 44, 49". On the other hand one very famous Paris firm who at least latterly undoubtedly finished many movements obtained from Japy, certainly received Japy *blancs-roulants* entirely innocent of any trademarks or inscriptions indicative of their real origin. How do we know? Because, as already noted, we have seen movements, which bear no name or trademark, but which are still in their original protective

---

(28) The late M. Henri Lengelle ("Tardy") in his *Dictionnaire* shows several Japy trademarks. One such mark is a cock. Another is an inverted horseshoe with the word "BEAUCOURT". Yet another has the word "Bocor" in two directions on a Maltese cross in a circle. Two others are variants of the monogram, but without the four stars. Another, in an ellipse, refers to "BARILLET INDEPENDANT". This modification was one of the firm's "rationalisations" introduced circa 1867.

VI Franche-Comté

paper wrappers stuck all over with Japy labels. Mr. Pitcher (whose reminiscences will be found in Appendix (c)) believes that Japy *blancs* were much inferior to those of Couaillet, and were produced far more cheaply. No doubt extreme cheapness was a feature of some of Japy's late manufacturing; but the firm's final productions were not necessarily inferior to those

Plate VI/21. *JAPY FRERES. The back of the movement of a clock similar to that shown in Plate VI/20. Note the main and sub-frames, and the twin barrels driving a common centre pinion. (By courtesy of Sotheby & Co.)*

eventually made by Couaillet. Observation leads to the belief that Japy made, although not necessarily at the same period, two quite distinct qualities of carriage clock. On the most interesting issue of cost, Sire, writing in 1870, devotes several pages and three folding plates to describing and illustrating Japy clock movement designs *à barillet indépendant* adopted a few years before. Plates VI/20, 21, 22 and 23 show these modified techniques as applied to *pendules de voyage*.

Plate VI/22. *JAPY FRERES. A side view of the movement of the clock shown in Plate VI/20. Sire, writing in 1870, described and illustrated this type of movement lay-out for which many advantages were claimed. (Private Collection. Photograph by P.H. Kohl)*

This particular example bears no identification mark of any kind. It is undoubtedly a Japy clock, as evident from the peculiar movement layout. In any case virtually identical pieces exist bearing Japy's trademark. The date of manufacture of the example under discussion must have been post 1867. The case is of rather poor quality, and the standard of work in the movement is inconsistent. For instance, the hands, dial, bell and escapement parts are all of the cheapest construction possible, while most of the wheelwork is reasonably well made. Another version of the same clock was produced having a music box in the base. Muston, writing in 1882, says that at that period it took only two hours at Badevel to make a clock from start to finish, and to have it cased and going. Reverchon, writing circa 1930, says that at one stage Japy Frères were invited to move their industry to Russia. This report may be believable in view of the

VI Franche-Comté

Saunier in the *Revue*, but also earlier by the *Bulletins* of the *Société d'Encouragement* and in *Dictionnaire Chronologique... de 1789 à la fin de 1820*. The reports do not appear always to agree, except in showing that watch and clock production at Beaucourt and Badevel was the envy of manufacturers everywhere. The great importance of Japy Frères et Cie as makers of carriage clocks must not be underestimated because this side of their business was dwarfed by many other activities. It is quite apparent that in horology as a whole they soon outstripped all the makers in Saint-Nicolas-d'Aliermont put together. The 18,000 platform escapements "destined for carriage clocks" and produced near Montbéliard in 1867 mentioned in Chapter III were a good indication of the production of *pendules de voyage* in the area at the time. According to Sire, writing in 1870 of the Montbéliard district, "In this area are found many

Plate VI/23. *JAPY FRERES. The two-plane, pin-pallet lever escapement with cone pivots used in the cheapest Japy carriage clocks, as shown in Plates VI/20-22. The lay-out of this escapement effects the required change in plane without the expense of a contrate wheel and pinion. (By courtesy of Sotheby & Co.)*

fact that craftsmen were recruited from Saint-Nicolas-d'Aliermont at the time of the Paris *Exposition Universelle* of 1900, as mentioned in Chapter IIII.

Frédéric Japy's great, great, great grandson Monsieur Albert Japy is still alive today. He was at one time *Director Général* of the firm. He says that by 1868 Japy had made 21,715,788 *ébauches* of watches, and that over 5,550 people were employed throughout the entire firm. He further says that the Badevel factory had already produced 1,960,805 clock movements. Figures supposed to represent Japy clock production at various periods are given not only by Muston and

Plate VI/24. *AUGUSTE L'EPEE. The founder of L'Epée & Cie, the world-famous makers of music boxes, and later of platform escapements. (By courtesy of Mme. Henry L'Epée)*

other workrooms occupied in the production of platform escapements for carriage clocks, one of the specialities of the Paris trade".

Some of the Badevel manufacturing books, dated January 1907 and covering *pendules de voyage* and *réveils Américains*, have miraculously survived. Notes concerning their contents are included in Appendix (d).

The firm of Frédéric L'Epée & Cie has already been mentioned in this book on a number of occasions.[29]

Plate VI/25. *SAINTE-SUZANNE. The music-box factory founded by Auguste L'Epée in 1839, as it appeared towards the end of the 19th century. Note the largely unspoiled countryside. (By courtesy of Mme. Henry L'Epée)*

Plate VI/26. *SAINTE-SUZANNE. Another view of the L'Epée factory which today makes platform escapements and is concentrated in the area near the old chimney. (By courtesy of Mme. Henry L'Epée)*

Plate VI/27. *L'EPEE & CIE. A keyless musical watch having pinned-barrel in place of the more usual pinned-wheel. Period late 19th century. (By courtesy of Mme. Henry L'Epée)*

This famous business was founded at Sainte-Suzanne in 1839 by Auguste L'Epée who had, according to Sahler, previously worked for Japy Frères. Plate VI/24 shows his portrait, while Plates VI/25 and VI/26 reproduce old views of the factory as it looked towards the end of the last century.

The world-wide reputation of the firm was at first founded upon music boxes which they continued to make until 1914. A large part of their business consisted in supplying other firms. This is perhaps the main reason why the name L'Epée is not better known today in connection with mechanical music. Another reason is that they did not usually sign their work, although the name did sometimes appear upon the lists of tunes attached to the lids of the larger boxes.

---

(29) The family crest and the business trademark depicts crossed swords, or épées.

L'Epée had an annual production by 1870 of approximately 2,000 large boxes, and some 40,000 small, and they employed 300 people. By 1900 the workforce had increased to 500, and at least 200,000 music boxes had been made. From 1850 the firm made musical watches and had also begun to produce the platform-escapements *(porte-échappements),* which have been sold in vast quantities and with great success ever since. Plate VI/27 shows a L'Epée musical watch. It is in a steel case, and its superiority and selling points evidently lay in its robust construction and in the good playing obtainable from a musical box type pin-barrel with sturdy comb and dampers, as opposed to the pinned-wheel construction which was almost the rule in "Swiss" musical watches made in the preceding half of the century. The L'Epée watch in its sensible case was probably also far less expensive than the earlier conventional key-wound musical watches. These were mostly shoddily-made from their backward-turning cylinder escape wheels to their low-quality gold cases, so thin that they "popped" like oilcans.

Eventually the demand for music boxes gradually became less, due to the advent of gramophones, and for a while part of the factory was turned over to making phonograph parts. In the same period, there was a surprisingly large demand for children's whistles *avec quatre touches,* while other lines which surprisingly continued to be needed were metronomes for setting music boxes to run at the correct speed and *claviers étalons* for tuning them. The *Revue Chronométrique* of 1878 quotes Muston as saying that one third of the music boxes made in the Montbéliard district were sent to England and America, one third to Germany and Russia, and the remainder to various other countries.

In 1945 the firm published its history under the title of *Causerie sur La Fabrication Des Boîtes à Musique à Sainte-Suzanne,* this being a transcription of a lecture given to the Société d'Emulation de Montbéliard on the 31st October 1942 by M. Henry L'Epée. In 1964 on the occasion of their one hundred and twenty-fifth anniversary, Frédéric L'Epée et Cie produced a brochure which included several illustrations of the older buildings at Sainte-Suzanne. Today L'Epée employs 450 people. Now, their very large production is confined to escapement platforms of a number of types in which they have cornered a large share of the world market. While many of these escapements are admittedly used in connection with horology, others are intended for use in industrial and scientific timers, and with specialist programming machinery. Originally the largest early customers for L'Epée *porte-échappements* were naturally Japy, Marti and Roux, their near neighbours, but at that time all three were largely unconnected. Further outlets were soon found in Paris and Saint-Nicolas-d'Aliermont. Here Couaillet, amongst others, was a regular customer. The annual output of platforms at Sainte-Suzanne was in excess of 18,000 per year before 1900.[30] Upstairs in the main offices at Sainte-Suzanne there are eight wall showcases of platforms showing at least some of the different models which have been made and required at one time and another. Prominent amongst many sizes, shapes and different arrangements are platforms with English-type pointed-tooth escapements. These were mainly for use in French carriage clocks destined for Britain. There is no mystery about

Plate VI/28. *L'EPEE & CIE. A peculiar platform escapement intended for carriage clocks and patented in 1879. (By courtesy of Mme. Henry L'Epée)*

---

(30) Megnin, 1909. See Bib. Many of these platform escapements were signed "A.L.", and also carried the L'Epée trademark of crossed swords.

VI Franche-Comté

such escapements, but many people wrongly believe that they must be partly English. One very strange frictional-rest escapement peculiar to L'Epée and shown in Plate VI/28 has twin identical 'scape wheels. These are made sometimes of brass and sometimes of steel, and they turn in opposite directions. The balance staff has milled in it a narrow, vertical slot (rather like that in the roller of a conventional duplex escapement) and the teeth of the wheels escape in turn through this slot giving impulse in either direction. L'Epée obtained a *Brevet* No. 131,336 of 23rd June 1879 for this "Echappement d'horlogerie à double roue". The L'Epée escapement seems to have been only half-successful as an alternative to the cylinder platform for use in carriage clocks. It could only ever have been intended for comparatively inexpensive pieces and is a far cry from the double-roller lever escapements which are now in production at Sainte-Suzanne.

Today the firm is owned and run by Madame L'Epée, who is the last of her family. Her husband, Henry L'Epée, great, great grandson of the founder died in 1964 at the age of 61. The present factory, close by the canal and the river Allan just outside Montbéliard, is a model of cleanliness and good organisation. It is equipped with the very latest and best automatic and electronic machines, besides employing advanced assembly and inspection techniques. The progress made by recent technology is everywhere apparent, and it is clear that no detail of anything anywhere in the place is unknown to Madame L'Epée and to her managers.

The L'Epée firm is the last of the old horological industries of Le Pays de Montbéliard. Its highly successful survival and transition in adapting its techniques continuously to changing world markets, is certainly a triumph for all concerned. In the *Revue Chronométrique* of October 1862, Pages 262-267 (Vol. IV) is an article by Muston entitled *De La Fabrication des Pièces à Musique dans Le Jura*. This gives a history of mechanical music in France generally, although it is mainly about L'Epée.

Although Japy Frères almost certainly made by themselves more *pendules de voyage* than did all the other firms in the Franche-Comté put together, Japy carriage clock movements were not of a design in any way typical of the predominant work styles of the area. It is true that their cheap pattern already mentioned and illustrated as employing a main and a sub-frame with two barrels was their own speciality, but their more expensive movements were almost indistinguishable from Saint-Nicolas-d'Aliermont work. We are thinking not only of the fairly conventional Japy carriage clocks often seen in the salerooms, but also of the *blancs-roulants* already mentioned as surviving exactly as supplied to be finished in Paris. These

Plate VI/29. FRANCHE-COMTE CLOCK. *Maker Michoudet, Foncine-le-bas. Date probably circa 1835-40. This clock goes for two weeks striking* grande sonnerie.

roulants could easily have been made by one of several makers in Saint-Nicolas so far as their design is concerned. Other earlier makers in the Franche-Comté, by contrast, produced carriage clocks which are characteristic of that area, differing sharply from the so-called Parisian work. The Franche-Comté clocks tend to have their external striking works on the back plates, and often they evince rather coarse finishing, combined with comparatively heavy wheels and

VI Franche-Comté

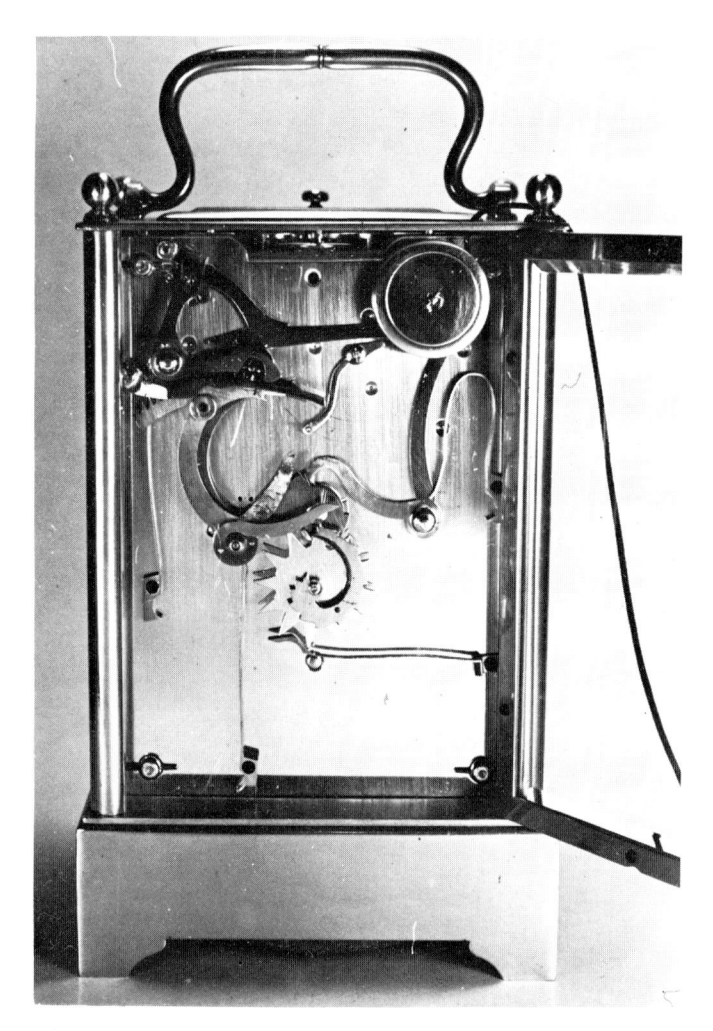

Plate VI/30. *FRANCHE-COMTE CLOCK.* The *cadrature* of the Michoudet clock shown in Plate VI/29.

Plate VI/31. *FRANCHE-COMTE CLOCK.* The Michoudet clock again. Note a movement layout very dissimilar from that favoured in Northern France.

pinions. Several views of a good example of a real Franche-Comté clock are shown in Plates VI/29, 30, 31 and 32. It is signed "Jn Marc Michoudet, Foncine-le-bas, JURA", and it is probable that it was made about 1835-40. The rather agricultural gilt-multi-piece case has plain handle, doors front and back, and glass on four sides as well as on top. The movement is wound from the front. The clock has centre seconds and alarm, and offers a choice of *grande* and *petite sonnerie* striking or silence, selected by a lever at the bottom of the dial. A single mainspring drives both going and striking trains for two weeks at one winding. A top button is provided for repetition. The rather heavy steel hands are intended to be set to time with the fingers. The external striking work is arranged on the back movement plate as is the spring barrel for the pull-wound alarm. Concealed beneath the base are the hammers and bells. The cylinder platform escapement is signed "FUMEY". No doubt this platform was obtained close to where the clock was made from the same Fumey family of horologists who had been in Foncine since the middle of the 18th century. The overall combination of features found in the Michoudet clock are not at all surprising considering that it is in the nature of a very much developed *Capucine,* a type of clock which almost certainly was produced in the part of France close to Switzerland. It is interesting to notice the great similarity in the striking work of the Michoudet clock as compared, for instance, with that of the *Capucine* illustrated in Plate I/33; but the family likeness really cuts much deeper than this. The Michoudet case is really not far removed from that of a *Capucine,* while the method of winding the alarm by pulling a cord on the top of the case is also a legacy from the same

VI Franche-Comté

Plate VI/32. *FRANCHE-COMTE CLOCK. The Fumey cylinder platform escapement of the Michoudet clock.*

ventional in later carriage clocks. The alarm mainspring remains in the old position inherited from the *Capucines,* but it is now wound with a key instead of by a pull cord. The external striking work maintains tradition by being placed on the back plate; but this clock strikes each hour twice, besides repeating hours at will and striking a single blow at the halves. As might be expected, the bell and hammers are in the base. The idea behind the double striking is to give a second chance to the listener who may not have heard properly the first time. This practice is very common in church clocks on the continent, particularly in Italy, but it is very unusual as applied to *pendules de voyage.*[31] The second striking takes place at two minutes after each hour. This anonymous Franche-Comté clock has a platform escapement with an uncut brass clockmaking traditions. Rather similar striking work was, of course, used in Switzerland in a number of portable clocks, including their rare very high quality *pendules de voyage;* but the finishing of this provincial French work is altogether inferior by comparison. Naturally the two clockmaking schools, being so close together, were to some extent influenced by each other, if not actually interdependent in the present context.

The clock in Plates VI/33 and VI/34 shows a continuation of the same style of work found in the Michoudet clock but carried one stage further. This later example, which is unsigned, has an engraved *Corniche* case, perhaps still a little quaint, but on the whole much more in conformity with Paris fashions. The sturdy manually-set hands are retained, but the alarm setting dial has moved to the position con-

Plate VI/33. *FRANCHE-COMTE CLOCK. An unsigned example, perhaps made 30 years after the Michoudet, but still characteristic of the work styles of the French Jura.*

---

(31) Double-striking is not, however, a surprising feature to find in a clock made anywhere near Morez!

153

VI Franche-Comté

Plate VI/34. *FRANCHE-COMTE CLOCK. The cadrature of the anonymous double-striking clock shown in Plate VI/33.*

and steel balance, right-angled lever escapement with single roller and pointed tooth escape wheel in the English style, but with a crescent-shaped counterpoise for the lever and with a double-ended coqueret to the balance cock. This type of escapement was made in great numbers by L'Epée. The finishing of this second clock, both train and escapement, shows considerably more refinement than the work in the Michoudet clock, probably because of a difference in date of perhaps thirty years.

In Chapter I was mentioned the fact that *Capucine* clocks may well have had their origins in the Jura; but a few more remarks relative to this matter may now be of interest to the reader. The famous Parisian clockmaker Antide Janvier (1751-1835) was born in Saint-Claude[32] the place where stood the Capucins monastery mentioned by Lebon in connection with the origins of clockmaking in the area. The *Tribune Chronométrique* says that Janvier received his grounding in horology from his father who was a farm labourer turned clockmaker; while Lebon in his study of horology in the Franche-Comté says that Janvier returned briefly to his old home in the years 1791 and 1792, and that he taught clockmaking while he was there. Lebon also gives the information that early in the 18th century Jean-Baptiste Dolard was the first to realise that clockmaking would offer a good and more secure outlet for the steel from the mills along the Bienne Valley. He is said to have taken over a number of forges and turned their production towards steel for clockmaking. The so-called *Comtoise* or *horloge de Comté* clocks — that is to say the tall-cased sort with the beautiful steel frames, seconds pendulums, upside-down-verge escapement and double-striking afforded by vertical racks — form no direct subject for this book, but they are peculiar to the Jura and were made there during the same period as *pendules de voyage*.[33] Pierre Dubois in one of his *Lettres... sur l'Horlogerie en Franche-Comté* (writing from Morez on 6th September 1852) mentions Saint-Colaude as one of the most industrious places in the French Jura, saying that there, as in Morez, they make clocks, turn-spits and musical boxes. Dubois says of carriage clocks that the ones now made in the Jura are different from the first travelling clocks made there, and that now they have balances instead of pendulums. Most of the escapements, he says, are cylinders; but a few escapements used in *pendules de voyage* made locally are levers or even have spring-detents. The reason for the preference for the use of the cylinder by the Jura manufacturers is that it is not only the least difficult to make, but also the least troublesome in use.

Dubois also speaks highly of the Japy *père*. He rates Frédéric Japy as the equal of Pons in transforming the manufacture of *pendules de cheminée*.

There is very much more to be learned about clockmaking in the Franche-Comté and in the Jura, especially as regards the finding and identifying of specimens of *pendules de voyage* made in this part of France.

---

(32) According to his Obituary in the *Tribune Chronométrique* pp. 24-29.
(33) *La Réparation des Pendules* by Jaquet and Gibertini deals specifically with the repair of *Comtoise* clocks. So does *Cours de Rhabillage de l'Horlogerie Comtoise* by Gabriel Moreau.

Chapter VII

Case Styles.
The Names, Shapes and Sizes of Standard French Carriage Clocks

CORNICHE. See Plate VII/14.

VII
Case Styles.
The Names, Shapes and Sizes
of
Standard French Carriage Clocks

The "One-Piece" Case. The "Multi-Piece" Case. Later Case Shapes. Distinctive Decorative Styles. Minor Case Styles. Decorative Panels. Carriage Clock Sizes. Conclusions. Carriage Clock Prices.

The qualification "standard" has been used in order to make it quite clear that this chapter is concerned mainly with those clocks made, at first in modest numbers but latterly upon a very considerable scale, during the one hundred or so years which saw for all practical purposes the beginning, heyday, decline and cessation of factory manufacture of *pendules de voyage*. The reader will already know about Breguet. He will also by now be aware that a few French travelling clocks are still finished in the traditional manner up to the present moment.[1] All this is beside the point. It is, however, beyond argument that most clocks, made during the period when the industry was active, largely conformed to a number of conventions as regards shapes and sizes of cases and also sizes of movements. It is with these standards that this chapter is concerned.

THE "ONE-PIECE" CASE

The "one-piece" (see Plates VII/1 & VII/2) is the earliest standard type of brass-and-glass case. A safe generalisation is to say that any clock having it is early, while examples found in cases comprised of many small component parts cover the whole of subsequent carriage clock history. A few such are early, but most are not. Conversely, a late clock will occasionally be found in a "one-piece" case.

"One-piece" cases were made in sizes from about four and a half to six and three quarters inches tall. They take their name from the fact that the strong, mitred and slotted brass sections which make up the glazed box (*boîte*) or frame are pinned and brazed to form one rigid unit. This in turn is screwed to a fairly massive cast brass base, which also serves to retain the glasses in their slots. The "one-piece" form of construction, by virtue of its stiffness, is a better arrangement than anything which followed. On the other hand it would have been far more expensive to produce and also does not lend itself to much variation in design. These two objections probably explain why the first style of case was soon abandoned. It was gradually superseded by the "multi-piece" cases, at first somewhat similar in appearance. These presently evolved into a large number of other shapes and sizes.

The first "one-piece" cases, dating from before 1830, did not have the hinged doors usually associated

---

(1) M. Pitou has retired since these notes were written.

VII Case Styles

Plate VII/1. *EARLY "ONE-PIECE" CASE. A style current from before 1830. This clock was made by Paul Garnier and is his No. 799, date circa 1834.*

Plate VII/2. *EARLY "ONE-PIECE" CASE. Date circa 1840. Striking and repeating movement of good quality, with alarm, lever escapement and plain balance. (Private Collection).*

with carriage clocks. Instead, the front and/or back glasses were made to slide in and out of the case vertically. The brass "box" frame was cut and slotted to allow the glass or glasses, which were attached to the appropriate top case rail or rails, to be removed. Sometimes the top rail was made in two parts, cut in the centre and hinging upwards at either end, to allow the removal of the glass alone. This arrangement is perhaps the earliest type of all. It suffers from the disadvantage that the glass becomes marked by finger and thumb prints, and also rattles abominably when the clock is moved. The last fault cannot occur when the glasses are attached to the well-fitted, mitred top units of the case itself.

Before leaving the "one-piece" case, it should be noticed that it is usually of really good quality and as a rule encloses a movement of more or less interest. Makers whose movements are found in "one-piece" cases include Lépine, Paul Garnier, Bolviller, Auguste, Jules and Japy. An early casemaker was L. Lange, whose name will be found struck on the base castings. Auguste, Jules and Japy certainly used Lange cases, sometimes plain, sometimes engraved, but often having embossed decoration. The period covered by "one-piece" cases is roughly 1826-1845.

THE "MULTI-PIECE" CASE

Plate VII/3 shows a transitional case, still resembling the clocks in Plates VII/1 and 2, but in fact made up of a number of pieces screwed together. This practice, as already observed, foreshadows the basic construction of all subsequent standard French carriage clock cases. Plate VII/4 illustrates the trend set by the clock in Plate VII/3, but carried one stage

VII   Case Styles

Plate VII/3. *EARLY MULTI-PIECE CASE. Date close to 1840. Maker Bolviller. Clock No. 2. This clock is described in Chapter II. The multi-piece cases both overlapped and succeeded the one-piece cases. (Private Collection).*

Plate VII/4. *LATER MULTI-PIECE CASE. Date probably mid-19th century. Maker Paul Garnier. Clock No. 2320. Two-plane escapement. Note the use of canted corners in a case of an altogether more modern appearance. The white enamel dial and trefoil hands also reflect a later taste. (By courtesy of E. Pitcher & Co.).*

further. In general, the earlier a case the larger the area of glass it will have. This fact, taken in conjunction with other features, provides quite a valuable guide to date.

The "multi-piece" case, except in its early forms, is quite dissimilar in appearance from those of the older "one-piece" conception. The *boîte* is fabricated from a mixture of brass castings and pressings held together by screws. The four vertical columns have milled slots to accommodate the bevelled glasses, which are further retained in position by the top case plate and by the base casting or pressing. The door, or doors, are hinged on pins top and bottom. This pattern determines the basic shape and assemblage found in all clocks made from the mid-nineteenth century onwards. While the best "multi-piece" cases are very good indeed, some of the cheapest are less satisfactory, being badly aligned and in consequence difficult to assemble. Against this must be set the fact that the carriage clock could never have become popular, much less have been made in very large quantities to sell for a reasonable price, with the old type of case. The cheaper, more flexible "multi-piece" construction offered new and previously undreamed of possibilities for variations in design.[2] The proliferation of patterns

---

(2) The report of D. Roussaille, delegate to the 1889 Paris Universal Exhibition on behalf of the Horological Chamber of Commerce at Lyon, mentions particularly this point.

VII Case Styles

evolved at one time and another between various makers was very great. All the same, most concerns followed a basic range of perhaps ten shapes. These are fairly well documented, if only in the now very rare surviving manufacturers' catalogues. Some makers (for instance Margaine) identified their clocks by number instead of by name. Whatever the reason, the fact remains that the names of even the commonest carriage clock models and shapes have largely been forgotten. In the heyday of the French carriage clock period, the top London stores, when offering and illustrating in their catalogues wide ranges of these clocks, normally did so without mentioning their names. This practice strongly suggests that correct names and designations were rarely used, and perhaps not even ever known in England except to the importers. The Catalogue "Guide to the Purchase of a Clock", published soon after 1900 by S. Smith & Son, illustrates no less than twenty-nine different clocks, but names very few. Many, if not all, of the clocks illustrated by Smiths in fact appear specifically named in the Duverdrey & Bloquel Catalogue of circa 1910. The same situation exists in respect of the Army and Navy Catalogue of 1907 where a dozen clocks of varying shapes are shown, but without a name between them. Accordingly, no apology is now necessary for listing, illustrating and discussing in these pages the standard models and shapes once used by many French manufacturers.

In this connection it should be made clear that carriage clock shapes, when used in the adjectival sense, are not given capital letters. They are correctly written thus:— *gorge, borne, cannelée, doucine,* etc. On the other hand, individual makers' models were customarily given capital letters in their catalogues:— for example *Galerie, Cavour, Acanthe. Zéro bis* or *Obis* is given a capital letter because it implies both a model and a shape used by many makers. Another model and shape, also in common use, was termed *Corniche.* Sizes, like models, were given capital letters. Thus we read *Mignonnette No. 3,* while the generic term for a miniature clock is a *mignonnette.*

LATER CASE SHAPES

To look through some of the old manufacturers' catalogues is an alarming experience. The varieties of cases and movements listed and illustrated by certain firms are quite bewildering. Anyone having the least experience of modern production technique wonders at once how it was ever possible, much less necessary, to offer so wide a choice. The fact that so many clocks were offered (and here we are thinking particularly of the Duverdrey & Bloquel Catalogue of circa 1910, and of a Couaillet Frères Catalogue dated 1914) shows not only the extent of the potential business but also the competition which must have existed. Duverdrey & Bloquel and Couaillet are perhaps extreme cases. The range of shapes offered in a typical Margaine catalogue, for instance, is far smaller. Many of the *boîtes* shown by Duverdrey & Bloquel are their own variations upon basic themes. They cannot be considered as standard cases, or at least not in the context of these notes. Furthermore, some Duverdrey & Bloquel models, as opposed to basic shapes, were

Plate VII/5. *BOITE JONC. Date circa 1910. A speciality of Duverdrey & Bloquel. Timepiece movement with alarm and with cylinder escapement. (By courtesy of Bryson Moore. Photograph by P.H. Kohl).*

VII Case Styles

Plate VII/C1. *ROCOCO. Made by Drocourt circa 1870. An important clock in very ornate style. The porcelain dial, side and top panels are finely decorated. "Giant" size. (By courtesy of E. Pitcher & Co.)*

Plate VII/C3. *CLOISONNE ENAMEL. The Anglaise Riche case has Corinthian columns. (By courtesy of E. Pitcher & Co.)*

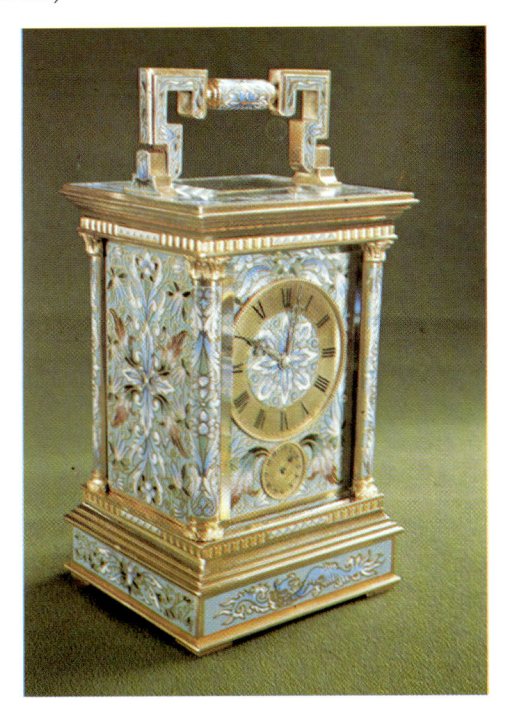

Plate VII/C2. *CLOISONNE ENAMEL. Date circa 1900. This style of enamelling originated in the Far East. Note an Anglaise Riche case with Corinthian columns. (By courtesy of E. Pitcher & Co.)*

Plate VII/C4. *LIMOGES ENAMEL. Date circa 1895. The case is Anglaise Riche with bamboo columns and with Limoges dial and side panels. (By courtesy of E. Pitcher & Co.)*

VII Case Styles

below the dial, was usually termed *riche*, in addition to its ordinary name, e.g. *Anglaise Riche*. Combined clocks and barometers were styled *jumelles* (twins). One Duverdrey & Bloquel model, with white "eyes" on a dark background, was called *Indienne* (Indian Girl).

The following names were those most commonly used by French manufacturers to describe the standard shapes of carriage clocks made during the last quarter of the 19th century up until 1914, and to some extent since.

Obis (short for *Zéro bis*, or "double nought")

This was the cheapest shape of case ever produced.

Plate VII/6. *BOITE CLASSIC. Offered by Couaillet as late as 1931-32. Duverdrey & Bloquel called an almost identical case* Boîte Colonnes. Obis *movement with alarm.* (Private Collection. Photograph by P.H. Kohl).

peculiar to this firm alone. Duverdrey & Bloquel had their factory at Saint-Nicolas-d'Aliermont and their specialities included *Boîte Jonc* (Plate VII/5), *Ovale Fantasie, Cubique Bombée, Chinoise, Turque, Mignon, Irlandaise, Dome, Colonnes Torses, Henri II, Gourmette Chaînette, Tresse, Victoria, Bouts Ronds, Six Colonnes, Double Corniche, Mauresque, Empire, Ecossaise*, to name but a few. Couaillet of Saint-Nicolas tended to give their own model names to what were otherwise standard case shapes. Plate VII/6 shows a pillared design given the model name *Classic* in a Couaillet Catalogue dated 1931-32. An almost exactly similar case was called *Boîte Colonnes* by Duverdrey & Bloquel. Any *boîte* with added ornament, particularly with vertically-slotted bands above and

Plate VII/7. *OBIS (Zéro bis) A model current from about 1880 to 1939. The cheapest standard case. A bread-and-butter line produced mainly in Saint-Nicolas-d'Aliermont. Height 5½ inches only. Note a typical* Obis *handle.* (By courtesy of E. Pitcher & Co.).

162

VII Case Styles

Plate VII/7 well shows its simplicity. Although it did not become current until towards the end of the 19th century, the *Obis* shape remains so fundamental to standard carriage clocks that it must be introduced at this stage. It was made in one size only, standing five and a half inches tall with handle up. It was the bread-and-butter line, produced soundly but inexpensively in very large quantities. The design is plain, with no decoration whatever, although relieved by the two-cut handle and by a simple moulding round the top. To save expense, the dials of the movements fitted into *Obis* cases were pinned directly to the front plates without the use of false plates, the motion work being accommodated between the movement plates instead of under the dials.[3] Redier of Saint-Nicolas obtained in 1857 a *Brevet* for enamel dials with no false plates. *Obis* clocks were mass-produced at Saint-Nicolas-d'Aliermont from about 1880 until the outbreak of the Second World War. It is important to make clear that the term *Obis* implies not only a case shape and a case size, but also a size of movement. At first makers used the *Calibre Obis* movement only in conjunction with the *Obis* one-sized case, as part and parcel of an inexpensive clock. Latterly, Duverdrey & Bloquel, and

Plate VII/8. GORGE. *The most expensive standard case used only by the best makers. It was current from before 1867 until after 1900. This particular clock was made by Jacot before 1867, and bears his number 825. (By courtesy of E. Pitcher & Co.).*

Plate VII/9. ENGRAVED GORGE. *Date circa 1890. Maker Margaine. Clock No. 2355 with hour and half-hour strike. (By courtesy of E. Pitcher & Co.)*

---

(3) The minute hands of *Obis* movements were retained in position by friction alone. They were simply pushed on to the centre pinions. A further invariable *Obis* economy was the absence of any stopwork.

VII   Case Styles

Plate VII/C5. *LIMOGES ENAMEL. Another Anglaise Riche* case with Corinthian columns used in conjunction with Limoges side panels. *(By courtesy of E. Pitcher & Co.)*

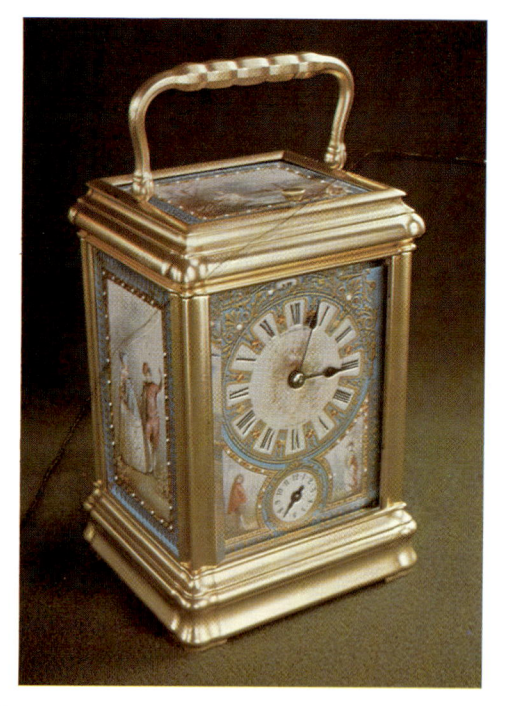

Plate VII/C7. *PORCELAIN PANELS. Date circa 1870-80. A fine clock in a gorge case with porcelain dial and panels. (By courtesy of E. Pitcher & Co.)*

Plate VII/C6. *GUILLOCHE ENAMEL. Period probably 20th century. The dial plate and side panels are enamelled in dark green over an engine-turned sunburst pattern. This clock is a miniature. The dial and hands appear to have been restored. (By courtesy of E. Pitcher & Co.)*

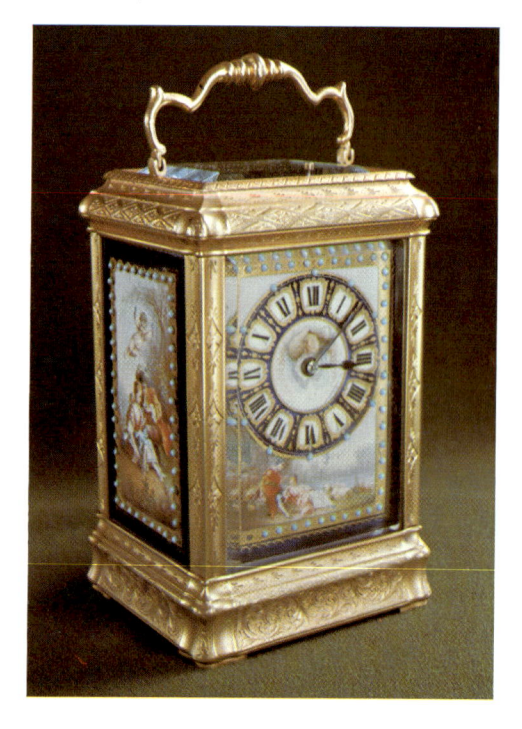

Plate VII/C8. *PORCELAIN PANELS. Date about 1870-80. Maker Margaine. Note a characteristic Margaine handle used so effectively in conjunction with an engraved gorge case. (By courtesy of E. Pitcher & Co.)*

VII Case Styles

certainly others, used *Obis* movements in a whole variety of inexpensive case styles, some larger and some smaller than the basic five and a half inches. Further details of *Obis* will be found under "**Carriage Clock Sizes**".

Gorge (grooved)

This very beautiful and restrained case shape was certainly current by 1867, continuing in use until after 1900. In contrast to the *Obis,* the *gorge* shape was never used for any but the finest clocks. It did, however, house most effectively the simple clocks of the top makers, for example Jacot timepieces. Plate VII/8 illustrates a typical *gorge* case. It will be seen that the base and pillars are deeply grooved. The top of the case is built in tiers, a mixture of convex and concave mouldings. The base is slightly concave at the four corners. The typical handle is moulded in five divisions, forming a sort of ripple pattern. Plate VII/9 shows an engraved *gorge* case. The dials supplied to complement *gorge* cases were usually plain white enamel of wonderful quality and clearness. Some clocks, however, had engine-turned metal dials, silvered or gilt, and with or without sunk circular enamel centres. Needless to say, *gorge* cases were used extensively by the best Paris firms. The cases were made in at least five sizes, and were always of very high quality. Unlike the cheaper styles of case, in which (if the truth be told) the

Plate VII/10. *CANNELEE. Current from circa 1885 onwards. A simplified version of* gorge. *By courtesy of E. Pitcher & Co.)*

Plate VII/11. *OVAL. Date circa 1870. Maker Samuel. Timepiece movement with two-plane escapement. See Chapter VIII. (By courtesy of Garrard & Co.)*

multiple parts at times fitted together more by good luck than good management, the *gorge* cases were accurately machined and also assembled with some degree of precision. Seemingly identical parts were not made interchangeable, but were steady-pinned and also numbered to ensure correct assembly.

VII Case Styles

Cannelée (fluted)

Plate VII/10 shows that while this case is similar to the *gorge* it is a simpler model. It was produced in much smaller quantities from about 1885. The *cannelée* case differs from the *gorge* in having an altogether harder top profile, and usually also a handle moulded in three divisions. It was used by the best Paris makers. Surviving examples will be found to house fine movements.

Oval

This attractive shape was introduced during the second half of the 19th century. Oval cases are usually found in three sizes:— *Mignonnette No. 3,*

Plate VII/12. ENGRAVED OVAL. *Date circa 1880. Size Petit-Zéro. Striking and repeating movement. This case has cloisonné side panels. (By courtesy of E. Pitcher & Co.)*

Plate VII/13. OVAL VARIANT. *Date probably post 1900. Note that the case is broader than usual. One maker called this shape Oval Bout. (By courtesy of E. Pitcher & Co.)*

Obis and *No. 3*. They occur in a considerable variety, often engraved, sometimes repoussé, and not infrequently with porcelain panels. Plate VII/11 illustrates a plain example and Plate VII/12 shows an engraved case with *cloisonné* enamel side and back panels. Plate VII/13 pictures an oval variant, height seven and a half inches. In one manufacturer's catalogue this shape is called *Oval Bout*. The case is broader than usual, decorated in relief with applied, tooled and gilt floral castings. The dial is in black and gold on a white background. It is worth remarking that with this clock it is not easy to see the time across a room. True oval cases have their shaped glasses set in fabricated bezels which are held in place by steady pins top and bottom. A few clocks were made with circular-section cases. Duverdrey & Bloquel certainly sold examples.

VII   Case Styles

miniature clocks, the standard sizes of *Corniche* and other semi-bread-and-butter productions, not miniatures, were *No. 1* (height six inches), *No. 2* (height six and a half inches) and *No. 3* (height seven inches).[4] These dimensions vary slightly with individual manufacturers. More will be said on this subject later. While *Obis* clocks were ordinarily limited to timepieces or timepieces with alarm, *Corniche* examples were rather more ambitious in scope, offering timepiece, timepiece with alarm, hour-and-half-hour-strike (with

Plate VII/14. CORNICHE. *A model current from circa 1880. More expensive and better made than* Obis. *Sometimes the base arches are un-ornamented. Note the typical* Corniche *handle. This particular clock, No. 18650, was made by Drocourt. (By courtesy of E. Pitcher & Co.)*

Corniche (cornice)

Now we return to a more modest shape of case, once again belonging to the mass-production era. Plate VII/14 shows a characteristic example. This model was produced in very large quantities in Paris, in the Franche-Comté and in Saint-Nicolas-d'Aliermont, before as well as after 1880. It was normally of better quality than the *Obis,* with a more elaborate moulding, or cornice, round the top. Apart from the

Plate VII/15. CORNICHE CARREE. *A rather squat variation of* Corniche. *The restricted height necessitates the use of a movement smaller than would be expected in a clock of this width. This model was probably introduced after 1900. (By courtesy of E. Pitcher & Co.)*

---

(4)   Margaine, at least, produced a *Corniche* alarm of no pretensions whatsoever and housed in a case of *Obis* size and quality. One Margaine catalogue also offers a *Corniche* in size *No. 0*. Its trade price was 40/- as timepiece, and 50/- with alarm. Delépine-Barrois produced two distinct qualities of *Corniche*.

VII Case Styles

or without alarm) and repeat (with or without alarm). Some *Corniches* even had *petite* or *grande sonnerie* striking. In contrast to the *Obis*, where, as already noted, the dial is attached directly to the front plate of the movement, the *Corniche* almost always has, behind the dial, a false plate, or dial-plate, joined to the front movement plate by four feet. This refinement, found normally in all types of carriage clock except *Obis,* leaves room for motion and striking work outside the plates. It also gives a clock greater depth.

Corniche Carrée (square cornice)

Plate VII/15 shows a carriage clock case shape which is really a variant of *Corniche*, although more squat in appearance. *Carrée* is also distinguished by heavy Roman chapters on a dial designed to meet the need for a clock which could be read very easily. The *carrée* shape was a speciality of several Saint-Nicolas manufacturers, dating from about 1890.

Plate VII/17. *CORINTHIAN COLUMNS. Corinthian columns were used mostly with Anglaise or Doucine cases. The example shown was almost certainly made in the early years of the present century. (Private Collection)*

Anglaise (English)

This case, Plate VII/16, was introduced about 1880. It is found in *Mignonnette No. 2, Obis* and *No. 3* sizes. The handle, top, pillars and base are all "squared". This severe style was marketed because the French believed that it would appeal to plain

Plate VII/16. *ANGLAISE. A style introduced about 1880. The French manufacturers believed that this shape would appeal to plain English tastes. Couaillet Frères offered a clock with exactly this case in their Catalogue of 1931-32, while Duverdrey & Bloquel showed a variant with a more ornate base arch in their Catalogue of circa 1910. (By courtesy of E. Pitcher & Co.)*

VII Case Styles

English tastes. However, as it is hoped has been made clear, virtually all French carriage clocks were made for sale in England anyway.[5]

Pillars

Plate VII/17 shows a clock having Corinthian columns, and which may be called typical of pillared designs as produced in quantity from about 1860. Pillared cases are, however, as old as carriage clocks. A-L Breguet was using *Empire* cases with pilasters by 1813, and among more modest makers, Lepaute showed a clock in a pillared case at the Paris Exhibition of 1827. Latterly pillared cases were made not only in many sizes, including miniatures, but also in a bewildering variety of designs, some having classical origins, but most of them stemming solely from the imaginations of their designers. The clock shown in Plate VII/18 is of pseudo-classical design, said to have been inspired by the neo-classical Madeleine in Paris. Various firms offered their own versions of this shape, most of them evincing alarming degrees of

Plate VII/18. *PEDIMENT-TOPPED. A pseudo-classical design said to have been inspired by the Madeleine in Paris. The case shown, of date circa 1880, is made of gilt brass; but some examples embody marble sections. (Private Collection)*

Plate VII/19. *PILLARS. Maker Margaine No. 6697. A very high quality clock using columns showing marked bamboo influence. (By courtesy of E. Pitcher & Co.)*

---

(5) Anyone doubting this statement needs only to see how many clocks he is able to find with the hand-setting and winding instructions in any language except in English. The American market for French carriage clocks was never great. What is more, they made their own. It seems that the only exhibitor of French carriage clocks at the Philadelphia Exhibition of 1878 was the Maison Breguet. Many other types of French clocks were, however, shown. (*Revue Chron.* Vol.IX, p.156).

VII Case Styles

Plate VII/20. *PILLARS. The rare use of free-standing columns. The glasses are set in separate ornamental bezels. Date post 1900. (By courtesy of E. Pitcher & Co.)*

Plate VII/21. *PILLARS. A twentieth century clock. Note the knurled, banded pillars used in conjunction with a semi-Anglaise case. The porcelain dial and side panels have applied butterfly wings used to colour the birds and insects. (Private Collection)*

artistic licence. Duverdrey & Bloquel called their model *Madeleine* and embellished its case with onyx. The clock illustrated has Ionic pilasters, and it generally has strayed very far from the Corinthian Order upon which the Madeleine is based. This style of carriage clock was current for a few years both before and after 1900. It is related to the "marble-mausoleum" type of French *pendule de cheminée* at present so much out of fashion. Plate VII/19 shows a clock of very high quality housed in a pillared case with columns showing marked bamboo influence. The maker was Margaine, and the clock was made after 1900. Plate VII/20 shows a clock with free-standing pillars, of no discernible origins and used in conjunction with a very ornate case. Once again this clock was made after 1900. The pillars in the clock shown in Plate VII/21 are best described as banded. The knurled pattern used for the bands makes it easy to believe that this clock was made at a date rather later than the other pillared variants illustrated.

VII   Case Styles

Louis XV Doucine (serpentine shape)

In this shape of carriage clock case the top and bottom sections of the front, sides and back are of an ogee or serpentine shape, an influence borrowed from furniture. Sometimes these cases have decorative friezes, superimposed above the curved portions of their cases as in Plate VII/22. Plate VII/23 illustrates a plain example. Both cases house *Obis* movements. *Doucine* cases were introduced before 1900 and may well have continued until World War II. They were a speciality of Duverdrey & Bloquel, but were also sometimes used for expensive clocks (see Plate V/25).

Plate VII/22. *LOUIS XV DOUCINE. A rather more elaborate doucine case having overlaid friezes in gilt repoussé work. Note that these comparatively inexpensive clocks have Obis movements.*

Plate VII/23. *LOUIS XV DOUCINE. This design was introduced shortly before 1900 and may well have continued until World War II. A serpentine frieze is used for the top and bottom case members. (By courtesy of E. Pitcher & Co.)*

VII   Case Styles

Plate VII/24. *CARIATIDES. Maker Jacot. An outstanding clock produced circa 1890. Jacot usually avoided ornate cases. This example is nevertheless characteristic, especially as regards the engraved dial motifs against a silvered background. (By courtesy of E. Pitcher & Co.)*

Plate VII/25. *CARIATIDES. Date post 1900. Maker Adolf Ollier. A miniature silver case, size* Petit-Zéro, *with cast, pierced and chased decoration. (By courtesy of E. Pitcher & Co.)*

Cariatides (caryatides)

Cariatides cases, as shown in Plates VII/24 and VII/25 perhaps have more in common with the pillared designs than with any other shape of case. The name is derived from the Greek and means a column in which is represented the upper half of a female figure, usually draped. In carriage clocks, these figures, either complete or tapered away, form the corner castings of the case. Some examples have the complete standing figures as in Plates VII/26 and VII/27. The corresponding male figures, called *Atlantes* (Atlas) or *Telemones,* are found less frequently. *Cariatides* were made in a very wide range of sizes in combination with a large variety of movements. Once again, this was a style used largely by the best Paris makers. It had been popular for clock cases long before *pendules de voyage* became current.

VII Case Styles

Plate VII/26. *CARIATIDES. Period last quarter of the 19th century. This clock is of exceptionally high quality with the figures finely chiselled and with the detail work very clearly defined. Note a top profile showing Chinese influence. (By courtesy of E. Pitcher & Co.)*

Plate VII/27. *CARIATIDES. Period late 19th century. Maker almost certainly Japy Frères at Badevel. A distinctive clock, but not of very high quality. The cast case details are not hand finished, and in consequence they lack the crispness found in more expensive clocks. (Private Collection)*

VII   Case Styles

Plate VII/28. OBLONG. *This inexpensive design was a speciality of Couaillet Frères, and was shown in their Catalogue of 1931-32. (By courtesy of E. Pitcher & Co.)*

Plate VII/30. CUBIQUE *Another late Couaillet style. These clocks were timepieces only, having* Obis-*type movements. (By courtesy of E. Pitcher & Co.*

Plate VII/29. OBLONG. *Maker Couaillet Frères. An oblong variant produced between the wars. This clock has an* Obis-*type movement. (By courtesy of E. Pitcher & Co.)*

Oblong

This very distinctive shape was a speciality of Couaillet Frères of Saint-Nicolas-d'Aliermont. Plates VII/28 and VII/29 illustrate examples. *Oblong* clocks have a height of four and a half inches and a width of three and a half inches. No other size was made. The shape was produced in large quantities after 1900. These clocks were usually timepieces only. Some examples may have had alarms. Usually the sides and back door were made of brass instead of being given bevelled glass panels.

Cubique (cubic)

Plate VII/30 shows an example. The height is four and a half inches, allied to a width of two and one eighth inches. This clock, as might be expected, was another Couaillet line. It was offered after 1900 as a timepiece only.

VII Case Styles

Plate VII/31. *BORNE. The milestone-shape, first used by A-L Breguet from circa 1813. The three clocks shown are by Adolf Ollier who revived the style in the early years of the 20th century. (By courtesy of E. Pitcher & Co. Reproduced from an Ollier Catalogue)*

Borne (milestone shape or "humpbacked")

See Plate VII/31. This shape, first used by Breguet, became a speciality of Adolphe Ollier of Paris after 1900. To anyone with an eye for the work of this particular maker, the examples shown are very characteristic. Note Ollier's use of arched tops with large carrying handles, and often in conjunction with silvered engine-turned dials. *Borne* cases cannot have been made in large numbers as they are rarely seen. Ollier sold the *borne* as a timepiece, or as a timepiece with alarm. Duverdrey & Bloquel offered a *borne* with a flat top.

DISTINCTIVE DECORATIVE STYLES

A few carriage clock cases do not seem to fall naturally into any of the basic shapes already mentioned. Good examples are:—

Plate VII/32. *BAMBU. Date just pre-1889. Maker Le Roy & Fils, in the time of G.H. Desfontaines. Clock No. 2374 F731. (By courtesy of Sotheby & Co.)*

Bambu (bamboo)

This very distinctive style, introduced during the final decades of the 19th century while bamboo furniture was still in vogue, needs no further description once the photograph is seen. Plate VII/32 shows

175

VII   Case Styles

Plate VII/33. *BAMBU AND ART NOUVEAU. Date circa 1900. Maker Drocourt. No. 20727. (By courtesy of Sotheby & Co.)*

Plate VII/34. *ART NOUVEAU. As first conceived, circa 1900. (By courtesy of Sotheby & Co.)*

an example. The clock seen in Plate VII/33 is really a *bambu* variant although it also owes something to *art nouveau*. The reader will notice that the case of this clock, which is signed "W. THORNHILL & CO", embodies influences drawn from *bambu, art nouveau,* and even from *chinoiserie*. Walter Thornhill & Co., whose business at 144-145 New Bond St. became a limited company in 1878, was at the same address until at least 1905. John James Thornhill, presumably the father of Walter, was Cutler to Queen Victoria. A firm with such a background would not have dreamed of importing inferior clocks, so it is not surprising that this example bears the name of Drocourt.

Art Nouveau

This mode of decoration was first shown to the world at the Paris Exhibition of 1900. W.A. Steward's book *Gold, Silversmithing and Horology at the Paris Exhibition, 1900,* published by Heywood, well shows the impact of the style. Plate VII/34 illustrates an unnamed carriage clock fully representative of the *art nouveau* idiom as first conceived. The piece is of very high quality and strikes *grande sonnerie*. Note the silver (or silvered) repoussé side panels, set off by a handle of entwined dolphins fit for any Master of the Fishmongers' Company! L. Leroy & Cie offered in 1904 a clock almost identical in appearance and

176

VII   Case Styles

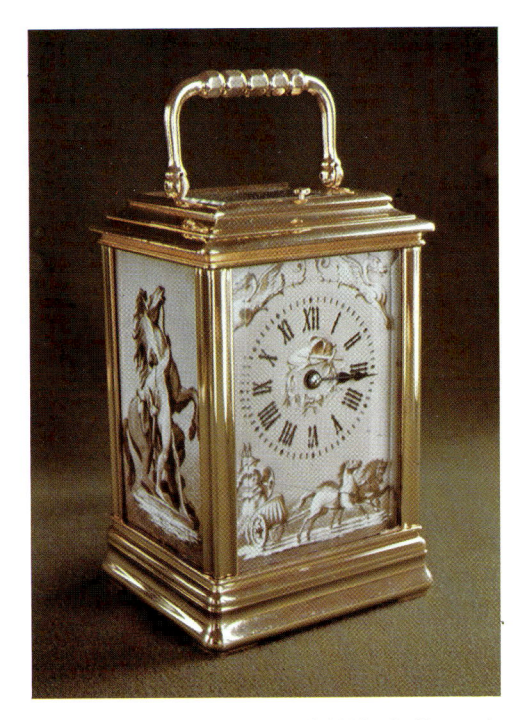

Plate VII/C9. *PORCELAIN PANELS. Date circa 1885. Panels celebrate* les Chevaux de Marly *by G. Coustou* père. Cannelée *case. (By courtesy of E. Pitcher & Co.)*

Plate VII/C11. *PORCELAIN PANELS. Period 20th century. The panels were made in Holland in the town of Delft. The clock is a* Corniche No. 1. *(By courtesy of E. Pitcher & Co.)*

Plate VII/C10. *PORCELAIN PANELS. Period 20th century. The panels depict farmyard scenes typical of the French countryside near the borders of Normandy and Picardy. Even the back door is decorated. This clock is a semi-miniature in the unusual size of* Petit-Zéro. *(By courtesy of E. Pitcher & Co.)*

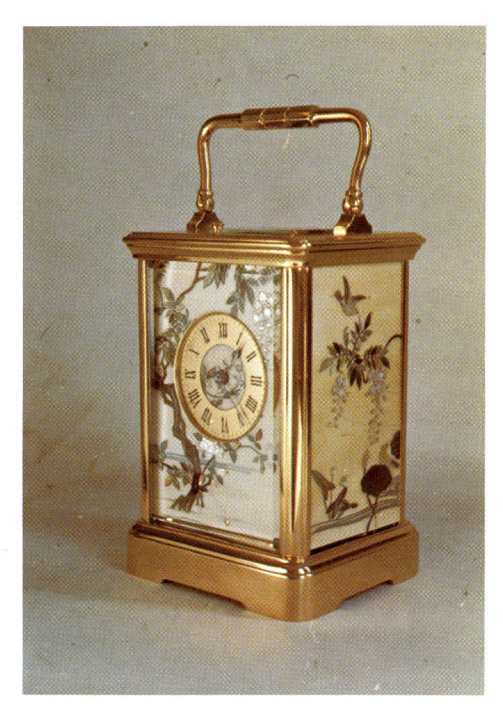

Plate VII/C12. *SHIBYAMA PANELS. Period 20th century.* Corniche *case. The panels were made in Japan of ivory decorated with applied birds, lotus flowers and wistaria executed in mother-of-pearl and other coloured materials. (By courtesy of E. Pitcher & Co.)*

VII   Case Styles

styled *Renaissance*. Plate VII/35 illustrates a clock reflecting the tastes and appearances of the early 1920's rather than those of the turn of the century. The shape of the case no longer suggests *art nouveau*, but its influence persists in the dial and side panels. Note the use of translucent enamel over engraved designs for the background, and of opaque enamel for the portraits of the "bright young thing" with her headband and Twiggy look. This particular clock strikes *petite sonnerie* and is a repeater. The *Obis* clock with *art nouveau* dial featured in Plate VII/36 is probably older than it appears. Certainly the model was current from well before 1900; while no manufacturer would have been likely to give an *art nouveau* dial to an inexpensive clock except at the height of a new fashion.

Plate VII/35. *ART NOUVEAU PANELS. Date circa 1920. Note panels still influenced by* art nouveau, *used in conjunction with an altogether more modern-looking case style, probably developed from a* Doucine. *The girl evinces none of the verve of the* art déco *maidens who were so soon to follow her. (By courtesy of Camerer Cuss & Co.)*

Plate VII/36. *ART NOUVEAU. Date shortly after 1900. A standard* Obis *clock, but with a black dial overlaid with stylised gilt flowers and foliage in high relief. (By courtesy of E. Pitcher & Co.)*

VII Case Styles

Boîte Chinoise (Chinese-look or Pagoda-topped)

Duverdrey & Bloquel offered the case style shown in Plate VII/37 in their catalogue of circa 1910. The firm made this shape as a *Mignonnette No. 2,* as an *Obis,* and as a combined clock with barometer, thermometer and compass. One or two other makers offered clocks in which the tops of the cases showed vaguely Chinese influence, as for example the *Cariatides* case shown in Plate VII/26, but the true pagoda cases were a speciality of Duverdrey & Bloquel.

Rococo (Baroque)

Plate VII/CI illustrates an extravaganza of a clock with a beautifully florid case in ormolu and coloured porcelain. The background of the panels is blue. The chapters and linings are gilt. The panel pictures are in soft colours. The dial shows a house, a bridge and a waterfall. This clock is so ornate, and so generally exaggerated in its conception and decoration, that the result is very impressive indeed. Probably a more *recherché* carriage clock case was never made. No amount of prejudice is able to rob it of its importance. Not surprisingly, the maker is Drocourt. The date of the clock is circa 1870. Plate VII/38 shows a less extreme example of the same type and period.

Plate VII/37. *BOITE CHINOISE. Date circa 1910. A speciality of Duverdrey & Bloquel of Saint-Nicolas-d'Aliermont. The clock shown has an* Obis *timepiece movement with the Lion trademark prominently displayed on the back movement plate. (By courtesy of Prestons Ltd. Bolton)*

Plate VII/38. *ROCOCO. Date circa 1880. Maker J.B. Beguin. A fine quality clock with the case cast and finished in high relief. The engraved dial plate has calendar apertures to show day and date. (Private Collection)*

179

VII Case Styles

Plate VII/C13. MARBLE PANELS. *Date circa 1900. Maker Drocourt. The panels are of Italian origin following the Florentine* pietra dura *tradition. (By courtesy of E. Pitcher & Co.)*

Plate VII/C14. MARBLE PANELS. *A side view of the clock with marble panels shown in Plate VII/C13. (By courtesy of E. Pitcher & Co.)*

Plate VII/C15. MINIATURE CARRIAGE CLOCKS. *Period before and after 1900. The largest clock shown is size Zéro, height five inches. (By courtesy of E. Pitcher & Co.)*

Plate VII/C16. MINIATURE CARRIAGE CLOCKS. *The smallest clock is a sub-standard size below* Mignonnette *No. 1. (By courtesy of E. Pitcher & Co.)*

VII *Case Styles*

MINOR CASE STYLES

At one time and another, there have been made carriage clocks, quite individual in appearance, yet difficult to classify. Good examples are the ivory-cased clocks and those small "square" examples made in silver.

Ivory Cases

We are inclined to think, with no evidence to support the notion, that these clocks may have been made primarily for dressing-cases or for dressing table sets. They are timepieces only with small unsigned movements marked "Made in France". The case of the example in Plate VII/39 is certainly at least partly British, since it has silver joints (hinges) bearing the English date letter for 1925; but the ivory part may well have come from the Dieppe area. The movement was undoubtedly made in Saint-Nicolas-d'Aliermont.

Square Cases in Silver or Pewter

The silver case of the semi-miniature clock illustrated in Plate VII/40 carries the English date letter for 1901. The height with handle up is approximately three and a half inches. On the back plate of the "works" is stamped "FRENCH MOVEMENT". There is no evidence which firm made the movements of these clocks, but it would be very safe to assume that they came from Saint-Nicolas-d'Aliermont. The *mouvement* follows *Obis* practice, scaled down, having internal motion work and no false plate for the dial.

Plate VII/39. *IVORY. Anno 1925. The ivory case (probably French) has silver hinges bearing the English date letter for 1925. The timepiece movement is French. One reason for the use of ivory would have been its comparative lightness. (By courtesy of E. Pitcher & Co.)*

Plate VII/40. *SILVER CASE. Date 1901. The timepiece movement is French. The case is English. (Private Collection. Photograph by P.H. Kohl)*

VII Case Styles

Such clocks were sold by smart shops in the West End of London until before the Second World War.[6] The silver cases were, of course, made in this country.[7] At a later period, Liberty & Co. imported similar movements, casing them in English pewter and selling them under the name of "Tudric". These clocks have copper-coloured chapter rings set against stippled backgrounds of blue and green ceramic.

DECORATIVE PANELS

Whilst it is admissible to show only a few of these highly desirable clocks within the compass of this book, their limitless variations in both materials and design, and often exquisite workmanship, ensures their continuing appeal. Such clocks never fail to create interest. Moreover, they attract many who could never be moved by anything purely mechanical. These decorative-panel clocks were made mainly in the last quarter of the 19th century. They were never numerous, but not surprisingly most examples seem to have survived.

Cloisonné Enamel. (Enamel fired into separate compartments, divided by thin metal walls).

The coloured illustration Plate VII/C2 shows a superb example of a *cloisonné* enamel clock. The case is decorated almost literally all over. The stylised, floral design is mainly in red, pink, blue, green and white. Another clock is seen in Plate VII/C3. This style of enamelling originated in the Far East.

Limoges Enamel

Plates VII/C4 & VII/C5 show a type of world-famous enamel which is extremely rare in clocks. The Limoges factory in the town of the same name south of Paris is still in business. In these enamel carriage clocks, the enamelling was often carried to the dial and not confined to the panels. The standard work on Limoges enamel is, or was, *L'Art de l'Email de Limoges* by Alfred Meyer, published in Paris in 1895.

Guilloche Enamel. *(flinqué)*

The term *guilloche* enamel implies coloured, translucent enamel, fired over a metal object previously decorated with engine-turning, or with an engraved design. This style of ornamentation was very popular for carriage clocks, especially in the smaller ones such as the example in Plate VII/C6. Watch case backs, bezels and even dials were sometimes given the same form of decoration, and so were the backs of ladies' hairbrushes, etc. The French verb *guillocher* means to engine-turn or to chase, and is usually applied to a precious metal.

Porcelain Panels

Carriage clocks with decorative panels of any kind were always expensive, and in consequence they were never made in large quantities. Among the most attractive are those having porcelain panels with painted scenes and figures. Happily, such clocks are not impossible to find, probably because they have always been treated as treasured possessions. The clock shown in Plate VII/C7 has "china" panels top, front, back and sides adorned with Italianate figures in red, blue, green, yellow and white. The sides are edged in turquoise blue, overpainted in a gilt looped pattern. Both panels and dial are decorated with half "stones", in alternate red and white. The dial, which bears the name of the vendor rather than that of the maker or producer, is centred with climbing roses. The date of this clock is circa 1870-1880. Plate VII/C8 illustrates a rather similar piece with painted panels front, back and sides. The sides show lovers and Cupid; and the dial a shepherd, shepherdess and sheep. The panels are edged in broad gilt lines, highlighted by applied blue split-beads set at equal pitch. The colours are soft. Note the unusual carrying handle. The maker was Margaine and the date probably circa 1870-1880. Plate VII/C9 shows a very rare clock indeed. Here the decorative panels are after Coustou's *Chevaux-de-Marly*. The predominant colours are sepia and white. This clock is unsigned but its date is circa 1885. Plate VII/C10 depicts a farmyard scene. The dial, sides and back of the case are set with panels in porcelain illustrating calm, restful scenes with ducks, geese and chickens. It should be no coincidence that exactly these views exist even now on all sides of Saint-Nicolas-d'Aliermont. The piece has neither name nor maker's mark. It is quite a modest clock in a

---

(6) See, for instance, Catalogues of Mappin & Webb.

(7) Another type of clock to appear in an all-silver case was a Swiss timepiece with added minute-repeating work based upon a pocket watch. Further notes on this ingenious artifice are given in Chapter X.

VII   Case Styles

Plate VII/41. *AFTER-PAINTED DIAL. This* Corniche *clock began life with a plain white dial!*

Plate VII/42. *CHINOISERIE PANELS. Date circa 1880. Maker Drocourt. This clock was intended to complement Chinese furniture and porcelain. The gorge case has a typical Drocourt handle. (By courtesy of E. Pitcher & Co.)*

simple case and offering hour-and-half-hour strike and repeat. Plate VII/C11 shows yet another attractive porcelain panel variant. Both clock and case are basically quite standard, but use is made of Delft china where normally there is glass. The design depicts pansies or violas. The movement has neither maker's name nor mark, but there is no doubt that it was made in Saint-Nicolas.

A number of French carriage clocks have come to light recently which were never made with painted panels, but which now have them. It is worth remarking that such panels are usually quite inferior, and that many of them, upon close scrutiny, will be found to be nothing more than transfers. The results are not unattractive; but a clock seen to have these recent decorative additions must not be confused with a genuine article. Plate VII/41 shows an overpainted dial originally sold as plain white enamel. The difference should be obvious. On the other hand, there exist modern blue replacement panels, embellished with gilt

183

VII Case Styles

decorations, which are well executed and in consequence are far harder to detect. *Caveat emptor*.

Porcelain panel clocks, on the whole, are a notable exception to a very accurate generalisation that most French carriage clocks were made for the English market. In fact many decorative-panel clocks were exported to the Far East, particularly to China. In recent years they have begun gradually to come back from the Orient to Europe by the "back door". The returning wanderers, dating from about 1880, often bear Hong Kong names on their dials.

Chinoiserie Panels on Copper

See Plate VII/42. Here a copper dial and side panels of the same material are enamelled in colour. The background is brown with figures, birds and Chinese junks overpainted in gilt and silver. The dial chapters are in the appropriate Chinese characters. Note that the silver painted chapter ring is not divided into minutes. The hands are gilt, and the brass case is polished and lacquered. Not unexpectedly, the maker was Drocourt, while the date of manufacture would have been circa 1880. A rare clock.

Shibyama Panels (Japan)

Plate VII/C12 shows a most attractive clock with ivory dial and side panels decorated with applied mother-of-pearl and other semi-hard and beautifully coloured materials. The result is delightful. It is something of a surprise to find such comparatively expensive decoration allied to a *Corniche* case, housing a fairly ordinary movement. The result, however, is most satisfying. The date is post 1900. There is no evidence that these clocks were intended for sale in Japan.

Marble Panels

This mode of decoration is rare in carriage clocks. We illustrate in Plate VII/C13 a very fine example where black marble with a coloured floral inlay is used for dial, sides and back. The sides of the case, as seen in Plate VII/C14 depict lily-of-the-valley, roses and blue flowers apparently of the Boragaceae family. The back concentrates on fuchsias. The panels almost certainly came from Florence. The date of the beautiful clock illustrated is circa 1900; but we have seen another, No. 1591 made by Jacot and sold by Klaftenberger, which may be accurately dated to the years between 1862 and 1867. This last clock is also interesting as having a particularly beautiful engraved *gorge* case, a style fairly new at the time.

CARRIAGE CLOCK SIZES

The standard French carriage clock was thoroughly established by the middle of the second half of the 19th century, with England the principal market for what in those times amounted to a very substantial production. In the interests of economy most manufacturers tended to adhere to a few standard sizes, although there were many exceptions, particularly towards the end of the century when demand was at its greatest. In the main, carriage clocks may be divided for convenience of reference into three size classes. These are "miniatures", "full-size" and "giants".

Miniature Carriage Clocks. (*mignonnettes,* or "Little Darlings")

There were three standard sizes of miniature carriage clock. The smallest was *Mignonnette No. 1,* which had a height of only three and three quarter inches with handle up. The next size was *Mignonnette No. 2,* four inches tall. The third and largest miniature was *Mignonnette No. 3,* which was four and a quarter inches high. Smaller clocks, sometimes very small indeed, were offered by individual makers. A few examples were also made in sizes between *Mignonnette No. 3* and the "full-sized" *Obis*. The commonest of these was the *Petit Zéro,* a semi-miniature of height four and five eighths inches, and the *Zéro,* height five inches. Before the *Obis* became so popular, the *Zéro* was the most common standard size.

A fact which will be found confusing is that certain makers, notably Duverdrey & Bloquel, in their Catalogue, instead of depending upon overall height, group all manner of cases into size categories as determined by their movements. Thus under the heading *Mignonnettes No. 1* are found five case shapes. Each shape will house an identical movement, while differing from the others in overall height. The explanation of this apparent contradiction and overlapping of sizes is that different designs in order to look right inevitably vary in their external proportions, while offering the same "box" inside. *Mignonnette No. 2* in the Duverdrey & Bloquel Catalogue appears in no less than six shapes of varying sizes. Each houses an identical movement, although larger than that

184

found in *Mignonnette No. 1*. Similar variations are found under each *Calibre*. Duverdrey & Bloquel, and others, offered several "bastard" sizes, such as *Zéro M* (five and a quarter inches) and *Zéro S* in four sizes from four and a half to five and a half inches tall. It is hoped that this question of sizes has not seemed to be over-laboured. Carriage clock size conventions were evidently elastic. As old catalogues come to light more accurate information will become available.

Almost every "full-sized" carriage clock had its counterpart in miniature. *Gorge, cannelée, Corinthienne, Corniche* and even *cariatides* appeared as *mignonnettes*. Even various examples having porcelain or enamelled dials and cases were offered in a scaled-down form. Plates VII/C15 & VII/C16 show particularly choice and uncommon selections. *Mignonnettes* will usually be found to be timepieces only, but occasionally they have alarms. Very occasionally an example will be found to strike hours and halves on a bell in the hollow base.[8] Rarest of all are the miniature repeating *grande sonnerie* clocks, such as the *Mignonnette No. 1* in pillared case found by Dr. Burnett at the end of 1973. The carrying handles of most of the very small cases are perforce rather large in proportion to the clocks. In consequence *mignonnette* handles tend to look slightly out of scale.

Full-sized Carriage Clocks

The term "full-sized" carriage clock embraces the vast majority of all those ever made. The commonest of these was the *Obis*, that most standard of all standard clocks. As the reader will know, the model *Obis* was by far the cheapest ever made, being mass-produced in Saint-Nicolas from about 1880 until the outbreak of the Second World War. *Obis* clocks were offered in two forms. The cheapest sort were timepieces only, while the rather more expensive clocks were timepieces with alarm. In order to save expense, the enamel-on-copper dials were pinned directly to the front plates of the movements, the motion work and alarm work (if any) being accommodated inside the movement plates.[9] *Obis*, unlike the *Corniche*, was never offered by the best makers. Its manufacture was largely confined to such Saint-Nicolas firms as Couaillet and Duverdrey & Bloquel who specialised in the mass production of run-of-the-mill clocks. It was not customary for either maker's name or mark to appear upon an *Obis* clock, although some of them were given serial numbers. As already explained, the standard *Obis* clocks were produced with characteristic inexpensive cases. This type of case, or rather its shape, was also called *Obis*. It was made in one size only, namely five and a half inches tall. Almost all *Obis* clocks were furnished with cylinder escapements having plain brass balances.[10] These cylinder platform escapements performed very well when the clocks were new.[11] Duverdrey & Bloquel, in their Catalogue of circa 1910, illustrated at least eighty case styles of varying sizes, but all having *Calibre Obis* movements.

Other "full-sized" cases in common use were *Calibre No. 1* (six inches tall), *Calibre No. 2* (six and a half inches tall), and *Calibre No. 3* (seven inches tall). Duverdrey & Bloquel, however, in their heyday offered some thirty-seven cases of varying sizes and appearances based on *Calibre No. 1*, seventeen on *Calibre No. 2*, and thirty-nine on *Calibre No. 3*. The Duverdrey & Bloquel selection ran the gamut of just about every known carriage clock shape and appearance, and plenty of "unknown" ones as well. Conspicuously absent from the array are the two most expensive standard cases, namely the *gorge* and the *cannelée!* The high-class makers using *gorge* and *cannelée* cases tended to stick rigidly to the standard sizes. *Calibres Nos. 1, 2* and *3* were by far the most usual sizes for *gorge* clocks, although this beautiful shape will also be found both in miniatures and in "giants".

Giant Carriage Clocks

Very few really large carriage clocks were ever made in France. When they were, the conventional size will

---

(8) Renneson filed a *Brevet* on 25th April 1868 in this connection. On 18th February 1869 Margaine recorded an alarm applicable *aux pièces d'horlogerie et spécialement aux pièces de voyage dites mignonnettes*. "Leroy" had a *Brevet* dated 15th February 1869 for *Application d'un réveil à sonnerie aux pendules dites mignonnettes*.

(9) The bells of *Obis* alarms were housed in the bases. There was no room for them anywhere else.

(10) Some, however, were sold with lever escapements.

(11) The current practice of exchanging cylinder platforms for modern lever ones is only admissible, if at all, when a cylinder is hopelessly worn. At best, a clock is to some extent spoiled by being given a modern escapement, although better timekeeping undoubtedly results.

VII Case Styles

be found to have been nine inches tall with handle up. We have noticed "giant" versions of four basic cases. These are early "multi-piece" (Plate II/37), *Corniche*, *Anglaise,* and above all *gorge*. A particularly good example of the latter shape was sold at Sotheby's in London on 14th December 1970. It has *grande sonnerie* striking and alarm, in addition to a calendar showing day of the week and date of the month. The maker was Jacot. Very occasionally the really large clocks were given coloured porcelain dials and side panels. Plate VII/43 shows a *gorge* "giant" clock beside a standard *No. 3* "full-sized" clock.

Plate VII/43. *"GIANT" CARRIAGE CLOCK. The clock in the* Anglaise *case (left) is seven inches tall. It is size No. 3 and thus it is the largest standard "full-sized" carriage clock. The clock in the* gorge *case (right) is a standard "giant", nine inches tall. (Private Collection)*

CONCLUSIONS

To sum up, the *Obis* and the *Corniche* models were by far the most commonly produced varieties of French carriage clocks. *Obis* and *Corniche* are, however, easily confused. In theory at least, *Corniche No. 1,* while maintaining very much the same external profile as the *Obis,* is deeper from front to back in order to allow room for a movement with a false plate. In fact no model of French carriage clock was ever made in so many different qualities as was the *Corniche*. The best examples of this model have very well made movements, and also substantial and well-proportioned cases. At the other end of the scale altogether it is possible to find examples of *Corniche* clocks which must have cost little more to produce than *Obis,* and which have very modest movements housed in thin-section cases, with *Obis*-type handles,[12] and with *Obis*-type un-ornamented base arches. *Corniche* design and quality depended upon the maker and upon whether a clock was *genre Paris* or simply a *pendule de voyage économique*. The presence or absence of a false dial-plate is by far the most reliable guide as to whether a plain cheap clock is a *Corniche* or an *Obis*. The handle in Plate VII/7 is typical of standard *Obis*, while that in Plate VII/14 is characteristic of standard *Corniche. Corniche Carrée* cases necessitate the use of movements smaller than would normally be found in carriage clocks of the same width.

Among clocks not produced upon such a large scale as *Obis* and *Corniche,* the opportunities for variation in design were enormous. The more elaborate a clock case, the more difficult it may be to classify. Individual makers strove to produce shapes and designs different from their competitors, and often gave names of their own to a particular model.

In general, for the standard types white enamel dials were preferred. These suited the cheaper and plainer cases, besides being comparatively inexpensive to produce. On the other hand, as it is hoped has already been made very clear, enamel dials of really high quality with finely-drawn Roman chapters were also largely used by top firms such as Jacot and Drocourt for their most expensive clocks. These pieces normally had *gorge* cases, which were as costly as they are restrained. The later white enamel dials had bold and heavy chapters, which in our view are far less elegant. Elaborate non-standard cases, whether for miniature or for larger clocks, tended to be accompanied by more fanciful dials. Ornate dials were also sometimes used in conjunction with quite standard cases in order to underline that a clock was better than usual. Clocks with *cloisonné* enamel panels are

---

(12) Standard *Obis* handles, unlike those of most other models, are "sprung" in place instead of being pinned. Their joints are not provided with stops to prevent their falling forward. Handle stops, latterly found on most clocks except *Obis*, were first employed when top repeat buttons became common.

VII Case Styles

often found with dials having gilt surrounds, either plain or engine-turned, or with gilt filigree centres, silvered chapter rings and Arabic numerals. Enamelled and similarly expensive cases were usually, but not always, reserved for expensive movements. Hands, in the case of the earlier makers, were most often "moon" (more properly called Breguet hands) or else they were trefoil (having three lobes). Spade and fleur-de-lys hands came later. Handles were made in very wide variety, usually (but not always) in keeping with the style of the case.[13] In general, *Obis* cases tended to have two cuts in the handle, *Corniche* two or three, *cannelée* three and *gorge* five. Plate VII/8 shows the handle usually found on the *gorge* cases of Jacot clocks. Note that the sides of the handle are more or less vertical when it is raised. Plate VII/42 shows a *gorge* case with a handle typical of Drocourt and having bowed sides. Plate VII/C8 depicts a handle favoured by Margaine for engraved *gorge* cases with decorative panels. We should expect a Margaine clock in a plain *gorge* case to have a handle like Drocourt's in Plate VII/42, and not one like the Jacot clock in Plate VII/8. It is probable that this shape was confined largely, if not entirely, to Jacot.[14] Square-section handles were favoured for *Anglaise*, *Corinthienne* and sometimes *Corniche* cases. Handles which were enamelled were invariably "flatted" to show off the work.

The variety of cases and dials produced in the various sizes was so great that no absolute rules will be found. Individual makers, in their catalogues, gave their own names to their own particular *boîtes*, alternatively called *cabinets*, within and without the basic types and sizes.

When aluminium was first developed in Paris, from about 1827, it was used for jewellery, and was just as expensive as silver. From 1854 aluminium was produced in bulk; that is to say, in the form of blocks instead of in fragments as previously. Louis Raby showed in Paris in 1855 a gold pocket chronometer of which the movement bars were made of aluminium. In 1886 new production methods made aluminium available cheaply everywhere. A fine *grande sonnerie* carriage clock, No. 15166, exists having an aluminium *gorge* case produced between 1889 and 1899. The

Plate VII/44. *CORNICHE. Another type of dial favoured for use in conjunction with Corniche cases. Period post 1900. (This plate is not mentioned in the text.)*

maker was "L. Leroy & Cie".

In the past year or two, paper dials have been used in the trade as replacements for badly damaged enamel-on-copper ones. The new dials, which perhaps are less objectionable than the old, when the latter are very badly cracked or patched, are often hard to detect. They do, however, tend to look rather like the photographs which they are, and also to have an "orange-peely" look when viewed along their surfaces.

CARRIAGE CLOCK PRICES

Thanks largely to the initial example set by Paul Garnier, French carriage clocks, other than special and complicated examples, were not unduly expensive

---

(13) The two manufacturers whose clocks showed the greatest diversity in handle design were Margaine at one end of the scale, and Duverdrey & Bloquel at the other.

(14) Needless to say a Jacot *gorge* handle will occasionally be found to have bowed sides.

VII Case Styles

once the industry became thoroughly established. The mass production methods pioneered by Japy at Beaucourt and Badevel, and copied at Saint-Nicolas-d'Aliermont not long afterwards, produced an amazing situation in which the cheapest clocks could retail in England for as little as a pound or two. This still allowed a profit for manufacturer, importer and retailer besides an acceptable living for the workpeople by the standards of the time. Even such grand London shops as Charles Frodsham & Co.,[15] whose overheads at 84 Strand would have been relatively high, were able to offer in their Catalogue of circa 1893 the following high-class French *pendules de voyage* at retail prices:—

Corniche	Timepiece, with lever escapement and compensated balance.	£6. 6.0d.
Corniche	Striking hours and half hours. Lever escapement with compensated balance.	£9. 0.0d.
Gorge	Striking hours and half hours and repeating hours. Lever escapement with compensated balance. Alarm. Best quality.	£15. 0.0d.
Gorge	As above, but in an engraved case. Superior quality.	£15.15.0d.

Although the clocks are illustrated and their functions clearly described, the words *Corniche* and *gorge* are nowhere mentioned.

In 1907 the Army & Navy Stores in London offered in their illustrated Catalogue the following French carriage clocks. These clocks, obtained from various sources,[16] ranged from the expensive to the mediocre; but they all had lever escapements, and struck on gongs where applicable:—

Corniche No. 2	Timepiece.	£3.16.0d.
Corniche No. 2	Striking hours and half hours.	£5. 1.0d.
Corniche No. 2	Striking hours and half hours, and repeating hours.	£5.18.0d.
Corniche No. 2	*Petite sonnerie*. Breguet spring.	£14. 1.0d.
Gorge No. 2	Timepiece.	£6. 2.0d.
Gorge No. 2	Striking hours and half hours.	£7. 7.0d.
Gorge No. 2	As above, but with repeat.	£8. 2.0d.
Gorge No. 2	*Petite sonnerie*. Breguet spring.	£15.12.0d.
Anglaise (6¼in. tall)	Timepiece.	£2.10.0d.
Mignonnette No. 1 (Corniche)	Timepiece.	£4. 7.0d.
Anglaise Riche (7½in. tall)	with Corinthian columns. Striking hours and half hours, and repeating hours.	£9.16.0d.
Square Silver Case	English silver case. Eight-day French lever movement.	£4.15.0d.
Doucine No. 2	Striking hours and half hours, and repeating hours.	£10. 5.0d.

Once again, neither shapes nor models were named.

---

(15) Mr. Donald de Carle recalls that Frodsham's liked to fit English platform escapements to the French carriage clocks which they sold. These escapements produced by a man called Rock, mainly from parts bought in Clerkenwell, are distinguishable by a completely circular cock taking in the 'scape wheel and pallets. A specially-made escapement would, of course, make a clock considerably more expensive. It is known that Frodsham's obtained some of their clocks from Drocourt.

(16) M. Pitou, who is mentioned in Chapter II, says that the Army & Navy Stores obtained at least some clocks from Jacot.

VII Case Styles

At the other end of the scale from the high quality carriage clocks sold by Frodsham's, and to a lesser extent by the Army & Navy Stores, those shown in the retail Catalogue of S. Smith & Son Ltd. and published from soon after 1900 under the title *Guide to the Purchase of a Clock*, were almost all obtained from the inexpensive mass production factory founded by Albert Villon in 1867, and which in 1910 became Duverdrey & Bloquel. It is most interesting to see the same clocks illustrated both in the *Guide* and also in Duverdrey & Bloquel's Catalogue published circa 1910. As might be expected the clocks shown in the Smith's *Guide* are identified by number only. Smith's did not perpetuate the names given to the clocks by the manufacturer. Amongst discernible models and shapes shown in the *Guide* are the following:—

Corniche	Timepiece with cylinder escapement.	£1. 1.0d.
	Timepiece with lever escapement.	£2. 2.0d.
Corniche	Striking hours and half hours on gong.	£3. 0.0d.
Corniche	As above, but with repeat.	£3.15.0d.

No sizes are mentioned.

Doucine No. 2	Timepiece in ornate case,	
	with cylinder escapement.	£1.17.6d.
	with lever escapement.	£2.12.6d.

A miniature timepiece clock with a height of some three inches (called *Boîte Marquise* in the Duverdrey & Bloquel Catalogue) cost £3.15.0d. with lever escapement. Various pillared designs, with movements striking hours and halves and repeating hours, cost from about £6 to £8.10.0d.

So far as Trade prices were concerned, the manufacturers provided their agents with catalogues, of which one issued by Margaine towards the end of the last century, and others from Duverdrey & Bloquel circa 1910 and Couaillet Frères in 1914, are perhaps typical. Margaine, in a Trade Catalogue, did not even mention their own name, simply contenting themselves with displaying the famous "Beehive" on the price list page! For that matter the same Catalogue did not name even one of the beautiful clocks illustrated. These pieces, which all had lever escapements, were simply identified by catalogue numbers. The first and cheapest clock offered was a *Corniche*. It was available in sizes *Zéro* and *Zéro M*, priced from £2.18.0d to £10.10.0d. This scale ran the gamut from timepiece to *grande sonnerie*. The price list was so tabulated as to ensure that each clock could be ordered with any one of ten different movements of progressively increasing complication. A second cheaper quality of *Corniche* was added to the Catalogue in red ink. It was available only as a timepiece, or as an hour and half hour striker, or as a simple repeater. *Corniche* was also offered in size *No. 0* as a timepiece at £2.0.0d, and as a timepiece with alarm costing £2.10.0d. A Margaine *gorge* (Catalogue No. 21) cost from £4.2.0d. to £11.14.0d. Shown but not listed was a repeater having centre seconds. The most expensive clock listed, costing £15.14.0d. for a "Superior Quality" *grande sonnerie*, was No. 42 in the Catalogue. It is difficult to describe, but had a very ornate case with *cariatides* and with cast top, base and handle decorated with formal designs in a sort of baroque style, most carefully finished in high relief. The base was centred with a mask, while the dial plate was overlaid with pierced strapwork representing foliage. The striking clocks struck upon gongs, and the prices quoted provided for ". . . clocks with ordinary enamel dials . . . delivered examined and timed in morocco cases". A list of extras included special dials and side panels, besides bevelled dial surrounds. *Grande sonnerie* clocks, with and without alarms, were offered in two distinct qualities. The one was described as "1st Qty", and the other as "Superior Qty".[17] The superior quality added one pound to the cost of any clock. Clocks not having *grande sonnerie* striking, (with the notable exception of "2nd Qty" *Corniches*), were

---

(17) Thus Margaine's cheapest clocks were described as "2nd Qty", their standard clocks as "1st Qty", and their best clocks as "Superior Qty". To avoid confusion the different sorts of movements were numbered 0 to 9, while the case numbers started at No. 11. See Chapter V where a page from this Catalogue is reproduced.

VII Case Styles

offered in "1st Qty" only. Margaine's price structures were no doubt little different from those followed by Jacot and Drocourt, although it may be that the two latter makers were rather more expensive. Jacot and Drocourt catalogues need to be found. The more complicated clocks, such as those having detent escapements, perpetual calendars, minute repeating, etc. must have been proportionately more expensive. It would be interesting to know the cost of such special pieces as those made by Oudin-Charpontier and by Bourdin. The prices of Breguet *pendules de voyage* were of course astronomic.

A noted wholesaler serving the horological trade and based in Clerkenwell used to be Grimshaw, Baxter and J.J. Elliott. One of their catalogues dated September 1912 shows a selection of inexpensive carriage clocks, apparently emanating both from Duverdrey & Bloquel and from Couaillet Frères. Couaillet's clocks, as shown in his Catalogue and Price List of 1914, were offered as *Qualité soignée* when they had cylinder platforms, and as *Qualité supérieure* when they had lever.

As a comment upon prices in general, it is worth remarking that the provision of lever escapements greatly increased prices of clocks at the lower end of the price range. At about the same period, when the cheapest *Corniche* timepiece could be retailed for a guinea and an *Obis* for even less, the weekly rent of a farm labourer's cottage ranged from about one shilling to half a crown.[18] Farm labourers earned some ten shillings a week. The wage of an ordinary watch repairer close to the turn of the century was £1.15s. a week, while clock repairers were paid rather less. According to the book *British Labour Statistics....* published by the Stationery Office in 1971, a fitter and turner earned £1.18.0d a week in London in 1900, and £2 in 1910. A bricklayer's labourer earned £2.10s. in 1900 and the same in 1910. By the same year, wages of agricultural workers had crept up to 15/4d. a week. A professional person, such as a solicitor, might have expected to earn a maximum of about £500 a year at the turn of the century. In contrast a County Court Judge was in receipt of £1,500 per annum by the year 1891.

On the whole only well-off and middle-class people bought carriage clocks.[19] The fact is that travelling clocks, like travel itself, were always in the nature of a luxury.

After this chapter had been set in type, a collector in France discovered a Delépine-Barrois Catalogue of carriage clocks. This Catalogue shows clearly that Delépine-Barrois, at least, styled as *cannelée* the shape of case which English importers have been brought up to call *gorge*. In the same Catalogue a case bearing strong resemblance to the one listed in this chapter as *cannelée* is styled *gorge*. The Delépine-Barrois Catalogue may or may not prove that the names of two case shapes have become confused over the years. Before being certain, a number of other manufacturers' catalogues will first have to be "un-earthed" and consulted!

---

(18) Flora Thompson *Lark Rise to Candleford*.

(19) What clocks did humble people have at about the turn of the century? They had American thirty-hour Ogee shelf clocks which were so reliable, inexpensive and well-distributed that one found a place in virtually every cottage in England. The Yankee clocks, with their lantern pinions and short trains, would run for years on end without any attention; and when one did stop, a feather dipped in the paraffin lamp and brandished about indiscriminately in the "works" usually sufficed to set it going again. Even today many such clocks are telling the time for elderly people in out-of-the-way places. The Yankee clocks were soon followed by even cheaper German and French alarms.

Chapter VIII

The Rarer French Carriage Clocks

DIGITAL DIAL. See Plate VIII/19.

VIII

The Rarer French Carriage Clocks

Alarm Work. Striking Work. Carriage Clock Trains. Calendar Work. Digital Dials. "Flick" Clocks. Unusual Escapements. Tourbillons and Karrusels. Four-Dial and Two-Dial Clocks. Bottom-Wind Clocks. Singing Bird Clocks, Musical Clocks, Sundial Clocks, etc. Year Carriage Clocks.

This part of the book could all too easily become unduly large. A satisfactory compromise seems to be to include such known peculiarities as will interest readers without depriving them of the pleasure of their own discoveries. In horology it is never wise to say that something was never done, or equally that something was always done, because the next day the exception will be found. This generalisation certainly applies to carriage clocks.

ALARM WORK

There is evidence that some of the earliest Italian clocks made before the year 1500 were provided with alarm mechanisms, while at a later date many German, French and English clocks and watches often had alarms almost as a matter of course. An alarm was a fundamental requirement in times of primitive lighting, and even our present ways of life have done little to lessen the need for such a device. It was natural that many carriage clocks, intended alike for journeys and for use upstairs and downstairs, should be provided with alarm mechanisms.

In its simplest form, the alarm carriage clock has a separate alarm train and mainspring, which must be wound whenever the "call" is required. A small 12-hour setting-dial determines when the device will "go-off". This exact arrangement is common to nearly all makers. It differs in principle from an ordinary modern "alarm clock" only in that, once set off, it cannot be silenced until the spring has run right down. The clocks in Plates VII/5 and VII/6 are of this type. At least one French manufacturer, however, offered a more refined variant. Plate VIII/1 illustrates the movement of a clock "signed" on the back plate "D.L.B.". It is of quite modest quality, being a timepiece only with *Obis*-like case and cylinder escapement; but it

Plate VIII/1. *TWENTY-FOUR HOUR ALARM MOVEMENT. This alarm is kept wound automatically by the going train. Maker Delépine-Barrois, Saint-Nicolas-d'Aliermont. The clock is housed in an* Obis-*like case, and stands only 4 inches high. (Private collection. Photograph by P.H. Kohl)*

VIII   *The Rarer Carriage Clocks*

has a 24-hour alarm, kept wound by the clock itself, and sounding automatically for a few seconds once a day so long as the clock is going. There is a lever to render the alarm inoperative. Altogether it is a great improvement upon the conventional arrangement.[1]

A small clock, having neither number nor maker's name, is illustrated in Plates VIII/2 and 3. It is unusual in having the alarm barrel planted on the right side as seen from the back.[2] Note also the unconventional arrangement by which the tail of the alarm click

Plate VIII/3. *THE MOVEMENT OF THE CLOCK IN PLATE VIII/2. Note the position of the alarm train. The bell is in the base. (Photograph by P.H. Kohl)*

Plate VIII/2. *MINIATURE TIMEPIECE ALARM. A good quality clock possibly made in the Franche-Comté. The lever escapement has double roller and the clock is provided with a calendar showing day and date. (Private collection. Photograph by P.H. Kohl)*

spring does duty as alarm setting detent. This clock is not a striker, and the bell is hung in the base below the movement.

One type of clock has as its alarm a musical box playing tunes. Marchand took out *Brevet* No. 127,663 in 1878 for such a clock. In the same year Ollier took out *Brevet* No. 116,868 for another variant. Several firms produced carriage clocks incorporating music boxes unassociated with any alarm work. In these clocks the tunes were either set off at intervals by the motion work, or else they could be played at will. See notes later in this chapter.

---

(1)   There was an American 8-day alarm clock made up on the same principle.
(2)   But see the Franche-Comté clocks shown in Plates VI/30 and VI/34 and also the Swiss clock in Plate X/10, all of which have alarm barrels in this position.

VIII The Rarer Carriage Clocks

A type of portable alarm included in this book only because it is old and also of travelling clock form is that shown in Plate V/26. Plate V/27 illustrates the movement. These clocks, produced circa 1840 at a time when the manufacture of carriage clocks was thoroughly established, were made only as cheap, thirty-hour pieces designed to employ obsolete watch movements. These movements, invariably full plate verges, were often old stock of Swiss origin[3] made out of date by the introduction of the Lépine caliper for watches at the end of the 18th century.[4] Sometimes these movements were taken from watches broken up for old gold when their appearance came to be considered too old-fashioned. The alarm mechanisms, usually planted in rectangular plates, were completely separate from the watch movements, and were "tripped" by means of a "floating" hour wheel, notched to release a spring loaded detent at a pre-set time. No bells were employed. Instead double-ended hammers hit the bottoms and one side of the clock cases. The clock illustrated is French, but an English counterpart exists. The English clocks were not usually provided with alarms. They were closely related to the so-called "cottage clocks", as they were to the so-called "Sedans".

Although alarm mechanism was so fundamental to carriage clocks, it was always an extra. Almost any model with or without other complications was available with or without alarm.

STRIKING WORK

Even the earliest ordinary French carriage clocks were usually hour and half hour strikers. This type of striking was a legacy from the everyday *pendules de Paris*. Many carriage clocks were also made to repeat at will. Both striking and repeating work tended gradually to become increasingly ambitious in the more expensive clocks, but on the whole the process was slow. Few makers, with the exception of Breguet, offered really complicated clocks until well into the second half of the century. Non-striking or timepiece carriage clocks did not appear in quantity until the advent of better lighting and a latter-day market for a less expensive clock created a demand for them.

Hour and Half Hour Strike

This is by far the most usual form of striking. The clocks strike the hour at each hour in passing. They also strike one blow at each half hour, usually on the same bell or gong. Hour and half hour striking in *pendules de voyage* is derived directly from *pendules de Paris* practice in that it employs conventional "warning" before the hour and half hour. Most early clocks which are not repeaters have this arrangement. When a clock is required to repeat, the "warning" type of striking work is not practicable, because when the clock is "on the warn" the striking train is temporarily rendered inoperative. To overcome this disadvantage, repeating carriage clocks employ a special type of striking work known as "knock-out striking". This enables the clock to be repeated at any time right up to the hour. It should be noted that once *blancs-roulants* were made in quantity "warning" striking virtually vanished. In order to standardise production as much as possible the same "knock-out" striking was also used in non-repeating clocks.[5]

Petite Sonnerie

Quarter chiming clocks are far less common than those with hour and half hour strike. At each hour *petite sonnerie* clocks strike the hour on one bell or gong of a low note. At a quarter past the hour they strike one double note, "ting-tang"; at half past, two "ting-tangs"; at a quarter to the hour, three "ting-tangs". The "tings" are sounded on a higher pitched bell or gong. *Petite sonnerie* clocks were customarily sold with repeat. When this type of clock is repeated, it usually strikes the last hour followed by the last quarter (i.e. it repeats *grande sonnerie*). Quarter striking clocks will mostly be found to be of good quality. They were never made in large numbers, and were considered special and exclusive.

Grande Sonnerie

At each hour the clock strikes the hour only. At each quarter it strikes first the last hour and then the

---

(3) Béliard, writing in 1767, makes quite clear that from 1750 onwards most French watches were in fact Swiss. See Chapter II.
(4) Lépine's improvement, which revolutionised watchmaking on the Continent, employed the so-called bar movement resulting in much thinner watches than were possible when two plates were used.
(5) Not always, however. Delépine-Barrois still used "warning" after 1910.

VIII *The Rarer Carriage Clocks*

appropriate "ting-tang" quarter or quarters. This pattern seems to be almost the unvarying practice in French work. Austrian and English clocks, on the other hand, customarily strike quarters first followed by the hour. The Vulliamy clock shown in Plate IX/13 strikes both hour and the four quarters at the hour. Swiss early carriage clocks customarily offer *grande sonnerie* striking during the night and *petite sonnerie* during the day. Such a clock is illustrated in Plate X/5. See Chapter X, Swiss.

Grande and *petite sonnerie* clocks as a rule may be silenced at will, and *grande sonnerie* may be reduced to *petite*.[6] Usually a lever under the base, or a set-square inside the back door, provides alternative settings. Very occasionally *grande sonnerie* clocks offer hours only as an alternative to full striking, instead of *petite sonnerie*. Yet others again were designed to strike the four quarters at each hour, after the hour had been struck. The famous maker Jacot sold some clocks of this kind. *Grande sonnerie* clocks were made in comparatively small numbers, mainly by the best manufacturers. It is unusual to find one not of really high quality.

Grande and *petite sonnerie* clocks use "knock-out" striking with the addition of a further rack to count the quarters. Descriptions of the actual mechanisms of striking and repeating clocks are not included here in order to keep the size of this book within some bounds. The basic principles are, however, described in a number of books, which will shortly be mentioned and which are readily available.

Half Hour Sonnerie

Variations on the *grande/petite* theme are by no means common, but they certainly exist. For example, Jacot made clocks now known as "half hour *sonnerie*". These pieces strike full *grande sonnerie* in passing; but only at the hour and half hour, and not at the quarters. In other words, they are silent at quarter-past and at quarter-to the hour. When repeated, "half hour *sonnerie*" clocks strike hours and quarters for each quarter. Alternative settings are not provided. One French clock of this type exists sold by Rossel et Fils, successor to J.F. Bautte et Cie, Geneva.

Westminster Quarters

The so-called Westminster quarters, familiar to everyone as the chimes of "Big Ben", are more properly termed "Cambridge Quarters". This familiar sequence is supposed to have been composed by Dr. Crotch in 1780, based upon an air by Handel. Lord Grimthorpe heard the chimes of Great St. Mary's in Cambridge and adopted them for the Great Clock at Westminster. Carriage clocks having Cambridge quarters are exceedingly rare. They are really a more expensive variant of *grande* and *petite sonnerie*. Plate VIII/4 shows a very fine quarter-chiming carriage

Plate VIII/4. WESTMINSTER QUARTERS. *This French* grande sonnerie *clock was made for Henry Capt of Geneva. The Westminster quarters are chimed on four small bells. The hours are struck on a gong. (By courtesy of Charles Terwilliger, New York)*

---

(6) The clock shown in Plate VIII/13 is unusual in that it strikes either *grande sonnerie*, or else the hours only, as alternative settings to "Silence". Richard & Co. of 24 Cannon Street published in the *Horological Journal* of November 1884 their Patent system for French quarter carriage clocks. In this arrangement first the hours and then the quarters were sounded in the normal way at each quarter, but at the hours the four quarters were struck first in order to warn the listener that an hour was about to be struck.

VIII The Rarer Carriage Clocks

clock. It offers *grande sonnerie,* sounding the quarters on four bells while striking the hours on a gong. The maker was undoubtedly French. The four bells are located between the movement plates. Note the porcelain dial and side panels. The four shields on the dial represent France, Italy, Germany and Great Britain. The arms on the side panels represent the Holy Roman Empire and the U.S.A. Plate VIII/5 shows a side view of the movement of another clock this time offering a modified Cambridge/Westminster chime on four bells. The quarters are sounded by two identical eight note sequences, thus allowing the use of a small pin-barrel. The bells are situated in the base. The pin-barrel, vertical hammer arbors, etc. may be seen in the photograph. The barrel is "pumped" out of the way laterally while the hours are being struck. All striking sequences are controlled by a large locking plate situated (improbably) under the dial.

Five Minute Repeaters

These clocks are decidedly uncommon, being much prized by collectors. Five minute repeating carriage clocks fall into two main types. The first type employs a form of striking work differing in principle but little from that of the normal *grande* or *petite sonnerie* already mentioned. This type of five-minute repeater, instead of striking hours and "ting-tang" quarters when the repeat button is pressed, strikes instead first the last hour and then up to eleven single blows each representing a five minute interval. The

Plate VIII/5. *SEMI-WESTMINSTER QUARTERS. A side view of a movement showing the pin-barrel with its associated vertical hammer arbors. The locking plate, which controls the striking work, may be seen under the dial. (Private Collection)*

Plate VIII/6. *FIVE MINUTE REPEATER. The side view of a two-button movement, showing the spring-loaded rack arbor.*

VIII   *The Rarer Carriage Clocks*

second type of five-minute repeating carriage clock employs two repeating buttons and differs considerably in its mode of operation. In effect it is little more than a double-acting hour-striking clock. All that is necessary is that the rack shall have two tails and be able to fall either upon an hour snail (conventionally mounted under the dial) or upon a five-minute snail (mounted on the back movement plate) as determined by whether a front or a back repeat button is pressed. The front button sets off normal hour repeating work. The rear button moves a spring-loaded rack arbor laterally, allowing the second rack tail to fall upon a five minute snail. The rack tails are so arranged that only one snail will be "read" at a time. The rack arbor is biased so the clock will strike normal hours and half-hours in passing. Plate VIII/6, shows the side view of the movement of a two-button clock, and Plate VIII/7, shows a rear view of the same movement.

Plate VIII/8. MINUTE REPEATER. *Jacot No. 3707. (Private Collection)*

Minute Repeater

These clocks combine minute-repetition with conventional *grande sonnerie* striking. When the repeat button is pressed, the clock begins by striking the last hour; then it sounds the last quarter, and finally on a third gong are rapidly counted out the number of minutes elapsed since the last quarter. Jacot's name appears upon some fine examples. Plate VIII/8 shows Jacot's No. 3707. The clock strikes *grande sonnerie* in passing. The striking work is provided with three racks and snails. Two take care of the hours and quarters in the conventional manner, while the third,

Plate VIII/7. FIVE MINUTE REPEATER. *A view of the back plate of the clock shown in Plate VIII/6 showing the five minute snail and rack tail.*

VIII   The Rarer Carriage Clocks

Plate VIII/9. *MINUTE REPEATER. The under-dial work of the Jacot clock No. 3707 shown in Plate VIII/8.*

mounted close to the pillar plate, controls the minute blows. Plate VIII/9 shows the *cadrature* of the clock in Plate VIII/8. Note not only the complication of the under-dial work, but also the size of the striking mainspring barrel. The case is a pillar variant. Incidentally, neither case, handle nor engine-turned dial is in any way typical of Jacot. The movement of this fine clock is provided with a superior lever platform escapement with compensation balance and overcoiled balance spring.

From just before 1900 until about 1930 the Swiss made small minute-repeating travelling clocks employing watch mechanisms to sound the hours, quarters and minutes. In these *montres pendulettes de voyage* the depression of a long plunger wound up the repeating train. A few Swiss minute-repeating travelling clocks made at the same period had *grande sonnerie* and minute-repeating trains wound once a week, and set in motion by a mere touch of a repeat button. The two systems were quite different, although the clocks looked very similar in shape and size. They are discussed further in Chapter X, Swiss.

Double Strike

The Franche-Comté clock already mentioned in Chapter VI and shown in Plate VI/33, strikes each hour twice; once at the hour, and once again two minutes later. This particular example is the only double-striking carriage clock so far seen by us, although other examples must certainly exist. Double-striking was popular on the Continent. In Italy many church clocks strike twice, while Morez clocks (otherwise called *Morbiers* or *Comtoises*) are particularly known for this peculiarity. They were made in the French Jura close to the border of France with Switzerland. Plate VI/34 shows the back of the double-striking carriage clock. In conformity with the practice of the region, all the external work is planted on the back movement plate. The case style suggests a fairly early clock. The hands are set to time with the fingers rather than by a key because the layout of the striking work leaves no room for a hand-set square on the back plate.

Ship's Bell Strike

These clocks strike nautical hours, as determined by the Watches kept at sea. Plate VIII/10 shows a very good example, said to have been made about 1880. The French circular-plate movement strikes on two bells, and employs a locking plate. There is a pull cord below the base for correcting the striking to agree with the hands when the locking plate falls out of step. Le Roy & Fils supplied just before 1900 a specially ordered size *Zéro gorge*-cased carriage clock with rack striking for nautical hours on gongs.

Striking Work in General

Broadly speaking, the first French production carriage clocks struck upon bells. By the middle of the century gongs were well on the way to becoming standard, and finally they superseded bells altogether. Sometimes clocks having bells were later converted to strike on gongs in order to conform with fashion. Such conversions, besides spoiling the clocks, were not wholly satisfactory because "bell" striking trains and "gong" striking trains should have quite different counts. Gongs require a much greater interval between blows than bells in order to sound right. To convert successfully from bell to gong necessitates a new warning wheel, with more teeth, besides a new fly pinion with fewer leaves. Bell flys turn on average 72 times per blow, whereas gong flys turn between 90 and 94 times per blow. Such niceties were usually unrealised and unrealisable by converters.

Detailed and well-illustrated explanations of carriage clock striking are available in several works currently in print. In view of this fact it seems pointless to repeat this information. Striking work generally is covered very ably by such authors as Haswell, Goodrich, De Carle, and Gazeley, the last two dealing specifically with carriage clock work. As for carriage clock repair, in addition to the last two authors mentioned, Thomas Schmith published in the *Horological Journal* of 1894-95 an excellent series of articles entitled *Examining French Carriage Clocks*.

CARRIAGE CLOCK TRAINS

While French clockmaking practices were always radically different from the traditions followed in England, those differences by the 19th century had resolved mainly into a continuing English pre-occupation with the use of fusees, in conjunction with massive frames and wheelwork, as opposed to the preference on the part of the French for light and delicate trains almost always driven by going barrels. No doubt both clockmaking traditions were based upon sound considerations in the eyes of their adherents. After all, there is much to be said for both of these basic designs. For the fusee it was argued that better timekeeping resulted; although as a matter of fact, unless maintaining power was also provided, fusee clocks often actually went backwards while being wound. Going barrel clocks, on the other hand, kept going during winding even if they were supposedly inferior as timekeepers. As for the trains, it cannot be denied that the delicately proportioned wheels and pinions used by the French in conjunction with hard pivots of small diameter in most of their domestic clocks, are superior at least in theory to the often coarse and heavy trains found in run-of-the-mill English work.

In eight-day fusee clocks the great wheel drives the centre pinion. The use of a going barrel, however,

Plate VIII/10. *SHIP'S BELL STRIKE. This clock strikes nautical hours. (By courtesy of Ernest R. Conover Jnr., Ohio, U.S.A.)*

necessitates the introduction of an intermediate wheel and pinion between the barrel and the centre pinion in order to obtain a similar overall gearing ratio, and hence the same duration of going. This fact accounts for the extra wheel and pinion found in most French spring-driven clocks between the great wheel and the centre pinion.

Carriage clock striking trains, as already noted, had differing numbers of teeth in their wheels and pinions dependent upon whether a clock was required to strike on a bell or on a gong. Obviously as time went by and bells fell out of fashion, trains which struck their blows rapidly ceased to be made altogether. This apart, the one all-important consideration governing the design of any striking train was that there had to be available amply sufficient power to ensure that the full striking facilities offered would be equal to incessant use during the going period of the clock.

It would appear that no French carriage clock train counts are anywhere readily available in tabulated form. For this reason it seems worth while to list the numbers of teeth used in the wheels and pinions of a few examples.

VIII The Rarer Carriage Clocks

Going Trains

Obis Eight-day Timepiece (Balance vibrates 18,000 times an hour).

	Wheel	Pinion
Going Barrel	90	--
Intermediate Wheel	56	12
Centre Wheel	64	12
Third Wheel	60	8
Contrate Wheel	80	8
Escape Wheel	15	8

Jacot Eight-day Petite Sonnerie No. 3157 (Balance vibrates 18,000 per hour).

	Wheel	Pinion
Going Barrel	90	--
Intermediate Wheel	72	14
Centre Wheel	64	10
Third Wheel	60	8
Contrate Wheel	80	8
Escape Wheel	15	8

This is a very typical train-count for a good quality clock.

Eight-day Strike/Repeat (Balance vibrates 16,230 per hour).

	Wheel	Pinion
Going Barrel	66	--
Intermediate Wheel	64	10
Centre Wheel	64	8
Third Wheel	64	8
Contrate Wheel	68	8
Escape Wheel	15	8

Eight-day Timepiece (Balance vibrates 15,600 per hour).

	Wheel	Pinion
Going Barrel	80	--
Intermediate Wheel	72	10
Centre Wheel	72	8
Third Wheel	60	6
Contrate Wheel	52	6
Escape Wheel	15	12

Striking Trains

Hour and Half Hour Strike and Repeat on a Bell

	Wheel	Pinion
Going Barrel	80	--
Intermediate Wheel	63	10
Pin Wheel (10 pins)	60	7
Locking Wheel	54	6
(1 pin and carrying the gathering pallet)		
Fifth Wheel	48	6
Fly	--	6

(The fly turns 72 times for each blow of the hammer).

Hour and Half Hour Strike and Repeat on a Gong

	Wheel	Pinion
Going Barrel	82	--
Intermediate Wheel	75	10
Pin Wheel (9 pins)	72	8
Locking Wheel	63	8
Fifth Wheel	63	7
Fly	--	6

(The fly turns 94.5 times per blow of hammer).

Petite Sonnerie and Repeat

	Wheel	Pinion
Going Barrel	90	--
Intermediate Wheel	72	14
Pin Wheel (10 pins)	80	8
Locking Wheel	60	8
Fifth Wheel	63	7
Fly	--	7

Drocourt Grande Sonnerie and Repeat No. 9325

	Wheel	Pinion
Going Barrel	120	--
Intermediate Wheel	104	14
Pin Wheel (18 pins)	90	12
Locking Wheel	84	10
(with two locking pins)		
Fifth Wheel	72	7
Fly	--	7

It is interesting that two clocks out of four chosen at random have "bastard" trains on their going sides. Perhaps cutting and sizing of pinions was not easy in the 19th century. It may be that carriage clock trains were sometimes dictated by what pinions were available.

The tremendous rise in value and popularity of clocks in general, and of carriage clocks in particular, during the last few years has prompted certain less scrupulous persons to try their hands at the turning of *petite sonnerie* clocks into *grande sonnerie,* (rather like changing a pumpkin into a carriage for a horological Cinderella). This back-street operation will convince the unsuspecting and ingenuous customer — but not for long.[7] The coach will soon need to be turned back into a pumpkin because the striking train will run down in much less than eight days, probably jamming the going train in the process. At first the customer may think that he has forgotten to wind the strike side fully, but soon the truth will dawn. If the repeat button is used to any extent, the crisis will occur far earlier. The converter is "wise" in his generation, but often his grasp of horology is so limited that he is incapable of appreciating the sheer impossibility of what he is trying to achieve. In fact 3,744 extra blows per week[8] are struck by a *grande sonnerie* clock over a *petite sonnerie.* This increment is not possible unless a clock is provided with a striking train specifically designed for the purpose, as shown in the train-counts. How may the purchaser of a *grande sonnerie* avoid being taken for a ride by a "pumpkin coachman"? Firstly, the customer must look at the striking train mainspring barrel. Unless it is much larger than the going barrel, and be seen to have many teeth of fine pitch, then suspicions should be aroused.[9] Secondly, almost without exception, *grande sonnerie* clocks are furnished with controls to enable the owner to vary the mode of striking at will (for example *"grande sonnerie", "petite sonnerie",* or *"silence"*). Thirdly, *grande sonnerie* requires the use of more than nine or ten pins in the pin wheel, and usually the number is eighteen. Fourthly, although this is less easily seen, *grande sonnerie* clocks very often use double-ended gathering pallets, and when this is done the locking wheel will be found to have two locking pins planted 180° apart. A few clocks lock on the tails of their gathering pallets, but when this is done they will still be found to have high train counts.

A final word of warning. A far subtler "conversion" recently perpetrated includes the provision of bottom controls. Then the only way of being sure of a clock is by the size of the mainspring barrel and by the train count. Unfortunately as rarer carriage clocks become ever more desirable, so malpractices of this kind will continue to be devised.

The clocks "butchered" as outlined above are usually good pieces, as originally made. The false *grandes sonneries* are all ex-*petites sonneries,* while a rash of short-winded repeaters have now been fabricated out of ordinary hour and half hour strikers. Need more be said?

CALENDAR WORK

Calendar mechanism is found in carriage clocks of all periods. Breguet used it, and it is mentioned here and there in this book in various clocks and in more or less complicated forms. At this moment, however, we are concerned with illustrating a few representative examples both standard and special.

Plate VIII/2, already mentioned, shows a small, late anonymous clock (five inches tall with handle up) and embodying a simple calendar. It shows the day of the week and date of the month. By "simple calendar" is meant that manual correction is necessary at the end of all months not having thirty-one days. This clock is also marginally unusual in that it has a lever platform, with cut brass-and-steel balance. Surprisingly also, the escapement has double roller, steel 'scape wheel and divided-lift. Simple calendar work, however, is found in various forms. For example Plate VIII/11 shows a fairly early clock giving both day and date. Plate VII/3 illustrates a fine quality and fairly early

---

(7) What the converter usually does is simply to bend a pin in the *cadrature* under the dial so that at each quarter not only the quarter rack but also the hour rack will be free to fall upon its respective snail. Then, if the striking barrel is provided with a stop-work, he throws that away in order to utilise the mainspring from fully-wound until its last gasp.

(8) Not counting the "dummy-blows" which are struck in certain clocks.

(9) A few *grande sonnerie* clocks will be found having strike mainspring barrels only marginally larger than those on their going sides. In such cases the additional strike capacity is obtained by the use of fairly high numbered wheels in conjunction with lower than usual pinion numbers, together with pin wheels having 18 pins and sometimes also double-gathering pallets.

VIII   The Rarer Carriage Clocks

Plate VIII/11. *SIMPLE CALENDAR WORK. Date circa 1845. Dial signed "BORRELL A PARIS". This clock has peculiar* petite sonnerie *striking work with no repeat and with Neuchâtel-type snail. (Private Collection)*

Plate VIII/12. *SIMPLE CALENDAR WITH MOON WORK. The* cadrature *is on the back plate suggesting Franche-Comté origins. (Private Collection)*

clock by Bolviller, but indicating date only. An interesting variation of a simple calendar is offered by the clock shown in Plate VII/38. Here the day and date appear through apertures in the dial. Yet other clocks, as for example that shown in Plate VIII/12, offer simple moonwork in addition to conventional calendars. The beautiful calendar clock by L. Leroy & Cie shown in Plate V/24 was probably exhibited in Paris in 1900. Note the "Empire" case. This style, although used by Breguet et Fils until 1970, was already old-fashioned by the end of the 19th century.

The so-called fly-back calendars are yet another arrangement found in carriage clocks. In these a pointer moves down a vertical scale, one day at a time.

At midnight on Sunday the pointer "flies back" to Monday in one movement. Plate VIII/13 shows the simplest form of this type of calendar. The clock, however, is very far from being ordinary. For a start it has *grande sonnerie* striking on gongs planted behind the dial. An alarm bell is mounted under the base. The last two arbors of the striking train have their pivots run in pierced jewels and so does the contrate pinion. A thermometer is also provided. Another example of a fly-back calendar clock has a maker's mark "V.W.B." in a heart. It is numbered 557. This clock has *grande sonnerie* striking with *cadrature* on the back plate. It also has alarm and thermometer. Perhaps the best and also the most complicated clock known to us having

203

VIII   The Rarer Carriage Clocks

Plate VIII/13. *FLY-BACK CALENDAR. This clock also offers thermometer and alarm. The* grande sonnerie *striking may be reduced to hours only, but not to* petite sonnerie. *(Private Collection)*

Plate VIII/14. *CALENDAR COMBINING DIALS AND FLY-BACK. Maker Boseet. Note thermometer, barometer and centre-seconds. The fly-back shows the day of week. The top dials indicate date, moon and month. (By courtesy of Robert M. Olmsted, New York)*

a fly-back calendar is that superb piece signed Boseet and illustrated in Plate VIII/14. In addition to the fly-back calendar, which indicates days of the week, subsidiary circular dials are provided for date and month. There is also a semi-circular moon dial, and both barometer and thermometer. Plate VIII/15 shows the under-dial work. The clock offers the further complications of *grande sonnerie* striking, centre-seconds, and minute-repeating. While almost certainly the piece was both finished and cased in Paris, the style of the movement suggests strongly the influence of the Franche-Comté, if nothing more. The *cadrature* is disposed upon the back plate in a manner similar to the clock shown in Plate VI/30. The central button found on the top of the Boseet clock is provided for repeating hours and quarters, and also for correcting the lunar work in conjunction with a selection lever. The right hand plunger (it is *not* a button) not only releases the quarter-chiming train, but also winds and releases a watch-type minute repeating mechanism. The left hand button is provided solely for manual correction of the calendar.

The pinnacle of calendar work is the perpetual calendar, that is to say one capable of catering for not only months of different lengths, but also with leap years. A few English, and some of Breguet's best pieces, had this refinement, but it is not usual in French carriage clocks. Plate V/23 illustrates a French clock with semi-perpetual calendar sold by Le Roy & Fils of Galerie Montpensier, Palais Royal, circa 1860

VIII   *The Rarer Carriage Clocks*

Plate VIII/15. *CALENDAR WORK. The under-dial work of the Boseet clock shown in Plate VIII/14. (By courtesy of R.M. Olmsted)*

and numbered 5693. The working of the calendar will be better understood by reference to Plate VIII/16. It will be seen that there is a one-year wheel at the bottom right hand corner, and that this wheel resembles a striking work locking plate. The five notches represent respectively February (the very deep slot nearest the bottom at the time when the photograph was taken) and continuing clockwise, April, June, September and November. This clock is not, however, provided with a four-year wheel and hence requires manual correction each leap year *(Bissextile)*. The effect of the notches, according to their depths, is to allow the change-over detent to perform a longer stroke than usual, thus advancing the date star wheel by increments appropriate to the short months. A clock by Lefranc of Paris and having a true perpetual calendar is shown in Plates VIII/17 and VIII/18. This piece, provided that it is kept wound, will indicate the correct date continuously. The four-year wheel

will be seen in Plate VIII/18 and is at eight o'clock to the left of the moon dial. There are four sets of five notches on the periphery of the wheel. Three sets are identical, but the fourth has a shallower slot for February. This particular clock has *grande sonnerie* striking. It is also provided with a separate train, wound once a week with the clock, to afford minute repeating. Note the difference between this system and the watch-type of minute repetition used in the Boseet clock already mentioned. It should be added that the Lefranc clock allows itself the luxury of three trains to achieve its minute repeating. The minute rack is visible in the top right hand corner of Plate VIII/18. The reader will remember that Jacot's clocks achieve the same result from two trains, using an extension to his well-tried *grande sonnerie* system.

Almost all of the Breguet *pendules de voyage* boasted calendars of one kind or another. In the best clocks these calendars were perpetual. Most were *à*

Plate VIII/16. *SEMI-PERPETUAL CALENDAR. The clock, No. 5693, was sold by Le Roy & Fils circa 1860. View of movement with the dial removed. See also Plate V/23. (Private Collection)*

VIII   The Rarer Carriage Clocks

rouleaux, that is to say with rollers showing through dial apertures.

Plate VIII/17. *PERPETUAL CALENDAR. A clock made towards the end of the 19th century by Lefranc of Paris. This piece, besides offering* grande sonnerie *striking, also repeats minutes. (By courtesy of E. Pitcher & Co.)*

Plate VIII/18. *PERPETUAL CALENDAR. A view of the Lefranc clock movement with the dial removed. (By courtesy of E. Pitcher & Co.)*

DIGITAL DIALS

A rare and unusual carriage clock is illustrated in Plates VIII/19 and 20. The layout is basically that of a conventional timepiece, but with the addition of a sub-frame between the main plates. This sub-frame accommodates an extra mainspring barrel and part of a subsidiary train driving the digital dials. The method of operation is ingenious. The extended seconds pivot of the going train (which carries the seconds hand) has machined in it a v-shaped slot through which a double-ended flirt, carried on the final pinion of the subsidiary train, escapes once each minute. The change of the unit digit occurs instantly and dramatically as the seconds hand reaches sixty, the tens-of-minutes and the hour indicators following suit at appropriate intervals. The provision of the additional train to drive the "motion work" ensures brisk operation without serious interference with the

VIII   The Rarer Carriage Clocks

Plate VIII/19. *DIGITAL DIAL. A special train is provided to free the going train of the task of moving the digits and also to ensure brisk operation. The time is 46 minutes past 11.*

Plate VIII/20. *DIGITAL DIAL. The movement of the clock in Plate VIII/19 showing the subsidiary train provided for moving the digits.*

going train. It is easy to provide for "jump-hours" from a single train; but "jumping-minutes" is another matter altogether because so little spare power is available high up in a going train. Clocks and watches having digital dials appeared in various forms during the 17th century, and they have been made at intervals ever since. Digital time dials, in common with calendar work, are usually progressed by going-trains working through a series of star-wheels and jumpers. Where this method of operation is employed the changeover is not instantaneous. Not only this, but the type of jolt which a travelling clock is likely to receive is often enough to cause a star-wheel to shift. Moritz Immisch invented a shock-proof star-wheel applicable to carriage clock strike and calendar work (see Lemaille in List of Names).

"FLICK" CLOCKS

It is probably debatable whether these small timepieces, which may perhaps be regarded as a variant of digital dials, should be included here. The main objections might be that they are normally not even of minimal carriage clock quality and were not usually made in France. Another is that since the flick-cards are easily deranged by "bumps", such clocks would be of little use for travelling. They are, however, included because they interest many people, besides being made in carriage clock form. Plate VIII/21 shows the type most commonly found. It goes for thirty hours, is of indifferent quality, and originally had a card calendar (now missing) on top of the case. The clock in Plate VIII/22 has a movement identical

207

VIII   The Rarer Carriage Clocks

stamped "FRANCE", and elsewhere are engraved "D" and the number 1557. The height of this "Flick" clock to the top of the handle is about eight inches. Note that the handle is characteristic of the less-good quality carriage clock. At the risk of appearing to state the obvious, it should be said that the upper cards are carried on the extended hour wheel pipe, the lower on the cannon pinion. Both the hour wheel pipe and the cannon pinion turn clockwise. The two retaining springs attached to the right hand pillar hold back the cards until they escape in succession. The minute cards flick each minute but the hour cards escape five times per hour (four repeats and one change). This method of operation allows much more accuracy of release for the hour numerals. The

Plate VIII/21. FLICK CLOCK. *The German type most commonly found. This clock goes for 30 hours only. It is wound by a fitted key below the base.*

Plate VIII/22. FLICK CLOCK. *A 30-hour German flick clock looking far more like a carriage clock.*

to that in the previous example but the case is far more carriage-clock-like. This type of clock was made both in Germany and in America,[10] before and after 1900. Plate VIII/23 shows a better product altogether. It goes for eight days, and has in the base a French drum-type movement, arranged with cannon pinion pointing upwards and with a lever platform below. It even has a cut bi-metal balance. The "works" are

---

(10)   E. Fitch, of West 147th St., New York, made such a piece in 1902, for which he took out a British Patent specification No. 20,371. It is described briefly and illustrated in the *Horological Journal* of October 1904, pp.24-25 (Vol. XLVII).

VIII   *The Rarer Carriage Clocks*

Plate VIII/23. *FLICK CLOCK. An eight-day French flick clock with drum movement in the base. (Private Collection. Photograph by P.H. Kohl)*

Plate VIII/24. *CHINESE DUPLEX ESCAPEMENT. This peculiar platform escapement is used in a centre-seconds clock bearing the trademarks both of Japy Frères and of Reclus. (Private Collection)*

top and bottom "fingers" ensure the correct alignment of the cards.

Flick clocks are not usually of good quality; but they are very far from being devoid of ingenuity.

UNUSUAL ESCAPEMENTS

In general, the only escapement found in a French carriage clock is either a cylinder or a lever. A few makers used other escapements, including Breguet who was always a law unto himself.[11] In any case deviants are not normally found in anything that could be termed "bread and butter lines". It is true that Paul Garnier in the first half of the 19th century employed an escapement of his own. His clocks are described elsewhere. He used an escapement based on that of De Baufre. Other early makers, such as Lépine, were inclined to use the duplex escapement (Plate II/42). At that time in France it was considered superior to most alternatives. A few late clocks, such as that by Japy Frères (Plate VIII/24) used the so-

---

(11)   Other makers associated with unusual escapements used in carriage clocks include Dejardin, Gontard, L'Epée and Meyer & Schatz, while Saunier wrote a whole article in the *Revue Chronométrique* of August 1872 and June 1873 concerned only with peculiar escapements intended for use in *pendules de voyage*.

209

VIII The Rarer Carriage Clocks

called "Chinese" duplex. This escapement, seldom seen except in the Swiss 19th century watches made in the Fleurier district for the Chinese market, was not employed from any notions of timekeeping superiority, but simply because it provided a simple method of "showing seconds" from a balance vibrating quarter seconds.[12] This particular clock was made by Japy Frères but was finished and sold by Reclus.

Chronometer (detent) escapements were but rarely used in French *pendules de voyage,* while the English and Swiss were distinctly inclined to put them in their best pieces. One French clock by Bolviller includes a detent escapement among a number of other peculiarities. It stands eight-and-a-quarter inches tall with handle up, has hour and half hour strike and repeat on bell, and plays one of two alternative tunes at twenty-five minutes past the hour. The case is remarkable for its deep relief casting in a bird motif. Messrs. Asprey of New Bond Street showed in their Exhibition in 1971 the very fine large carriage clock shown in Plate V/12. It is signed simply "Berthoud Paris". It has a pivoted-detent escapement, train remontoire and up-and-down dial. Another good French chronometer carriage clock, this time with a spring-detent escapement, appeared at Sotheby's on 16th October 1972, Lot 189. It bore the Le Roy 13 & 15 Palais Royal address and the serial No. 4082.[13] The escapement of this clock is comparatively advanced in design. There is also a good compensated balance and a steel helical spring with terminal curves.

A very curious carriage clock, for which most extravagant claims were made by its perpetrator, was "Samuel's Vertical". It is described in the *Revue Chronométrique* April 1873, pages 392-395, and also in more detail by Berlioz in *L'Horlogerie, Rapports Sur l'Exposition Universelle de 1878.* Plates VIII/25 and VIII/26 show the escapement. The exterior of the clock is shown in Plate VII/11 as illustrating a standard

Plate VIII/25. *SAMUEL'S VERTICAL ESCAPEMENT. A two-plane escapement developed from that of Sully. Note the combined balance cock and potence. (By courtesy of Garrard & Co.)*

Plate VIII/26. *SAMUEL'S VERTICAL ESCAPEMENT. Note the integral balance cock and potence; also the unusual form of springing. With the change of plane taking place in the escapement itself, no contrate wheel is required. (By courtesy of Garrard & Co.)*

---

(12) In pendulum clocks, a *coup-perdu* escapement performs precisely the same function, showing "seconds" from a half-seconds pendulum.

(13) A misprint in the Catalogue made nonsense of the address which in fact reads on dial "LE ROY & FILS, PALAIS ROYAL GALERIE MONTPENSIER 13 & 15, H<u>GERS</u> DU ROI", and on the side of the movement plate "No. 4082 LE ROY & FILS, H<u>GERS</u> DU ROI A PARIS".

VIII The Rarer Carriage Clocks

oval case. Samuel's escapement in fact is nothing more than yet another variant of the two-plane design of the type associated with Sully. Notes on two-plane escapements will be found in Chapter II and in Appendix (a). Despite the fact that Samuel applied jewelling to the working surfaces of his escapement, and also used a compensated balance in conjunction with a spring of unusual form (said to be endowed with mysterious isochronal properties), it is difficult in the context of the year 1873 to be unduly impressed by this escapement. However, Berlioz in his description enters into paeans of admiration for the clock and for its escapement. He even goes so far as to state, in all seriousness, that these pieces were "... intended to be used in case of accident as marine chronometers". Elsewhere in connection with the case, the same author claims that the clock is watertight. He says "The *assemblage* is perfectly closed, inaccessible to dust, to humidity, and could if need be stand up to immersion in water without a single drop penetrating inside". While it is possible that some Samuel clocks were indeed water-resistant, the case of the example illustrated has no such properties. As for the chronometric propensities of Samuel's escapement, it seems likely that the reports of both Saunier and Berlioz are gross exaggerations, if not indeed utter nonsense.

In 1791 Robert Robin of Paris had produced an escapement which locked upon a lever, but in which impulse was given to the balance at each alternate vibration in the manner of a chronometer. Throughout the 19th century many variants of this hybrid escapement were tried, including a number used in pocket watches sold in France by such distinguished artists as A-L Breguet and his pupil Louis Raby, besides by others such as Rochat the Swiss, who worked for Audemars. Most of these escapements were really in essence pivoted-detent chronometers. They show great diversity in lay-out. Seemingly indiscriminate use was made of chronometer, duplex or pointed-tooth 'scape wheels, in conjunction sometimes with lever-escapement-like fork-and-roller, and at other times with chronometer-like horn-and-passing-spring. Even the function or functions of the "pallets" or "locking-stones" carried upon the "levers" or "detents" varied enormously within the type. In general,

however, these hermaphroditic essays were chronometer/lever variants, locking either single or double wheels upon "lever" pallets without impulse faces. As a result impulse was imparted in the manner of the chronometer or duplex escapement.[14] Plate VIII/27 illustrates the platform of a clock signed "Meyer & Schatz A PARIS", and having an escapement described enigmatically in Sotheby's Catalogue of 26th April 1971 as a "Rubisola-duplex". The reader may be able to discern the composite features of a duplex wheel used in conjunction with a "lever". Such escapements are rare in carriage clocks, and this particular instance may represent an attempt to achieve a "dead"

Plate VIII/27. *DUPLEX PIVOTED-DETENT VARIANT. This escapement employs a conventional duplex 'scape wheel. The long teeth are locked by a pair of "pallets". The impulse is imparted by the short teeth to an impulse-pallet carried in a roller. (Private Collection)*

---

(14) No attempt has been made to explain and illustrate the *modus operandi* of well-known escapements because the basic information is readily available everywhere. As for rare escapements, a great many of even these are to be found described in such un-rare books as Saunier's *Treatise on Modern Horology* and Chamberlain's *It's About Time*.

locking in a "frictional-rest" escapement. The clock, which has an early "one-piece" case, was made circa 1835-40. It is of good quality, being very much on a par in every way with Garnier's early *pendules de voyage.* It seems possible that this clock was marketed in competition with Garnier's business and that the manufacturers (who were unlikely to have been Mayer & Schatz) sought an escapement both cheaper than a chronometer and better than a cylinder.

Another unusual escapement used in carriage clocks during the second half of the century was one devised by Gontard. This escapement, while it is certainly related to a "Robin", must be considered as an attempt to improve a pivoted-detent chronometer escapement in order to make it more suitable for pocket watches or carriage clocks. Fig. VIII/1 shows the escapement

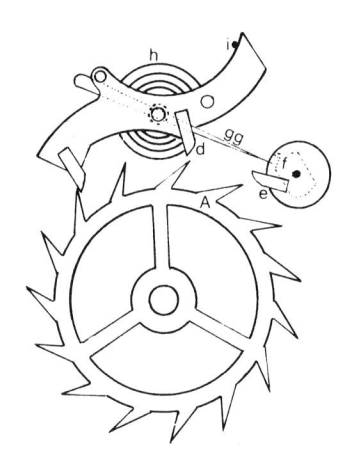

Fig. VIII/1

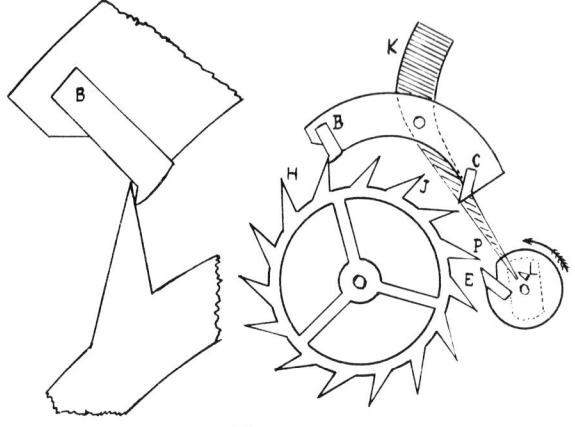

Fig. VIII/2

as first devised, while Fig. VIII/2 illustrates an improved version produced only months after the first. Gontard's original design is described and illustrated in the *Revue Chronométrique* of October 1855 on pages 51-54, while his modified design appears in the same journal on pages 255-256 of the December 1856 issue. As originally conceived by Gontard the unlocking of the escapement is performed by the action of the discharging pallet 'f' upon a discharging detent 'gg' which is pivoted independently on the "lever" (or pivoted-detent) which is kept to a pin 'i' by a volute spring 'h'. Impulse is given by the 'scape wheel 'A' upon an impulse pallet 'e'. The escapement is returned at the appropriate moment to the locked position by the action of a 'scape wheel tooth upon an exit-pallet 'd'. The volute spring 'h' also allows the discharging pallet 'f' to deflect and pass the tip of 'gg' on the return (idle) vibration of the balance. Readers familiar with chronometers and with "Robins" of various descriptions will see how interestingly this escapement differs. An example of Gontard's first escapement is shown in Plate VIII/28. It is

Plate VIII/28. GONTARD'S ESCAPEMENT. *A rare and expensive pivoted-detent escapement designed for use in carriage clocks. (Private Collection)*

VIII The Rarer Carriage Clocks

used in a fine carriage clock signed "Delmas, PARIS". Both the *roulant* and the case of this piece were probably obtained from Jacot. The appearance of the clock, which has a *gorge* case, is virtually identical to that shown in Plate VII/8. The platform bears Gontard's trademark prominently displayed. It is interesting to note that Gontard uses an index in conjunction with a helical spring. In this respect his thinking is in line with that of the Swiss, and at variance with English chronometer carriage clock practice. According to the *Revue*, Gontard's second version of his escapement was devised specifically for carriage clocks. Saunier says in his *Treatise* that Jacot thought well of it. While the action is very similar to the first design already described, Gontard evidently considered that the volute spring was unnecessary and made the discharging detent 'K' (Fig. VIII/2) integral with the "lever" (or pivoted-detent). In this arrangement the locking pallet 'B' is shaped and set so as to provide a very large amount of "draw", thus ensuring safe locking. On the return (idle) vibration of the balance the discharging detent 'K' momentarily deepens the locking *en passant*. What happens is that the locked 'scape wheel tooth slides further along the resting face of the pallet, returning to its draw-retained locked position as soon as the discharge pallet 'L' releases the tail of discharging detent 'K'. The later design was said not only to be easier to make, requiring little more exactitude in manufacture than a cylinder, but above all able to maintain a heavier balance in a carriage clock than ordinary detent escapements, which cost far more money and were very fragile into the bargain. The *Revue* of 1855 already mentioned is very scathing about lever escapements, saying that when they are badly made they lack the ability to keep a sensibly-sized balance moving properly, and are the despair of those who make them.

At the other end of the scale altogether are found a few clocks of no pretensions yet having unusual escapements. A good example is the Japy escapement shown in Plate VI/23. It is described in Chapter VI. Another lever variant is employed in the "R.E.D." travelling timepiece shown in Plate VIII/29. The clock is of fair quality but was not made at all according to the accepted *pendule de voyage* format. It was, however, provided with a travelling box and made so as to be completely portable. Plate VIII/30 shows the escapement. The first escapement of the type was probably that attributed to Thiout some time prior to

Plate VIII/29. "R.E.D." TRAVELLING TIMEPIECE. *This clock has a two-plane lever escapement, and also employs a fusee. The middle section of the case is glazed to show the movement. The clock stands nearly 6 inches tall with handle, and is 4¼ inches in diameter. (Private Collection)*

1741. It was intended for use in conjunction with a pendulum. Many variants for use with both pendulums and balances were subsequently made. Some are illustrated by Gros in *Les Echappements* and yet others by Tardy in his booklets dealing with watch escapements. In a carriage clock the advantage of a two-plane escapement is that it makes a contrate wheel unnecessary while at the same time allowing the balance staff to be planted vertically. In an "R.E.D." clock the locking of the escapement takes place on the fronts of the 'scape wheel teeth in the normal way. On the other hand impulse is given alternately to the pallets by the sides of the wheel. The construction of both case and movement are as unconventional as the escapement. The nickel-plated case consists essentially

VIII The Rarer Carriage Clocks

Plate VIII/30. *"R.E.D." TWO-PLANE LEVER ESCAPEMENT. The escapement of the clock shown in Plate VIII/29. An example exists dated 1886.*

Plate VIII/31. *V.A.P. ESCAPEMENT. The solid steel pallets are similar to those found in* pendules de Paris *but with the provision of slight "draw".*

of front and back circular plates, separated by six pillars and with the movement visible sideways through a cylindrical glass. The movement goes for eight days and departs from normal French practice in having fusee and chain. The fusee has only ten turns. In this clock there is no arbor turning in an hour, and in consequence there is no centre wheel and pinion. These peculiar clocks were current by 1886. A dated example has been found bearing the address "Hampton House, Hereford".

The movement of a clock rather similar in quality to those mentioned above, is shown in Plate VIII/31. Clocks of this type have the initials "V.A.P." in an oval on their back plates. The maker is otherwise unidentified. These clocks also bear the legend "Made in France". The "V.A.P." mark is most commonly found not on carriage clocks but on circular-plated timepiece movements, cased either in plain brass tambours or in round ebonised wood cases standing on small plinths. The dials of such pieces are usually made of glazed paper. Their escapements are the same as that used in the carriage clock movement illustrated, but they are mounted on the back plates without the use of contrate wheels. In this escapement the pallets are almost the same as those of a dead-beat pendulum escapement, but having slight "draw". This inexpensive and unsophisticated escapement is in fact most satisfactory and maintains a good balance arc. "V.A.P." carriage clocks, unlike their small mantel clocks, cannot have been made in large quantities.

VIII   The Rarer Carriage Clocks

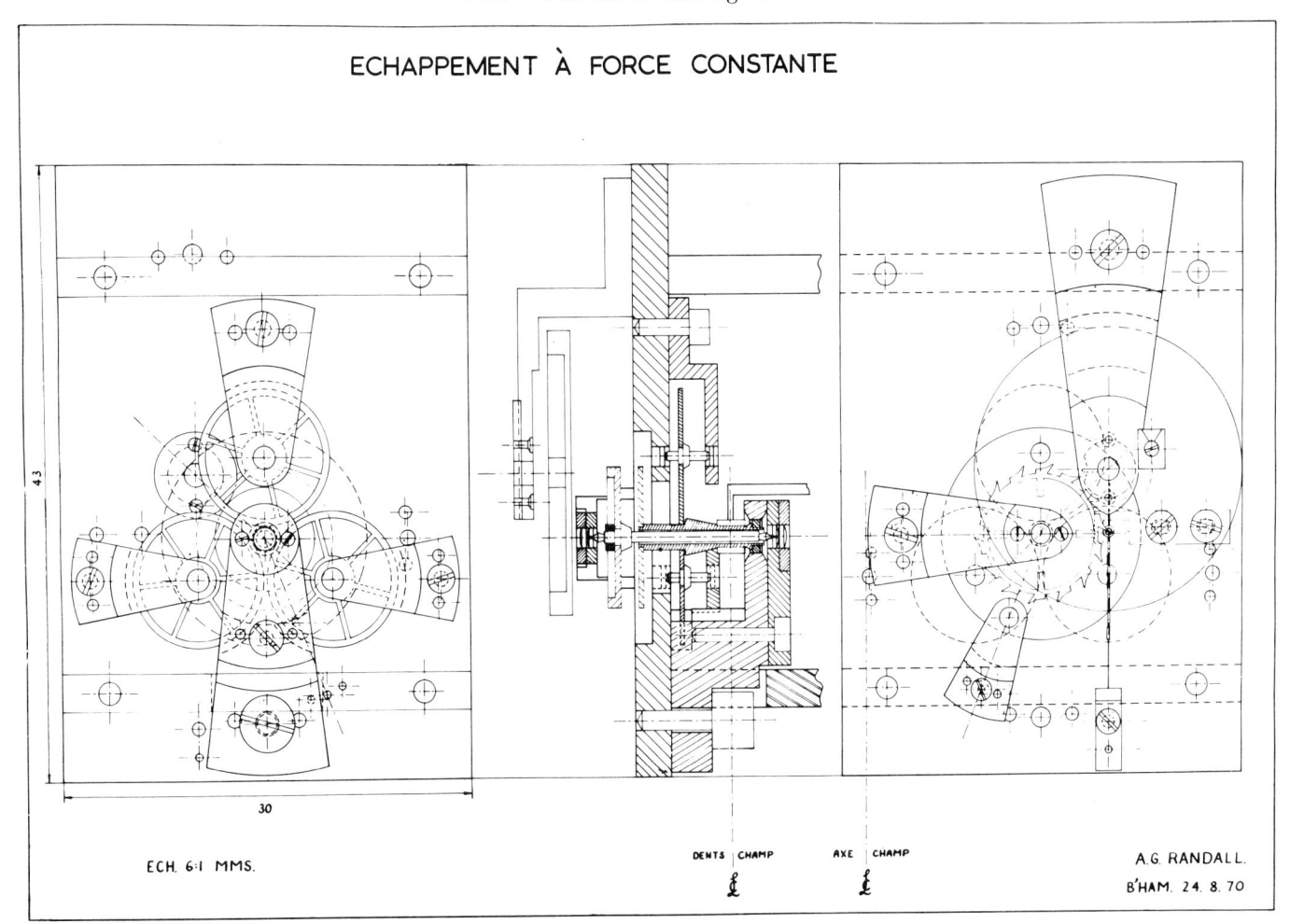

Plate VIII/32. *RANDALL/THEUVILLAT CONSTANT FORCE ESCAPEMENT. Projection drawings showing the platform seen from above (right), a sectional elevation (centre) and the platform seen from below (left). (By courtesy of A.G. Randall)*

Apart from their escapements they are so unremarkable that they were probably finished from standard *blancs-roulants* and adapted to accommodate the large and unusual *porte-échappement*.

Before leaving unusual escapements it is worth noting that today, when there has been a much-needed revival in the manufacture of interesting and unusual clocks and watches, Mr. Anthony Randall of Birmingham has developed and made the new constant-force platform escapement of which three views are shown in Plate VIII/32. This platform is intended for use in portable clocks and is based on one originally devised by his Swiss father-in-law. This constant-force escapement in its developed form is perhaps technically the most perfect ever devised. It ensures that both the energy required from the balance and spring assembly to perform the unlocking, and also the amount of energy imparted to the balance, will remain constant over a very extended period of use. In the Randall/Theuvillat escapement a spring-detent is used in conjunction with two escape wheels separated by a spiral remontoire spring. One wheel is locked by the detent and gives impulse to the balance, while the other only receives impulse from the train. A double pallet system enables the 'scape wheel, which imparts impulse to the balance, also to lock and unlock the 'scape wheel associated with the remontoire. The remontoire is rewound at each alternate vibration of the balance after impulse has been given. The whole mechanism is based upon a platform with all the functions visible. Plate VIII/33 well shows the beauty and also the sophistication of the arrangement, while Plate VIII/34 gives a detailed close-up view of the underside. Not only will this escapement ensure very great accuracy

VIII   The Rarer Carriage Clocks

Plate VIII/33. RANDALL/THEUVILLAT CONSTANT FORCE ESCAPEMENT. *The platform showing the arrangement and layout of the parts. Note the Guillaume bimetallic balance and the spiral balance spring with terminal curve. The remontoire spring lies above the upper escape wheel. (By courtesy of A.G. Randall)*

Plate VIII/34. RANDALL/THEUVILLAT CONSTANT FORCE ESCAPEMENT. *A view of the underside of the platform of the escapement shown in Plate VIII/33. Three anti-friction wheels support the hollow pinion arbor. The arbor of the upper escape wheel passes through this hollow arbor and there are separate and independent pierced jewels and endstones for the pivots. (By courtesy of A.G. Randall)*

from a mechanical timekeeper, but it will also enable any number of additional complications to be attached to a clock without impairing its potential accuracy.

The illustration in Plate VIII/35 shows a mass-produced remontoire platform escapement of Swiss manufacture and made during the present century. It is very ingenious and employs a conventional lever escapement driven by a volute spring which is re-wound frequently.

TOURBILLONS AND KARRUSELS

The terms "Tourbillon" and "Karrusel" are customarily used fairly loosely to describe the two basic types of revolving escapement carriages which were designed for use in pocket watches. The tourbillon was the invention of A-L Breguet and his *Brevet* for it is dated 26th June 1801, although he made examples in the last ten years of the 18th century. The karrusel, on the other hand, was invented almost a hundred years later by Bahne Bonniksen in 1894. He published a booklet on the subject in 1905.[15]

In any watch controlled by a normal balance and spring system, the effects of gravity acting upon the system will vary depending on the position of the watch. When the plane of the dial is horizontal, whether it is facing upwards or downwards, the effect of gravity is minimal. However, in any other position, and especially in those where the plane of the dial is vertical, the effect of gravity varies depending on the position. Variations in timekeeping result because the centre of gravity of the moving balance and spring cannot be maintained on the axis of the balance however carefully the watch is made. Although this limitation was partly known before the time of A-L Breguet, it was his genius which produced a most elegant and efficient solution, which he called the tourbillon. Breguet mounted both his balance and

---

(15) From 1895 karrusel watches achieved great success at Kew, and at a later date many were purchased by the Admiralty as a result of the annual trials of deck watches.

VIII   The Rarer Carriage Clocks

Plate VIII/35. *CONSTANT FORCE LEVER PLATFORM. The 'scape wheel is driven by a small volute spring, re-wound at each beat of the escapement.*

spring system as well as the escapement itself on a circular carriage which rotated about its own axis. In such an arrangement the effects produced by gravity in the vertical positions of the watch were "averaged out". However, even this ingenious system left one further serious problem. It was the difficulty of equalising the rates of going of the watch both in the horizontal and in the vertical positions. The variations in rate were caused mainly by differences in balance pivot friction.

In a normal carriage clock the balance and spring system is mounted as in a watch when its dial is horizontal; and so gravity produces no effect on timekeeping, unless of course the clock is laid on its side. However, if the balance and spring systems are mounted on the back plate of a clock in order to avoid using a contrate wheel[16] then gravity will affect the system and will produce changes in timekeeping whenever the balance amplitude varies (because of the running down of the mainspring and of imperfections in the gearing) or if the clock is put in a different position.

Breguet's tourbillon carriage was usually arranged to turn once a minute, although improved examples were subsequently made by Breguet and by others in which the carriages rotated far more slowly, for instance once in every six minutes, or in longer periods as in Taylor's type made circa 1903. The karrusel was a similar device, but in this arrangement the carriage rotated either in 52.5 minutes or else in 34 minutes depending upon whether or not the watch was required to have a centre-seconds hand. Another detail difference between tourbillons and karrusels was that while the carriages of the former had pivots top and bottom karrusel carriages were unsupported at the top. However, there is a far more fundamental difference than this between tourbillons and karrusels. It is one, moreover, which does not seem to have been clearly stated in readily available horological literature. The difference is that, in the case of a tourbillon, the driving torque to the escapement is transmitted via the rotating carriage, which, of course, has to start and stop very rapidly. In a karrusel, on the other hand, the drive to the escapement is totally independent of the rotation of the carriage.

Although A-L Breguet was the inventor of the tourbillon, he appears to have made use of it only once in a carriage clock. This is the example shown in Plates II/2 and 3.[17] On the other hand, Nicole Nielsen in 1914 made at least seven identical tourbillon carriage clocks. One of these clocks is illustrated in Plates IX/81, 82 and 83.

---

(16) As implied many times in this book, contrate wheel gearing is never wholly satisfactory and is the most frequent cause of stoppage in carriage clocks.

(17) It is worth remarking in this connection, if only as a matter of interest, that Breguet also made one very beautiful tourbillon mantel clock, No. 1252, sold in 1814 to the Prince Regent. This piece, which rests upon a massive square marble base under a glass shade, is not of course in any sense a carriage clock or even a *pendule portative*. It was obviously made in order to demonstrate Breguet's tourbillon. Three photographs of this clock will be found in Sotheby's Sale Catalogue of 1st June 1964. The same piece is also illustrated by Tardy in *La Pendule Française*.

VIII The Rarer Carriage Clocks

Revolving escapements have occasionally been placed on the tops of the frames of carriage clocks in the position normally occupied by standard fixed escapements. When this was done the rotation of the carriage could be seen and appreciated through the top glass of the case. While the planting of a revolving escapement on top of a clock, thus requiring the retention of a contrate wheel, was perhaps motivated more by a desire for "mystique" than for any practical reason, such an arrangement would still help to equalise the going of the clock in different positions. Plates VIII/36, 37 and 38 illustrate such a clock by Auguste and one moreover in which the carriage turns in about 43 seconds. This curious hybrid piece transmits the drive to the escapement through the carriage as in a tourbillon. At the same time the carriage has no supporting top pivot, and in this respect the arrange-

Plate VIII/36. *REVOLVING ESCAPEMENT. This "giant" clock, which is not numbered, is signed on the dial and on the movement "AUGUSTE A PARIS". (Private Collection)*

Plate VIII/37. *REVOLVING ESCAPEMENT. The movement of the Auguste clock shown in Plate VIII/36. (Private Collection)*

ment resembles that of a karrusel. While almost certainly unique, such a system would not be likely to produce satisfactory results on account of the rapid rate of rotation. The carriage is also unrealistically heavy, so that the inertia inseparable from its starting and stopping makes for a very low mechanical efficiency. There is no evidence when the piece was made, but this clock bears such strong resemblance to the Leroy giant clock shown in Plate II/37 as to suggest a date of manufacture of circa 1830. The revolving escapement carriage, although it has been re-made, could just conceivably be an original feature.

It is interesting that neither Bonniksen nor anyone else apparently ever thought of trying a karrusel-like arrangement for use in a carriage clock, for which purpose it might have answered rather well.

VIII   *The Rarer Carriage Clocks*

Plate VIII/38. *REVOLVING ESCAPEMENT. The escapement of the Auguste clock turns in 43 seconds. (Private Collection)*

What has been said in these pages about tourbillons and karrusels is, of course, very much an over-simplification of an extremely complicated subject. Readers wishing to know more will find rewarding reading in the chapter in Haswell's *Horology* entitled "Epicyclic Trains".

Well worth mentioning, although not strictly speaking either a tourbillon or a karrusel, is the device used in the American carriage clock shown in Plate XII/31. This ingenious mass-produced travelling clock incorporates the "rotating-movement" system used by the Waterbury Company of Connecticut in their so-called "Long-Wind" watches, Series A-E.

FOUR-DIAL AND TWO-DIAL CLOCKS

In 1878 Guilmet took out *Brevet* No. 119,862 for a four-dial clock. Plate VIII/39 cleverly shows simultaneously three views of a four-dial clock which has a plate under the base reading "PATENT. BREVETE S.G.D.G. BLUMBERG & CO. LIMITED, PARIS AND LONDON". This carriage clock stands seven inches tall with handle up and is a timepiece only, going eight days. A similar clock is signed on the dials "H RY MARC", but below the base on a pinned-on elliptical plate in raised lettering is the Blumberg inscription already mentioned. Both clocks have circular white enamel dials, set in rectangular engine-turned surrounds. They are wound and set from the bottom through shutters which reveal inverted lever platforms. The escapements have flat balance springs and uncut bi-metal balances. A central contrate wheel drives four sets of motion work through straight-cut pinions.

Besides four-dial carriage clocks there exist a few two-dialled examples. It is sometimes said that these were intended for "partners' desks"; but the theory sounds like wishful thinking on the part of the antique trade. An anonymous two-dial clock under scrutiny as these pages are written is plainly of Saint-Nicolas-d'Aliermont manufacture. It uses a timepiece movement, suitably modified and housed in a *Corniche No. 1* case. The movement is wound from below the base where squares for setting the hands and for remote control regulation are also accessible. The escapement is a cylinder and the whole clock is of modest quality.

BOTTOM-WIND CLOCKS

Bottom-wind clocks were a speciality of Le Roy & Fils of 13 & 15 Palais Royal, Paris, and 57 New Bond Street, London. These "keyless" clocks were made during the last quarter of the 19th century. While most of them were "full-sized", including examples having *grande sonnerie* striking, a scaled down bottom-winding mechanism was also made for semi-miniature clocks.

In the Le Roy bottom-winder, winding is accomplished by means of a sturdy, permanently fitted turn-key housed in the recessed base. The going and striking trains are wound alternately by turning the "key" about 45 degrees first in one direction and then in the other. The "key" has a contrate wheel squared on to its arbor. The contrate teeth engage a floating wheel, the axis of which moves in a slot. The floating wheel winds the going train directly through a wheel squared on to the barrel arbor; and winds the striking train via

VIII   The Rarer Carriage Clocks

Plate VIII/39. *FOUR DIAL CARRIAGE CLOCK. The timepiece movement is wound and set from below the base. (By courtesy of Charles Terwilliger, Bronxville, New York. Photograph by Vincent Pollizzotto)*

an idler pinion to reverse the direction of rotation (both barrels need to be wound in the same direction). Plate VIII/40 gives an idea of the layout of the mechanism. Plates V/25 and VIII/41 show a fine Le Roy clock and also its fixed winding "key". The purpose of the bottom-wind clocks is to avoid the troublesome necessity of a separate key and also of opening the clock, besides offering the advantage of winding both trains in a single operation. Care, however, is still necessary in use to ensure smooth and unharmful engagement of the mechanism. In this connection it is worth remarking that the Le Roy device also avoids those accidents which occur when worn or badly-fitting keys are used, or even sometimes when clocks are wound carelessly. Every repairer sees clocks with broken pinions and also teeth stripped out of mainspring barrels caused by slipping keys. It has already been mentioned in Chapter V that the origins of the Le Roy bottom-winding clocks are still unexplained. It is also necessary to say that the inscription "Patent No. 9501", which is found engraved on the turn-keys, does not apparently refer to any British patent. On the other hand the number does not seem to relate to any French *Brevet* either. Here is a small mystery which for the time being at least must remain unexplained. It has already been stated that Le Roy carriage clocks emanated from several different sources. Those examples having bottom-winding seem without

220

VIII   The Rarer Carriage Clocks

Plate VIII/40. *LE ROY BOTTOM-WIND. A view of the back plate of a "bottom-wound" clock showing part of the winding work. (Private Collection. Photograph by P.H. Kohl)*

Plate VIII/41. *LE ROY BOTTOM-WIND. A Le Roy clock, No. 9487, showing the permanently-fitted winding "key".*

exception to have been made prior to the year 1900. Perhaps significantly, the 57 New Bond St. address, with or without the Palais Royal, will usually if not always be found on bottom-winding clocks.

Another interesting invention intended to improve winding is that of Henri Lioret of Paris, *Brevet* No. 122278 of 1878. Lioret's winding work is far superior to the Le Roy system in that the train of wheels connecting the fixed winding key to the barrel arbors is continuously engaged. Many modern travel alarm clocks still employ this exact arrangement, as (with modifications) do many two-train keyless pocket watches. Lioret positions his key at the back of the clock.

SINGING BIRD CLOCKS, MUSICAL CLOCKS, SUNDIAL CLOCKS, ETC.

The so-called singing-bird clocks are an interesting *pendule de voyage* variant and perhaps evince more variant than *de voyage*. Three quite different examples are shown in Plates VIII/42, 43 and 44. In the first clock, which is surmounted by a sort of tall parrot cage, the bird briefly sings, flutters, and opens and closes its bill at every hour as the clock goes. It is also possible to make the bird sing by pressing a button. The bellows and "Swanee" whistle are housed in the base of the clock, the singing-bird train being accommodated between the movement plates, together with trains for going, strike and repeat, and alarm. The clock has a cylinder escapement. A second example with a clock in a "one-piece" case but with a bird below in an early "multi-piece" case is signed "BOLVILLER A PARIS". The bird sings at will

VIII   *The Rarer Carriage Clocks*

Plate VIII/42. *SINGING-BIRD CLOCK. The bird sings every hour or at will. (By courtesy of Madame Ducatez, Paris)*

Plate VIII/43. *SINGING-BIRD CLOCK. The bird sings "by request" only and is not set-off by the clock. Maker Bolviller, Paris. This clock stands 12 inches tall and still retains its original leather travelling box. (By courtesy of Gerry Planus)*

by means of a lever (bottom right) and is not connected to the clock. The tall travelling box of this clock survives. A third clock, made at a rather later date by or for Japy Frères, and having duplex escapement and centre-seconds hand, is housed in a half-ornate, half-plain "multi-piece" case. The bird in this instance perches above the clock in its own private jungle of greenery, while the singing mechanism is housed in the base of the clock itself. Bird song replaces the usual cacophony of a conventional alarm. It is overwhelmingly likely that all three clocks were produced in Eastern France. Not only do they look the part, but there is no evidence that musical boxes, singing birds, etc. were ever made in Saint-Nicolas, or for that matter anywhere in the north of France. The Franche-Comté, on the other hand, was noted for this sort of work, and so of course was Switzerland.

Another Bolviller clock, this time having a musical box in the base playing alternately two different tunes at twenty-five minutes past every hour, is shown in Plate II/45.

VIII   *The Rarer Carriage Clocks*

Plate VIII/44. *SINGING-BIRD CLOCK. The bird sings in place of a conventional alarm. This clock was made or sold by Japy Frères, and has a duplex escapement. (Private Collection)*

Plate VIII/45. *CARRIAGE CLOCK WITH UNIVERSAL EQUINOCTIAL SUNDIAL. The magnetic compass set in the top of this carriage clock allows the sundial to be oriented correctly. The latitude scale provides settings appropriate to London and to various parts of India. (By courtesy of private owner in Rhodesia)*

A clock right out of the ordinary run of travelling pieces is seen in Plate VIII/45. This example stands eight and a quarter inches tall with handle up, and its features include a sundial complete with its associated magnetic compass and latitude scale. The latitudes of London, Delhi, Benares, Lucknow, Calcutta and "Lahors" (sic) are engraved on the top of the clock to ensure correct latitude setting. The original owner, according to family tradition, was a Briton serving in the Indian army during the last century. On the right hand side of the clock case is a thermometer. This clock, which is also unusual in having the escapement set in the dial, was first described by Mr. S.M. Baker in the transactions of the Antiquarian Horological Society of March 1966. A rather similar clock, bearing the name "R. HOLDT, PARIS", has a sundial and compass together with instructions for their use, as well as Fahrenheit and Centigrade thermometers. In this instance a normal movement is used with top escapement platform.

223

VIII   The Rarer Carriage Clocks

Plate VIII/46. COMPASS & THERMOMETER CLOCK. (By courtesy of N. Bloom & Son Ltd.)

Plate VIII/47. "BUBBLE" CLOCK. A striking clock with the escapement planted on top of the case under a glass shade. (By courtesy of Gerry Planus)

Plate VIII/48. "BUBBLE" CLOCK. A close-up view of the escapement of the clock shown in Plate VIII/47. (By courtesy of Gerry Planus)

Another clock having its escapement set in the dial is the pattern typified by Plate VIII/46. Some of the glass chapter rings found in these and other French carriage clocks may well have been imported from Germany. The top compass and circular thermometer were presumably provided for the reassurance of the pompous Victorian traveller!

Plate VIII/47 shows a clock having the escapement visible through a top "bubble". The whole style of the piece suggests a fairly early date. Plate VIII/48 illustrates the escapement.

YEAR CARRIAGE CLOCKS

Very few carriage clocks were made going for a year, and most of these which were made appear to have emanated from Saint-Nicolas-d'Aliermont. Plate VIII/49 shows an anonymous factory-made year carriage clock probably dating from the early years of

224

VIII The Rarer Carriage Clocks

Plate VIII/49. *YEAR CARRIAGE CLOCK. This anonymous piece was in all probability made in Saint-Nicolas-d'Aliermont early in the present century. (By courtesy of Martin Hutton Antiques, Battle)*

the present century, and which was first illustrated in *Antiquarian Horology*, March 1966. The reader will notice that the clock is housed in a very plain case, basically of the *Obis*-type but being made unusually deep from front to back. In this clock the (theoretical) year's duration of going is achieved by the employment of a very large and deep mainspring and mainspring barrel used in conjunction with a ludicrously long train of wheels and pinions terminating in a slow-beating lever escapement. The train layout is peculiar. No arbor turns in an hour, and so there is no centre wheel. The barrel-bottom and the barrel cover embody ball bearings introduced with the object of reducing friction. One clock of this description was tested. Perhaps predictably it not only proved to be a very indifferent timekeeper, but also it was reluctant to keep going for many months on end. It is probable that as a production model the type may well have been a failure and that, in consequence, it was soon abandoned. It is certainly a rarity, and its interest makes up to some extent for an overall lack of quality.

At the other end of the scale altogether is the year clock by Saulet of Paris shown in Plate VIII/50. This piece was probably made circa 1820-30. The Saulet clock achieves its required duration of a year by the use of two very large going barrels driving a long train incorporating a remontoire. No doubt the fluctuations in torque through the train during the course of a year are comparatively great. The clever maker wisely decided to make a virtue of necessity. He accordingly placed a half-minute re-winder on the front movement plate, making the working visible through an aperture in the dial. Plate VIII/51 shows a close up view of the remontoire mechanism. The remontoire is re-armed every thirty seconds, this operation being reflected by

Plate VIII/50. *YEAR CARRIAGE CLOCK. Maker Saulet, Paris. Note the remontoire visible through an aperture in the dial.*

225

VIII   The Rarer Carriage Clocks

the minute hand. The escapement of this clock is that of a pivoted-detent chronometer. While it is of exactly the right period and style, circumstantial evidence suggests that it has been re-instated. This year clock is wound from the back by means of a very large key. The winding hole shutter bears the inscription "Saulet à Paris", and is the only place where the clock is signed. The winding square is protected by a circular boss round which are engraved the words "Remontez une fois par an". The winding square is connected to the twin barrels by means of a reduction gearing. There is no opening door in the case. The hands are set by means of a small key inserted through a hole in the right-hand side panel, which is provided with a sliding cover to help exclude dust. The silvered dial is drawn in regulator fashion, that is to say with separate hour and minute circles. Not surprisingly no seconds dial is provided. Probably the visible remontoire, occupying the usual seconds-hand position, was considered more desirable; and in any case the time is shown accurately every thirty seconds by the flirting forward of the minute hand. The remontoire is most ingenious and subtle, affording a continuous drive to the escape pinion.

Plate VIII/51. *YEAR CARRIAGE CLOCK. Close-up view of the remontoire mechanism of the clock by Saulet shown in Plate VIII/50.*

English Carriage Clocks

Chapter IX

English Work and Workmen

JAMES McCABE No. 2927. See Plate IX/63.

IX

English Work and Workmen

Introduction. Difficulty of Establishing Dates. Carriage Clock Production – Some Celebrated Workmen. The Coles. B.L. Vulliamy. J.R. Arnold. The Dents. Thomas Earnshaw II. Charles Frodsham and Family. Webster. Whitelaw. Barwise. Desbois. Thwaites & Reed. Barraud and Successors. Sir John Bennett. James McCabe and Family. The MacDowall Family. Victor Kullberg. Usher & Cole. Bridgman & Brindle. S. Smith & Son. Jump. C.R. Hinton. Philip Thornton. Col. Quill and J.S. Godman. Nicole Nielsen. Thomas Mercer. Other Makers. French Fear of English Competition. Huber.

INTRODUCTION

English carriage clocks are normally far larger and heavier than French ones, so much so that no one could be blamed for wondering whether they were ever seriously intended for taking on journeys.[1] However, it then becomes hard to explain the existence of the beautifully-made and fitted travelling boxes invariably supplied with the best English clocks. The very design of these boxes shows a marked concern for the safety of the contents while in transit, by means of padded and baize-lined bolsters, locks, robust carrying handles, and brass-bound corners and edges. Moreover provision is usually also made to enable the time to be seen at any stage of a journey. Certainly large travelling clocks would not have been considered at all incongruous in the spacious days of massive luggage. In Edwardian times even moderately well-off families thought nothing of taking with them twenty or thirty pieces of baggage when going away for a short visit.[2] Equally, no one who has looked at John Leech's *Pictures of Life and Character*, taken from *Punch* of 1842-1864, could find any difficulty in seeing English carriage clocks against a Victorian background. In earlier times, grander families undoubtedly carried about with them even more of the artifacts

Plate IX/1. *A LATE VICTORIAN LADY'S DRESSING CASE. This beautifully-made piece of hand luggage is covered in morocco leather and furnished with Bramah's patent lock. It contains 78 items in ivory and silver, the latter bearing the London date letter for 1889.*

---

(1) The distinguished horological historian, the late Professor D.S. Torrens, expresses this doubt in an article published in the *Horological Journal*, August 1946, under the title *Carriage Clocks and Chronometers*.

(2) Plate IX/1 shows an opened lady's dressing-case. There are seventy-eight different items, all made of silver-gilt and/or ivory. Every female member of a well-off Victorian or Edwardian family would have had one of these; just as each daughter would have had her own lady's-maid. The beautifully-equipped 18th and 19th century travelling knee-writing desks also give a good idea of the type of portable furniture which people took wherever they went almost as a matter of course.

IX English Work and Workmen

of their day. R.E. Ball and T. Gilbey in *The Essex Foxhounds*, 1896, paint a clear picture of what at least one country squire took with him in the second half of the 18th century when he travelled between his estates in Essex and Berkshire.[3] Mr. Archer was not, in fact, particularly eccentric by the standards of the period; but in any case the first carriage clocks made in England from soon after 1820 were modest in both size and weight. They were entirely suitable for carrying from place to place, and they would certainly have been used for the purpose for which they were designed in a period when church clocks, even in neighbouring villages, often differed by as much as three quarters of an hour.[4]

Fine English carriage clocks were always made in small numbers, and then usually only to special order. They had expensive movements employing fusees and chains, and at no period did they become in any sense a "popular" item. English carriage clocks were bought by the "carriage trade", and were never intended to compete in the same market as the far cheaper French products. While French *pendules de voyage* were made in a wide range of qualities, the English carriage clocks were far more evenly matched as regards presentation and execution.

The early clocks sold by concerns such as the Coles, Vulliamy and Desbois had robust lever escapements and, needless to say, they would have travelled happily. After the middle of the century, famous firms like the Dents, Frodsham, and later Victor Kullberg, made great use of the chronometer escapement. No doubt by this time carriage clocks were used less for travelling than in the drawing room, smoking room or study, but all the same those sold were still provided with fitted outer boxes. Where chronometer escapements were used, they were designed with short detents and light balances. Such clocks, provided they were treated with reasonable care, were probably not nearly so delicate as might be supposed. In other words the escapement proportions, as in the case of pocket chronometers, undoubtedly made some allowance for the possibility of an occasional twist or bump. Having said this, it must be re-emphasised that a chronometer carriage clock was an expensive investment, if not indeed a "status-symbol". At a time when many country houses boasted regulators or box chronometers, a good portable piece would not have been purchased by someone who had no intention of looking after it.[5] In Victorian and Edwardian times, one or two top-flight firms offered chronometer carriage clocks as their best travelling pieces,[6] suggesting lever escapement clocks to customers wishing to pay less. In the North of England, in places like Harrogate, York or Newcastle-upon-Tyne, the best "London-made" clocks would have been available. In Liverpool a few carriage clocks were finished locally by firms like Roskell. These clocks, like the marine chronometers from the same district, show excellent if characteristic work. However, they do not usually equal the sheer panache of finishing found in a best London clock, especially when workmen really set their minds to the task.[7] Across the Irish Sea in Dublin, shops like West's may have sold both London and Liverpool clocks. A few comparatively inferior London clocks certainly exist, but no unkind criticism is implied in this observation. These somewhat coarser pieces, apart from mostly being early for carriage clocks, would have cost in the first place a

---

(3) "First the coach and six horses, with two postillions, coachman, and three outriders; a post-chaise and four post-horses, phaeton and four followed by two grooms, a chaise marine with four horses carrying the numerous services of plate — this last was escorted by the under-butler, who had under his command three stout fellows; they formed part of the household, and all were armed with blunderbusses. Next followed the hunters with their cloths of scarlet trimmed with silver and attended by the stud-groom and huntsman; each horse had a fox's brush tied to the front of the bridle. The rear was brought up by the pack of hounds, the whipper-in, the hack-horses, and the inferior stablemen. In the coach went the upper servants, in the chariot the eccentric master's wife . . . . or . . . . she accompanied Mr. Archer in the phaeton, he travelling in all weathers in that vehicle, wrapped up in a swan's-down coat."

(4) Abbott, *Treatise on the Management of Public Clocks*... (1837). According to E. Beckett Denison in *A Rudimentary Treatise*, local time was still kept in 1850 in some places in the West of England.

(5) Very many Victorian and Edwardian gentlemen suffered from an obsession with punctuality which amounted almost to a mania. These were the customers for whom chronometer carriage clocks were made.

(6) Clocks sold by "James McCabe" were almost certainly an exception. Although of the highest quality, they seem always to have had lever escapements, and this despite the reputation of the firm for marine and pocket chronometers.

(7) This class of finishing was called "Exhibition work". Carriage clocks, with their bevelled glass side panels, offered unrivalled opportunities for showing off to perfection polished and spotted movement plates, pinions hollowed and glossed, wheels with six crossings polished and screwed to their collets, screws and steelwork finished "white" or blue; or, in a word, movements beautifully executed in every detail.

IX English Work and Workmen

fraction of the price of those in the Dent or Frodsham class. The fact that the cheaper clocks are still going today shows that they were good value. They are, however, "cast" in an altogether coarser mould than the élite pieces. It is long overdue for someone to pinpoint these differences. There is much pleasure and satisfaction to be found in learning to differentiate between clean and inferior case castings, vigorous or shapeless mouldings, fine or poor gilding, good or bad hands, dial-engraving, etc.; to say nothing of the infinitely subtle differences between movements and escapements finished to chronometer standards, as opposed to those examples of everyday clockmaking which amount to no more than good ordinary work.

DIFFICULTY OF ESTABLISHING DATES

As a rule it is exceedingly difficult to determine the dates of manufacture of English carriage clocks, for the following reasons:—

(1) Unlike watches, English carriage clocks do not usually have cases made of either gold or silver. They lack the provenance afforded by casemakers' marks and by the conclusive date-letters of London or other Assay Offices.

(2) Serial numbers found on clocks are not as a rule confined to a carriage clock series. Sometimes one number series embraces an entire stock sequence of clocks, watches and chronometers. Messrs. Dents of Pall Mall confirm that this was the case with their clocks and watches. Dents further say that the numbers used on goods sold from the Strand address were unrelated to those from Cockspur Street. Bearing in mind the permutations of proprietorship and of address used by several distinguished firms throughout their long careers, clearly it is unwise to rely only upon numberings as a means of endeavouring to establish dates.

(3) While the addresses of firms are usually recorded upon their clocks, it is not always known when premises were first occupied or when they were finally vacated. In this connection the information contained in the London Trade Directories is sometimes helpful.

(4) On the whole, the conventional stylings of carriage clocks as regards cases, dials and movements remained little changed throughout the ninety or so years during which these clocks were made. Every "maker" tended to obtain his cases, dials and hands from the same sources, although not necessarily in the same combinations. What is more, most London "makers", at any given period, obtained their "rough" movements from one of several specialist suppliers. Afterwards more often than not they employed the same combinations of workmen for all the processes which followed.

(5) With the notable exceptions of the workbooks of Vulliamy and of Victor Kullberg, no manufacturing records of any significant producer of carriage clocks appear to have survived. The Vulliamy books, however, cover most of the period 1797 to 1834, and show that the firm sold carriage clocks from 1826. The workbooks of Kullberg are almost complete. They cover the serial numbers of everything made from No. 740 of circa 1870 to 9978 of 1943. Clocks, watches and chronometers are, however, indiscriminately mixed as regards numbering.

In order to understand better the position outlined in paragraph (4) above, it must be explained that one of the more disconcerting results of the shared producers of cases, dials, "rough" movements, platform escapements, etc., is that it is possible to find early Victorian carriage clocks virtually identical in all respects and yet bearing different names. For instance, exceedingly similar clocks, made circa 1845 exist signed "James McCabe", "Charles Frodsham", "James Whitelaw", "Robert Roskell", "Barwise" and others.[8] On the other hand, Vulliamy's clocks from the first example of 1826 always look distinctive. From soon after 1843 Dents of Cockspur Street produced chronometer carriage clocks in the Gothic style. They also favoured exceedingly plain and elegant cases, often with double carrying-handles, and some-

---

(8) It would seem that the final dates given by Britten and Baillie for 19th century makers are not necessarily correct in every case. Occasionally, for one reason or another, a name continued but has not appeared in the Lists. The use of the name "Robert Roskell", for instance, apparently did not cease in 1830. There was a son of the same name. The name "James McCabe" (sometimes spelt Mc Cabe) certainly appears on pieces made long after his death in 1811, and this despite other McCabe business names used by successive members of the family until 1883. Mr. C. Clutton brings this point out clearly in *Watches*, p.140.

times made of polished nickel. A characteristic feature of many Dent carriage clocks was the persistent use of circular white enamel dials set in engraved, or engine-turned, or pierced-appliqué surrounds. By 1849[9] canted corners had appeared, but they were on the whole more characteristic of later clocks. At first the cants often had hooked and ogee decoration, but latterly almost all had applied reeded ornament repeated on the handles. On the other hand, "humpbacked" cases, made in emulation of Breguet, occur throughout the entire period when English carriage clocks were made. James Ferguson Cole almost certainly produced the first in 1823. John Roger Arnold was also early in the field. Jump sold clocks of similar appearance for a few years from about 1883, and so did Brock of George St., Portman Square. Just before 1914 Nicole Nielsen made to special order about six highly complicated "humpbacked" clocks, all apparently for one customer. In the early part of the present century Louis Desoutter produced superficially similar clocks, while much more recently Philip Thornton of Great Hayward has completed clocks of the same shape. Finally, as recently as 1963, Colonel H. Quill and Mr. J.S. Godman together devised and made yet another example having a perpetual calendar employing a modification of the system described by A.L. Rawlings.[10] All these clocks are, however, quite different mechanically, besides which it so happens that their dates are well known to within a few years.

CARRIAGE CLOCK PRODUCTION – SOME CELEBRATED WORKMEN

The actual methods of carriage clock production, particularly as regards the parts relating to movements and escapements, were so sub-divided that many hands went into the making of a major, even if small, assembly such as a platform escapement. The London workman John Travers (1849-1937)[11] who was very well known in the trade as a "chronometer escapement-maker", in fact performed a mixed sequence of operations, known at the time as "making the pivoting". Travers' chronometer work consisted of pivoting and polishing the escape pinion, mounting the 'scape wheel on its collet, and also in making completely the balance staff and its rollers. His versatility later included the cutting and finishing of chronometer 'scape wheels and the making of fusees.[12] His obituary says plainly that "he worked for every British manufacturer of note and for many well-known foreign firms". In his youth Travers was trained by the first Thomas Mercer, the well-known chronometer maker. In his old age, and as late as 1936,[13] Travers was still doing chronometer work for the celebrated firm of Kullberg, then owned and run by the late S. Lundquist. Travers' workbooks, which were in existence in 1938, showed that he completed upwards of 20,000 "pivotings" and a wide variety of other chronometer parts during his long working life. His association with Kullberg lasted for over fifty years. Although Travers' work was usually confined to marine chronometers, he would have been employed, when the need arose, in chronometer carriage clock work. It is worth remembering that when a carriage clock was to be given a chronometer escapement, then quite a different team of craftsmen, more highly paid than usual, were engaged in the processes of its finishing. On the other hand, a clock having a lever escapement would, apart from its platform, have been clockmakers' work. All the lever platforms appear to have been derived from a few common sources, and it is safe to assume that there were specialists in the art of making them. There is certainly no doubt that the lever escapement was far more practical than the chronometer for use in a widening market, especially if used for travel. A

---

(9) Moore's pattern book of this period shows carriage clock cases having canted corners; but, conversely, it also offers pillars, Gothic, castings with figures and fruit in high relief, *art nouveau* fifty years too early, besides the plain cases mentioned above as being used by so many makers circa 1845-50. A catalogue published by G.E. Frodsham in 1889 still offers four carriage clock case styles shown by Moore.

(10) *Science of Clocks and Watches*, 1944.

(11) Travers' Obituary will be found in the *Horological Journal* of February 1938, page 25.

(12) An unusual combination of tasks, since the skills demanded were so very dissimilar. Clever clock and watchmakers today often face very unenviable combinations of work. They are expected to embody much of the lost knowledge and "know-how" of numberless dead craftsmen, who themselves seldom made more than one part.

(13) His last billheads read:– "JOHN TRAVERS (LATE OF LONDON), 7, Spencer Square, Ramsgate, Kent. Chronometer Escapement Maker, Scape Wheel and Fusee Cutter".

IX English Work and Workmen

knock which would destroy a chronometer might well be survived by a lever. However, given the best attention[14] it is certain that a chronometer clock would have kept consistently better time. The two famous firms, Vulliamy and James McCabe, both of whom produced carriage clocks to very high standards, appear to have made them only with lever escapements. It was, of course, the improvements made to the English lever escapement in its more developed forms from about 1820 which made carriage clocks a practical proposition for manufacture. It is fairly certain that all the first carriage clocks had lever escapements, and this moreover at a time when the chronometer was considered paramount for timekeeping. Only at a slightly later stage did firms like the Dents, the Frodshams and Kullberg introduce it into some of their best pieces for "seconds-hunters".

Other workmen, famous in the trade, some within living memory, were Henry Mann "the great pivoter", William Sills "prince of finishers", and George Abbott, whose detents were probably the finest ever made. "Old" George Abbott "black-polished" his detents all over, including the spring. According to D.S. Torrens, Abbott was paid an extra 1/6d. per detent for his tour de force. This was in the period when a two-day marine chronometer of the very best quality still could be produced for about £17, and commanded a trade selling price from £21 to £25, except in the case of a special piece with auxiliary-compensation balance. These prices, which had become current by the year 1855, remained almost unchanged for perhaps 50 years.[15] Chronometer clocks, unless they had striking trains, cost about the same to produce. Top-class London-finished carriage clocks, like box chronometers, were often made nearly if not quite to "Exhibition" standards. Second-class London finishing and Liverpool work were proportionately less impressive and expensive. The difference between the best and second-class work, although highly significant by absolute standards, was such as to be undetectable by most people. From about 1869, several of the principal escapement makers (according to Professor Torrens) agreed among themselves to divide the work into two parts, namely "pivoting" and "planting", and to specialise in doing one or the other of these. This arrangement made for slightly lower prices and for higher efficiency. Presumably the main work involved in producing a platform escapement for a carriage clock would, therefore, after this period have been condensed into four main areas of responsibility. The first was "making the pivotings", the second the planting of the parts on the platform, the third the jewelling, and lastly the springing and timing. It is hoped that the preceding pages may have helped to show how many different permutations of trades went into the making of any particular unit or assembly. To further illustrate a situation which today seems almost unbelievable a few further comments, applicable to horology in general, may help the reader.

Prescot in Lancashire was the home of the "rough" watch and chronometer movement trade, and also many tools were made there. In 1825 there were thirteen different trades in the town, all connected to some extent with horology.[16] By 1863 there were twenty-eight different trades in Prescot, each occupying from one to thirty-five firms.[17] It is true that of these the vast majority were concerned with watches. As has been noted, very many workmen just made literally one item only, often something quite simple like one type of screw. There is a record somewhere (possibly in connection with one of the Arnolds) where it is stated that a certain artisan who made beautiful balance screws would not make a detent banking-screw (or it may have been the other way round). The late Frank Mercer, giving a lecture before the British Horological Institute on the 24th November 1955, mentions a workman called Tom Foster, whose job in connection with chronometers

---

(14) English chronometer clocks, unlike levers, could not be "regulated" except by a watchmaker. On the other hand people likely to own expensive clocks usually had in operation "wind-and-maintain" contracts, involving a weekly visit from their clockmaker.

(15) Torrens Op. Cit.

(16) According to Baines' *History, Directory and Gazetteer of the County Palatine of Lancaster* published in 1825.

(17) See Hogg's *Directory* mentioned in the Bibliography. The most notable local makers of chronometers and chronometer parts (and therefore presumably when occasion demanded it of "rough" movements for chronometer carriage clocks) were Richard Doke, wheelcutter, and Jonathan Welsby, pinion maker, who were never surpassed. Other famous firms were I. & C. Webster, Hewitt Bros., Taylor Bros. and Joseph Preston & Sons, each of whom was skilled and employed a small staff. *(Horological Journal, January 1926)*.

was to make the motion wheels. He also made the up-and-down mechanism, but he could not drill a cannon pinion. Another man, Tom Howard, made beautiful barrels and barrel arbors, but could not read or write. The lecturer said "I don't think he had any gauges except a piece of steel with a couple of slots in it".[18] Horology, however, is not to be mistaken for a simple art. The bibliography of the subject is enormous, and for the past five or more centuries the continuous necessity of devising better methods of time-keeping and of time-determination has occupied some of the best "mathematical practitioners" of the principal nations. From the 18th century, if not before, the horological industries produced more specialised tools and appliances than did any other trade or profession until comparatively recent times. The prototypes of most modern machine-tools may be seen in the "engines" developed originally in connection with the manufacture of clocks and watches.[19] Perhaps the most obvious examples of this derivation are the screw-cutting lathes derived from fusee-engines, milling machines descended from wheel and pinion engines, press-tools developed from punches, and there are very many others. Three illustrated catalogues of horological tools were published by "Wyke", "Ford, Whitmore and Brunton", and by "Stubs" respectively in England from soon after 1770. Tools were also illustrated in Rees' *Cyclopaedia* of 1819-1820, to say nothing of numberless superb plates found in the more important French books and encyclopaedias.

J.F. COLE AND THOMAS COLE

The earliest English carriage clock of which the date is known is almost certainly the example made by James Ferguson Cole in 1823, the year of the death of Breguet.[20] Plate IX/2 shows the clock, which is of the beautiful silver-cased "humpbacked" design originated by Breguet in or about 1813. This Cole clock is superb. To say that it is in most respects equal to the finest *pendules de voyage* of A-L Breguet

Plate IX/2. *JAMES FERGUSON COLE. The silver "humpbacked" clock of 1823. This piece, which has a perpetual calendar, also shows the times of sunrise and sunset. (By courtesy of the British Museum)*

is not an overstatement. The inspiration for the clock is due entirely to him, but at the same time it is a fact that English work at its best was better finished than anything ever produced on the Continent. J.F. Cole, on account of his watches, has been called "the English Breguet"; but while no-one anywhere ever equalled A-L Breguet in ingenuity and inspiration, Cole and many others of the 18th/19th century English school could have taught even the great French genius something in terms of sheer execution. The clock is enigmatically

---

(18) In other words a "go or not-go" gauge.

(19) D.S. Torrens *Nail and Cork.* (*Horological Journal* Feb. 1938). This article is so informative that it is quoted in full in Appendix (e) together with notes entitled *Rule of Thumb.* Both were written to correct the impression that in the old days horologists worked by haphazard methods and without the aid of proper equipment.

(20) Chamberlain, the noted American writer, in his book *It's About Time*, 1941, gives a very charming account of his visit to Cole's son in 1924.

IX English Work and Workmen

inscribed "D",[21] but fortunately the date letter for London 1823 is struck clearly both on the top of the case and on the door. There is no doubt whatever about the date of manufacture. Plate IX/3 shows the movement. The reader will see that this is signed in flowing script "James Ferguson Cole, 3 New Bond St., London". It would be easy to write a whole chapter about this one clock, but it is necessary to keep the description short. The escapement is an "underslung" lever platform. All the "lift" is on the steel 'scape wheel of which the teeth are both drilled and slit in order to retain oil. There is a double-roller, a brass-and-steel compensation balance, and a "white" steel spring with overcoil.[22] This clock "D" of 1823 also has *grande sonnerie* striking and repeat on two curved steel gongs of rectangular section. There are "circular" striking racks, again in the manner of Breguet. One mainspring drives both trains. There is a going barrel with great wheels at either end. Returning to Plate IX/2, attention is drawn to the fly-back perpetual calendar working in a sector showing up to

Plate IX/3. *JAMES FERGUSON COLE. The movement of the 1823 clock. (By courtesy of the British Museum)*

Plate IX/4. *JAMES FERGUSON COLE. A carriage clock, cased in the style of Breguet, having simple calendar, grande sonnerie striking, and incorporating a musical box which may be set off at will. (By courtesy of Lady Prestige. Photograph by George Daniels)*

---

(21) Possibly representing serial number 4.

(22) The year 1823 is very early, in terms of English work, for these features. Quite apart from any other consideration, the disadvantage of a single-roller lever escapement (that of nearly stopping the balance by friction should a "knock" bring the safety dart into contact with the roller) should have led to its earlier demise in England. As it was, the double-roller only became general well into the last quarter of the 19th century, with Kullberg advertising that he at least always used it.

IX English Work and Workmen

31 days, with subsidiary dials for day and month. Elsewhere are sunrise and sunset dials, seconds and moon. According to the *Horological Journal* of 1859, Cole made several other such clocks, but it is not clear how many, since the writer speaks in the same breath of Cole's complicated watches. However, a similar clock was shown at the *Five Centuries of British Timekeeping* Exhibition in London in 1955.

Plate IX/4 shows another J.F. Cole piece, this time a 30-hour with cylinder escapement. Unfortunately it bears neither date nor number. The case is in the Breguet "Empire" style. This clock offers *grande sonnerie* striking and repeat by a lever underneath the base.[23] Plate IX/5 shows the musical work which may

Plate IX/5. *JAMES FERGUSON COLE. A view of the musical work under the base of the clock shown in Plate IX/4. (By courtesy of Lady Prestige. Photograph by George Daniels)*

be played at will and which is not set off by the clock.

Another "humpbacked" clock, this time in a silver-gilt case with the London date letter for 1825, bears the joined names of J.F. Cole and of his brother Thomas Cole. The clock is very much like the 1823 piece already described, having similar escapement,

Plate IX/6. *THOMAS COLE. A travelling desk clock made circa 1850. Note the key-drawer just above the dial. This clock is provided with a beautifully-fitting leather travelling box having apertures for reading both mean time and moon dials. (Private Collection)*

perpetual calendar, etc., this time in combination with *petite sonnerie* striking and repeat and with a pull-wound alarm. Unlike the 1823 Cole "humpbacked" clock, this example bears no identifying letter.

Thomas Cole is remembered today chiefly for the very ornate, beautifully engraved and gilt "desk clocks" and also "strut" clocks in which he specialised. These travelling pieces appear in many forms including even candlestick-clocks complete with their travelling boxes; but they were certainly supplied mainly for sale by others, notably by Payne of 165 New Bond Street, by Hancock, and by Hunt &

---

(23) According to Mr. Daniels the clock originally had a pull-repeat, but in the end it was accidentally pulled off a mantelpiece. Sotheby's Catalogue of 21st May, 1973 illustrates the clock complete with its travelling box.

IX English Work and Workmen

Roskell, to name but three.[24] Plate IX/6 shows a travelling desk clock signed "THOS COLE". Rather more common are his "strut" or "easel" clocks. Plate IX/7 illustrates a fine example. "Strut" clocks usually stand six or seven inches tall and mostly they have engraved silvered dials set in gilt framings and admirably complemented by fleur-de-lys hands. These last were a speciality of the Pendleton family in Prescot.[25] A serious defect in many of T. Cole's otherwise admirable clocks, is to be found in their lever escapements. These are usually planted on their sides on the back plates (the worst position known) and to say that they are not good would be to flatter them. Similar single-roller escapements survive in great numbers in inferior English lever watches, but why Cole should have used them in his superior little clocks will remain a mystery for ever.[26]

The *Illustrated Catalogue* of the Great Exhibition of 1851 published by *The Art Journal*, page 111, contains a steel engraving of an inkstand calendar clock by Thomas Cole. It is designated as "a compendium for the writing table, made and contributed by Mr. Cole of Clerkenwell". A more typical example of the work of this maker, and for that matter of early Victoriana, is impossible to imagine. The clock and calendar are housed in the lid of the inkstand. When the compendium is open pen tray, inks, sand, calendar, time, etc. are simultaneously offered under the hand of the scribe.

Plate IX/8 shows a carriage clock signed "Hancock" but with Thomas Cole's name on the movement. Thomas Cole also made miniature eight-day carriage clocks. One collector in Surrey has two examples. The first clock, only 2½ inches tall, strikes the hours. It is wound at the front for the going train, and at the rear for striking. The pointed-tooth lever escapement with overcoiled balance spring is planted on the back plate, as are the striking rack and snail. A lapis lazuli stone is set in the top of the case. While the clock bears the name "HANCOCK" below the dial, inside it is signed and dated "Constructed and made by Thos. Cole, LONDON, 1862". In appearance this clock is very similar to the full size carriage clock shown in Plate IX/8. The second Cole miniature stands 3¼ inches tall including handle and is wound from the

Plate IX/7. *THOMAS COLE. A strut or easel clock. These portable boudoir clocks, usually timepieces only, are as a rule provided with two means of support. The first is an easel-type back prop, which tilts a clock slightly backwards. The second is a two-position turnbuckle foot which holds a clock nearly vertical. Both supports are independent of each other and stow away flat, leaving a very slim clock. Strut clocks were always provided with travelling boxes. (By courtesy of Evans & Evans, Alresford).*

---

(24) Sometimes these clocks bear only the vendor's name; but often Thomas Cole's name is hidden on some part of the totally-enclosed movement or on the case in some place which is normally quite inaccessible.

(25) In March 1935 the late Professor Torrens visited Peter Pendleton, the last of the line, for several days on end. He (Torrens) made sketches of Pendleton's workroom and of all his tools; also notes "Things I watched him do".

(26) T. Cole perhaps did not always employ poor escapements planted in unfavourable positions. One "strut" clock, chiming Whittington quarters on gongs, and striking hours, employs a Savage two-pin escapement with compensation balance and planted on top of the movement. This particular piece is a bedroom clock. The chimes are so subdued that they would never awaken even the lightest sleeper, while still being just audible. No Cole signature has yet been discovered anywhere on this clock. However, the overall style strongly suggests his work despite the fact that most, if not all, of his clocks seem to have been timepieces.

IX   *English Work and Workmen*

Plate IX/8. *THOMAS COLE/HANCOCK. A timepiece carriage clock with lever platform escapement and with silvered panels either side of the case, one with a thermometer and one with a hand-set perpetual calendar. The back movement plate is signed "C.F. HANCOCK", but Cole's name is engraved on the left-hand side panel deliberately concealed by the applied framing. (By courtesy of Evans and Evans, Alresford)*

back. It is a timepiece and has a lever escapement, once again on the back plate, but this time not having an overcoil spring. The side and top panels are in malachite. As in all Cole clocks, the case is finely engraved all over. This second example bears the name "R. & C. Garrard & Co., Panton Street, Haymarket, London. Made by T. Cole".

J.F. Cole's[27] Obituary is to be found in the *Horological Journal* of February 1880, where attention is drawn to his interest in a number of modified forms of the lever escapement. This particular subject is well covered by Chamberlain in *It's About Time*. Thomas Cole's Obituary appears in the *Horological Journal* of February 1864. He is described and much praised "as a maker of portable clocks". Beautiful though the Thomas Cole clocks often are, it is nonetheless true that his mechanical work was never in the same class as that of his much more famous brother.

Mr. John Hawkins of Sydney, Australia, who has collected the work of Thomas Cole for some years, is of the opinion that in later life Thomas may well have concentrated his own abilities more upon the cases than upon the movements of the clocks which passed through his workshop. These cases are superb by any standard. For sheer quality they surpass not only the best French work but also the gilt metalwork of such makers as B.L. Vulliamy, or Weeks of Coventry St., both of whom were far better known.

The construction of all Thomas Cole clocks is based upon a central casting. These central castings will be found to be completely flawless. They are precise in their definition, having all edges absolutely square and clean. The applied mounts are also cast. They are built up in layers, pierced so as to give a perfection of line unobtainable by direct casting. Each layer is engraved to a standard not excelled by any working silversmith engraver in the 19th century. A characteristic of Cole engraving is often that the background is cut away leaving the pattern in relief. This method alternates, and also deliberately contrasts, with more standard methods of line engraving. The results are very impressive indeed. A further feature of Thomas Cole's clocks is that the dial chapters are nearly always painted and are not engraved. In the context of so much engraving, such a method is unexpected.

Mr. Hawkins is hoping to publish in 1974 a specialist book on the work of Thomas Cole. The above observations on cases and engraving are due to him.

B.L. VULLIAMY

Vulliamy probably did not make many carriage clocks, but he was early in the field. One of his workbooks[28] preserved at the British Horological Institute shows that on 13th June 1826 he sold to Lord Yarborough a small balance-controlled clock with "... detached lever escapement, compensation curb... in a red Sanderswood case..." The movement of this clock, which was numbered 873, was made by Holmden, whose name appears regularly in the workbooks mainly in connection with carriage clocks. In the same year similar clocks, numbered 883 and 886 respectively, were sold to the Marquess of

Plate IX/9. VULLIAMY NO. 1052. *An hour striking carriage clock sold in 1834 and having knock-out repeating, single-roller pointed-tooth lever escapement, plain balance and compensation curb. The case is veneered in rosewood and the travelling box is made of solid mahogany. (By courtesy of Dr. David A. Fermont)*

---

(27) By the way, he was named after James Ferguson, the great Scottish astronomer.
(28) The workbooks cover clock Nos. 297 to 496 and 746 to 1067. Mr. Geoffrey de Bellaigue, Deputy Surveyor of the Queen's Works of Art, has found in the Public Record Office early papers relating to the activities of the Vulliamys from about 1800-1815/20. These papers deal almost exclusively with the firm's ornamental work, such as chimney pieces, candelabra and inkstands. There are, however, invoices relating to the sale of clocks. In 1854 the Vulliamy business was acquired by Charles Frodsham.

IX English Work and Workmen

Worcester and to the "Duke" of Leicester. In 1827 No. 887 was made for Count Warrinzon. In 1829 No. 882 was sold to Lord Leicester, and in the same year Count Warrinzon was again a customer, purchasing

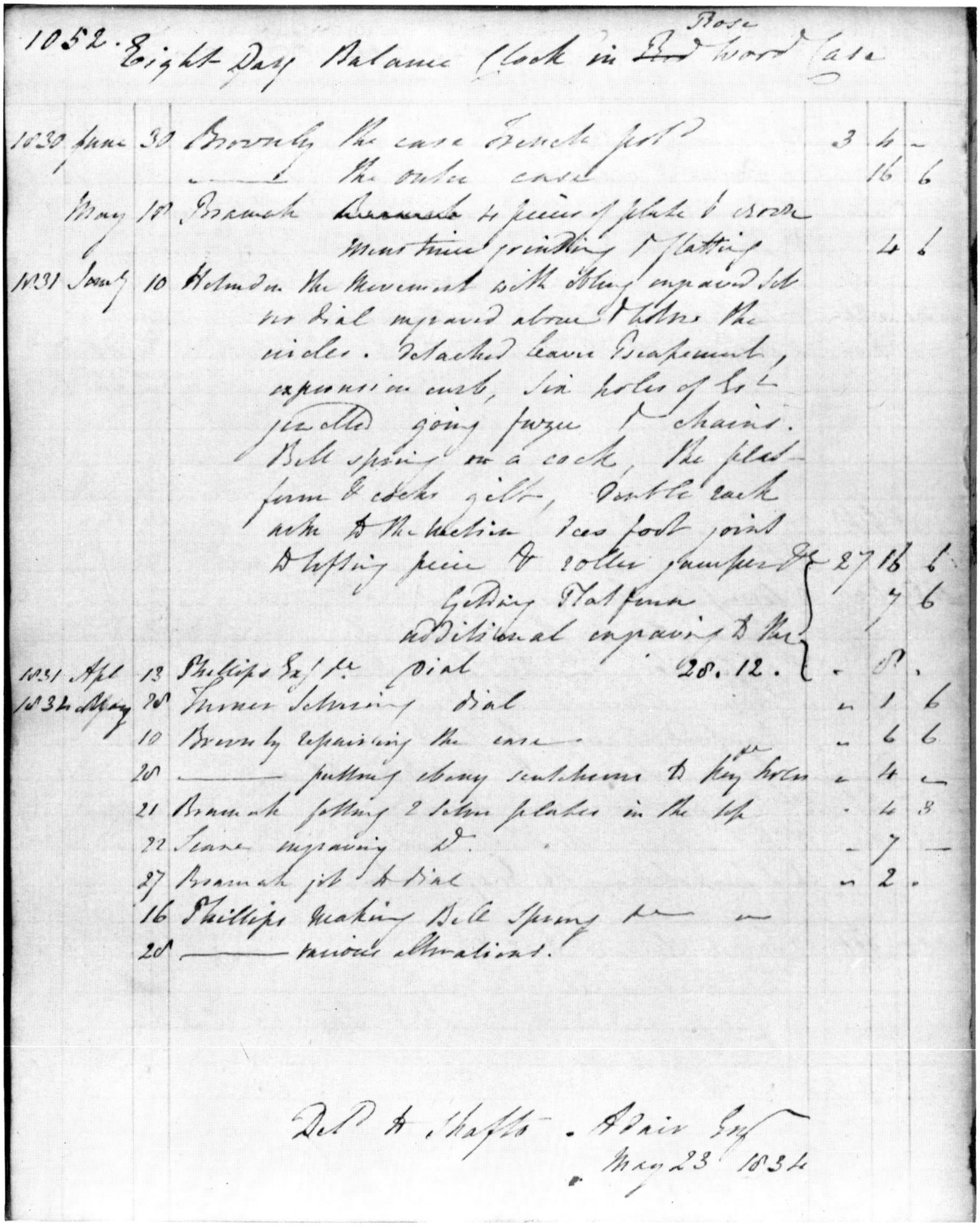

Plate IX/10. *VULLIAMY'S WORKBOOKS. The entry with reference to the carriage clock No. 1052 sold to Shafto Adair. (By courtesy of the British Horological Institute and of Percy G. Dawson)*

Nos. 1016 and 1017. All of these clocks were housed in wooden cases, either of rosewood or of "redwood" (presumably mahogany), and all of them had lever escapements. Numbers 1016 and 1017 are recorded as having compensation curbs. Plate IX/9 shows clock No. 1052 in a rosewood case. Its date is 1834, and it is very typical of Vulliamy carriage clocks in general. It cost £34.2s.3d. to produce, and was sold to Shafto Adair Esq. in May of that year at a figure not revealed. Once again, the movement was made by Holmden, as were all those mentioned between 1826 and this date. Plate IX/10 reproduces the entry in Vulliamy's workbook relating to the Adair clock No. 1052. The handwriting of the entry is not always easy to decipher, but it reads approximately as follows:—

The clock No. 1052 strikes the hours and has a pull knock-out for repeating. The escapement is a single-roller type, used in conjunction with a plain balance and a compensation curb. Vulliamy no doubt considered this combination adequate in the circumstances. It was certainly rather old-fashioned. The brass balance has two arms and four quarter screws. A slow train is used, the balance vibrating quarter seconds.

Early Vulliamy carriage clocks all have wooden cases. Probably the use of this material was dictated by the sound, practical consideration of reducing weight as much as possible. Vulliamy, of all people, would have been in the position of being able to supply superb metal cases had he wished to do so. (see Bibliography). His wooden cases are, however, both

1052.   Rose
        Eight Day Balance Clock in R̶e̶d̶ Wood Case
1830 June 30   Brownly the case French pol^d 3   4   -
                "   the outer case   16   6
       May 18   Bramah B̶r̶a̶m̶a̶h̶ 4 pieces of plate to cover
                       mens time grinding & flatting   4   6
1831 July 10   Holmdn. the Movement with oblong engraved
                silver dial engraved above and below the
                sides. Detached leaver Escapement,
                expansion curb, pin holes of Es^t
                jewelled going fuzee & chains.
                Bell spring on a cock, the plat-
                form & cocks gilt, Double rack
                (?) to the (?) (?) foot joint
                to lifting piece & roller jumper etc.   ⎰ 27. 16. 6
                       Gilding Platform   ⎨   7. 6
                       additional engraving to the   ⎩
1831 Apl 13   Phillips Esq. dial   28.12.   8. -
1834 May 18   Turner Silvering dial   1   6
         10   Brownly repairing the case   6   6
         28   "   putting ebony scutcheons to key
                                          holes   4   -
         21   Bramah filling 2 silver plates in the top   4   3
         22   Scane engraving etc.   7   -
         27   Bramah job to dial   2
         16   Phillips making Bell Spring etc.
         28   "   various alterations

            Del^d· to Shafto Adair Esq.
                    May 23 1834

IX English Work and Workmen

beautiful and unmistakable. While they are not all the same, they tend to have certain features in common. These include round-section carrying handles (usually silvered), and also secret, sliding, chamfered top panels beneath which the escapements are to be seen under glass. Other features common to the Vulliamy cases are the peculiar and very well-made locks which they invariably embody. It is probably reasonable to believe that these may have been made by Bramah, with whom Vulliamy was associated. Certainly they are nothing like the later famous "Bramah locks". The wooden cases are superbly made, demonstrating a standard of clean, vigorous work which seldom has been excelled.

The carriage clocks listed in the workbooks up to 1834 all struck the hours, but none of them repeated.

It is true that No. 1052 is provided with a pull-cord for repeating the hours; but this device, usually incorrectly termed a "pull-repeat" in the trade, only knocks out the rack hook of an ordinary striking train in order to make a clock repeat the last hour. At risk of stating the obvious, this arrangement was never intended for continual use, and the striking train will run down long before the going if much use is made of such a device. By contrast, clocks provided with repeat-buttons will be found not only to have striking trains of different counts (allowing for many more blows than usual) but also detents for holding up the striking trains while the racks are falling. The fundamental difference between the two arrangements, although seldom acknowledged, is very considerable.

Plates IX/11 and IX/12 show another clock No.

Plate IX/11. *VULLIAMY NO. 1361. A typical Vulliamy carriage clock striking hours and housed in a rosewood case. Note the Strike/Silent lever below the dial. (Private Collection).*

Plate IX/12. *VULLIAMY NO. 1361. The movement of the clock shown in Plate IX/11. This clock has a lever escapement and compensated balance. (Private Collection)*

IX English Work and Workmen

1361 in a rosewood case. It is likely to have been made a few years before 1840. Plate IX/13 illustrates a fine and highly unusual Vulliamy three-train, thirty-hour carriage clock No. 1420 with *grande sonnerie* striking. Other than clocks made by J.F. Cole, this appears to be one of the very few wholly English carriage clocks extant with this type of striking. There are three bells; two for the ting-tang quarters, which are struck first, and one for the hours. There is no repeat. The short duration of going was probably dictated more by lack of space than any other reason. The movement, which is superb in the sense that the finishing is almost up to that of a good chronometer of the same period, fills the case so completely that there would have been no room for an extra wheel and pinion in each of the three trains. This clock is typically English in having a 14,400 train. In other words its balance vibrates four times per second. Plate IX/14 shows the lever escapement, which is rather more developed than those found in earlier clocks. Scratched upon the movement out of sight is the inscription "Made by J. Couttes, 1838". It is not known how much importance should be attached to this message, which if correct would have meant a very sudden large increase in clock production over that shown in the workbooks. Plate IX/15 illustrates the movement. The cranked winding key shown in Plate IX/13 is not original, but makes the necessary daily winding a pleasure instead of a tedious performance.

Plate IX/13. *VULLIAMY NO. 1420. Date circa 1838. A very rare* grande sonnerie *clock going for thirty hours. The mahogany case is veneered in ebony. (Private Collection, Canada.)*

Plate IX/14. *VULLIAMY NO. 1420. The escapement platform and balance of the clock shown in Plate IX/13. (Private Collection, Canada.)*

245

IX English Work and Workmen

Plate IX/15. *VULLIAMY NO. 1420. A side view of the three-train movement of the clock shown in Plates IX/13 and IX/14. Note the short arbors and the overall high quality. Scratched on this movement are the words "Made by J. Couttes, 1838". (Private Collection, Canada)*

Vulliamy Carriage Clock Numbers found in Work Books

Serial No.	Date Sold	To:
873	1826	Lord Yarborough
882	1829	Lord Leicester
883	1826	Lord Worcester
886	1826	Lord Leicester
887	1827	Count Warrinzon
915	1827	Speaker of House of Commons
1016	1829	Count Warrinzon
1017	1829	Count Warrinzon
1019	1829/30	No customer's name
1042	1835	W. Garnier
1043	1831	Count Warrinzon
1051	1852	Customer not known (date of final delivery)
1052	1834	Shafto Adair Esq. (End of Books)

Other Known Clocks

1361	perhaps	c.1837	Still in purchaser's family.
1420	"	c.1838	Customer not known
1422	"	c.1838	Customer not known

All the clocks listed above have wooden cases, except for No. 1422 which has a metal case. No. 1420, the clock with *grande sonnerie* striking, is veneered in ebony. The others have cases of rosewood, redwood, or "red Sanderswood" (probably the same as redwood).

JOHN ROGER ARNOLD

The carriage timepiece shown in Plates IX/16 and IX/17 is interesting because it is signed "Arnold 84 Strand" and is numbered 543. Dr. Vaudrey Mercer, whose important book *John Arnold & Son, Chronometer Makers, 1762-1843* was published in 1972, says that in the course of his researches he has encountered Arnold house clocks addressed 84 Strand and numbered respectively 435, 500 and 635.[29] When were these pieces made? Bearing in mind both the signature and address found alike on all four clocks, then only two periods of manufacture seem to be possible; namely before 1830 or after 1840. It is known that the earliest date at which Arnold was at 84 Strand was 1821. From 1830-1840 he had E.J. Dent as his partner at 84 Strand and work sold during this period carried both their names. From 1840 until his death in 1843 Arnold was once again on his own at 84 Strand. Common sense suggests that 635 pieces would have been altogether too many to have been produced during the short period of life remaining to John Roger Arnold after the partnership was dissolved. It is far more likely that the clocks in question date from the period 1821-1830. A production of some 600 clocks spread over about nine years would seem to be entirely statistically probable. Of course it is always just

---

(29) No. 435, which is a four-glass mantel clock in a mahogany case, is illustrated in *Antiquarian Horology*, December, 1963, page 128.

IX *English Work and Workmen*

possible that Arnold may have abandoned a clock number sequence from 1830 and then have gone back to it after 1840. At present there is no evidence; but Dr. Mercer is collecting serial numbers of Arnold clocks so that eventually it should be possible to date them accurately.

THE DENTS

Although much has been written about the Dent family[30] it still seems necessary to include here a few new facts in addition to those which are already well known. Obviously it is not practicable to attempt to write sensibly about a long succession of undated Dent carriage clocks without trying at least to fit them into some sort of sequential pattern relating to the known facts. For this reason it has seemed necessary to compile a brief Dent history before offering a select list of clocks with some estimates as to their dates.

The family business was started by the famous Edward John Dent (1790-1853). The early period of his life, and particularly the years prior to 1814 and up until 1830, are obscure. It seems however that circa 1810 E.J. Dent worked with Richard Rippon,[31]

Plate IX/16. *JOHN ROGER ARNOLD NO. 543. Date of manufacture probably circa 1828. Timepiece. "Humpbacked" case in gilt brass. (By courtesy of Arthur Lister, London)*

Plate IX/17. *JOHN ROGER ARNOLD NO. 543. The movement of the clock shown in Plate IX/16. Note the position of the lever escapement. (By courtesy of Arthur Lister, London)*

---

(30) For instance by Chamberlain in *It's About Time*, and by Arthur Tremayne in *London's Great Family of Horologists*.
(31) A very fine eight-day box chronometer by Richard Rippon of Long Acre, London and numbered 126 was sold at Sotheby's on 18th October, 1971. Lot 89. This chronometer would have been made after 1816 and before 1827.

247

IX   *English Work and Workmen*

Plate IX/18. *ARNOLD & DENT. This clock, which is not numbered, was made between 1830 and 1840. It stands just over seven inches tall with the handle up, strikes and repeats hours, and has a lever escapement. (By courtesy of Evans and Evans, Alresford)*

IX   *English Work and Workmen*

Plate IX/19. *ARNOLD & DENT. The movement of the Arnold & Dent clock shown in Plate IX/18. Note a very typical English movement back plate. (By courtesy of Evans and Evans, Alresford)*

IX English Work and Workmen

whose widow Elizabeth he was eventually to marry. Historians are generally agreed that E.J. Dent later worked for both Barraud and Vulliamy. In 1814 E.J. Dent made a standard astronomical clock for the Admiralty, besides chronometers for determining longitude for the Colonial Office African Expedition. In 1829 he made the chronometer No. 114 which won the sum of £300 from the Admiralty.[32] Between 1830 and 1840 E.J. Dent was in partnership with John Roger Arnold. The clock shown in Plates IX/18 and IX/19 is signed "Arnold & Dent, London" and therefore it cannot have been made later than 1840. The year of its manufacture may indeed have been closer to 1830. After all, Breguet used this style of case from soon after 1800. Front-winding, as used by Arnold and Dent in this clock, is a most sensible feature. Unfortunately it is rare in English carriage clocks except for those by Vulliamy. The Arnold and Dent clock strikes hours on a gong, a practice common enough in English carriage work from the beginning.[33] In France, with the exception of *pendules de voyage* made by Breguet, bells were used for most of the earlier examples. The Arnold and Dent movement is typical of English work throughout the whole of the period when carriage clocks were made.

E.J. Dent, who traded as "E.I. Dent", started in 1840 the famous family business which has continued ever since. He was awarded the contract for the Westminster clock ("Big Ben"), completed in the time of his stepson Frederick. One fact, however, which appears to be far less well known, is that for a time there existed two totally separate Dent establishments in competition with each other and that eventually the two re-merged. The table of Dent dates and addresses which follows was deduced from various London Post Office and Trade Directories, and from correspondence with Mr. Daniel Patrick Buckney (great great grandson of the first Thomas Buckney) and with Dr. Vaudrey Mercer. Daniel Patrick Buckney, who joined the firm of E. Dent & Co. in 1934, was Managing Director from 1956 to 1968. The Buckney family had a very long connection with the Strand side of the Dent firm, although not all of them worked in it. Patrick's father had his own watchmaking business which was later absorbed into Dent.

Dent Dates and Addresses

1814	Dent, E.J. ("E.I. Dent")	
1830	Arnold & Dent	84 Strand.
1839	Arnold & Dent	84 Strand.
	"Chronometer, Watch and Clockmakers to the Queen".	
1840	Arnold and Dent parted.	
1841	Arnold, J.R.	84 Strand.
	"Watch and Chronometer maker".	
1841	Dent, E.J.	82 Strand.
	"Appointed Chronometer Maker to Her Majesty by Special Warrant".	
	"Appointed Chronometer Maker to H.R.H. Prince Albert by Special Warrant".	
	(Dent Advertisement, 1879. See Bib.)	
1843	Dent, E.J.	82 Strand.
1844	Dent, E.J.	82 Strand and 33 Cockspur St.
1846	Dent, E.J.	82 Strand, 33 Cockspur St. 34 & 35 Royal Exchange.
1852	Dent, E.J.	61 Strand. 33 Cockspur St. 34 & 35 Royal Exchange.
1853	E.J. Dent died, leaving his business to be divided between his stepsons, Frederick William Rippon and Richard Edward Rippon, on condition they took his name.[34]	

---

(32) This piece was sold at Christie's salerooms in London on 7th June 1972.
(33) Vulliamy carriage clocks are a notable exception. Clocks sold by the Dent family will mostly be found to employ gongs for hours and bells for quarters.
(34) Frederick William was to take over the Strand shop, the Royal Exchange shop and also the clock and compass factory in Savoy Street. Richard Edward received the Cockspur St. shop and also £10,000 towards stock. Both continued to use the name "Dent" on their respective shops. This caused trouble because Frederick claimed that Richard was obtaining credit for E.J. Dent's inventions and also for the Great Clock.

IX English Work and Workmen

1853	Dent, F.W. (formerly F.W. Rippon)	61 Strand, 34 & 35 Royal Exchange. Factory Savoy St.	1853	Dent, R.E. (formerly R.E. Rippon)	33 Cockspur St.
			1856	R.E. Dent died and his widow Marianna Frederica Dent (formerly Rippon; née Cowslade) continued the business using his name.	33 Cockspur St.
			1858	Dent, M.F.[35]	33 Cockspur St.
1860	F.W. Dent died. The firm continued under the control of F.W. Dent's brother-in-law, the first Thomas Buckney, and F.W. Dent's sister Mrs. Amelia Lydia Sophia Gardner (née Rippon). They adopted the new title of "E. Dent & Co."	61 Strand, 34 & 35 Royal Exchange.			
1862	E. Dent & Co.	61 Strand, 34 & 35 Royal Exchange.	1862	Dent, M.F.	33 & 34 Cockspur St.
1872	E. Dent & Co. Advertisement front cover *Horological Journal*, December 1872:— "NOTICE. E. Dent & Co. caution all Engravers, Dial Painters, and others, against engraving or marking upon any piece of work their name and addresses, or any such names or addresses as shall bear a fraudulent resemblance to theirs, without a written order signed by themselves. Any person giving such information as shall lead to a conviction under the Trades Marks' Act, will be handsomely rewarded". Addresses given are: 61 Strand, 34 Royal Exchange and factory Savoy Street, London. Advert. claims "Manufacturers of Watches, Chronometers, Astronomical and Turret Clocks to Her Majesty...."		1872	Dent, M.F. Advertisement in *Horological Journal*, December 1872:— "The Astronomer Royal in his Report to the Admiralty (13th August, 1870), on 40 Chronometers entered for the annual competition, says of M.F. Dent's Chronometer 'This is the finest chronometer we have ever had on trial'..." Advert. claims "Chronometer, Watch and Clock Maker to the Queen..."	34 Cockspur St.
1876	E. Dent & Co. Application filed for triangular trademark with the word "DENT" within and with the words "TRADE MARK" over the pediment.[36] A new number sequence was begun starting at No. 38,000.				
1887	E. Dent & Co.	61 Strand, 4 Royal Exchange.			

---

(35) It seems that from 1858 M.F. Dent, who had previously traded as "Dent", agreed to waive all claims to E.J. Dent's past achievements and to trade as "M.F. Dent". This she and later her successors did until 1921 when the two companies merged.

(36) In 1879 a booklet with 28 pages and illustrations was published by E. Dent & Co. advertising the adoption of the trademark and listing the "Principal works executed by the House since its foundation". This pamphlet was produced in order to further differentiate between E. Dent & Co. and M.F. Dent. On every page is stated that no instrument numbered higher than 38,000 is genuine unless it bears the Dent triangular trademark. Addresses given are: 61 Strand, 34 and 35 Royal Exchange, Factory Gerrard Street.

IX English Work and Workmen

1897	E. Dent & Co. Ltd.	61 Strand, 4 Royal Exchange	
		1916 (?) Dent, M.F.	28 Cockspur St.
1921	E. Dent & Co. Ltd. (M.F. Dent merged with E. Dent & Co. Ltd.)	28 Cockspur St. 4 Royal Exchange.	
1937	E. Dent & Co. Ltd.	41 Pall Mall, 4 Royal Exchange.	
1940	Watchmaking firm of Daniel Buckney absorbed.		
1944	The shop at 4 Royal Exchange was closed leaving the 41 Pall Mall address for E. Dent & Co. Ltd. (The firm is still in business managed by Mr. K.A.G. Butcher)	41 Pall Mall.	

DENT CARRIAGE CLOCKS
EXAMPLES PRESUMED TO HAVE BEEN SOLD FROM THE STRAND ADDRESSES IN THE LIFETIME OF E.J. DENT

SERIAL NO. & ESTIMATED DATE	NAME & ADDRESS AS SHOWN ON CLOCK	SPECIFICATION	ILLUSTRATIONS
517 (1844)	Dent, 82 Strand London	Timepiece. Single-roller lever escapt. Gold balance with flat spring. Height approx. 6½in. Brass case, bronzed and gilt. Engraved dial surround.	Clock: Plate IX/20 This clock was sold at Sotheby's on 19th June 1972, Lot 48.
693 (1845)	Dent, London	Chronometer. Timepiece. Patent balance. Plain gilt case with deep well top. Pierced and engraved overlaid dial surround.	Clock: Plate IX/21
1302 (1850)	Dent, London (not known whether any further address)	Chronometer. (?Timepiece). Very attractive engraved and gilt case with deep well top. Engraved dial surround.	Clock: Plate IX/22
1503 (1852)	Dent, 61 Strand London	Timepiece. Lever escapt. with compensation balance. Plain gilt case with well top. Engraved dial surround.	Clock: Plate IX/95 (extreme right top row)

IX   *English Work and Workmen*

DENT CARRIAGE CLOCKS
A SELECT LIST OF SOME KNOWN EXAMPLES ALL PRESUMED TO HAVE BEEN SOLD FROM 33 COCKSPUR ST. DURING AND AFTER THE LIFETIME OF E.J. DENT

SERIAL NO. & ESTIMATED DATE	NAME & ADDRESS AS SHOWN ON CLOCK	SPECIFICATION	ILLUSTRATIONS
12683 (1845-47)	Dent, London	Chronometer. Strike and repeat on a gong. Dent patent balance. Plain, rather narrow gilt case.	Clock: Plate IX/23 Balance: Plate IX/24 Plate IX/25
14880 (1849-50)	Dent, London	Chronometer. Quarter strike on bells. Hours on gong. Patent balance. Polished nickel case. Fold-flush 2-piece handle. Engraved dial.	Clock: Plate IX/26
17966 (1853)	Dent, London	Chronometer. Timepiece. Patent balance. Engine-turned dial. Very plain case with reeded handle.	Clock: Plate IX/27
20338 (1855)	Dent, 33 Cockspur St. London	Chronometer. Strike/repeat on gong. Patent balance. "Gothic" case.	Appearance as No. 21245
21245 (1856)	Dent, 33 Cockspur St. London	Chronometer. Timepiece. Standard chronometer balance. Gothic case.	Clock: Plate IX/28
21574 (1856)	Dent, 33 Cockspur St. London	Chronometer. Strike/repeat on gong. Standard chronometer balance. "Gothic" case. Height: 8in. with handle up.	Escapt: Plate IX/29 Clock: Plate IX/95 (extreme left top row)
22006 (1857)	Dent, 33 Cockspur St. London	Chronometer. Strike/repeat on gong. Balance with auxiliary compensation. Height 9in. with handle up. Plain gilt case with deep well top.	Clock: Plate IX/30 Escapt: Plate IX/31
23083 (1858)	Dent, 33 Cockspur St. London	Chronometer. Timepiece. Standard chron. balance. Plain case with fold-flush 2-piece handle.	Case style as No. 14880; but case gilt brass with engine-turned dial.

IX   *English Work and Workmen*

SERIAL NO. & ESTIMATED DATE	NAME & ADDRESS AS SHOWN ON CLOCK	SPECIFICATION	ILLUSTRATIONS
23715 (1860)	M.F. Dent, 33 Cockspur St. London. 23715 (on movement) Dent, Chronometer Maker to the Queen, 33 Cockspur St. London. 13715 (on dial)	"Giant" clock. Perpetual calendar. Equation of Time and moon. Quarters struck on eight bells. Hours on gong. Chronometer escapement. Patent balance. 3 trains, all with fusees. The going train drives the calendar via the motion work.	Dial: Plate IX/32 Back: Plate IX/33 Side: Plate IX/34 An inscription on this clock records that it was presented to a parson in Wanstead in August 1864; but there is no evidence when it was made.
24128 (1862)	M.F. Dent, Chronometer Maker to the Queen, 33 & 34 Cockspur St., Charing Cross, London.	"Giant" clock. Perpetual calendar. Equation of Time and moon. Quarters struck on eight bells. Hours on gong. Chronometer escapt. 4 trains, 3 with fusees. The fourth train, with going barrel, relieves the going train of the task of moving the calendar. Weight approx. 58 lbs. Height: 18½in. with handle up.	Overall appearance very similar to No. 25262. An inscription on the base of this clock records that it was presented to H. Custance in October 1879 by the Duke of Hamilton.
25262 (1865)	M.F. Dent, Chronometer Maker to the Queen, 33 & 34 Cockspur St., Charing Cross, London.	"Giant" clock. Perpetual calendar. Equation of Time and moon. Quarters struck on eight bells. Hours on gong. Chronometer escapt. 4 trains, 3 with fusees, and the calendar train with a going barrel. Weight approx. 60 lbs. Height: 19in. with handle up.	Clock: Plate IX/35 Movement: Plate IX/36 Plate IX/37 This clock was sold at Christie's 26th October, 1971, Lot 24.
30478 (1888) date known	Not stated in workbook.	Described in workbook as "a gilt, lever carriage clock, 8-day, striking on gong, made for Daniel Buckney".	
32571 (1898)	Dent, 33 Cockspur St. London.	Lever escapt. Strike/repeat (also strikes one blow at half hours to compete with French clocks) on gong.	Clock: Plate IX/38

IX   *English Work and Workmen*

Plate IX/20. *DENT NO. 517. Date probably circa 1844. A fairly early timepiece carriage clock with lever escapement and sold from 82 Strand. (By courtesy of Meyrick Nielson of Tetbury Ltd.)*

Plate IX/21. *DENT NO. 693. Date circa 1845. Chronometer timepiece with Dent's patent balance. (By courtesy of Christie, Manson & Woods)*

IX   *English Work and Workmen*

Plate IX/22. *DENT NO. 1302. Shown top left of photograph. Date circa 1850. Chronometer carriage clock, presumed to be a timepiece. Note the very attractive engraved case and dial surround. (By courtesy of N. Bloom & Son Ltd., London.)*

IX English Work and Workmen

Plate IX/23. DENT NO. 12683. Date circa 1845-47. Chronometer carriage clock with Dent's patent balance, and striking and repeating hours. The dial plate is overlaid with pierced and engraved appliqué work. (By courtesy of Garrard & Co. Ltd.)

Plate IX/24. DENT NO. 12683. The balance of the clock shown in Plate IX/23, showing Dent's patent balance having his "staple-shaped" compensation pieces. (By courtesy of Garrard & Co. Ltd.)

IX   English Work and Workmen

Plate IX/26. DENT NO. 14880. Date circa 1849-50. Chronometer carriage clock with Dent's patent balance and striking hours on a gong and sounding quarters on bells. No repeat. The beautiful case is made of polished nickel. The handle is made in two pieces which fold away when not in use flush with the top of the case. (Private collection)

Plate IX/25. DENT NO. 12683. Another view of the balance of the clock shown in Plate IX/23. (By courtesy of Garrard & Co. Ltd.)

Plate IX/27. DENT NO. 17966. Date circa 1853. Chronometer timepiece with Dent's patent balance. Note the introduction of an engine-turned dial used in conjunction with a very plain case and reeded handle. (Private Collection, New York.)

258

IX English Work and Workmen

Plate IX/29. *DENT NO. 21574. Date circa 1856. An escapement view illustrating a standard mid-19th century brass-and-steel chronometer balance, having circular weights and with two small weight screws in addition to the timing nuts at the extremes of the crossbar. This type of balance is as normal in Dent chronometer carriage clocks as the patent type. (Private Collection)*

Plate IX/28. *DENT NO. 21245. Date circa 1856. Chronometer carriage timepiece having a standard balance and housed in a beautiful Gothic-revival case. (By courtesy of Asprey & Co. Ltd.)*

Plate IX/30. *DENT NO. 22006. Date circa 1857. A chronometer carriage clock, striking and repeating hours on a gong. (By courtesy of Asprey & Co. Ltd.)*

IX   English Work and Workmen

Plate IX/31. DENT NO. 22006. *A view of the escapement platform of the clock shown in Plate IX/30. Note that the balance has auxiliary compensation. See also on the extreme right of the plate the vertical brass screen protecting the escapement from the air currents occasioned by the fly. (By courtesy of Asprey & Co. Ltd.)*

Plate IX/33. DENT NO. 23715. *The back of the movement of the clock shown in Plate IX/32. Note that while the movement bears the number 23715, the dial is numbered 13715. This discrepancy is discussed in the text. (Private Collection. Photograph by courtesy of Alfred Stradling, Cirencester Ltd.)*

Plate IX/32. DENT NO. 23715 (also numbered 13715). *Date circa 1860. This "giant" chronometer carriage clock has Dent's patent balance and offers quarter striking on eight bells, perpetual calendar, Equation of Time and moon. (Private Collection. Photograph by courtesy of Alfred Stradling, Cirencester Ltd.)*

IX English Work and Workmen

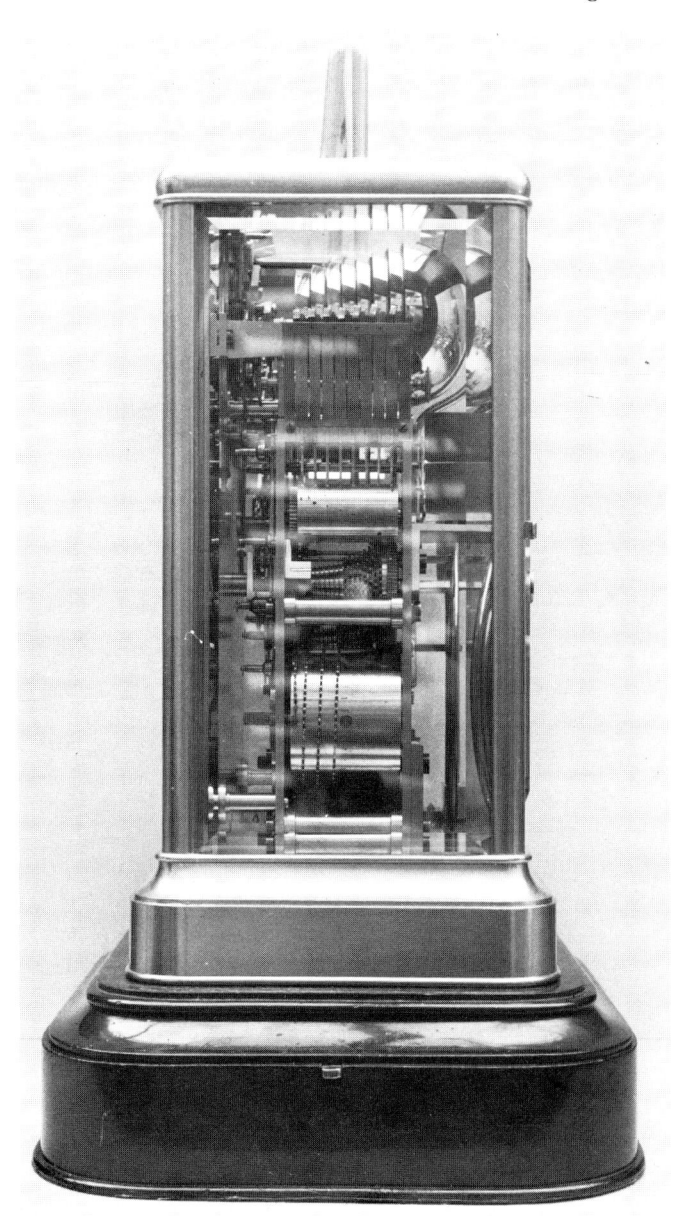

Plate IX/34. DENT NO. 23715. A side view of the clock shown in Plates IX/32 and IX/33. Note the turntable base enabling a clock weighing some 60lbs. to be rotated through 180° once a week for winding. (Private Collection. Photograph by courtesy Alfred Stradling, Cirencester Ltd.)

Plate IX/35. DENT NO. 25262. Date circa 1865. "Giant" chronometer carriage clock with Dent's patent balance, quarter striking on eight bells, perpetual calendar, Equation of Time and moon. Compare with the standard French gorge cased carriage clock (size No. 3, 7 inches tall with handle up) shown in the same photograph. (By courtesy of the late Commander C.A.K. Norman and of E. Dent & Co. Ltd.)

Plate IX/36. DENT NO. 25262. A view of the movement, looking into it from behind and downwards. The bells have been removed in order to show the general layout. (Acknowledgement as Plate IX/35 above)

261

IX   *English Work and Workmen*

Plate IX/37. DENT NO. 25262. *A side view of the movement of the clock shown in Plate IX/35. Note at bottom right the small subsidiary train, terminated in a fly, and used to move the calendar work. (By courtesy of the late Commander C.A.K. Norman and of E. Dent & Co. Ltd.)*

Plate IX/38. DENT NO. 32571. *Date circa 1898. Carriage clock with lever escapement striking hours and half hours and repeating hours on one gong. Interestingly enough, this late clock is signed simply "Dent". (Private Collection, Spain)*

It is apparent from the few surviving Dent workbooks, now at Pall Mall, that a single series of numbers was used for both clocks and watches emanating from Cockspur Street (but not for box chronometers). It would further appear that this number sequence, started in about 1844 or soon afterwards, began with No. 12000. The only clock in the whole sequence of which the date is known with certainty is No. 30418, made for Daniel Buckney in 1888.[37] One "giant" carriage clock was undoubtedly shown in the London International Exhibition of 1862, but the Official Catalogue does not mention its serial number. It is perhaps reasonable to assume that the Exhibition clock was No. 24128, if only because of its more developed perpetual calendar and also the use of the address "33 and 34 Cockspur Street". It seems improbable that No. 23715 would have been used for exhibition purposes, displaying as it does conflicting

---

(37)   It seems strange that Daniel Buckney, whose family alliance was with the Strand firm, should have ordered a carriage clock from Cockspur Street.

262

IX   *English Work and Workmen*

serial numbers and signatures.[38] The higher of these two serial numbers is the one most likely to be correct. The smaller serial number is so far out of period as to be almost certainly the result of an engraver's error. The fact that the two later "giant" clocks Nos. 24128 and 25262 are provided with small subsidiary trains for moving their perpetual calendars suggests that the first type may have given trouble; if not by stopping the clock at least by changing its balance amplitude and thus spoiling its rate. The "giant" clock No. 23715 is provided with a massive mahogany carrying box furnished with a leather strap in the manner of a marine chronometer outer travelling case. At least some of the standard-sized Dent "Cockspur" clocks were given French-style travel boxes covered in morocco leather and furnished with straps and with Bramah locks.

Dent carriage clocks bearing the addresses 82 or 61 Strand are far rarer than any from Cockspur Street. So far as is known only two Strand examples, conclusively addressed, have appeared in recent years. The first clock, which is addressed 82 Strand, is numbered 517 and it is shown in Plate IX/20. The other clock, numbered 1503 and addressed 61 Strand, is shown in Plate IX/95. Both these clocks, according to our reasoning, were made in the lifetime of E.J. Dent and before the division of the family business. We have yet to see a clock which we believe to have been sold from 61 Strand during the time of Frederick William Dent. It has already been said that the series of numbers used on Cockspur Street clocks and watches appears to have begun with No. 12000. It would appear, on the other hand, that E.J. Dent's clocks (but not watches) as sold first from 82 Strand and latterly from 61, carried serial numbers beginning at one. In addition to the clocks No. 517 and No. 1503 already mentioned, Nos. 693 and 1302 are also still in existence. So far as we know both are only signed and addressed Dent, London.

A final Dent clock which must be mentioned, although it does not fall into any number sequence or design pattern already mentioned, bears a serial number "1" and is shown in Plates IX/39 and IX/40. This clock, which offers *grande sonnerie* striking, is almost certainly continental in origin,[39] despite

Plate IX/39. *DENT. This* grande sonnerie *clock, bearing the serial number "1", strikes Westminster quarters on four gongs. It also has a chronometer escapement. (By courtesy of Asprey & Co. Ltd.)*

manifestly English finishing and also an English chronometer platform escapement. The clock strikes Westminster quarters on four gongs, a selection lever in the base offering alternative settings. The

---

(38) The dial is signed "DENT CHRONOMETER MAKER TO THE QUEEN 33 Cockspur St. LONDON 13715", while the movement is signed "M.F. DENT 33 Cockspur St. LONDON 23715".

(39) Perhaps in much the same manner as many perpetual calendar, minute-repeating watches were produced as *ébauches* in Switzerland but were finished, escaped and cased in England.

IX English Work and Workmen

Plate IX/40. DENT. The movement of the grande sonnerie *carriage clock shown in Plate IX/39. This clock employs a fusee in the going train, while the hour and quarter trains are driven by going barrels. (By courtesy of Asprey & Co. Ltd.)*

movement employs a fusee for the going train, while the striking and quarter work are driven by going barrels. This clock, which has a very high quality movement, presents something of an enigma. The style of case, movement and dial suggests a somewhat late date of manufacture. A similar clock exists bearing the serial number "2".

That the Dent family also sold entirely conventional French-made carriage clocks is clear from Chapter II. In fact, the Cockspur Street stockbooks of 1860-65 show that the firm imported large quantities of *pendules de voyage* from Vologne, from Raingo, and from Soldano. The last supplied clocks a dozen at a time. Sotheby's auction catalogue for 27th March 1972 illustrates (Lot 6) a typically French "giant" carriage clock in a *gorge* case. Almost certainly it was supplied by either Jacot or Drocourt. It is signed on the dial "DENT, 33 Cockspur St. LONDON". The piece is No. 8533 and stands eight and a half inches high.

Before leaving the Dent family it is necessary to reiterate that their best carriage clocks represent top class English work. Nothing better of the same kind was ever made anywhere. There is no direct evidence of the sources used by any of the Dents for "rough" movements of carriage clocks, but the few surviving workbooks in Pall Mall show that Glover supplied movements for marine chronometers during the years 1866-71, while Richard Doke provided others in 1874 and 1875. The name of Joseph Preston does not appear in the books until 1890, but from then onwards he supplied all the chronometer "rough" movements up until 1917. It is overwhelmingly likely that Glover, Doke and Preston would also have supplied movements for chronometer carriage clocks when they were required, and no doubt would have been willing to supply both the Cockspur Street and the Strand firms. As to E.J. Dent's patent balance, this appears in Specification No. 9302 of 1842 and it also figures in the *Nautical Journal* of the same year, where it is illustrated both as the "S" shape shown by Gould in figure 72a, and in the simpler form like a letter "U" lying on its side, as seen in Plate IX/24. E.J. Dent called this last arrangement "staple-shaped".

Readers who have studied the list of Dent Cockspur Street clocks may have questioned the large temporary increase in production which seems to have occurred from soon after 1850. Perhaps the reason was related to the upsurge in national prosperity which began soon after the Battle of Waterloo, reaching a peak at the time of the Great Exhibition of 1851. No doubt the Exhibition, at which E.J. Dent made a massive showing which included the Kings Cross Station clock set up on scaffolding, influenced sales and prestige both before and after the event.

Charles Allix will be grateful to receive from readers notes of serial numbers, signatures and addresses found upon both dials and back movement plates of "Dent" English carriage clocks. Of particular interest are "Strand" clocks and those bearing presentation dates and inscriptions which will help to provide further information for any future edition of *Carriage Clocks*.

THOMAS EARNSHAW II

Plates IX/41 and IX/42 show a carriage clock signed "Thos Earnshaw, 119 HIGH HOLBORN". On the basis of this name and address the clock, which has a duplex escapement, should not have been sold later than 1845. It has an engraved pillared case with bun feet very like one design shown in Moore's Pattern Book of 1849 which is mentioned in the Bibliography. To judge from the photograph, the movement of this Earnshaw clock is of most unusual construction, having the barrel in a sub-frame and being surmounted by a large escapement platform.

Plate IX/41. *THOMAS EARNSHAW II. Timepiece carriage clock with duplex escapement and with going barrel. Sold prior to 1845. (By courtesy of Percy G. Dawson)*

The late Malcolm Gardner made researches at the Guildhall Library in London after the last war, coming to the conclusion (despite the dates given by Britten and Baillie) that Thomas Earnshaw II, who was the son of the famous Earnshaw, remained at 119 High Holborn until at least 1842, in which year he apparently occupied both this address and No. 87 Fenchurch Street. Malcolm Gardner's unpublished notes read as follows:—

1842	119 High Holborn also 87 Fenchurch Street.
1845	2 Brewer Street, Clerkenwell.

Plate IX/42. *THOMAS EARNSHAW II. The movement of the Earnshaw clock shown in Plate IX/41. (By courtesy of Percy G. Dawson)*

1852	48 St. John Street, Clerkenwell.
1854	Same address, but ceased trading as a manufacturer, the original business being discontinued.
1867-71	Trading from St. John Street as a watch balance maker.
1872	JAMES EARNSHAW, compensation balance maker at Alfred Street, City Road.
1874	The same but from 104 St. John Street, Clerkenwell. These premises were then taken over by Henry Sainsbury & Co. watch and clock manufacturers with Earnshaw still in part possession.
1875	London entries cease.

It seems likely that Thomas Earnshaw II was established on his own account as a balance maker prior to his father's death in 1829, and subsequently ran both businesses until 1854, when the chronometer manufacturing was discontinued. Thus it is most probable that round this time the remaining stocks were sent to auction.

For a further 20 years THOMAS EARNSHAW III continued in business as a balance and escapement maker, and is reputed to have retired to Brighton, where he died.

265

CHARLES FRODSHAM AND FAMILY

Charles Frodsham's carriage clocks are far rarer than either those made by his distinguished contemporaries the Dents, or the fine examples produced in Victorian times under the name of James McCabe. It is reasonable to assume that most Frodsham carriage clocks have survived; and yet it has proved difficult to find examples with which to illustrate this chapter. Before looking at any pieces, it would be as well to consider briefly their background.[40]

There is evidence enough that the Frodsham family had their origins in the village of the same name on the Mersey, while there are also numerous references to Frodshams in the early Prescot and Huyton parish registers from 1580. A branch of the family came to London in the early 18th century.

William Frodsham (1728-1807) was the first member of the family who is known for certain to have been engaged in horology[41] although it is thought that his father, also William, may have been in the employment of Justin Vulliamy. William Frodsham founded the firm which was to continue, under various members of the family and at different addresses, until the present day. He opened his shop at 12 Kingsgate Street, Bloomsbury, although the date is not certain. He had previously been a watch jeweller and had been very friendly with Thomas Earnshaw from whom he had learned this art in about the year 1780. William took his son, also William, into partnership and traded under the name of "Frodsham & Son". This son[42] died before his father; therefore when William Frodsham (the one born in 1728) died in 1807, the business which he had founded passed directly to his younger grandson John.[43] The trading style then became "John Frodsham". In 1809 John took as a partner a chronometer maker called Baker, the two of them trading under the name of "Frodsham & Baker". The business remained in Kingsgate Street until 1823 when the firm moved to 31 Gracechurch Street. In 1849 John Frodsham died and his son George Edward inherited the business. He traded as "G.E. Frodsham". It is not clear what became of Baker. G.E. Frodsham's Catalogue published in 1889 (see Bib.) uses the words "Established 1796". The meaning of this statement is not clear. It is possible that after G.E. Frodsham's death the remaining assets of his business were assimilated by Parkinson & Frodsham, because the London Trade Directory for 1902 says simply "G.E. Frodsham, 31 Gracechurch Street, see Parkinson & Frodsham", while no entry appears for G.E. Frodsham for the year 1903.

William James Frodsham, the elder grandson of the William born in 1728, having learned watchmaking in the family workshop, began business on his own account. In 1801 he entered into partnership with William Parkinson at 4 Change Alley, Cornhill, and they established there the firm of "Parkinson & Frodsham". They mainly made chronometers and were very successful. When Change Alley was rebuilt they took fresh premises in the City, finally removing to 5 Budge Row where the firm was to remain until 1947.

On 1st January 1804, William James Frodsham married Hannah Lambert. They had eight children. Four of the sons were apprenticed to Parkinson & Frodsham. Two of these, William Edward and Henry, who were apparently twins, were articled on the same day, 13th January 1823. William Edward joined the Clockmakers' Company in 1847. Henry went to Liverpool and made chronometers and watches

---

(40) There exists an unpublished Frodsham history drafted soon after the second World War and instigated by the late Philip Clowes. Unfortunately this manuscript was never completed, and much further work and checking remains to be done.

(41) His horological standing was very considerable. In 1763 his services were retained by the Board of Longitude in connection with Harrison's "Timekeepers", while in 1804 he gave evidence in favour of Thomas Earnshaw against J.R. Arnold's claims.

(42) Britten states that this William married Alice, the granddaughter of "Longitude" Harrison"; but Col. Quill, in the course of extensive research made during the preparation of his very important and thorough monograph on Harrison, was unable to discover any evidence of such a granddaughter.

(43) John (1785-1849), the younger grandson of William Frodsham (1728-1807), presumably inherited the business because his brother William James (1778-1850) already had his own business prior to 1801 and from that year was a partner in Parkinson & Frodsham. It seems that Britten confused this John with his nephew born in 1812, while the Clowes history appears to be wide of the mark in stating that William James was the second son of William and Alice.

IX English Work and Workmen

similar to those for which that City was famous. We have the dial of his two-day chronometer No. 2202. Ironically enough, it seems that the Parkinson & Frodsham business was inherited by George, the eldest son, who had never been apprenticed. He was joined by his younger brother John who had been apprenticed in 1826 and who became a member of the Clockmakers' Company in 1840. The remaining indentured son was the Charles who was to become the most famous member of the Frodsham family. He was born on 15th April, 1810, was educated at Christ's Hospital and was apprenticed to his father on 14th July, 1824. In spite of the fact that the family business was thriving, Charles, already a successful chronometer maker, chose not to remain with it but started on his own in 1832. His early business addresses are not known; but he probably worked first from his private house at 5 Navarino Terrace, Hackney, and after his marriage in 1834 from Barossa Place, Islington. Having been established for five years he decided that it would be in his interest to open a shop in the City and accordingly in 1837 he rented premises at 12 The Pavement, Finsbury. A year later he moved to larger premises at 7 The Pavement. On the death of John Roger Arnold in 1843, Charles Frodsham acquired the business at 84 Strand. Clocks made for at least ten years after this date will more often than not be found to bear signatures such as "Arnold and Frodsham" or "Arnold, Charles Frodsham", although the two were never in partnership. The eminent Charles Frodsham soon acquired another business, namely that of Vulliamy, in 1854. In 1861 he bought the watchmaking concern of William Johnson, formerly with Grimalde. Charles Frodsham was not only by far the most best-known member of the famous clockmaking family, but he was also the one whose working life embraced the main years of the carriage clock era.

According to the Clowes notes Charles Frodsham had two sons, John Mill and Charles Mill.[44] John broke with tradition and did not follow a horological career. Instead he became a doctor. Charles Mill was apprenticed to his father on 24th May, 1852, becoming a member of the Clockmakers' Company in 1863.[45] After 1897 Charles Frodsham & Company were at 115 New Bond Street, and from 1920 until 1942 they were at 27 South Molton Street. Bombed out of Molton Street, they were briefly at Mandeville Place, subsequently moving to 62 Beauchamp Place and then to 173 Brompton Road, where they remained until the summer of 1973. At the time of writing these notes, the firm was seeking new premises.

When one considers the number of changes of address made during a long life by the business started by Charles Frodsham, to say nothing of the bombing incident during the last war, it is not surprising that none of their early clock records remain. It would certainly appear, however, that at all stages entirely separate number sequences were used for clocks, watches and chronometers respectively. Charles Frodsham & Co. eventually owned the firm of Nicole Nielson, whose superb carriage clocks are discussed elsewhere in this chapter. Today, in 1973, Charles Frodsham & Co. is managed by Mr. F.L. Thirkell. Previously Philip Clowes was Managing Director, and before him Mr. R.B. North. Nielson became a Director towards the end of the last century. Overleaf will be found a partial Frodsham Family Tree followed by tables showing the addresses occupied at various dates by the different members of the family.

---

(44) Harrison Mill Frodsham, who was on the Council of the British Horological Institute besides a regular contributor to the *Horological Journal*, was almost certainly another son. He used the address 84 Strand (vide *Horological Journal*, December 1877). Mr. Thirkell of Charles Frodsham & Co. Ltd. is fairly sure that Harrison Mill, who died as recently as circa 1920-21, ran the firm for a number of years after the death of the famous Charles Frodsham.

(45) A beautiful box chronometer exists, No. 3/33085, having a very complicated auxiliary balance with cross-wires and bearing the name C. Mill Frodsham Jnr., and the words "Elève de Chas Frodsham.".

IX English Work and Workmen
A Partial Frodsham Family Tree

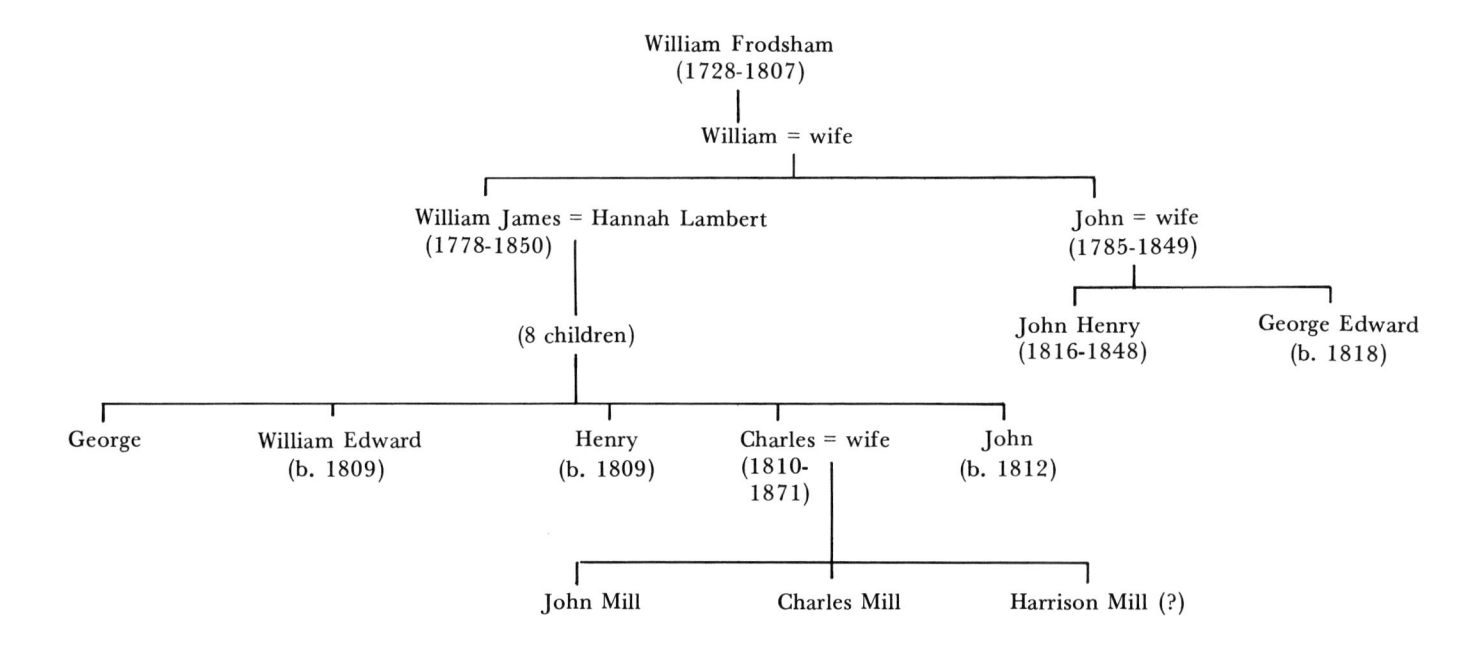

```
William Frodsham
  (1728-1807)
       |
   William = wife
       |
  ┌────┴─────────────────────────┐
William James = Hannah Lambert    John = wife
  (1778-1850)                    (1785-1849)
       |                              |
  (8 children)                  ┌─────┴──────┐
       |                    John Henry    George Edward
                            (1816-1848)    (b. 1818)

┌──────┬──────────┬──────┬────────┬──────┐
George  William   Henry  Charles = wife  John
       Edward    (b.1809) (1810-          (b.1812)
       (b.1809)           1871)
                            |
                ┌───────────┼──────────────┐
            John Mill   Charles Mill   Harrison Mill (?)
```

Frodsham Styles, Dates and Addresses

Pre 1790	William Frodsham	12 Kingsgate Street, Bloomsbury.
1790	William Frodsham & Son	12 Kingsgate Street, Bloomsbury
1807	John Frodsham	12 Kingsgate Street, Bloomsbury.
1809	Frodsham & Baker	12 Kingsgate Street, Bloomsbury.
1823	Frodsham & Baker	31 Gracechurch Street.
1849-1901	G.E. Frodsham	31 Gracechurch Street.
1902	Company probably absorbed into Parkinson & Frodsham.	

1801	Parkinson & Frodsham	4 Change Alley.
1891	Parkinson & Frodsham	16 Queen Victoria Street, E.C.
1893	Parkinson & Frodsham	35 Royal Exchange.
1896	Parkinson & Frodsham	15B Royal Exchange.
1907	Parkinson & Frodsham	5 Budge Row.
1940-1947	Parkinson & Frodsham (G.C. Harris)	5 Budge Row.

1832	Charles Frodsham	5 Navarino Terrace, Hackney.
1834	Charles Frodsham	Barossa Place, Islington.
1837	Charles Frodsham	12 The Pavement, Finsbury.
1838	Charles Frodsham	7 The Pavement, Finsbury.
1843	Charles Frodsham	84 Strand.
1893 (July 25th)	Charles Frodsham & Co.	84 Strand.
1897	Charles Frodsham & Co.Ltd.	115 New Bond Street.
1920	Charles Frodsham & Co.Ltd.	27 South Molton Street.
1942 (?)	Charles Frodsham & Co.Ltd.	Mandeville Place.
1942	Charles Frodsham & Co.Ltd.	62 Beauchamp Place.
1947-1973	Charles Frodsham & Co.Ltd.	173 Brompton Road.

The reader will remember that clocks sold by Charles Frodsham from 84 Strand for at least ten years after 1843 more often than not carried the names of both Arnold and Frodsham, used in the form of such signatures as "Arnold, Chas Frodsham" or "Arnold and Frodsham". Evidently, however, there were some exceptions. An excellent example is a mahogany bracket clock No. 655 illustrated by Dr. Mercer. This piece is signed "Arnold 84 Strand" on the dial, while on the back plate of the movement the name of Charles Frodsham appears in conjunction with the same address. It is probably reasonable to

IX English Work and Workmen

conclude that this clock was taken over by Frodsham when he acquired Arnold's business in 1843. It may be that the serial number of the clock continues a sequence begun by Arnold from 1821 when he first started at 84 Strand.[46] Another house clock No. 863, this time in a mahogany "four-glass" case, is signed CHA^S FRODSHAM 84 STRAND" on the dial, but "Arnold, Charles Frodsham" on the movement. On the other hand a brass-cased carriage clock No. 841, which will be discussed later, has Charles Frodsham's name alone on both dial and movement, although it was apparently sold at the time when most clocks were signed "Arnold, Charles Frodsham". Much more work remains to be done before it will be possible to be sure about the relationship (if any) between clocks possibly belonging to one number sequence but bearing any of the names already mentioned. Charles Allix will be most grateful to receive from readers particulars of any house or carriage clocks by Arnold, Arnold and Dent, Dent, Arnold and Frodsham, or Frodsham which they may own or encounter.

Dr. Mercer's book relates how, soon after the death of John Roger Arnold, Charles Frodsham bought the Arnold business from the executor Richard Steele. Although Arnold was over seventy when he died, and although it is probable that the business had declined during those three years since the departure of E.J. Dent, it is important to appreciate that the name "Arnold" was the most famous in the world in the context of chronometers. The acquisition of the business was altogether a shrewd and sensible move on the part of Charles Frodsham, and this despite the formidable competition of E.J. Dent only a few feet away.

Plates IX/43 and IX/44 are of great interest because they show a clock signed simply "CHA^S FRODSHAM 84 STRAND LONDON", when in the ordinary course of events it should have been sold bearing Arnold's name as well. Perhaps it was brought by Frodsham from Finsbury Pavement? The piece is numbered 841 and it is manifestly a fairly early clock. Its date is probably very close to 1843, and this despite a somewhat late-looking dial and hands. Mechanically this clock is very much on a par with the one by Arnold

Plate IX/43. *CHARLES FRODSHAM NO. 841. An hour striking carriage clock probably dating from 1843 or earlier and having a lever escapement. (By courtesy of P.G. Dawson)*

and Dent illustrated in Plates IX/18 and 19 and with the Desbois clock shown in Plates IX/55 and 56.

Another clock, this time signed "Arnold CHA^S FRODSHAM LONDON 860" is shown in Plate IX/45. It is as different from No. 841 as chalk from cheese. Here is shown an entirely conventional mid-19th century carriage clock. It has a typical Frodsham dial with white enamel chapter ring superimposed upon an engraved background and set off by "Vulliamy" hands.[47] The probable date of the clock No. 860 is circa 1844.

---

(46) After all, Arnold's reputation was based on marine chronometers and precision watches. He may not have sold many clocks.
(47) Another type of dial favoured by Frodsham was that with a pierced and engraved overlay as in Plate IX/43. This mode of decoration was much favoured by B.L. Vulliamy, who habitually used it on the back plates of his watches.

IX *English Work and Workmen*

Plate IX/44. *CHARLES FRODSHAM NO. 841. The movement of the clock shown in Plate IX/43. Compare with Plates IX/19 and IX/56. (By courtesy of P.G. Dawson)*

Plate IX/45. *ARNOLD, CHARLES FRODSHAM NO. 860. Date circa 1844. Lever escapement. Note a case style also found in clocks sold by Dent, McCabe, Barwise and Whitelaw. The self same decorative handle was also favoured by the Dents of both Strand and Cockspur St. and by McCabe. Need more be said about shared sources of supply? (By courtesy of Meyrick Nielson of Tetbury)*

The next clock which we are able to illustrate is No. 1080 which bears Charles Frodsham's name alone. It is shown in Plate IX/46. It has a lever escapement and also strikes and repeats hours. Its unusual features include not only a centre-seconds hand but also twin up-and-down dials showing the state of wind of both the going and the striking mainsprings. On the dial appear the words "CLOCK MAKER TO THE QUEEN". It would seem that Charles Frodsham was able to take over Vulliamy's Royal Appointment with his business in 1854[48]; but in any case this year was the first in which he described himself as "Clock Maker to the Queen". This being so, the clock, No. 1080, cannot have been made before 1854. In all probability it was at this period that the use of the Arnold name was finally discontinued.

In view of the very considerable reputation enjoyed by Charles Frodsham as a maker of both marine chronometers and of chronometer watches, it seems surprising that the only example of a chronometer carriage clock by him offered for sale in recent years seems to have been that sold by Sotheby's on 9th December, 1968. It is numbered 1089[49] and, like the lever clock already described, will have been made very soon after the year 1854. It is inscribed on the

---

(48) While it was in Frodsham's interests to use Arnold's name with his own for a number of years after 1843, it appears that he never once used that of Vulliamy.

(49) In the auction catalogue the number appears as "1098", but this is a misprint.

IX *English Work and Workmen*

Plate IX/46. CHARLES FRODSHAM NO. 1080. Date post 1854. Note the seconds hand and also the unusual feature of up-and-down dials for both going and striking trains. (By courtesy of Asprey & Co. Ltd.)

Plate IX/47. CHARLES FRODSHAM NO. 1089. Date soon after 1854. This clock has a chronometer escapement and also strikes and repeats hours. The up-and-down dial relates to the going train only. (By courtesy of John D. Altmann, Australia)

back plate "Cha^S Frodsham, CLOCK MAKER TO THE QUEEN, 84 Strand, LONDON 1089", and it offers hour striking and repeat in addition to having an up-and-down dial relating to the going side only. The photograph, Plate IX/47, does not begin to do justice to a beautiful clock which is quite as attractive as the Dent "Strand" and "Cockspur" chronometer clocks already illustrated. A customer wishing to buy such a clock soon after 1850 certainly had the choice of three really admirable alternative sources of supply, there being little if anything to choose between them in terms of quality and execution. It would seem, however, that Frodsham was the only one of the three to provide his clocks with up-and-down dials.

Plate IX/48 shows a curious hybrid drum-shaped timepiece clock. It is neither a chronometer, which perhaps at first glance it superficially resembles, nor yet is it a carriage clock in the ordinary sense of the term. It is signed "CHA^S FRODSHAM & CO. 84 Strand, 1487. Makers to the Queen". Plate IX/49 shows the movement, from which it will be seen immediately that the whole layout and scheme of planting is totally different from that of a chronometer. The fusee has its large end towards the back plate, and the escapement (detailed view Plate IX/50)

271

IX English Work and Workmen

Plate IX/48. *CHARLES FRODSHAM NO. 1487. Date post 1893. This type of travelling timepiece clock goes for eight days and is provided with an up-and-down dial. (By courtesy of G. Manning)*

is a lever platform set into the tops of the frames. The address "84 Strand" in conjunction with the "& Co." proves conclusively that this clock must have been made after 1893, and indeed most examples of this type will be found to have the Bond Street address. We are at a loss to explain why the serial number is no higher than No. 1487. In the absence of any more sensible theory, it is suggested that when Charles Frodsham's business became a Company in 1893, another number sequence was started for clocks beginning once again at No. 1000. On this basis a lever timepiece carriage clock No. 1814, bearing the inscription "Cha^S Frodsham, Clockmaker to the Queen, 84 Strand London" (sold Sotheby's, 21st May 1965, Lot 40) is likely to belong to the same number sequence as the Charles Frodsham clocks 1080 and 1089. We have not seen No. 1814, but from the catalogue description should be very much surprised if it were not far earlier than the drum-shaped clocks.

WEBSTER

R.W. Webster & Son of Cornhill sold, circa 1840, a type of miniature timepiece travelling clock apparently peculiar to that firm. An example is shown in Plate IX/51. It is numbered 74 and it was illustrated by both Mr. Percy Dawson and by Dr. Bruce McLean in articles mentioned in the Bibliography. These Webster clocks have lever escapements. They stand only two inches high in their mahogany cases, which have brass edgings but no handles. The outer travel-boxes have backward-opening tops.

Plate IX/49. *CHARLES FRODSHAM NO. 1487. The movement of the clock shown in Plate IX/48. Although the movement of this piece resembles superficially that of a two-day chronometer, it is in reality quite different. Note the lever platform escapement set into the tops of the frames. (By courtesy of G. Manning)*

IX *English Work and Workmen*

Plate IX/50. *CHARLES FRODSHAM NO. 1487. The escapement of the clock shown in Plates IX/48 and 49. (By courtesy of G. Manning)*

Plate IX/51. *WEBSTER NO. 74. Date circa 1840. A miniature timepiece travelling clock with lever escapement and standing 2ins. high. (By courtesy of Evans & Evans, Alresford)*

WHITELAW. BARWISE

The illustration Plate IX/52 shows a top quality hour striking and repeating clock signed "JA^S WHITELAW, EDINBURGH". Apart from the fact that this piece may conceivably have been "finished" in Edinburgh, a more typical example of good London 19th century work could scarcely exist.[50] The un-engraved case is finished a bronze colour instead of the far more usual gilding. This is a man's clock. It was intended for a Smoking Room, Gun Room or Study.

Another clock, really very similar although finished and decorated altogether differently, is signed "BARWISE, London". It is illustrated in Plate IX/53. This is a Drawing Room clock. The case and dial are floridly engraved. The case gilding is unusually bright. The appearance of this piece is as extrovert as the Whitelaw is restrained! Plate IX/54 shows the lever platform escapement of the Barwise clock. Another Barwise carriage clock was sold at Sotheby's on 27th March 1972, and it is illustrated in their catalogue (Lot 5).

DESBOIS

While the Desbois family was neither particularly notable nor prolific as a maker of carriage clocks, it is

---

(50) The Whitelaw clock has a brass-bound mahogany travelling box of the type associated with McCabe clocks and of which an example is illustrated in Plate IX/64.

IX English Work and Workmen

with the handle up. Unfortunately, while Desbois ledgers and day books are preserved in the Guildhall Library, no workbooks appear to have survived. Positive dating is therefore impossible. In any case, such records as are available do not go back before 1854.

The movement of the clock in question is really little more than that of an English bracket clock, scaled down to suitable proportions for a travelling piece, and then furnished with a balance-controlled escapement. Indeed the movement was probably designed to be fitted into a four-glass case, and in other circumstances would have ended life as a pendulum mantel clock. The same observations apply equally to the clock by Arnold and Dent already described. In the case of the Desbois clock, it is the platform escapement, perhaps more than any other feature, which

Plate IX/52. *JAMES WHITELAW. An un-numbered clock made circa 1845 and being almost certainly London work. This clock strikes and repeats hours and has a lever escapement. (Private Collection)*

nevertheless a fact that the name appears on some quite early examples. Such a clock is shown in Plate IX/55. It is un-numbered and signed on the back plate "Desbois, London". Note the combination of French and English influence apparent in the case, dial and hands. The clock strikes and repeats the hours on a bell. Both going and striking trains employ fusees. In many respects the movement is very like that of the Arnold and Dent clock mentioned previously. Plate IX/56 shows the movement of the Desbois clock. A hand-setting indicator is provided on the back plate to allow for setting to time without the trouble of looking at the dial. The clock stands nine inches tall

Plate IX/53. *BARWISE NO. 376. Date circa 1845. This clock strikes hours, but does not repeat. It is 10¼ins. tall and has a lever escapement with helical balance spring. (Private Collection)*

IX   English Work and Workmen

Plate IX/54. *BARWISE NO. 376. The escapement of the clock shown in Plate IX/53. (Private Collection)*

Plate IX/55. *DESBOIS. An un-numbered clock, probably made circa 1845, and striking and repeating on a bell. (By courtesy of Sotheby & Co.)*

provides some sensible clue as to its date. Plate IX/57 shows the plain brass balance and the brass balance bridge used in conjunction with a single-roller lever escapement. The pattern is wholly consistent with those escapements found in watches made between 1830-40, although by the latter date most pieces of any pretension were given compensation balances. Despite the primitive balance and escapement used by Desbois, we do not consider that the clock would have been made much before 1845. The piece was provided with an outer travelling box of rudimentary design, made of an inexpensive wood, and having a top lid and carrying handle. A circular aperture was made in the front to show the dial, but no shutter was provided.

Desbois evidently covered a very long period as a maker of carriage clocks. At least two examples, clearly made close to 1900, appeared in the London sale rooms in 1971. The first well-known member of the Desbois family was Daniel, 1800-1840. His son Daniel died in 1885. The grandson, Edwin, died in 1917. The great grandson, Ernest Desbois, died in 1951. Ernest's daughter, Miss Kate Desbois, ran the firm for many years in the present century.

THWAITES & REED

The firm of Thwaites & Reed is one of the very oldest in the business, having been established in Clerkenwell since early in the 18th century. The firm was started by Aynsworth Thwaites in Rosoman Row

275

IX English Work and Workmen

in about the year 1735. It continued under the direction of Aynsworth's sons Benjamin and John, who appear to have entered into partnership with Reed in 1808. The Thwaites firm, in the 18th and 19th centuries, was probably the most prolific in England. They made turret clocks and also domestic clocks of all descriptions, although seldom under their own name. They specialised in supplying other firms whose names appeared on the finished work.[51]

Plate IX/58 shows a fine striking carriage clock with lever platform escapement and with perpetual calendar, signed on the dial "THWAITES & REED LONDON". This piece is not numbered but it was

Plate IX/57. *DESBOIS. The escapement of the clock shown in Plates IX/55 and 56. (By courtesy of Sotheby & Co.)*

probably made circa 1850. Note a case very similar to the Dent chronometer carriage clock No. 14880, for which a similar date is suggested and which is illustrated in Plate IX/26.

BARRAUD AND SUCCESSORS

The well-known firm of Barraud, which later became Barraud & Lund, then Lund & Blockley, and finally Herbert Blockley (successors to Lund & Blockley) produced a few carriage clocks in their time. Cedric Jagger (see Bib.) mentions four examples, in-

Plate IX/56. *DESBOIS. The movement of the clock shown in Plate IX/55. (By courtesy of Sotheby & Co.)*

---

(51) See *The Maker Behind the Clock* by Geoffrey Buggins. Thwaites turret clocks, which often bear names such as "Holmes" or "Perigal" are very characteristic and instantly recognisable. Less obvious are the domestic clocks, although these will more often than not be found to be signed in block letters "THWAITES" on the front plate and concealed by the dial.

IX   English Work and Workmen

Plate IX/58. THWAITES & REED. *Date circa 1850. A striking carriage clock, apparently not numbered, with lever platform escapement and with a perpetual calendar. The handle is in two pieces and folds away flush into the top of the case. (By courtesy of Asprey & Co. Ltd.)*

Plate IX/59. BARRAUD & LUND NO. 1795. *Date circa 1855. This piece strikes hours and also has a lever escapement; but its main interest lies in the fact that its case shape anticipates the French gorge design which does not seem to have been current much before 1867. (By courtesy of Sotheby & Co.)*

cluding No. 1323 which he states cannot have been made earlier than 1844, No. 1386 of circa 1845, No. 1690 of circa 1850, and No. 1794 of circa 1855. On 27th March 1972 Sotheby & Co. sold clock No. 1795 "BARRAUD & LUND" for which is deduced a date of 1855. It is shown in Plate IX/59. Here is an untypical English carriage clock, if ever there was one. The case and handle anticipates the French *gorge* style, while the dial is in the manner of its English contemporaries. This clock is not without an interest of its own because the French, so far as we are aware, did not produce *gorge* cases until almost thirty years later. Mr. Dawson illustrates a massive clock signed "LUND & BLOCKLEY, 46 PALL MALL, LONDON" in a multi-pillared case in the form of a temple or triumphal arch, approached by steps. The author suggests a date of about 1880. See Plate IX/60. This clock, which sports a mahogany travelling box, has a glass dome in place of a top carrying-handle. So much for one definition of a carriage clock! Herbert Blockley was the star pupil of the British Horological Institute Classes and was well-known by 1893. Among his specialities were the so-called "Explorer's watch",

277

Plate IX/60. *LUND & BLOCKLEY. This enormous "carriage clock" was made circa 1880. Apparently it has no serial number. (By courtesy of Percy G. Dawson. Clock formerly in the possession of W. Hodson)*

having not only screw back and bezel, but a screw-on captive cap for the "winder".

SIR JOHN BENNETT

In 1851 the Great Exhibition Catalogue listed John Bennett, Parkinson & Frodsham, E.J. Dent, Charles Frodsham and William Payne & Co. of 163 New Bond St. as showing carriage clocks.[52] The last four makers are mentioned elsewhere so that only Bennett needs further comment, beyond remarking that the clock shown by Payne was probably made by Thomas Cole. Bennett of Cheapside (or Sir John Bennett as he became in 1872) was an interesting character, well-known and thoroughly unpopular in the trade. The *Horological Journal* of August 1897 is particularly uncharitable towards him in an obituary. John Bennett was born at Greenwich, beginning in business on his own in Cheapside in 1843. In 1851, according to the *Journal*, he advertised at the cost of £750 in the back of the Great Exhibition Catalogue "... and this venture seems to have ensured his successful business career. Of a peculiar personality, he may be said to have been deficient in veneration and dignity, but with abounding assurance and self-esteem...." Elsewhere the same writer continues "The accident of holding the office of Sheriff when the Prince of Wales visited the City in 1872, led to his receiving the honour of knighthood". According to the *Horological Journal*, poor Bennett's chief sin, other of course than his success, was his public criticisms of the modes of manufacture used in Clerkenwell, "... together with his posing as the saviour of the English trade..."! The knighthood, too, was evidently a cause of much jealousy, for one finds the *Horological Journal* of September 1878 saying of the Paris Exhibition "Bennett of Cheapside, — perhaps we are irreverent and should say Sir John Bennett, for that title can be seen in various degrees of boldness in four or five places on the stand — shows clocks, watches and chronometers, an ordinary shop stock, a great part of which is not of British manufacture at all".

A further indication of the success of Bennett may be inferred from the *Horological Journal* of March 1889 where a short announcement reads: "It is proposed to form a limited liability company to take over the business of Sir John Bennett, the price to be paid being £90,000, of which £31,900 are for goodwill. The prospectus states that for the last ten years the net profits have averaged £8170 per annum. However justly manufacturers may complain of hard times it is pretty clear from this that distributors have been doing very well, though it is doubtful if any other watch-dealer than the redoubtable Knight has of late years been able to pay income tax on net profits of over £8000 per annum".

JAMES McCABE AND FAMILY

James McCabe died in 1811. In other words he had nothing whatever to do with the very fine English

---

(52) In the 1862 London Exhibition only Bennett, Dent, Parkinson & Frodsham, Charles Frodsham and Payne seem to have exhibited carriage clocks out of approximately 130 English firms.

IX   *English Work and Workmen*

carriage clocks found bearing his name. These were produced by successive members of his family who carried on the business. Britten gives various McCabe addresses, always in the City. Of these the best known were 97 Cornhill (or 97 Royal Exchange, which meant the same thing) occupied from circa 1804, and later 32 Cornhill where the business was to remain until it closed in 1883.

McCabe carriage clocks will usually be found in one or other of two kinds of case. The first style of case, which is typified by a clock No. 2841 and shown in Plate IX/61, is little different from pieces found bearing various other names and all being made in the years close to 1845. It has a lever escapement with a plain balance, and it also strikes and repeats hours. A

Plate IX/62. *JAMES McCABE NO. 2841. The top of the movement of the clock shown in Plate IX/61 showing the lever escapement platform. (Private Collection)*

top view of the movement is shown in Plate IX/62. The second type of case found in McCabe clocks, and incidentally one of the finest examples which we have ever seen, is well illustrated by the quarter-striking clock No. 2927 shown in Plate IX/63. These larger clocks, standing 9¾ inches high with handles up, have cases with canted corners embellished with vicious downward-turning "beaks" and equally sharp lower ogee mouldings. Usually the cases are left plain and are gilded, but sometimes they are engraved all over as in the illustration. The difference in cost between plain and engraved cases, even in the days of cheap labour, was always very great. A particularly attractive feature of No. 2927 is that the gilt and engraved dial has been finished a slightly different colour from

Plate IX/61. *JAMES McCABE NO. 2841. Date probably circa 1845. This clock strikes and repeats hours and has a lever escapement. (Private Collection)*

IX English Work and Workmen

Plate IX/63. *JAMES McCABE NO. 2927. Date uncertain but almost certainly post 1850. This clock strikes hours and quarters and has a lever escapement with a fine compensation balance. (Private Collection)*

the case in order to make a contrast. Also worth noticing are the fleur-de-lys hands, almost certainly made by the Peter Pendleton of the day.[53] No. 2927, in common with all McCabe carriage clocks seen by us, has a lever escapement. This particular example has a compensation balance, and the flat balance spring is so well made and so accurately centred that the clock was able to continue going and keeping time for many years with no top balance pivot.[54] It is perhaps worth remarking that a similar circumstance in the case of a clock furnished with a chronometer es-

capement would almost certainly have resulted in some far more expensive damage. Perhaps the firm of McCabe knew very well what they were about when they provided their travelling pieces with lever escapements. Before leaving this particular clock, it is worth saying that it is finished throughout to similar standards to the chronometer clocks sold by the Dents and Frodshams. Plate IX/64 shows clock No. 3354 complete with its mahogany travelling box with sunk handle and brass corners. These boxes, which are beautifully made, were originally supplied with all McCabe carriage clocks and may be peculiar to this maker. The inside of the box is lined with green baize. The back door is fitted with a lock, and at the front of the case is a rising-and-tip-over shutter to allow the dial to be seen on journeys.

Serial numbers of McCabe clocks known still to exist range from 1733, which is a lever timepiece with a seconds dial and standing only 4½ inches tall, to 3376, which is a quarter striking clock with an engraved dial and housed in a pillared case. Other known clocks are numbered 2141 (a lever timepiece in a French-type case), 2777, 2994, 3156 (which strikes quarters) and 3373 (which strikes and repeats hours). The existence of no less than three clocks striking the quarters among a mere handful of examples suggests that the McCabe firm may have concentrated upon this type. The numbered brass-bound mahogany travelling box of No. 2770 still exists but it has lost its clock.

No McCabe workbooks have survived, but it would appear that clocks had a number sequence of their own. While it is probably reasonable to assume that the clock 2841 was made at about the middle of the 19th century, it is difficult to attempt to date the other examples without having any idea how many clocks were produced a year. The style of the clock No. 2927 is shown exactly in Moore's pattern book of 1849, except that there the dial has a square top and not the arch which is almost certainly rather later in date.

Many McCabe clocks were exported to India where the firm had a thriving agency. A number of carriage clocks are still to be found in that country, particularly

---

(53) The Pendleton family, hand makers in Prescot, Lancs., were particularly noted for their fleur-de-lys hands, the finest ever made.
(54) History does not relate whether the top pivot was worn away from years of going without oil, or broken off as the result of an accident.

IX *English Work and Workmen*

Plate IX/64. *JAMES McCABE NO. 3354. This clock strikes and repeats hours and has a lever escapement. The beautiful mahogany travelling box, with brass corners and edges and with a front shutter to show the time, is typical of those supplied with McCabe carriage clocks. Note that the winding key carries the serial number of the clock. (By courtesy of Ronald A. Lee)*

in Calcutta. These "Indian" clocks include some in wooden cases, while it is known that one exists housed in a very ornate case executed in Bengali silver. This particular clock has a musical box in its base. When McCabe clocks were exported to Pondicherry, then the instructions for winding, setting, etc. were always in French.

THE MACDOWALL FAMILY

Charles MacDowall (1790-1872) was an inventive Yorkshireman. He is best remembered by horologists today for his "Helix Lever" clocks, those strange month-going skeleton timepieces first made in about 1830, and which have 25-turn fusees to say nothing of the helical gearing used throughout their trains.[55] C. MacDowall is also remembered for a "single-pin" escapement intended for pendulum clocks. In this arrangement a crank makes a half-turn at each excursion of the pendulum in either direction.[56] An example was shown at the Great Exhibition of 1851, while E.B. Denison illustrated it in the later editions of his book. Denison pointed out that MacDowall was the independent inventor, rather than a first inventor, of most of his brain-children. Another of Charles MacDowall's developments was a practical method of making twist drills from pinion wire, in much the same fashion as the old Prescot workmen in the 19th century made simple lead-screws for wheel-engine indices using the pinion wire then so readily to hand in Lancashire.

Joseph Eden MacDowall, possibly a son of Charles, patented in 1838 an escapement in which the impulse is imparted by means of a helix. One application of this escapement is found in a two-day box chronometer signed "J. Eden MacDowall's PATENT London" and bearing the serial No. "1". It is described and illustrated by G.A. Rose and R. McV. Weston in *Antiquarian Horology* December 1964. Another related piece has survived. It is a portable clock movement,

---

(55) This type of construction is briefly described and illustrated by Royer Collard in *The Skeleton Clock*. The "Helix Lever" clocks are usually addressed "Leeds" or "Wakefield".

(56) Chamberlain draws attention to patent No. 13587 of 1851 in the name of Charles MacDowall and relating to this sort of escapement. This is also noted by Saunier in his *Treatise*.

281

almost certainly intended for a carriage clock. Plate IX/65 shows the dial of this piece, while Plates IX/66 and 67 show the movement with its curious sub-frame in which run the escapement and some of the train. The escapement is essentially that shown in the draw-

Plate IX/65. *J. EDEN MACDOWALL. The dial of a portable clock having Macdowall's patent escapement. Note the variant of the spelling of the surname. (By courtesy of the British Museum)*

ing on a patent specification No. 7874 of 15th November 1838 ("MACDOWALL, Joseph Eden"). It is not difficult to believe that this somewhat crude clock movement is earlier than the Chronometer No. 1. Presumably various pieces and arrangements were tried at about the time when the patent was first filed. It will be noticed that an extra wheel and pinion are required in the clock train because of the rapid rotation of the final mobile.

Charles MacDowall's obituary, which shows his portrait, was written by John James Hall and published in the *Horological Journal* of September 1873, where the "Helix-Lever", "single-pin" escapement

Plate IX/66. *J. EDEN MACDOWALL. The movement of the portable clock. Note the sub-frame carrying the escapement and some of the train wheels. (By courtesy of the British Museum)*

and "spiral drill" are all mentioned, together with a surprisingly large number of different London addresses. J. Eden MacDowall is not mentioned at all by J.J. Hall; but the fact of the former's preoccupation with helices alone makes some direct link with Charles seem inevitable. Charles was born in 1790 and, assuming that he married young, he could well have had a grown-up son helping him by 1838. It would surely have been too great a coincidence for C. Macdowall and J.E. not to have been somehow connected?

VICTOR KULLBERG

Victor Kullberg! What a name to conjure with, however ungrammatically! Kullberg was the most

Plate IX/67. *J. EDEN MACDOWALL. A view inside the sub-frame showing train wheels and part of the escapement. (By courtesy of the British Museum)*

famous of English Victorian chronometer makers at the time when the art reached a peak both in terms of beautiful execution and of performance.

Kullberg, according to his obituary,[57] was born in Wisby on the Swedish Island of Gothland in 1824. He was apprenticed at the age of sixteen "to a watchmaker who had just at that time produced the first satisfactory chronometers made in Sweden". The obituary is at pains to point out that "Watchmaking, as carried on in Sweden in those days, was an excellent school for the learner, the trade not being subdivided there as it is now in more advanced watchmaking centres". Later Kullberg worked for Urban Jürgensen in Copenhagen, coming to London in time for the 1851 Exhibition. He soon obtained employment specialising in marine and pocket chronometers and also making watches with his patent keyless fusee work from about 1856. He quickly reached the top of his profession, being known for many inventions and especially for his chronometer balance incorporating an auxiliary device to compensate for extreme temperatures. His first chronometer success was in the Greenwich Trials of 1862. Thereafter his name was often at the top of the list. The *Horological Journal* recorded his achievements for many years on end. The list of the awards which he received both in England and in Paris, La Havre, Naples, Trieste, Besançon, etc. is "endless".

As noted at the beginning of this Chapter, the Kullberg workbooks have survived virtually intact and covering the period 1870 to 1943. D.S. Torrens acquired these Kullberg records (now in the Guildhall Library) on the death of S. Lundquist.[58] To the last Lundquist continued to finish a few chronometers,[59] besides doing deck watch work during the War. Torrens, never one to give out fulsome praise, always contended that the last Kullberg (Lundquist) chronometers maintained the firm's traditional standards, even though in the end Lundquist was compelled in his old age to make or finish parts which had become "dead trades".

The Kullberg firm, despite a total production of some ten thousand pieces, mainly watches and box chronometers, made very few carriage clocks. The list on the next page is compiled from the workbooks:—

---

(57) See the *Horological Journal* of August 1890. Further notes on Kullberg will be found in *Invention and Inventor's Mart*, October 17th, 1885. According to Dr. Torrens' notebooks the late George Abbott, the detent maker, remembered old man Kullberg... "Mr. Kullberg. Oh! very clever man and a very fine workman used to working at the board, at watches". Kullberg was regarded as a god in the trade. He was the first to make polished detents and he introduced diamantine.

(58) Mr. Lundquist's obituary will be found in the *Horological Journal* of May 1947.

(59) As might be expected, Lundquist obtained his "rough" chronometer movements from Joseph Preston and Sons of Prescot. Latterly the firm, which was established in 1840, consisted only of old Harry Pybus, who worked alone and did everything. He died on 27th May 1952, and many of his tools are now in the Liverpool Museum. He was the last English movement-maker. His workplace was described and illustrated by F.A. Bailey in 1953 (see Bib.). Joseph Preston & Sons described themselves as "Marine Chronometer and Watch Movement Manufacturers, Manufacturers of Chronometer and Keyless Going-barrel Movements in all Styles and Sizes, also Pocket Chronometer, Repeater, and Keyless Fuzee Movements &c. Keyless Conversions. Wheel-cutting, Wheels, Pinions, Fuzees and Barrels to any size. Movements to old Case and Private Callipers. Regulator Trains. Clock Movements."

IX English Work and Workmen

Victor Kullberg Carriage Clocks

Serial No.	Date Delivered	Brief Description
2973 (this number seems to have been altered later to 3973	1876 (Dec.)	Chronometer with auxiliary compensation. Gilt brass case. Leather outer box. Sold to Mr. J. Ogilvy who was a very regular and faithful customer of V.K.
4013	1881	Chronometer in silver case weighing 93½oz!! Silver dial and silver key. Leather outer box. Made for Bond & Son of Boston. Case was originally specified to be of 18-carat gold! Total production cost £81.6s.6d.

Kullberg died in the year 1890

Serial No.	Date Delivered	Brief Description
4052	1899	Large chronometer with quarter striking on four gongs. Gilt brass case. This was probably the clock mentioned by J. Tripplin (see Bib.) as having been shown in the Paris Exhibition of 1889. It was finally sold to Mr. J. Ogilvy.
4854	1890	Lever. Small pretty gilt case. For Mr. J. Ogilvy.
5296	1895	Lever. Eight-day. Sold to Mr. J. Ogilvy for Mrs. Huth.
6078	1897	Lever. Eight-day. Going barrel. Case with columns.
6079	1897 (May)	Chronometer. Two-day. Spotted frames. A small clock in gilt pillared case. Made for Mr. J. Ogilvy. (See Plate IX/68).
6499	1898	Small lever timepiece with spotted frames. Finished "in chronometer style". Silver-plated case. Sold to Mr. Simpson Benzie.
6540	1909 (May)	Chronometer. Gilt brass case. Silver dial. Sold to "Dunn Esq." (probably the astronomer).
6554	1921	Chronometer. Gilt brass case. Spotted plates. In morocco outer box. Sold to E.J. Barker, Esq. See Plates IX/69, 70, 71 and 72.

It would be most interesting to know what became of the chronometer carriage clock No. 4013 in the massive silver case. The original purchaser had instructed the makers that this clock should be "in every sense a *Gem*", and no doubt it was. The original order, as already noted, specified a case of 18-carat gold but, according to D.S. Torrens, "This was too much for London even in those spacious days, and a com-

Plate IX/68. VICTOR KULLBERG NO. 6079. Date 1897. A miniature two-day chronometer timepiece carriage clock standing approximately 3 inches tall with handle up. (Private Collection, Berkshire)

284

IX English Work and Workmen

promise was effected with a case of solid silver. The other appointments were in keeping with the case..."

The firm of Kullberg obtained most of their "rough" movements from Joseph Preston of Prescot as already noted. The workbooks show the progress of each clock, watch or chronometer during its manufacture, together with the names of the workmen and the prices charged. In addition, order books, cash books, stock books, outworkers' job-books, and duplicate correspondence books etc. between them paint a very clear picture of the running of the firm.[60] The reader will be interested in the following entries from the workbook of 1897 in respect of the small two-day chronometer carriage clock No. 6079, which is shown in Plate IX/68. The entries are exactly as recorded.

John Ogilvy Esqr   May 18th 97

6079 Small carriage Clock. case 1½ dial fluted columns. with ornamented capitol and movem't 2 day chron move't fusee	Morrisson July 27th 94	11. 10. 0	
Balance	Smith & Son Aug 17th 94		7. 0
Escapement	Wilde Sept 21st 95	4. 5. 0	
Jeweling escape holes	Milligan Aug 19th 95		10. 6
Finishing	Lawrence Dec 14th 95	3. 0. 0	
Jewelling 3^d 4th	Milligan Nov 95		10. 0
New fusse	Preston Dec 21st 95		2. 6
Cotton spring	Oct 31st 95		1. 6
Chain	Jenkins March 5th 96		2. 3
Polishing & Spotting	Dean W May 26th 96	1. 0. 0	
Naming dial V Kullberg & figures	Abbott Oct 6th 96		8. 6
Hands steel spades up & down	Hood Oct 12th 96		4. 0
Balance Spring	Romberg Sept 96		6. 0
Exam'ing new setting arbor recolleting crown wheel collet spring etc.	Lundquist Oct 28th 96	4. 0. 0	
Hands an other set	Hood Dec 9th 96		2. 0
Gilding clock case	Johnson Dec 14th 96	1. 10. 0	
Engraving the dial	Abbott Dec 8th 96		8. 6
Russia Leather case	Robinson March 10th 97		18. 6
Gilding Key	Johnson March 11th 97		
" Bushing for key holes	" March 23^d 97		1. 0
Making key for winding and dust cap	Lundquist Feb 22^d 97		8. 0
Drilling hole in glass & making bushing	Lundquist March 22^d 97		10. 0

Preston July 19th 95 fitting screws to pillars 9^d
Preston Oct 10th 95 crown-wheel 1/6
   "   hollow centre-pinion Oct 29th 95   2/-
jewelling crown wheel pivot with endstone June 96   2/-
Preston fusee wheel Dec 24th 96   1/-
Lundquist stoning up the brass-case for gilding
       and recleaning movement   £1. 10. 0.

---

(60) The Kullberg workbooks etc. are now in the Library of the Clockmakers' Company at the Guildhall.

Other men who did chronometer work for Kullberg included W. Sills No. 2 who was born in 1840 and died in 1901. He spent all his days in his workshop, starting at 7-7.30 a.m. and working till 10 at night, except on Saturdays when he knocked off at 6 p.m. and "went marketing". He "used to do a little bit of work on Sunday mornings", as did old Kullberg and many of his contemporaries. Sills' main rival in chronometer finishing was Wicks. Other contemporary finishers were Bench, Bird, Lynch, Lycett, Ducker, Silke and Pape. Sills' father, William (Bill) Sills No. 1, "the General of the Green Lanes", came to London from Coventry in the 1830's. Originally a watch finisher, he took up chronometer work and manufactured a few on his own account; but he did not follow up the manufacture, and eventually specialised in chronometer finishing. His own chronometers he "finished out" including the escapement pivoting and springing. His abilities were not confined to the chronometer business; he was an accomplished poacher and a great fighter! His advent to London was precipitated by his becoming involved in a poaching affray. His two accomplices were arrested and sent off to Van Diemen's Land, while he made off and "padded the hoof" to London where he became lost in the crowd. He was never very fond of work, being too much a sportsman and liked exploring the lanes and fields of Tottenham with his dog.[61]

Kullberg's address before 1870 was 12 Cloudesley Terrace, Islington. He then moved to 105 Liverpool Rd. His name, like that of the great Breguet, lost no impact whatsoever after his death. Victor Kullberg had one son, William, but his association with the business was brief. The concern was continued most ably by Victor Kullberg's two nephews, the cousins George and Peter Wennerstrom. They were joined by other Swedish, Danish and Finnish workers including Mr. Lindquist and Mr. Sanfred Lundquist. Old George Wennerstrom had two sons, George and William, who later joined the business. Lundquist, trained by Peter Wennerstrom, became a partner about 1930. He was the final proprietor.

A direct living link with the old Kullberg firm is Mr. H.C. Young, who was always known as "Horace". He was at Liverpool Rd. from 1920 until 1939. Mr. Young, also trained by Peter Wennerstrom, married Peter's daughter, Lillian Florence Marjorie in 1930. From 1940 to March 1973 Mr. Young worked for H.M. Chronometer Department. Today he is always

Plate IX/69. *VICTOR KULLBERG NO. 6554. Date 1921. An eight-day chronometer timepiece carriage clock standing $8\frac{3}{8}$ inches tall with handle up. (Private Collection, New York)*

known as "Chris". He lives at Herstmonceux and has a profound knowledge of the Kullberg business.

Plates IX/69, 70, 71 and 72, received after these notes were written, illustrate a beautiful eight-day chronometer carriage clock No. 6554 which amazingly enough was completed only in 1921, a period when very few firms indeed could possibly have undertaken this standard of work.

---

(61) D.S. Torrens' notebooks. Prof. Torrens obtained his information from E. Sills, the grandson of W. Sills No. 1. Unlike his father W. Sills No. 2 was very serious-minded, to say nothing of being a highly-skilled workman.

IX English Work and Workmen

Plate IX/70. *VICTOR KULLBERG NO. 6554. A side view of the movement of the clock shown in Plate IX/69. A reversed fusee is used. Mr. Chris Young actually remembers this chronometer clock being made. It was finished by George Wennerstrom II. (Private Collection, New York)*

Plate IX/71. *VICTOR KULLBERG NO. 6554. The platform escapement of the clock shown in Plates IX/69 and 70. Note Kullberg's balance with auxiliary compensation, and also the beautiful palladium spring with its carefully executed terminal curves. The detent and 'scape wheel are glossed all over. Better work was never done. (Private Collection, New York)*

Plate IX/72. *VICTOR KULLBERG NO. 6554. A further view of the platform escapement of the clock shown in Plates IX/69, 70 and 71. (Private Collection, New York)*

USHER & COLE

A famous firm on record as having made chronometer clocks was Usher & Cole of St. John's Street, Clerkenwell. Steward, writing of the Paris Exhibition of 1900, page 45, mentions that Usher and Cole exhibited "A fine eight-day chronometer clock, silvered dial, with seconds, and up-and-down circular dial . . .". He does not say specifically that this was a carriage clock; but unless it was enormous it was certainly something of this nature, whether it had a top handle or not. Usher and Cole began business in 1861. The Cole family came from Ipswich. Curiously enough, Mr. J.F. Cole, the last proprietor of the firm, had no connection with James Ferguson Cole already mentioned. In 1958 the firm was assimilated by Camerer Cuss & Co. and in 1961 J.F. Cole and T.P. Cuss published a history of Usher & Cole under the title *A Watchmaking Centenary*.

BRIDGMAN & BRINDLE

The name Bridgman & Brindle of 45 Haymarket, London appears upon the very fine English carriage clock, with mean and sidereal trains, illustrated in Plate IX/73. Although little is on record about the partnership, the work produced was of an exceedingly high order to judge from the only two pieces which

287

IX   *English Work and Workmen*

Plate IX/73. BRIDGMAN & BRINDLE. *This complicated carriage clock, having lever escapement and showing both mean and sidereal time, was made for G. Dunn, the astronomer, in 1901. (Private Collection)*

Plate IX/74. BRIDGMAN & BRINDLE. *The back view of the movement of the clock shown in Plate IX/73. Note the peculiar location of the escapement, and also the disposal of the twin movements, one above the other. (Private Collection)*

seem to be known.[62] The carriage clock, made for Mr. George Dunn the astronomer, carries the date letter "aiza" corresponding to 1901. Plate IX/74 shows the movement. The clock has a lever escapement with a "block balance" and also an oddly-shaped balance spring rather suggestive of those used in France by Motel. Note how the escapement platform, instead of being placed on top of the frames, is set into the spotted back movement plate. There are two time trains, each with fusee and maintaining work.

Mean time is shown on an upper regulator-type dial, while sidereal time is indicated on the lower dial provided for that purpose. On the back plate of the movement are two seconds dials, one for each train. The trains may be set to time either simultaneously or individually. This clock is almost certainly unique, and without doubt it was made to the client's own specification.

Bridgman was always known as "Gentleman Bridg-

---

(62) The two Bridgman and Brindle pieces are the carriage clock illustrated and also a superb gold chronometer pocket watch, having two volute balance springs, coiled in opposite directions and each terminated in a resilient stud. This watch was first described in the *Horological Journal* of 1900. Since that time W.J. Gazeley has written about it in the *Watchmaker, Jeweller and Silversmith* of August 1953. The article has been reproduced in the *American Horologist and Jeweler* in September 1971. This watch, like the carriage clock, was made for Mr. George Dunn. It is dated "aizz" which corresponds to 1900.

IX English Work and Workmen

man" in the trade. He was probably the clock maker who appears in the Trade Directories as working at 18 Mount Street in 1881. It is also likely that he is the same Bridgman who was the inventor of a tool for measuring lever escapement angles and which was described in the *Horological Journal* in 1876. It would further appear that he worked for Charles Frodsham & Co. at 84 Strand from 1883 to 1895. Brindle appears to have worked from 196 Beresford St., S.E. from 1885, and from 206 Beresford St. from 1889. The London Trade Directory for 1896 gives the joined names of Bridgman and Brindle at 45 Haymarket. They are no longer listed separately.

S. SMITH & SON

S. Smith & Son traded from No. 9 Strand only a few doors from Charing Cross Station. The frontage of the shop was modest, but it concealed one of the largest horological showrooms in London. The firm also had premises at 13 Soho Square, besides in Clerkenwell and in Coventry. According to a Smith's Catalogue *Guide to the Purchase of a Clock*, published soon after 1900, the business was first established in about the year 1830. The firm still continues as Smiths Industries Ltd. S. Smith & Son were famous for high class English clocks, watches and chronometers until the 1914-1918 War saw the end of most hand-made horology. Plate IX/75, which is reproduced from their Catalogue, shows one of their English carriage clocks. The cheapest version, an eight-day timepiece with lever escapement, cost £28 at about the turn of the century. Alternative movements were available. A clock striking hours and half hours on a gong and repeating hours cost £37.10s.0d. A clock striking both hours and quarters and also repeating cost £47.10s.0d. Movements having chronometer escapements were also available. The Catalogue was at pains to point out that "These Clocks are quite suitable for Mantel Clocks".

JUMP

Plate IX/76 illustrates yet another example of the type of "humpbacked" clock originated by Breguet,

Plate IX/75. *S. SMITH & SON. Date circa 1900. An eight-day English timepiece carriage clock standing six inches tall with the handle up. Note the up-and-down dial showing that the clock has run for five days. Smiths sold their clocks complete in leather-covered travelling boxes.*

and which has remained in favour ever since. This un-numbered clock, which was produced by the famous firm Jump of Mount Street,[63] has a silver case bearing the date letter for 1901; but according to Col. H. Quill, the first Jump Breguet-type carriage clock was made by 1883 if not before. In connection with this clock, Col. Quill has a letter from Mr. A. Hayton Jump, who designed the last dozen of these clocks, which

---

(63) Richard Jump was apprenticed to Vulliamy on 7th November, 1825, and Joseph Jump on 5th November, 1827. They founded the firm.

IX English Work and Workmen

C.R. HINTON. PHILIP THORNTON. COL. QUILL AND J.S. GODMAN

During the second World War Mr. Hinton, working for Malcolm Webster, made a silver "humpbacked" clock which was sold to Sir John Prestige. In the immediate post-war years Mr. Philip Thornton of Great Hayward, Staffordshire, completed several fine

Plate IX/76. *JUMP. A perpetual calendar timepiece clock, the silver "humpbacked" case bearing the date letter for 1901. This clock was formerly in the collection of Sir John Prestige. (By courtesy of Asprey & Co. Ltd.)*

Plate IX/77. *JUMP NO. 115. A timepiece carriage clock in a gilt brass case with silver or silvered dial. Not all Jump clocks were numbered. (By courtesy of Christie, Manson & Woods)*

were all different.[64] He is the great grandson of R. Jump and he writes "The first of these clocks was made for a Lord Ashburton of the Victorian period before I was in the business. It cost the firm a load of money in time and trouble, for so many men had to make so many parts (the man that made the hands couldn't do anything else, etc. etc.). My father presented the bill to Lord A with trembling hands and apologised for the high charge. Lord A took the bill and wrote out a cheque at once *for double the amount charged on bill!!!* and expressed his appreciation". An example of an altogether different type of Breguet-inspired clock made by Jump is shown in Plate IX/77.

silver-cased "humpbacked" clocks with perpetual calendars. The first example was commissioned by Charles Frodsham & Co. Ltd. and was presented to the Queen Mother. Plate IX/78 shows either this clock, or else another in the same series. Of particular interest is its perpetual calendar mechanism.

Probably the last "humpbacked" English carriage clock to be made up to the present time was com-

---

(64) These clocks were almost certainly made by the famous London workman W. Barnsdale. The Barnsdale workbooks, now in the possession of the Clockmakers' Company, show that Barnsdale supplied in 1905 a "humpbacked" carriage clock in shagreen-covered case with silver mounts. The clock offered perpetual calendar and moon, and it had a double roller escapement with a Breguet spring. It cost Jump £60.10s. Barnsdale supplied another "humpbacked" carriage clock to Jump in 1906.

IX English Work and Workmen

pleted in 1963. It has a rhodium-plated silver case and is shown in Plate IX/79. It was conceived by Col. H. Quill, D.S.O., and by Mr. J.S. Godman. It is based upon a Thornton clock, but it has been given a new chronometer-type train by Mr. Godman, while Col. Quill has redesigned the calendar. He took as his model the plan proposed by A.L. Rawlings in *The Science of Clocks and Watches* first published in 1944. This calendar has run for ten years without making a mistake. The mechanism is planted at the back of the movement and may be seen in Plate IX/80. The movement, which has a lever platform escapement, carries the serial number "1". The travelling box has been as carefully constructed as any part of the clock. Col. Quill made this box. It was covered in soft green morocco leather and lined with silk velvet by Mr. Stanley Bray of Sangorski & Sutcliffe, the world-famous London book-binders.

NICOLE, NIELSEN

This famous firm was noted for superb watches, and more particularly for tourbillons. They made very many different standard patterns of watches, but their speciality was the production of one-off complicated pieces to special order. The heyday of Nicole, Nielsen was the last decade of the 19th century. Their watch factory was not only very advanced for the times but

Plate IX/78. P. THORNTON. *A silver-cased timepiece carriage clock with perpetual calendar and made soon after the second World War. This clock, or one like it, was presented to the Queen Mother. (By courtesy of Philip Thornton, Esq.)*

Plate IX/79. H. QUILL & J.S. GODMAN. *Clock No. 1. This timepiece clock, completed in 1963, has a special type of improved perpetual calendar. The chronometer-type train was "hobbed" throughout by Mr. J.S. Godman and the clock is based upon a Thornton movement. (By courtesy of Col. Quill)*

IX   *English Work and Workmen*

Plate IX/80. *H. QUILL & J.S. GODMAN. The rear view of the movement of the clock illustrated in Plate IX/79 showing the disposition of the perpetual calendar mechanism on the back movement plate. This calendar mechanism is Col. Quill's improved version of the type first proposed by A.L. Rawlings in 1944. (By courtesy of Col. Quill)*

IX *English Work and Workmen*

Plate IX/81. *NICOLE, NIELSEN NO. 11553. Date circa 1914. A tourbillon carriage clock with grande sonnerie striking. Up-and-down dials are provided for both going and striking trains. The clock stands 5½ inches tall with handle up. (By courtesy of Richard Good)*

it was also mechanised to a degree then undreamed of elsewhere in England. The *Horological Journal* of June 1889 describes the premises in Soho Square, London, with notes upon some of the machines and the operations which they performed. Nothing seems to be known of the origins of either Nicole or of Nielsen, but latterly Nielsen was closely connected with Charles Frodsham. After Frodsham's death in 1871, Nielsen became a director of Charles Frodsham & Co. Ltd. The latter firm in turn eventually owned the Nicole, Nielsen business, which only finally ceased manufacture circa 1935.

In 1914 Nicole, Nielsen produced seven remarkable English tourbillon, *grande-sonnerie* carriage clocks, probably numbered consecutively, which are quite unlike the work of any other maker. Plate IX/81 shows one of these clocks, No. 11553, complete with its burr walnut brass-bound travelling box. According to Mr. Richard Good[65] the clocks were all commissioned by one client, who then gave six of them away to girl friends in exchange for small favours. Another example No. 11558 was sold at Christie's on 26th October 1971.[66] Richard Good, who has worked upon No. 11553, says that mechanically its movement is not that of a clock but rather pure watch-work suitably increased in size and driven by large barrels and fusees of chronometer style and finish. Good shows an under-the-dial view which admirably supports his contention. This illustration is reproduced in Plate IX/82. The back of the movement is seen in Plate IX/83 in which the tourbillon carriage and bridge are plainly visible. Nicole, Nielsen no doubt had sound technical reasons for incorporating tourbillons in these carriage clocks. This question is discussed at some length in Chapter VIII. The Nicole, Nielsen tourbillon clocks have silver cases with engine-turned dial plates of the same precious metal. The beautiful white enamel dials were almost certainly made by Willis[67] while the fine hands are likely to have been the work of the Pendletons of Prescot. There are up-and-down dials for both going and striking trains.

THOMAS MERCER LTD.

In 1972 it once again became possible to buy a brand-new chronometer timepiece carriage clock.

---

(65) Author of *Two Masterpieces by Nicole Nielsen*, published in the *Horological Journal*, 1969.
(66) It was re-sold in Zurich at the Galerie Am Neumarkt on 19th October the following year.
(67) He was the best of the latter-day London watch dial makers.

IX English Work and Workmen

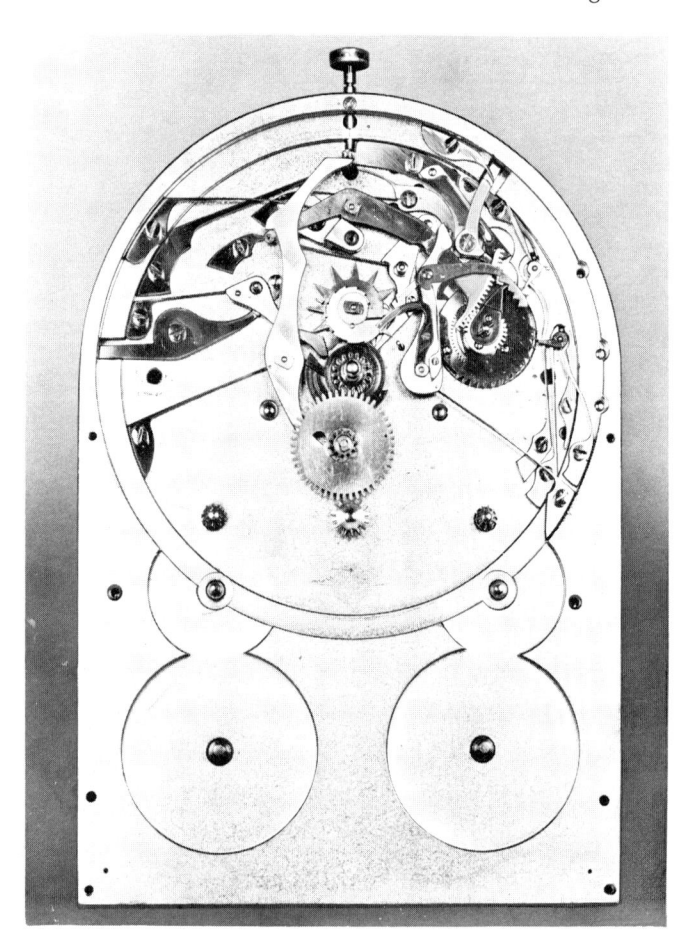

Plate IX/82. *NICOLE NIELSEN NO. 11553. An under-the-dial view of the clock shown in Plate IX/81. The work is in watch-making rather than clock-making traditions. (By courtesy of Richard Good)*

Plate IX/84 shows the very model. It is made by Thomas Mercer Ltd. of St. Albans. The present proprietors of the firm are Gurney and Tony Mercer. They are the grandsons of the founder, who died in 1900. Their father, Frank Mercer, was a very well-known horologist who died as recently as 1970. The clock in the photograph is number 1173, and with its camera-type travelling box and its tapered gilt case it represents a skilful blend of traditional and modern styling. Several very beautiful variants of this clock, having solid silver cases in completely modern idioms, were shown for the first time by the firm at the Basle Fair in 1973. One of them is illustrated in Plate IX/85.

The Mercer firm still manufactures in quantity two-day and eight-day marine chronometers as well as a type suited to surveying. The present factory at St. Albans has a staff of 430. Besides making chronometers it specialises in the manufacture of instruments for linear measurement.

OTHER MAKERS

Readers will encounter a number of other English carriage clocks which would have been included here

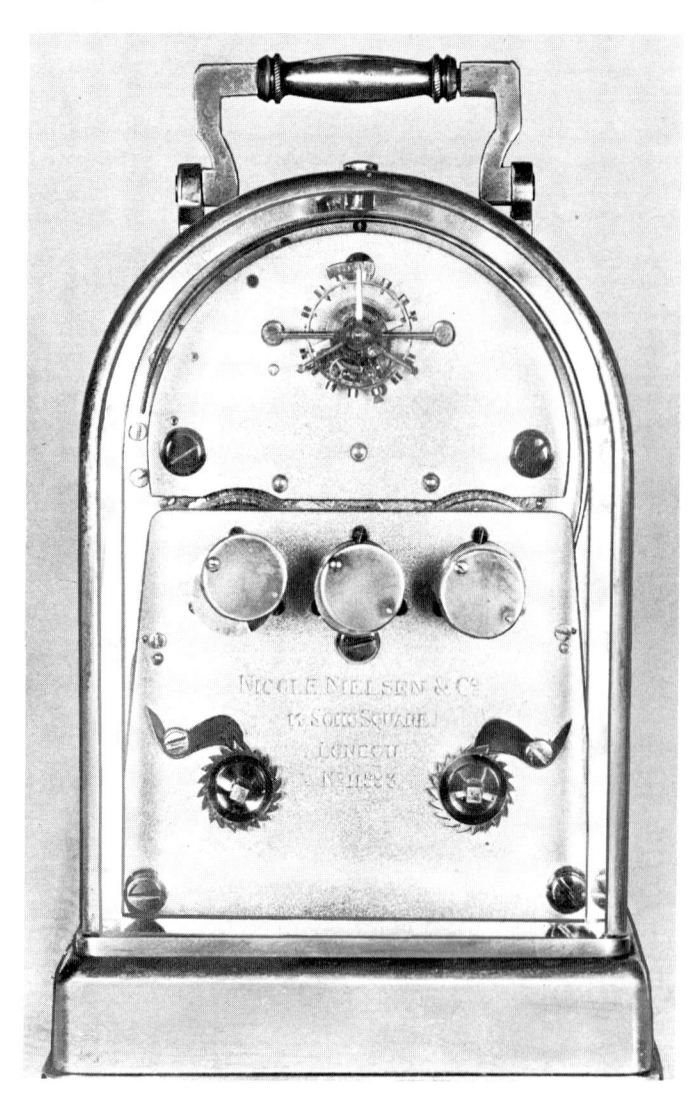

Plate IX/83. *NICOLE NIELSEN NO. 11553. The back of the clock shown in Plates IX/81 and 82. The fusees are just visible between the main and sub frames. The tourbillon carriage rotates in sixty seconds. (By courtesy of Richard Good)*

if only space had allowed and time were of no account. Names which come to mind include "Walter Yonge, Strand", "Blundell", "Connell, Cheapside", and "Jas. Gowland, 52 London Wall", while there remain a few distinctly important makers such as Losada, examples

IX *English Work and Workmen*

Plate IX/84. THOMAS MERCER NO. 1173. *A brand new English chronometer timepiece carriage clock. The serial number 1173 continues a special sequence begun at No. 500 in 1922. (By courtesy of Thomas Mercer Ltd., St. Albans)*

of whose work have simply eluded us. J.E. de Losada was a Spaniard[68] who worked in London from 1835 and who became famous for both clocks and watches. His best-known address was 105 Regent Street, where his nephew continued the business until circa 1890.

Two clocks, very different from each other, are now included more or less as an afterthought since both are of some interest. The first clock, which is anonymous, is shown in Plates IX/86, 87 and 88. It carries a presentation inscription for the year 1856. The escapement is a "patent lever". The second clock is a timepiece and the movement was made by William Johnson of 54 Threadneedle Street. The business of this very good maker of watches and chronometers was acquired by Charles Frodsham in 1861, and thus the clock may be presumed to have been sold prior to this date. The reason for its inclusion here is that its case is apparently French and what is more it is covered with by far the finest engraving which we have ever seen on a carriage clock. The case is signed "Faucherre S.C.", and no doubt this man was well known. Plate IX/89 illustrates the clock, while Plate IX/90 shows a close-up view of one of its side panels.

Plate IX/85. THOMAS MERCER. *One of the very beautiful silver-cased chronometer clocks first shown at the Basle Fair in 1973. This particular model is known as the "Buckinghamshire". (By courtesy of Thomas Mercer Ltd.)*

FRENCH FEAR OF ENGLISH COMPETITION

The reader will by now have seen more than enough photographs of 19th century English carriage clocks to realise how very different they always were from the French national product. As already observed, the two manufacturing countries were scarcely in competition. While it is true that the French produced some exceedingly expensive *pendules de voyage* and the English some modest carriage clocks, it is still a sensible generalisation to say that the French clocks were on the whole cheap and English ones dear. This undoubted fact adds interest to a report found in the

---

(68) David Glasgow, author of the much respected book *Watch and Clockmaking*, 1885 and 1897, worked for Losada.

IX *English Work and Workmen*

Plate IX/86. ANONYMOUS CLOCK. *An anonymous timepiece clock bearing a presentation inscription for the year 1856. (Private Collection)*

Revue Chronométrique and dealing with the Paris Exhibition of 1878. Under the heading of *Horlogerie Anglaise* appear the following remarks:— "Selon des renseignements particuliers que nous avons lieu de croire exacte, il se fait peu de pendules de cheminées en Angleterre; mais, depuis quelques années, il s'y construit un grand nombre de petites pendules portatives; genre de fabrication qui aspire à remplacer de ce pays nos produits français designés sous le nom de *pièces de voyage"*. A free translation might read:— "According to certain reports, which we have no reason to doubt, few mantel clocks are made in

Plate IX/87. ANONYMOUS CLOCK. *A side view of the timepiece clock shown in Plate IX/86. (Private Collection)*

296

IX English Work and Workmen

Plate IX/88. *ANONYMOUS CLOCK. The "patent lever" platform of the clock shown in Plates IX/86 and 87. (Private Collection)*

England, although during the past few years they have produced there very many small portable clocks, made with a view to supplanting in England our national products known under the name of *pièces de voyage*". It is a mystery which clocks the French reporter could possibly have had in mind. It does not seem reasonable to suppose that Saint-Nicolas and Paris would have been worried even at a later date by the existence of English "French-style" carriage clocks like that shown in Plate IX/91, if only because they must have been not only expensive to produce but also quite unlikely to have been made in anything but small numbers. The clock in question bears the serial number 2004. On the backplate is a trademark consisting of a tasselled cushion with the initials "EWS", surmounted by a crown and supported by an hourglass. Below on a serpentine "ribbon" appear the words "MACHINE MADE". Plate IX/92 shows a further view of the case and also part of the movement. Note the very large going barrel having 148 teeth and which drives an intermediate pinion of 12 leaves and carrying a wheel of 80 teeth. The centre pinion has 10 leaves and the centre wheel 80 teeth. The barrel turns in 98 hours and a stopwork allows two turns only of the mainspring to be brought into use. A side view of the movement is shown in Plate IX/93. The escapement is a ratchet-toothed lever, having single roller, and a flat

Plate IX/89. *WILLIAM JOHNSON. Date prior to 1861. This timepiece clock is housed in an engraved French case of truly outstanding quality. (Private collection)*

Plate IX/90. *WILLIAM JOHNSON. A close-up view of one of the side panels of the clock shown in Plate IX/89. The engraver is Faucherre. (Private Collection)*

297

IX   *English Work and Workmen*

Plate IX/91. ENGLISH CLOCK IN FRENCH IDIOM. *The maker of this timepiece clock, which stands 5¼ inches tall, was "E.W.S." The case is gilded and the engraved dial is silvered. The top handle does not fold down. (Private Collection)*

Plate IX/92. ENGLISH CLOCK IN FRENCH IDIOM. *A further view of the clock shown in Plate IX/91. Private Collection)*

balance spring. This clock still retains its original travelling box, covered in dark red leather and lined in velvet.

W.J. HUBER LTD.

The year 1972 saw in England the manufacture for the first time of a "French" carriage clock! It is made by W.J. Huber Ltd. of Hatton Garden, London and an example is seen in Plate IX/94. The clock stands 5½ inches high and is a reproduction of an *anglaise* made by Couaillet Frères of Saint-Nicolas-d'Aliermont as is evident from their 1931-32 Catalogue. The new Huber clock has a timepiece movement made as a reproduction of a typical Saint-Nicolas *Obis*. The movement has been given a modern lever platform escapement and in consequence keeps very good time indeed. The Huber reproductions are at least as well made as were the originals, and their manufacture to sell at a most sensible retail price is a creditable achievement at the present time.

The first clocks produced carried the initials "W.J.H." in an ellipse on the back plate, but they were not numbered. After about 150 clocks had been made a number sequence was introduced starting with "150"

IX   *English Work and Workmen*

Plate IX/93. ENGLISH CLOCK IN FRENCH IDIOM. *A side view of the clock shown in Plates IX/91 and 92. Note the French influence apparent in the movement which, however, is finished very much to English watch standards. (Private Collection)*

Plate IX/94. HUBER NO. 1706. Date 1973. *A reproduction French timepiece carriage clock made in London. (By courtesy of W.J. Huber Ltd.)*

and well over a thousand clocks have been produced to date. At approximately serial number 1000, the initials on the movement changed to "A.C.G." (A.C. Gibson). The future plans of the firm include the manufacture of a *mignonette* and also a striking clock.

299

IX *English Work and Workmen*

Plate IX/95. ENGLISH CARRIAGE CLOCKS. *A very choice collection. Top left to right, Dent No. 21574, Dent No. 1, McCabe No. 2927, Vulliamy No. 1420 and Dent No. 1503. Below left to right, James Abbott No. 103 (perhaps more of a mantel clock), Barwise No. 376, Desbois, Jas. Whitelaw, not numbered. (By courtesy of various private owners and also of Asprey & Co. Ltd.)*

IX English Work and Workmen

B.L. VULLIAMY. *(By courtesy of the Clockmakers' Company, London)*
Benjamin Lewis Vulliamy (1780-1854) was "... the son, grandson and great grandson of watch and clockmakers.." and was the last of his line to conduct the famous family business, which was assimilated by Charles Frodsham in 1854. B.L. Vulliamy's most recent biographer (S. Benson Beevers in Antiquarian Horology, March 1954) not only emphasises Vulliamy's great personal integrity but also outlines his main horological achievements. Benson Beevers is at pains to point out that Vulliamy's modifications to important old clocks were made at a time when "antiquarian horology" did not exist in its present sense, and were therefore not nearly as reprehensible as may now appear.

E.J. DENT. *(By courtesy of the Clockmakers' Company, London)*
This is the best known portrait of Edward John Dent (1790-1853), who always called himself "E.I. Dent". His best memorial is the Great Clock at Westminster, despite the fact that it was not completed in his lifetime and that it later became the subject of much acrimony. E.J. Dent was by far the most distinguished member of his family. He was very inventive and one of his major interests was the improvement of marine chronometers. He devised a number of balances of which perhaps the best remembered today is his Patent Balance of 1842 affording secondary compensation and subsequently sometimes called in the trade "Dent's hurdygurdy".

IX English Work and Workmen

SIR JOHN BENNETT. (By courtesy of The City of London, Guildhall Library)

The Guildhall Library contains many different portraits of Bennett, including a number of cartoons. Also preserved are many articles taken from newspapers and periodicals and dealing with the part played by Sir John Bennett (1816-1897) in the life of the City in Victorian times. The Guildhall Library possesses a beautiful coloured printed invitation card for a garden party held on 26th June 1875 at Sir John Bennett's home, The Banks, Mountfield, Sussex.

VICTOR KULLBERG.

This unfamiliar portrait of Victor Kullberg (1824-1890) appeared in Invention and Inventors' Mart, 17th October 1885. Kullberg's marine chronometers were considered the best in the world. They held a most enviable reputation as timekeepers due to Kullberg's improved balances, to his auxiliary compensation and to his advanced springing techniques. During Kullberg's lifetime his chronometers were top of the list in the Greenwich Trials on no less than eight occasions, namely the years 1864, 1872, 1881, 1882, 1883, 1885, 1888 and 1889. After Kullberg's death his business was most ably continued under his name by his nephews the Wennerstroms. They achieved many more chronometer successes, the name "Kullberg" regularly heading the lists showing the results of the annual Royal Observatory Trials.

Other Carriage Clocks

Chapter X

Swiss Carriage Clocks

Fritz Courvoisier
1799—1854
d'après un dessin de H. Fischer

FRITZ COURVOISIER, ("The Commandant"). Perhaps the best known portrait, showing the uniform and the arm-band which he wore on 29th February, 1848. Fritz Courvoisier was both the leader and the hero of the Neuchâtel revolution.

X
Swiss Carriage Clocks

The Courvoisier Family and Other Early Makers. Later Clocks sold by Bautte, Moulinié and Henry Capt. The Mathey-Tissot Montres Pendulettes de Voyage of circa 1900. The Stauffer Family. D. Elffroth's Clock.

Swiss carriage clocks fall into four main categories:—

1. The early true Swiss carriage clocks which are so rare that they are hardly ever seen, and which appear to have been made only in or near La Chaux-de-Fonds in the canton of Neuchâtel. The name Courvoisier is the most important in connection with such pieces.

2. Later clocks which, while they bear Swiss names and have peculiarities not normally associated with French carriage clocks, nevertheless have the appearance of being based upon French *blancs-roulants*. An example of this type of clock will be shown signed "J.F. BAUTTE & Cº à Genève". The exact origin of these pieces is something of a mystery; but almost certainly they are far more French than Swiss.

3. Entirely conventional French carriage clocks made in the second half of the 19th century and bearing Swiss names. Of these, the examples most commonly found are either inscribed "Henry Capt, Geneva", or else "Moulinié à Genève". While it is possible that clocks bearing such names may have been given their finishing touches in Geneva, any examples seen by us are manifestly of standard French origin.

4. The small *montres pendulettes de voyage*, first finished in the years close to 1900 by firms such as Mathey-Tissot of Les Ponts-de-Martel, and of which manufacture continued for perhaps thirty years. These clocks were often repeaters, employing for that purpose part of the mechanism of a watch.[1]

The foregoing complicated and confused situation, while difficult to summarise briefly, need puzzle no-one who has read the earlier parts of this book.

Plate X/1 shows one of the earliest true Swiss carriage clocks. It is signed on the dial "COURVOISIER ET CIE".

Plate X/1. *COURVOISIER ET CIE. A very early true Swiss carriage clock, possibly made soon after 1811. This clock has a lever escapement. (Private Collection)*

---

(1) The business slump experienced everywhere in the 1920's, and the by then rapid spread of electric lighting in houses, were probably the two factors which finally "put paid" to the manufacture of repeating clocks and watches.

307

X Swiss Carriage Clocks

The Courvoisier family, one of the oldest associated with the horological industry of La Chaux-de-Fonds, played by far the most important part in the history of early Swiss carriage clocks. According to Chapuis[2] the great Swiss historian, the main family businesses were "Courvoisier et Cie" (1811-1845), and "Courvoisier Frères" (circa 1845-1882). In the year 1842[3] Frédéric-Alexandre Courvoisier *(dit* "Fritz") left the main family business in order to work on his own account. "Courvoisier et Cie" was dissolved in 1845, and Henri-Louis and Philippe-Auguste revived the family concern under the name of "Courvoisier Frères". In about 1852 a new firm of the same name was formed under Henri-Edouard, Louis-Philippe and Jules-Ferdinand. This business continued after the death of Henri-Edouard and until the year 1882. The Courvoisier family tree, appearing on page 316, is compiled from the text of Chapuis' *Histoire de la Pendulerie Neuchâteloise.* The Courvoisier pedigree shows the different generations associated with clocks and watches, besides giving the names of the horological families with whom they intermarried. Notes on the early business history of the Courvoisier family and their link with Captain Louis Robert will be found in Chapter I. The reader will remember that pre-*pendules de voyage* exist signed both "Robert" and "Robert and Courvoisier".[4]

Returning to the clock in Plate X/1, the reader will remember that it is signed "COURVOISIER ET CIE". This signature, according to Chapuis, was first used in 1811,[5] and there is no reason why such a piece should not have been made within a few years of this date. The wooden and glass case, veneered in mahogany, is decidedly primitive and fragile. It is held together by tenons, mortices and wood screws, and only thin picture glass is used for the front, back and side panels. The overall height is 9¾ inches with the handle up. On top of the case is a round window with gadrooned edging. There is a top repeat button. The squared feet of the case are gilt in common with its other furniture. The movement of the clock is shown in Plate X/2. The piece goes for eight days, offering calendar, alarm and also *grande sonnerie* striking and repeating. The reader will notice that the movement is supported by the dial. The *modus operandi* of the striking work

Plate X/2. *COURVOISIER ET CIE. The back view of the clock shown in Plate X/1. Note the external* cadrature *and typical Swiss back-to-front bells with steel hammers. (Private Collection)*

---

(2) *Histoire de la Pendulerie Neuchâteloise, 1917,* and *Pendules Neuchâteloises (Documents Nouveaux) 1931.*

(3) The date is confirmed by the book *La Chaux-de-Fonds, Son Passé et Son Présent, 1894.* This work contains a biography of "Fritz" Courvoisier by Dr. Farny. He drew much of his basic information from an earlier biography written by Fritz's son Paul-Frédéric and subsequently published by Jeanneret in 1863. In 1947 appeared Chapuis' *Fritz Courvoisier, 1799-1854, Chef de la Révolution, Neuchâteloise* containing yet further biographical details.

(4) See Chapter I where such a clock is described. It is illustrated in Plate I/36.

(5) Op. cit.; but this fact is not brought out in the book *The Swiss Watch* by the same author, where the impression is given that *raison sociale* "Courvoisier et Cie" was not used before 1816. Probably Chapuis, in his latter work, was offering précis rather than an exhaustive history.

X Swiss Carriage Clocks

is most interesting, differing not only from French and Austrian work, but also from the practices found in the slightly later Swiss carriage clocks. The *cadrature* is, however, planted on the back movement plate in the manner which the reader will already have learned to associate with the area of the Jura, and which is also typical of Swiss work. In the Courvoisier clock almost all this external work is made of steel, while the back-to-front arrangement of the twin bells with their associated steel hammers is characteristically Swiss. With these remarks in mind, the reader may wish to compare Plate X/2 with Plates VI/30 and VI/31. It comes as no surprise to find that the Swiss carriage clock drives both its going and its striking trains from a single barrel. The clock offers a choice of *grande sonnerie, petite sonnerie* or silence as determined by a lever accessible in the top left-hand corner of the dial. When the clock is set for *grande sonnerie*, and is left to strike and chime by itself in passing, then it strikes at each hour the hour followed by the four quarters. A further peculiarity is that "tang-tings" and not "ting-tangs" are struck.[6] At the quarters in passing the clock strikes the hour followed by the quarters. When, however, the control lever is set for *petite sonnerie* the clock will strike "tang-ting" quarters in passing, but at the hour it will strike the hour, and then the four quarters. When the repeat button is pressed the clock will follow the same *grande sonnerie* pattern already described. The hammer tails are raised, not by a conventional pin-wheel, such as one expects to find in a domestic clock, but rather by a cam-wheel reminiscent of tower clock practice. "Knock-out" striking is employed, the "knock-out" being achieved by a downward-acting hammer.

The platform lever escapement of the clock is contradictory because in certain respects it shows comparatively advanced design, while in others it remains rudimentary. A very short straight-line lever is used in conjunction with a divided-lift wheel, plain steel balance, and a steel balance spring with an overcoil. The pallets and 'scape wheel are mounted below the platform. The escapement is primitive in that it employs unjewelled steel pallets, having curved locking faces, being arcs struck from the pivoting point of the lever. There is no draw. The absence of this feature is

Plate X/3. *CUGNIER LESCHOT. Another early true Swiss carriage clock, probably made circa 1815. The signature is hidden by the bezel, as is the single winding square for both going and striking trains. Note the resemblance to certain of the early Breguet clocks, for example No. 2020 as seen in Plates II/4 and 5. (By courtesy of G.C. Del Vecchio, Milan)*

demonstrated in practice by the fact that the 'scape wheel does not recoil on unlocking. Essentially the form of the pallets is little different from the early design used by Breguet.[7] There are thirteen teeth in the 'scape wheel, of which 3½ teeth are embraced by the pallets. In contrast to the unsophisticated form of the pallets, the safety action is quite advanced. A roller of comparatively small diameter is used. It has a curved passing-hollow while the lever horns are almost exactly the same shape as those favoured by Massey. It is also worth mentioning that the escapement is banked by a single pin, planted vertically in the lever just behind the fork, and limited in its movement by

---

(6) The "tang" bell is also used for striking the hours.

(7) Tardy in *Les Echappements de Montres* illustrates this pallet form in Fig. 237 on page 126.

X Swiss Carriage Clocks

Plate X/4. *CUGNIER LESCHOT. The movement of the clock shown in Plate X/3. Compare with Courvoisier Plates X/1 and X/2, and Breguet Plates II/4 and II/5. (By courtesy of G.C. Del Vecchio, Milan)*

a slot cut in the platform proper. At the bottom of the dial is a lever by means of which the simple calendar may be indexed forward one day at a time for purposes of correction.

Another Swiss clock, made in very much the same period as the wooden-cased Courvoisier & Cie, is an un-numbered example signed "CUGNIER LESCHOT" and which is illustrated in Plates X/3 and X/4. The resemblance of this piece to some of the early Breguet *pendules de voyage* is very striking indeed. The Cugnier Leschot clock stands just under eight inches tall and it offers the alternatives of *grande sonnerie* or *petite sonnerie* striking on two gongs, besides having an alarm which sounds on a bell in the base. The escapement is a lever. The single winding square common to both going and striking trains is concealed by the bezel below chapter '6'. The signature "CUGNIER LESCHOT" is divided on either side of the winding square. Behind the bezel at chapter '12' is a control lever for *grande* or *petite sonnerie* and at chapter '3' is another lever offering *sonnerie* or *silence*. In its present state the clock runs twelve to thirteen days if set to *silence*, but only four to five days if the full striking is used. The clock still has its original wooden outer travelling box.

Plate X/5 illustrates a rather later and exceedingly fine Swiss carriage clock, this time bearing the name of Frédéric Courvoisier[8] and possibly made as early as about 1830.[9] It has a chronometer escapement, while the striking is of that kind peculiar in carriage clocks to the few true Swiss examples made in the canton of Neuchâtel. In this tradition *grande sonnerie* is struck throughout the night, and *petite sonnerie* during the day. The change-over occurs automatically each morning and evening at eight o'clock. This particular clock also offers repetition, alarm and calendar. It is to be found in the Musée d'Horlogerie at La Chaux-de-Fonds. The clock still has its original leather-covered travelling box with double-opening front doors. It is wound from the front, the other end of the same key serving to set the hands for meantime, for calendar and for alarm. Engraved in block capital

---

(8) As already mentioned, the books *La Chaux-de-Fonds. Son Passé et Son Présent* and *Histoire de la Pendulerie Neuchâteloise* both contain biographies of Frédéric-Alexandre Courvoisier. The writers make it plain that, while "Fritz" received more commercial horological training than practical, he nevertheless attended a horological school run in the canton of Neuchâtel by Henri-Louis Maillardet and Charles-Frédéric Klentschi. At a later period "Fritz" Courvoisier travelled extensively on behalf of the family business, representing the house in many countries. He married in 1826, thereafter enjoying about three years with his wife before becoming involved in the by then serious local political troubles. He rose to be a Lieutenant-Colonel in the *carabiniers* and was temporarily banished from the canton circa 1831. After the death of his father Louis Courvoisier in 1832, "Fritz" re-constituted the firm "Courvoisier et Cie" with his two brothers Henri-Louis and Philippe-Auguste. In 1842 "Fritz" left the family business in order to set up on his own account in La Chaux-de-Fonds. He specialised in the sale of watches made in the Neuchâtel mountains. He was both the leader and the hero of the Neuchâtel revolution of 1848.

(9) Repair dates for 1864 and 1867 are scratched on the back plate.

X *Swiss Carriage Clocks*

Plate X/5. *FREDERIC COURVOISIER/C.F. KLENTSCHI. A true Swiss carriage clock, probably made as early as 1830, and having chronometer escapement and peculiar Swiss striking. Height approx. 7ins. with handle up. (By courtesy of the Musée d'Horlogerie à La Chaux-de-Fonds)*

Plate X/6. *COURVOISIER/KLENTSCHI. The very complicated movement of the clock shown in Plate X/5. This clock strikes in passing grande sonnerie at night and petite sonnerie during the day. Engraved on the back plate but hidden by the bells is the signature "C.F. KLENTSCHI NO 645". (By courtesy of the Musée d'Horlogerie à La Chaux-de-Fonds)*

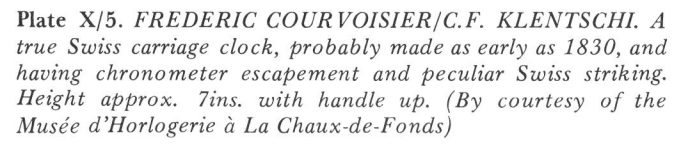

letters on the back of the clock are the words "ECHAPPEMENT LIBRE A RESSORT LEVEE REPOS ET DEGAGEMENT EN RUBIS DIX TROUS EN PIERRE SPIRAL ISOCHRONE BALANCIERE COMPENSE" (sic.). On the top of the case at the front is engraved "Frédéric Courvoisier CHAUX-DE-FONDS EN SUISSE". However, engraved on the back plate of the movement, underneath the bells and quite hidden by them, is the inscription "C.F. KLENTSCHI NO. 645".[10] Plate X/6 shows a back view of the movement with its peculiar *cadrature* and bells. Plate X/7 gives a close-up view of the escapement platform. Note how Swiss practice most sensibly avoids the use of free-springing, providing in this clock an index and means of regulation, even though the balance has a helical spring. Normally clocks, watches and chronometers which are free-sprung, especially those with helical springs, need skilled attention to bring them to time, involving adjustments made on the balance itself. The pivoted-detent escapement of the Courvoisier/Klentschi clock is also exceedingly different from those seen in most continental work. As a rule pivoted-detent escapements, or *bascule* escapements as they are sometimes called,

---

(10) Klentschi was born in 1774 and died in 1854. He not only worked for the Courvoisier family, but also ran with Maillardet a horological school which was attended by "Fritz" Courvoisier. He was a noted clockmaker in La Chaux-de-Fonds.

311

X Swiss Carriage Clocks

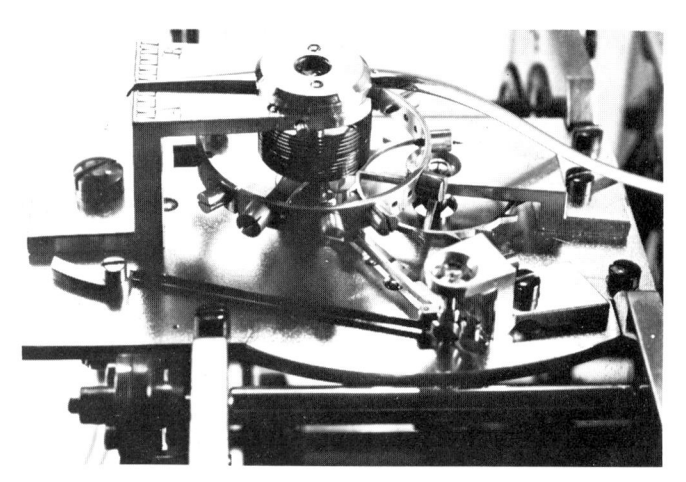

Plate X/7. COURVOISIER/KLENTSCHI. *The beautiful and very Swiss chronometer escapement of the clock shown in Plates X/5 and X/6. Note the long pivoted-detent with a straight spring instead of the usual continental volute spring. The index is an unusual but sensible feature as applied to a domestic chronometer. (By courtesy of the Musée d'Horlogerie à La Chaux-de-Fonds)*

are controlled by small volute springs similar to the balance springs of watches. This example has its detent under the control of the long straight spring seen in the foreground. The "vital statistics" of the clock, already mentioned as being prominently engraved on the outside of the case, are concerned only with the charms of its escapement. The words of the inscription convey, if somewhat obscurely, that the clock has a detached escapement controlled by a spring, and that the discharging pallet, locking stone and impulse pallet are jewelled in rubies. Furthermore six pivot holes and four endstones associated with the escapement are also jewelled, while the escapement boasts a compensated balance with isochronous spring. The striking work of this Swiss *pendule de voyage* is particularly interesting. As already noted, the clock strikes *grande sonnerie* at night and *petite sonnerie* during the day. Such striking is typical of the true Swiss carriage clocks made in the canton of Neuchâtel. It is, of course, possible to reverse the night and day sequences by setting the clock forward twelve hours; but it is abundantly clear from Exhibition reports that the intention was for the great striking to be reserved for the hours of darkness. A lever is provided to suppress the striking altogether. It does not offer a choice of *grande* or *petite sonnerie*. The clock, however, strikes full *grande sonnerie* whenever the top repeat button is pressed. The calendar

requires correction for all months not having thirty one days.

Chapuis, in *Histoire de la Pendulerie Neuchâteloise*, 1917, shows three views of a chronometer carriage clock which looks as if it was made c.1830. It is signed on the top of the case "Frédéric Courvoisier CHAUX-DE-FONDS SUISSE". Chapuis states firmly that the piece was made by Frédéric himself. The clock is not No. 645 which has just been described although it is very similar. Fortunately Chapuis includes a photograph of the top of the clock, showing not only slightly different inscriptions, but also a compensation balance rather simpler than that of No. 645 now in La Chaux-de-Fonds museum. The two clocks also have quite different *étuis* or travelling boxes. Chapuis states that "Fritz" himself worked upon a clock

Plate X/8. FREDERIC COURVOISIER. *A true Swiss carriage clock, No. 1609, with lever escapement and Swiss striking. Date circa 1830. (Private Collection)*

X *Swiss Carriage Clocks*

which accompanied him on all his travels ("... Fritz Courvoisier travailla lui-même à la pendule qui l' accompagna dans tous ses voyages.."). This piece would almost certainly have been a *pendule de voyage*. While it is very tempting to theorise upon when such a clock may have been made, there seems to be no direct evidence. "Fritz" Courvoisier could scarcely have "finished" a complicated *grande sonnerie* chronometer carriage clock while under training with Klentschi; although it is stated by Chapuis that Frédéric "finished" two clock movements during this period. A possible theory might be that Frédéric Courvoisier carriage clocks were made during the years 1826-1829, this being almost the only period of his working life, other than his apprenticeship, when he would have had time to make anything. It seems fairly improbable that Frédéric would have signed clocks with only his own name while he was a member of "Courvoisier et Cie", while after 1842 his business was largely confined to watches. In any case, the clocks in question were manifestly made well before this date.

Another carriage clock having a lever escapement and signed "FRED COURVOISIER, CHAUX-DE-FONDS, 1609" is shown in Plate X/8. The movement and striking work are almost identical to those of the two chronometer clocks already mentioned. The straight-line platform lever escapement, as seen in Plate X/9 has a single roller. The pallets are jewelled with rubies, while the steel 'scape wheel has a large amount of "lift" on the teeth. The plain balance has a flat spring and is oversprung. The movement, see Plate X/10, is engraved "ECHAP.T LIBRE A VISIBLES TROIS" (sic) on the left, and on the right "ANCRE

Plate X/9. *FREDERIC COURVOISIER. The lever escapement of the clock shown in Plate X/8. (Private Collection)*

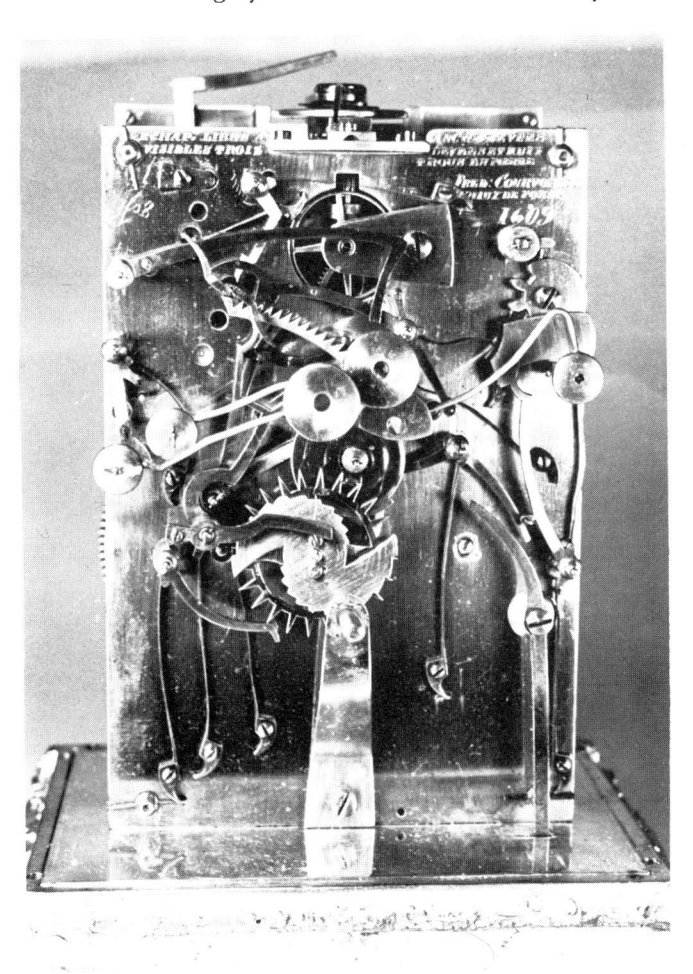

Plate X/10. *FREDERIC COURVOISIER. The very interesting and characteristically Swiss movement of the clock shown in Plate X/8. (Private Collection)*

X *Swiss Carriage Clocks*

Plate X/11. *AUGUSTE COURVOISIER. A true Swiss carriage clock, No. 2119, with chronometer escapement and made for the Turkish market. Date circa 1830. (By courtesy of David Wakefield)*

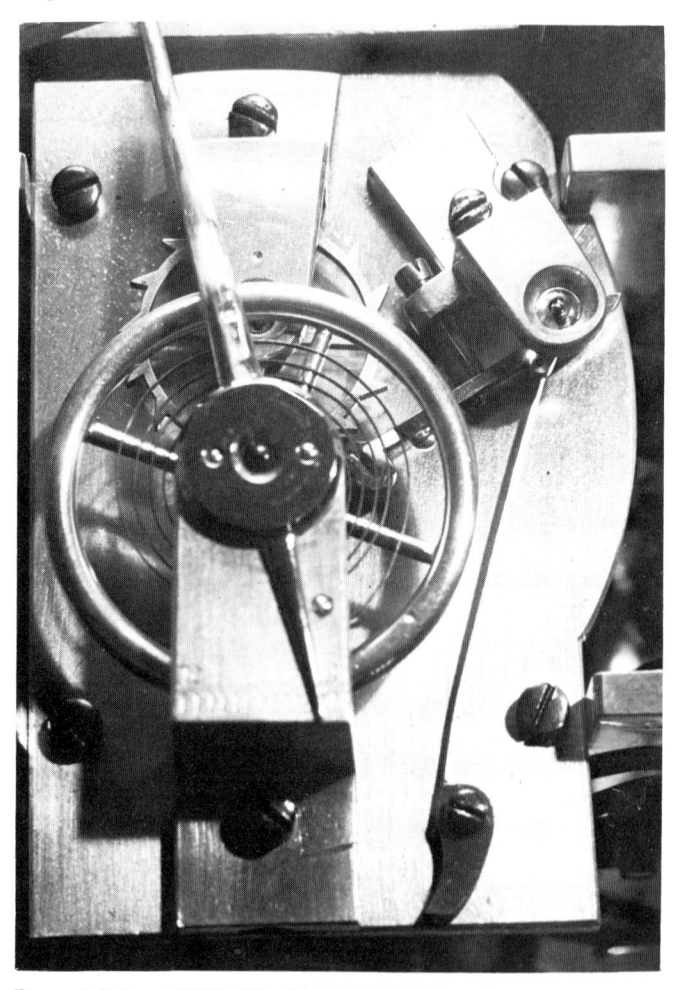

Plate X/12. *AUGUSTE COURVOISIER. The comparatively simple chronometer escapement of the clock shown in Plate X/11. Compare with Plate X/7. (By courtesy of David Wakefield)*

LEVEES. LEVEES ET HUIT TROUS EN PIERRE". The inscriptions once again refer to the escapement. Although jewelling is used throughout, only the balance holes have endstones. The whole escapement is very much more "modern" in its conception than that of the early Courvoisier et Cie wooden-case clock already described.

Plates X/11 and X/12 show a further chronometer carriage clock. This example was made in very much the same period as the clocks already described. The back plate of the movement is signed "Aug^{te} Courvoisier et Comp^{e} CHAUX-DE-FONDS", No. 2119. The peculiar Swiss striking work is once again used, and once more the escapement is that of a pivoted-detent chronometer. A plain balance is over-sprung with a flat spring. A very interesting feature is the use of Turkish chapters on the dial. This particular clock was found in Istanbul in the covered bazaar in 1970, and it is clear from both chapters and from dial signature that the piece was made for the near-Eastern market. A further peculiarity is the use of the Islamic calendar.[11] Like two of the clocks already mentioned,

---

(11) This differs from the normal 31 day calendar in having 30 divisions instead of 31. The Islamic year is divided into months alternately having 29 and 30 days. Manual correction of the calendar is thus required every other month. Chapuis in *Technique & History of the Swiss Watch ...* pp 139-140, makes abundantly clear the extent of the Swiss trade with the Near East, and especially Turkey from as early as 1592. At the beginning of the 19th century the firm of Robert & Courvoisier, later Courvoisier & Cie, to say nothing of Bautte of Geneva, carried on a very substantial watch export trade to the whole of the Ottoman Empire. Most of the large and important Swiss firms were represented in this trade.

X Swiss Carriage Clocks

this example has its original travelling box. This is similar to the one illustrated by Chapuis, having a backward-opening top with side hasps secured by latches. Engraved upon the top left-hand side of the movement back plate is the inscription "ECHAPPEMENT A RESSORT ET HUIT TROUS", and on the right-hand side of the same plate appears the inscription "LES LEVEES EN RUBIS". These shortened and colloquial inscriptions say that the chronometer escapement has jewelled holes, while only the balance holes have endstones. The words "LES LEVEES EN RUBIS" mean simply that all working parts of the escapement (that is the discharging pallet, impulse pallet and locking stone) are all in ruby.

The reader will remember that Philippe-Auguste Courvoisier, born in 1803, was a younger brother of "Fritz" and of Henri-Louis. Auguste married Adèle Jacky, and he was a member of "Courvoisier et Cie" and later of "Courvoisier frères". On this basis the clock may reasonably be assumed to have been made between the years 1832 and 1845. In 1832, as already explained in a footnote, Auguste helped his brothers Henri-Louis and "Fritz" to re-constitute the firm of Courvoisier et Cie following the death of their father Louis Courvoisier, while in 1845 the *raison sociale* became Courvoisier frères. The signature "Aug.te Courvoisier et Comp.e CHAUX-DE-FONDS" would seem more likely to have been used in the former period than in the latter. The main and very real difference between the chronometer clock signed "Frédéric Courvoisier" and the example bearing Auguste's name is that the entire escapement of the latter is very much more of utility standard than that of the former. It could be argued that the cruder escapement is earlier, or equally that the product was cheapened somewhat as time went by. Another valid point might be that a relatively inferior escapement would have been considered quite good enough to send to Turkey. Whatever the facts, the quality of the chronometer escapement of the Auguste clock No. 2119 is about equal to that of the lever escapement of the "Fred Courvoisier" No. 1609. There is no evidence at present what relationship, if any, there is between the respective serial numbers as found on any of these Courvoisier clocks.

A clock produced at a rather later date than those already discussed is illustrated in Plates X/13 and X/14.

It is signed on the back plate "P. Girard". In most

Plate X/13. *P. GIRARD. A later Swiss carriage clock with a lever escapement. This, or a similar piece, was exhibited at the London Great Exhibition of 1851. (Private Collection)*

details it is very similar to the series of pieces already mentioned. This fact is in no way surprising, especially as it is known that Girard did work for the Courvoisier family sometime after 1820. The striking work of the Girard clock, with its day and night alternating *grande* and *petite sonnerie*, is identical in character and in operation to the Courvoisier examples. The lever escapement is of a rather more sophisticated design than that used by Fred Courvoisier, while the use of gongs for the striking and the altogether later appearance of the case, dial and hands suggest a period of closer to 1850 than 1830. In fact Peter Girard is on record as having exhibited a clock of much the same specification at the London Great Exhibition of 1851. (See Alphabetical List of Names) The clock under

X Swiss Carriage Clocks
Part of the Courvoisier Family Tree

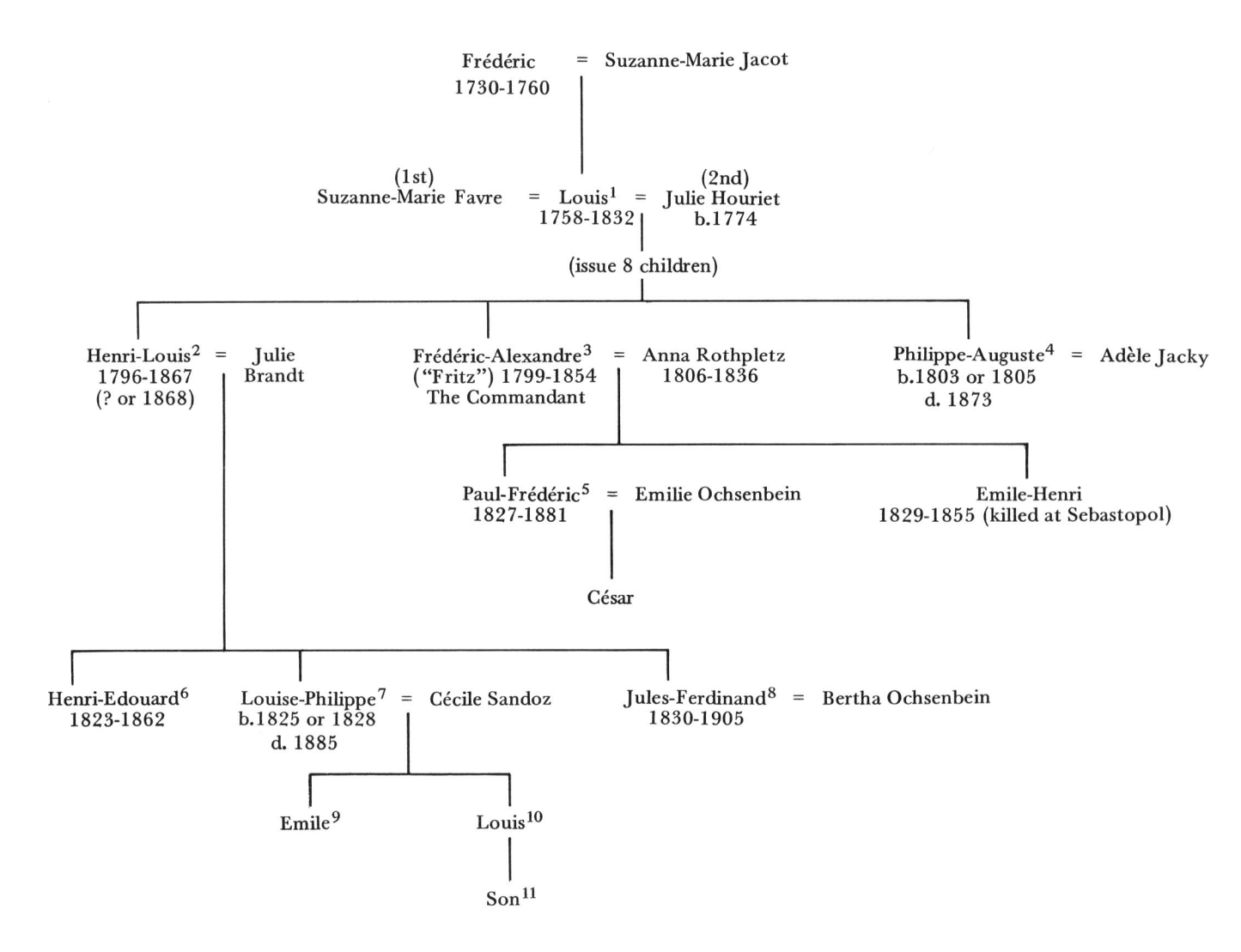

```
                        Frédéric      =    Suzanne-Marie Jacot
                        1730-1760
                            |
        (1st)                                       (2nd)
   Suzanne-Marie Favre  =   Louis¹   =        Julie Houriet
                           1758-1832              b.1774
                            |
                    (issue 8 children)
   ┌────────────────────────┼────────────────────────────┐
Henri-Louis² = Julie   Frédéric-Alexandre³ = Anna Rothpletz   Philippe-Auguste⁴ = Adèle Jacky
1796-1867     Brandt   ("Fritz") 1799-1854   1806-1836         b.1803 or 1805
(? or 1868)            The Commandant                          d. 1873
                                   |
                    ┌──────────────┴──────────────┐
                Paul-Frédéric⁵ = Emilie Ochsenbein      Emile-Henri
                1827-1881                                1829-1855 (killed at Sebastopol)
                    |
                  César
   |
   ┌──────────────┬──────────────────┐
Henri-Edouard⁶   Louise-Philippe⁷ = Cécile Sandoz   Jules-Ferdinand⁸ = Bertha Ochsenbein
1823-1862        b.1825 or 1828                     1830-1905
                 d. 1885
                     |
                ┌────┴────┐
             Emile⁹     Louis¹⁰
                           |
                         Son¹¹
```

Key to Courvoisier Family Tree

1. Josué-Robert, with his son Louis, founded "Robert Josué et fils". On 1st May 1781 the firm became "J. Robert et fils et Cie". At its head was Capt. Louis Robert and Louis Courvoisier. Louis Robert's widow continued the firm from July 1787 under the *raison* "J. Robert et fils, Courvoisier et Cie". She was assisted by Aimé Robert, Louis Courvoisier, Jean-Pierre Robert, and Florian Sandoz. In 1805 the firm became "Robert, Courvoisier et Cie", and in 1811 "Courvoisier et Cie".

2. Member of "Courvoisier et Cie", and later of "Courvoisier frères".

3. "Courvoisier et Cie". Own business from circa 1842.

4. "Courvoisier et Cie" and later of "Courvoisier frères".

5. *Pendulettes de voyage* with Villeumier.

6. "Courvoisier frères", as re-constituted circa 1852.

7. "Courvoisier frères", the new firm.

8. "Courvoisier frères", the new firm.

9. "Courvoisier frères", the new firm.

10. "Courvoisier frères", the new firm.

11. "Courvoisier et fils". (After retirement of Emile. Date uncertain).

X *Swiss Carriage Clocks*

most likely basic source of supply. It is made plain by Chapuis,[12] however, that Maillardet, Klentschi and Girard dealt directly on behalf of the Courvoisier business with small workshops scattered about the canton of Neuchâtel. In other words, every one of these Swiss carriage clocks, in common with both French and English examples, were very largely the product of a complex out-worker industry involving many people. All the same, it seems certain that the true Swiss carriage clock was only ever made in relatively small quantities, and that it became extinct from soon after 1850.

Plates X/15 and X/16 show a very different carriage clock representing the second type mentioned at the

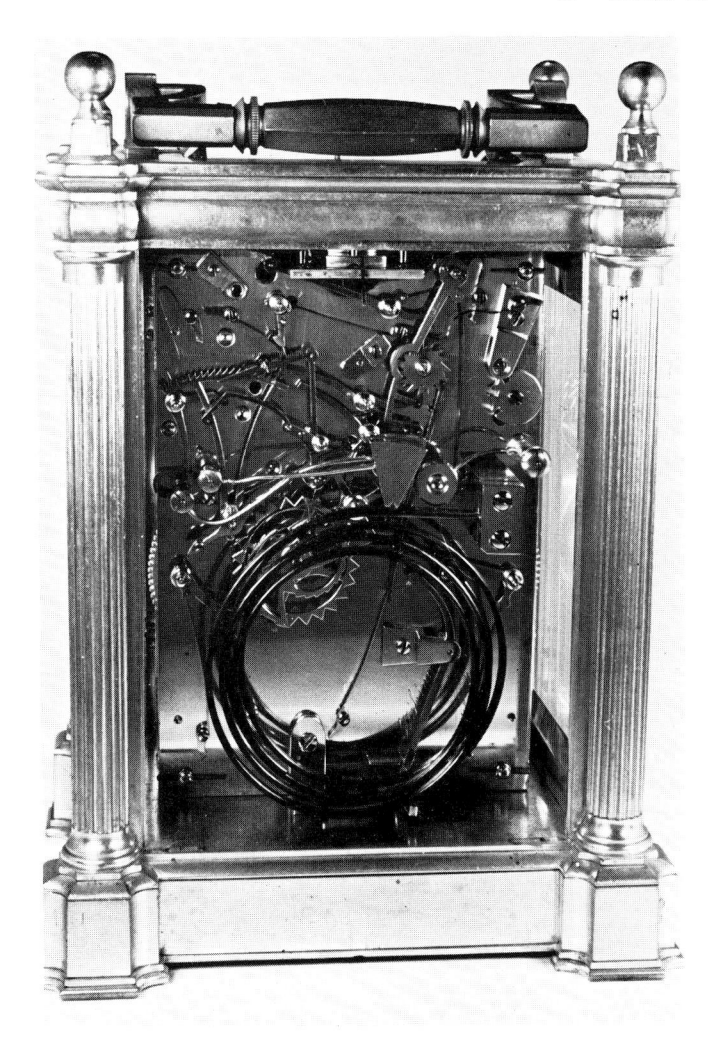

Plate X/14. *P. GIRARD. The back view of the clock shown in Plate X/13. Note the characteristic Swiss hammers and the introduction of gongs. (Private Collection)*

discussion may even be the very piece. Its case has a decidedly English flavour, particularly as regards the handle and top finials, and may even have been all or partially produced in this country. Plate IX/28 in the chapter dealing with English clocks illustrates the resemblance.

The characteristically Swiss carriage clocks, such as those described above and bearing such names as Klentschi, Fritz or Auguste Courvoisier, Courvoisier et Cie, Girard or Borel, are so alike as to leave little doubt that many of their parts must have had the same origin. In this connection the Klentschi and Maillardet families perhaps seem to have been the

Plate X/15. *J.F. BAUTTE & CO., GENEVA. This carriage clock is almost certainly more French than Swiss. Date circa 1860. (By courtesy of Major General Denis Redman)*

(12) Op. cit., 1917.

317

X Swiss Carriage Clocks

Plate X/16. *J.F. BAUTTE & CO., GENEVA.* A back view of the clock shown in Plate X/15. Note the unusual blanc upon which this clock is based. (By courtesy of Major General Denis Redman)

beginning of this chapter. The signature "J.F. BAUTTE & C⁰ à Genève" appears both on the dial and on the bottom of the back plate. The clock strikes and repeats the hours on a bell and is also provided with an alarm. The curious lever platform escapement has an uncut balance and yet it is provided with a very elaborate overcoil, obviously formed with great care. The well-made movement is unusual in having the back plate cut open and the intermediate wheels and pinions of both trains made visible and run in a common horizontal bar. There are also peculiar mainspring clicks provided with double-ended cocks and with "C" shaped springs. Unlike the Courvoisier pieces, this clock is wound through shutters in the back door. With its peculiar escapement, white enamel dial, rather tall and narrow case offset by an exuberant handle, it seems to represent an amalgam between French and Swiss tastes and influences. In fact a very similar clock signed "LE ROY ET FILS H$\underline{ERS}$, PALAIS ROYAL, G$\underline{\text{IE}}$ MONTPENSIER 13 & 15, PARIS", was sold at Sotheby's on 13th December 1971, Lot 37. This clock had exactly the same style of "bar" movement, "C" click springs and back winding. The movement undoubtedly has the same basic origins as the "Bautte" example. The clocks also have exceedingly similar cases and dials.

In addition to the peculiar carriage clocks mentioned above, Bautte's name also appears upon a number of entirely standard French pieces. One such clock signed on both dial and movement "J$\underline{n}$ F$\underline{ois}$ Bautte & C$\underline{ie}$ A GENEVE" is a characteristically Paul Garnier piece.

Jean-François Bautte (1772-1837) was not only very well known in Switzerland, but he also had a shop in Paris, which apparently continued in business well after his death. His speciality was watches, and more especially those of the decorative kind. Chapuis speaks of him as being a merchant rather than workman, and mentions a post-chaise which Bautte used during the French Revolution to conduct his business between Paris and Geneva in kind rather than in cash. Bautte was succeeded by Rossel et Fils.

Continuing to the third type of carriage clock bearing Swiss names, Plate X/17 shows a very attractive small example on the dial of which appears "MOULINIE GENEVE".[13] While this piece could have been finished in Switzerland and possibly has Swiss hands, the feeling given by both movement and case is that the clock is French, made perhaps 1845-50. The same applies to another clock bearing a similar inscription and shown in Plate X/18.

A few "Geneva" clocks, manifestly French beyond all question, are found signed "Henry Capt. Genève". See Plate VIII/4. They are always fine pieces, but appear to have absolutely standard French *pendule de voyage* format and finish. Henry Capt was advertising carriage clocks in at least one New York newspaper at the turn of the century. They were described as "of

---

(13) Mr. E. Jurmann of Arundel bought in 1972 an obvious Garnier clock, with his escapement; but engraved on the back plate was "Moulinié à Genève". On later examination this clock was found to have an "H.L." movement, this further proving its French origins.

X Swiss Carriage Clocks

Plate X/17. *MOULINIE, GENEVA. A small carriage clock made circa 1845-50, and almost certainly French. (Private Collection)*

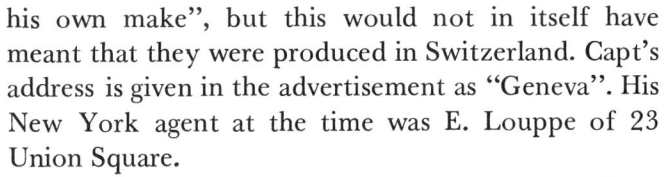

Plate X/18. *MOULINIE, GENEVA. Another carriage clock having French origins. (By courtesy of Dr. Mull)*

his own make", but this would not in itself have meant that they were produced in Switzerland. Capt's address is given in the advertisement as "Geneva". His New York agent at the time was E. Louppe of 23 Union Square.

Evidence that no real carriage clock industry existed in Geneva during the second half of the 19th century is not lacking in the works of contemporary writers. For instance, the famous Adrien Philippe[14], at a time when horological business was extremely bad, saw fit to advise the manufacturers and workmen of Geneva to take up the manufacture of *pendules de voyage*. Had these clocks already been in production, at least on any scale, such advice would surely not have been necessary. If anyone was in the position to know, it was Philippe. Indeed, he was not only a partner in the world-famous watch firm of Patek et Philippe, but he was also the expert sent by the Swiss to report on the clocks, watches and chronometers exhibited in Paris in 1878 by their manufacturing rivals, including the Americans. What Philippe actually said translates to read: "Concerning the *pendules de voyage*, it occurs to us that the following information should be given to the manufacturers and workers of Geneva. As the political situation and the threat of war have brought about unemployment, it seems that there could be an element of activity which would give new resources to our industry. The work is very little different from that of pocket-watchmaking and it is evident that a workman of even mediocre ability, providing that he

---

(14) *Etudes sur l'Horlogerie à L'Exposition de Paris, 1878*, p.74, published by the *Journal de Genève* in the same year.

is intelligent, will quickly adapt himself to the relatively easy practice of this kind of horology. We put forward this suggestion to those who both desire and look for a new branch for the horological industry at Geneva". These remarks, which appear to have fallen upon deaf ears, surely contain nothing to suggest that any carriage clock industry existed at the time in Geneva! Before leaving this interesting subject, other quotations may be pertinent. The *Revue Chronométrique,* speaking of *L'Horlogerie à l'Exposition de 1878*[15], says "L'industrie de la Suisse, en ce qui concerne l'horlogerie, est concentrée tout entière dans la production de la montre et de certains genres d'outils". This may be translated "The Swiss industry, in so far as horology is concerned, is confined exclusively to the production of watches and of certain kinds of tools". Even as concerns Geneva watches, Lebon in *Etudes. .sur L'Horlogerie en Franche-Comté,* printed in Besançon in 1860, says in his letter written from Geneva dated 6th August 1862 words which translate to read:— "One thing I say to you, Sir . . . is that there are not, in the true meaning of the term, real manufacturers of horology in Geneva. Suppose, for example that one goes to the house of M. Bautte, or that of M. Mercier or to a number of other *horlogers* calling themselves manufacturers, one does not find the smallest sign of any factory. They will show you beautiful watches, well decorated and reasonably well made, but that is all. If you ask one of these manufacturers where his factory is, he will reply that it is in this and that part of the town, but he will be careful not to offer to take you there, because the fact is that it does not exist". ("car c'est un fait qu'elle n'existe pas"). Lebon goes on to say that the Geneva *fabricant* buys *ébauches* easily from Brassus (canton de Vaud) or from Fontaine-Melon (canton of Neuchâtel) or even from MM. Japy (Haut-Rhin). Then he finds one worker to make the escapement, another to perform the finishing, and then gets hold of all sorts of other miscellaneous pieces until the watch is complete. It must, however, be said that Lebon mentions with praise a few real Geneva makers such as Patek & Philippe, Vacheron and, with a few reservations, Soldano.

The final type of Swiss carriage clock, which appeared just a few years before the present century, was the *montre pendulette de voyage,* finished and sold by the house of Mathey-Tissot at Les Ponts-de-Martel in the canton of Neuchâtel.[16] These clocks were smaller than the "full-sized" *pendules de voyage* made in France. They went for eight days and were made in *qualité soignée* and *qualité extra soignée,* having stopwork for the barrels, lever escapements and often quarter, five-minute or even minute-repeating work based on watch movements supplied by Hahn of Le Landeron, Lugrin at l'Orient de l'Orbe and by Le Coultre of Le Sentier.[17] The Mathey-Tissot illustrated Catalogue of 1905 shows no less than twenty-three *Pendulettes-Mignonnettes,* besides clock cases, interchangeable movement parts and fold-flat travelling clocks of the type sometimes called *calottes.* The case styles on the whole were individual to the firm, although there are found familiar names like *Corniche, Indienne,* and *Anglaise.* Other names included *Empire craquelé, Empire rayon de gloire, Empire moiré,* and *Anglaise craquelée.* Plate X/19 is reproduced from the Mathey-Tissot Catalogue. The cases were mostly made of silver and were semi-miniatures, although a few more conventional-looking brass cases were offered. The silver cases mostly showed marked *art nouveau* influence. Some were inlaid *(niello),* some were plain polished silver, others "crackle" finish, some sunray, some "hammered", some lined and some like ribbon *(moiré).* Not a few had decorative patterns in high relief *(ciselure très riche).* One example had an inlaid case. Others had coloured enamel fired over engine turning *(guilloché).* Another clock boasted a case apparently with *cloisonné* enamel, and yet another had the contradictory name of *Anglaise Chinoise.* As the name suggests, the real difference between *montres pendulettes de voyage* and Swiss carriage clocks of the early type is that the former are far more watch-like in their conception. They do not usually strike in passing, although some of them are repeaters. As the

---

(15) August 1878, p.120; usually found bound in Vol. X.
(16) Guinard and Golay were the main importers to England.
(17) In the U.S.A. is a travelling clock, reputedly a minute-repeater with *grande sonnerie,* bearing the name of the "Geneva Clock Company". This may be some type of *montre pendulette de voyage.*

X Swiss Carriage Clocks

Grandeur naturelle. — Natürliche Grösse.
Natural Size. — Grandezza naturale.

N° 5

N° 6

Modèles déposés.
Gesetzlich geschützt.
Designs protected.
Modelli depositati.

———

Se finissent en bronze doré ou vert.
Se font aussi en argent massif.

—

Herstellung in vergoldeter oder grüner
Bronze oder in massivem Silber.

—

Finished in gilt bronze or green.
Also made in massive Silver.

—

Si finiscono in bronzo dorato o verde.
Si fanno anche in Argento massiccio.

Empire craquelé.
Gehämmerte Empire-Uhr.
Empire, hammered.
Impero «craquelé».

Empire rayon de gloire.
Empire-Uhr mit Glorienstrahlen.
Empire, Sun's Rays.
Impero raggio di gloria.

Plate X/19. *MATHEY-TISSOT. Typical* montres pendulettes de voyage. *Plate reproduced from a Mathey-Tissot Catalogue issued in 1905. (By courtesy of Etienne Ch. Mathey)*

reader will know, conventional striking and repeating carriage clocks have two trains. In such clocks the striking trains are released by the going trains at appropriate intervals; or, in the case of repeaters, when top repeat buttons are pressed. Most *montres pendulettes de voyage,* on the other hand, are basically timepieces. They have no clock-type striking trains, although they embody the repeating mechanism of a watch in order to be able to repeat quarters, five-minutes or minutes as the case may be. In this kind of clock, when the user wishes it to repeat, he depresses a long plunger on the top of the case. In so doing he winds the spring of a watch repeating mechanism located under the dial.

The movement of another type of eight-day Swiss silver-cased *pendulette de voyage,* very similar in outward appearance to the Mathey-Tissot pieces and also to the French clock with an English case shown in Plate VII/40, is illustrated in Plate X/20. This clock strikes *grande sonnerie* in passing and is also a minute-repeater with alarm. These clocks, according to M.

X Swiss Carriage Clocks

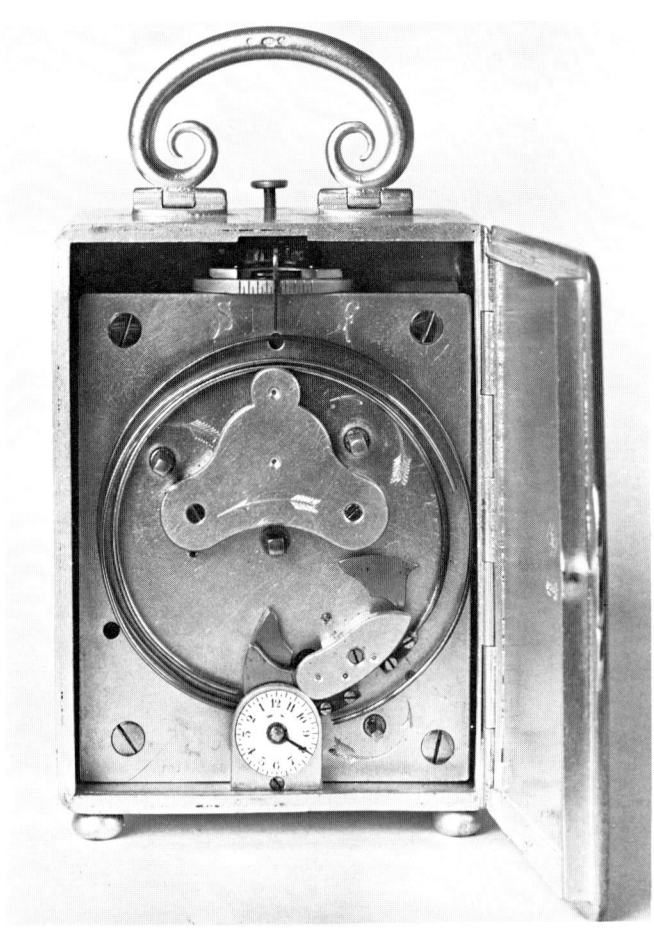

Plate X/20. *STAUFFER FAMILY.* Montre pendulette de voyage *with* grande sonnerie *striking and minute repeating. These clocks were made circa 1930 by the Stauffer family and by out-workers in the Vallée de Joux. (By courtesy of Meyrick Nielson of Tetbury Ltd.)*

Etienne Ch. Mathey of Mathey-Tissot, were made about 1930 for M. Onésime Stauffer of Les Ponts-de-Martel. M. Stauffer took out several patents for striking and repeating mechanisms which included minute, five-minute and quarter repetition. A few clocks from the same area had perpetual calendars. The movement *ébauches* were produced in artisan workshops in the Vallée de Joux. The platform escapements were obtained and the finishing completed by the Stauffer family. These *montres pendulettes de voyage* were sold in large numbers in smart cities all over the world. Those found in England today mostly bear the names of fashionable West End shops, and in New York Messrs. Black, Star and Frost specialised in their sale.

An isolated instance of a miniature Swiss carriage clock, made almost throughout by one man towards the end of the 19th century, is the piece produced by M. D. Elffroth which achieved much acclaim at about the time of the Paris Exhibition of 1878.[18] The clock which stood only 61mm. tall (about two and three-eighths inches) had a wooden case, struck *grande sonnerie* in passing, repeated, besides having a calendar showing the days of the week, the date of the month, name of the month and the phase of the moon.

In his second book about Neuchâtel clockmaking Chapuis includes a small chapter entitled "Pièces de carrosse, pendules d'officier, pendulettes, etc.". This chapter is interesting because it illustrates not only a number of portable pieces of exactly the type discussed in Chapter I of the present book but also a real carriage clock still styled *pendule d'officier* and the *calibre* of a round-plated portable striking clock made about 1840 and styled by Chapuis *pendule de voyage*. This emphasises the overlapping of the various types. A further point is that Chapuis mentions that Charles-Frédéric Klentschi in association with Constant Borel was producing from 1818 movements for *pendules d'officier* on behalf of Robert et Courvoisier. Chapuis goes on to say (translated) "... they made also some travelling pieces of another sort more in the nature of mantel clocks in as much as they were covered by glass shades (!), something not much use for moving about". He quotes a letter from Blavet of Paris to Louis Courvoisier which says "Your firm asks me for three glass shades for small portable clocks, models which I have already supplied. Please will you send me the size quickly, as I have kept nothing which gives it to me". This point is mentioned as linking up with the very un-portable *pendules portatives* mentioned in Chapter II in connection with Paul Garnier.

---

(18) *Journal Suisse d'Horlogerie*, 1876-77, and also Adrien Philippe in *Etudes*, already mentioned, pp.134-135.

Special Offer

Join the Antique Collectors' Club and receive your first issue of our magazine 'Antique Collecting' and our book catalogue FREE

For further details see other side.

To: Antique Collectors' Club Membership Service, Freepost, 5 Church Street, Woodbridge, Suffolk IP12 1BR, United Kingdom.* Please enrol me as a member of the Antique Collectors' Club. I understand I will receive the first issue of the magazine and a book catalogue free. I enclose a cheque for UK — £15.95 Overseas — £17.95 USA — US$35.00 Canada — CAN$45.00 OR please charge my credit card VISA / ACCESS / AMERICAN EXPRESS / DINERS CLUB (delete where appropriate).

Card no. ☐☐☐☐☐☐☐☐☐☐☐☐☐☐☐☐ Expiry date day ☐☐ month ☐☐ year ☐☐

Signature_____ Name_____

Address _____

Town_____ County_____

Postcode_____ I understand my membership fee will, if requested be repayable to me any time up to 14 days after receipt of my first magazine.

If you do not wish to take up our offer, but would like further details of the Club please tick here. ☐ This offer is not available to existing members.)

*NO STAMP REQUIRED IF POSTED IN GREAT BRITAIN
CHANNEL ISLANDS OR N. IRELAND

We invite you to join the Antique Collectors' Club
— the key to intelligent collecting —

The Antique Collectors' Club was formed in 1966 by a group of keen collectors. Since then the Club has grown to a membership of over 12,000 worldwide. Members of the Club receive our magazine **Antique Collecting** published every month except August.

Antique Collecting provides you with the information needed to collect profitably and intelligently, and helps you avoid costly mistakes.

In every issue you will find:

* practical, authoritative and detailed articles on collecting including current price trends, written by experts, on subjects not discussed elsewhere.
* comprehensive listings of venues and dates of Antiques and Collectors' Fairs and Auctions throughout the United Kingdom, plus Saleroom and Auction Features.
* details of over 100 regional and overseas clubs — an easy and pleasant way to gain information is to meet other experienced collectors at regional clubs; their addresses are published in the magazine. Collectors and would-be collectors can meet together to discuss their collections, listen to lectures and improve their knowledge.
* news of seminars and conferences for members on their favourite collecting subjects.

ANTIQUE COLLECTORS' CLUB BOOKS AND PRICE GUIDES

The Dog on the Spine is the Sign of Quality

Books and price guides published by the Antique Collectors' Club are hailed by the press, collectors and the antique trade as the best available. Their depth of specialist information is unsurpassed.

The magazine gives you advance information on forthcoming titles (about 12 every year). They are available through your local bookseller. We will send you a free book catalogue with your subscription to the Club.

Receive your first issue of **Antique Collecting** and the Antique Collectors' Club book catalogue free. Simply fill in your name and address on the form overleaf, and post it today. You don't need a stamp (if posted in Great Britain, Channel Islands or N. Ireland).

Chapter XI

Germany and the Austro-Hungarian Empire

A FINE AUSTRIAN CARRIAGE CLOCK. Date circa 1840. This rare broad-based example, which stands only 4½ins. tall, retains its original travelling box. The fold-flat handle and the style of the engraving are both typical of Austrian work of the period. The clock has a duplex escapement and offers grande sonnerie *striking with repeat. (By courtesy of Camerer Cuss & Co.)*

XI

Germany and the Austro-Hungarian Empire

Vienna's Cultural Influence upon Surrounding Territories. Small Austrian and German Travelling Clocks. Viennese Carriage Clocks. Later German Alarms.

Chapter I touched upon the first spring-driven clocks of the early 16th century, and how from them gradually evolved in Europe that progression of travelling pieces culminating in the true carriage clocks of the 19th century. Towards the end of the very long intervening period there appeared many small travelling clocks of the type which in this book have been called "pre-*pendules de voyage*". A few French and Swiss examples have already been illustrated; but clocks of similar intent were also current at the same period in those Eastern European countries which formed at the time part of the Austro-Hungarian Empire. Most of these travelling pieces were made in Vienna; but examples enough will also be found bearing the names of such cities as Berlin, Dresden, Pressburg (Bratislava), Budapest, Warsaw, etc. All are so very similar, both mechanically and in their mode of decoration, as to leave no doubt that they stem from one clockmaking tradition, if not all indeed from Austria. Whatever the facts, it is certain that the culture of Vienna impressed itself upon the adjacent territories and along the river Danube.[1]

Austria was a kingdom which brought its own sometimes-bizarre touch of eccentricity and romanticism to horology. Apart from the French, it is doubtful whether any clockmaking nation in history ever ranged over such a repertoire of differing appearances.

The explanation no doubt lies in the peculiar political history of Vienna, in its extraordinary geographical disposition, and in the nature and temperament of the people. Like the waltzes of Strauss, the Viennese clocks seem to embody an "... incarnation of that kind of hedonism which has become proverbial in all lands: the Viennese spirit, that rare amalgam of lively temperament and slight sentimentality, of joie de vivre and nostalgia, of inborn ease and nonchalance".[2] More seriously, it must be pointed out

---

(1) Just as the French provincial makers followed Paris, while all England copied London. Vienna itself of course looked to Paris, then the cultural centre of Europe, for much of its own inspiration. French fashions were sometimes followed, sometimes varied and sometimes ignored.
(2) These clever words belong to Dr. Mosco Carner.

XI Germany

that Austria and South Germany, having once led the entire horological world, became almost totally eclipsed by England and France towards the end of the 17th century. It is scarcely surprising, then, to find that clocks made after that period often assumed alongside their native peculiarities decorative styles borrowed from other clockmaking countries.[3] The Austrian travelling clocks are both mannered and unpredictable in their appearances; yet the most dissimilar examples all possess some indefinable common denominator which at once sets them together and at the same time apart from all others. Austrian travelling clocks were not usually very expensive in the first place. As a rule it will be found that they have a duration of only thirty or forty hours, and almost invariably they are provided with alarms. Some are quarter clocks, while others offer *grande sonnerie* striking. In the latter case it was the Austrian practice for quarters to be struck before the hours. While most Austrian travelling clocks will be found to have verge escapements, later examples exist having cylinder escapements, or occasionally such interesting variants as the double-wheel duplex.

Plates XI/1 and XI/2 show an Austrian travelling clock of the pre-*pendule de voyage* type and made circa 1760-1770 by Johann Collman of Türnau. While it does not look in any way like a carriage clock, it is certainly a step in the right direction, having a balance-controlled escapement and offering pull repeat and alarm. The next Austrian clock, which is shown in

Plate XI/1. *AUSTRIAN TRAVELLING CLOCK. Date circa 1760-1770, signed "Johann Collman à Tÿrnau" (Türnau). This clock goes for 40-hours, and has pull repeat and alarm. (Documentation: photograph collection of H. von Bertele)*

Plate XI/2. *THE MOVEMENT OF THE COLLMAN CLOCK. The balance and escapement are under the top bell. (Documentation: photograph collection of H. von Bertele)*

---

(3) The most obvious examples are the "English" Viennese bracket clocks of the 18th century, and in the 19th century the "French" Empire mantel clocks and also the superb complicated skeleton clocks inspired by the best Paris makers.

326

XI Germany

Plate XI/3. *AUSTRIAN TRAVELLING CLOCK. Date circa 1775-1785, signed "Joseph Ruetschmann in Wien". Brass case with rococo ornament. Enamel dial with floral paintings. Goes 40 hours. Quarter striking and alarm. Side glasses are introduced for the first time. (Documentation: photograph collection of H. von Bertele)*

Plate XI/4. *GERMAN TRAVELLING CLOCK. Signed "Fridrich Tiede, Därkchm". Mid to late 18th century. (By courtesy of Messrs. S. Lampard & Son, London)*

Plate XI/3, represents a positive advance in the progression with glassed sides making their first appearance. This clock is signed "Joseph Ruetschmann in Wien" and its date is circa 1775-1785. It offers quarter striking and also alarm. Note the "romantic" nature of the rococo decoration in contrast to the vaguely "English" and also "Dutch" influences apparent in the previous example.

A shape of case which must have been a leading European fashion during the last half of the 18th century is well illustrated by two clocks bearing German addresses and shown in Plates IX/4 and IX/5. The first piece is signed "Fridrich Tiede, Därkchm".[4] It goes

---

(4) Alternative spelling "Darkehmen". This place is in (former) East Prussia, between the Seesker Heights and Instersburg, now under U.S.S.R. administration.

327

XI Germany

1787; and so there is no doubt about its date.[5] Note the use of a white enamel dial and steel hands unprotected by either bezel or glass. The central alarm setting ring is numbered from one to twelve, while the top subsidiary dial is used for regulation. Another clock in an almost identical case, but lacking its original movement, was sold by Messrs. King & Chasemore at Pulborough early in 1972. It is signed "C.H. Weisse, Dresde". Both clocks were probably made by Christian Heinrich Weisse who died in 1793.

The next clock illustrated in Plate XI/6 was made

Plate XI/5. *GERMAN TRAVELLING CLOCK. This piece, which is provided with an alarm, carries a presentation inscription for the year 1787. Note a shape of case which must have been the leading European fashion at the time. Clocks of this type had fusees on their going sides, but employed standing-barrels for the other trains. (By courtesy of Marouf, Düsseldorf)*

for thirty hours, and it is possible that Fridrich Tiede was a member of that family which later became so famous in connection with marine chronometers. The second clock, which is signed "Weisse, Dresden" bears an engraved presentation inscription for the year

Plate XI/6. *AUSTRIAN TRAVELLING CLOCK. Date circa 1790-1800, signed "Carolus Hofmann in Wienn" (sic). Brass case. Plain white enamel dial with subsidiary dials for strike/silent, etc. Quarter striking. Goes 40 hours. (Documentation: photograph collection of H. von Bertele)*

---

(5) This inscription translates to read: "Hans Casper Keck von Schwartzbach, Capt. Saxonian Chürassiers(?), bought me for 90 Guilders in the year 1787, and I am henceforth to remain with the eldest of the Schwartzbach family as a memento."

XI Germany

Plate XI/7. *AUSTRIAN TRAVELLING CLOCK. Date circa 1800, signed "Philipp Fertbauer in Wien". Pear-shaped brass case. Hour strike and alarm. Goes 40 hours. (Documentation: photograph collection of H. von Bertele)*

Plate XI/8. *THE MOVEMENT OF THE FERTBAUER CLOCK. Note the balance and scale for regulation. (Documentation: photograph collection of H. von Bertele)*

just before the turn of the century, probably circa 1790-1800. It is signed "Carolus Hofmann in Wienn". How indigenous this clock is to Vienna! Plates XI/7 and XI/8 show a piece made in about 1800 by Philipp Fertbauer of Wien. Although this clock goes for only forty hours at a winding it is mechanically well on the way to being what came to be termed a carriage clock. The positions and form of the clickwork for the winding, the central bell, the hammer, and even the position of the balance, foreshadow what later often became standard practice. This particular clock also undoubtedly anticipates the ordinary 19th/20th century cheap German alarm. An unpalatable truth is that at times a hairsbreadth line divides the so-called "carriage clock" from the so-called "alarm". Their functions have always been similar, and more often than not they are

329

XI Germany

"sisters under their skins". The very unusual "lion" travelling clock shown in Plate XI/9 is on exhibition in the Château des Monts museum at Le Locle in Switzerland. It was made about 1815-1820. Only

Plate XI/9. *AUSTRIAN "LION" TRAVELLING CLOCK. Date circa 1815-1820. (By courtesy of the Musée d'Horlogerie, Château des Monts, Le Locle.)*

Plate XI/10. *AUSTRIAN TRAVELLING CLOCK. Date circa 1815-1825. (By courtesy of Ritter, Vienna). The eight clocks shown in Plates XI/1-10 could very well have been featured in Chapter I; but it seemed a pity to separate them from their peculiar national context.*

marginally later in date is the circular clock supported on the backs of two crouched lions as seen in Plate XI/10. Professor H. von Bertele and other Viennese experts have suggested a date between 1815 and 1825 for its manufacture. In a way this clock may be said to mark the end of an era, after which Austrian travelling clocks on the whole tended to look rather more like the conventional *pendules de voyage* as made in France. This incipient trend is well illustrated by the

330

XI Germany

Plate XI/11. *AUSTRIAN TRAVELLING CLOCK. Date circa 1820-1825. Goes 40 hours. Unsigned. Unusual 2-wheel escapement. (By courtesy of Ritter, Vienna)*

Plate XI/12. *AUSTRIAN TRAVELLING CLOCK AND AUSTRIAN CARRIAGE CLOCK. Both clocks were made by Franz Gutkaes of Vienna in the period circa 1825-1830. The clock on the left is characteristically Viennese in appearance and has* grande sonnerie *striking. The clock on the right was made in emulation of Breguet. It has off-set seconds, calendar for month and date, alarm and strike/silent lever. (Documentation: photograph collection of H. von Bertele)*

Plate XI/13. *AUSTRIAN CARRIAGE CLOCK. Date circa 1830. Not signed. Goes 40-hours. Quarter strike. Skeleton movement with steel plates. Silver case, engraved and partly pierced to release sound. (Documentation: photograph collection of H. von Bertele)*

square-looking clock in Plate XI/11, having a date of perhaps 1820-1825.

Plate XI/12 illustrates two utterly dissimilar clocks which were both produced by the same maker, Franz Gutkaes of Vienna, at exactly the same period, circa 1825-30. The example on the left remains unrelentingly "Austrian", while everything about the clock on the right is made in emulation of the then highly fashionable firm of Breguet et Fils. These two entirely different Austrian carriage clocks perhaps underline

XI Germany

better than any words some of those strange contradictions which characterise Austrian work. The extraordinary clock shown in Plate XI/13 once again demonstrates the Austrian temperament and love of the ornate. The clock was made circa 1830, and it is housed in an engraved silver case. The movement has skeletonised steel plates and strikes hours and quarters. It goes for forty hours only.

By about the year 1840 Viennese makers were at last regularly producing carriage clocks of conventional appearance. Plate XI/14 shows such a clock

Plate XI/15. *AUSTRIAN CARRIAGE CLOCK. Date circa 1850-1860. Maker Rettich of Vienna. Goes eight days with grande sonnerie striking, calendar and alarm. (By courtesy of Ritter, Vienna)*

Plate XI/14. *AUSTRIAN CARRIAGE CLOCK. Date circa 1840. Maker Philipp Happacher of Vienna. Goes 30-hours. Grande sonnerie, alarm and repeat. Case engraved all over to a very high standard. The carrying handle folds away invisibly into the top of the case in "military-chest" fashion. (Private collection)*

made by Philipp Happacher of Vienna. It goes for thirty hours and has *grande sonnerie* striking, alarm and repeat. Clocks of this type were often made with duplex escapements. The next photograph, Plate XI/15, shows an example of a true Austrian carriage clock in its fully developed form as produced circa 1850-1860. This clock, going for eight days, is signed "RETTICH IN WIEN" and it offers *grande sonnerie* striking, calendar and alarm. The heavy cast case is well made and finished, the dominant decorative motif being deep

332

XI *Germany*

Plate XI/16. *AUSTRIAN CARRIAGE CLOCK. Date after 1860. Maker A.W. Mayer of Vienna. Goes eight days.* Grande-sonnerie, *repeat and alarm. (Photograph P.H. Kohl, Richmond Beach, Washington, U.S.A.)*

Plate XI/17. *MOVEMENT OF THE MAYER CLOCK. Side view. Note the use of main and sub-frames. (Photograph: P.H. Kohl, Richmond Beach)*

fluting. Here is an Austrian carriage clock which, in its own right, might compete on equal terms with the French national product.[6] Its movement is very similar to that of a final Austrian clock which is illustrated in Plates XI/16, 17, 18 and 19. This eight-day *grande sonnerie* piece with repeat and alarm was made by A.W. Mayer of Vienna in about 1860, if not later. The Mayer clock stands seven and five-sixteenths inches tall with the handle up. The gilt case is to such an extent "multi-pieced" that to re-assemble it is a nightmare of the highest order. Nothing really fits, and the case tends to disintegrate the moment that the four bun feet "nuts" are unscrewed. The movement is radically different from that of a French *pendule de voyage,* being throughout characteristically Viennese in conception. The movement plates are

---

(6)   F. Effenberger of Vienna exhibited carriage clocks at the Paris Exhibition of 1867, obtaining an Honourable Mention.

333

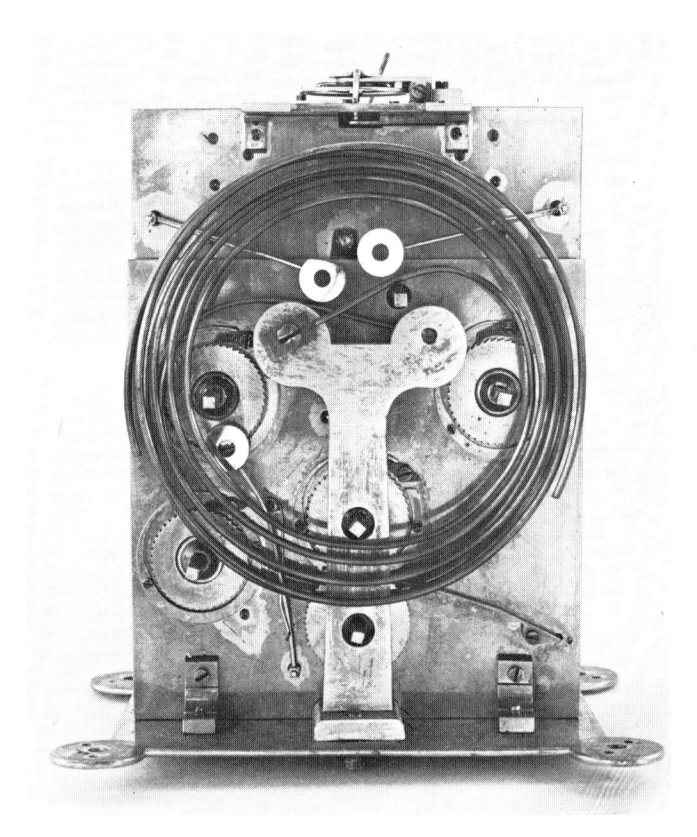

Plate XI/18. *MOVEMENT OF THE MAYER CLOCK. Rear view showing central gong standard and disposal of the trains. (Photograph: P.H. Kohl, Richmond Beach)*

Plate XI/19. *CYLINDER PLATFORM ESCAPEMENT OF THE MAYER CLOCK. (Photograph: P.H. Kohl, Richmond Beach)*

close together, and both a main and sub-frame are used. The three trains are respectively for going, *grande-sonnerie* striking and alarm. The striking follows Austrian convention, the quarters being struck before the hours.[7] Instead of a ting-tang for each quarter, a single ting per quarter is struck on a high note wire gong, followed by a low note for the hours. Four quarter notes are struck before each hour. There is a repeat-button. The movement planting is divided between the main frames, which are used both for the mainspring barrels and for the complete alarm train, and a sub-frame carrying the remainder of the trains. The cylinder platform escapement is comparatively crude, having plain steel balance, flat balance spring and steel 'scape wheel. The balance cock is screwed to the platform, but is not steady-pinned. It is intended to be movable about its foot screw for the purpose of adjusting the depth of the escapement. The lead-screw provided to assist this "performance" may be seen close to the cock foot. Note the style of the clickwork and hammers. Austrian clocks often have peculiar four-vaned flys, but in this example they are of the two-bladed form. In general, Austrian carriage clocks were in no way equal to French, although some fine examples exist. As already noted few of even the best Viennese clocks go for eight days.

---

(7) French *grande sonnerie pendules de voyage* almost always strike the hours before the quarters, although there are some exceptions. The English *grande sonnerie* clock by Vulliamy (Plate IX/13) strikes the quarters followed by the hours.

XI Germany

Plate XI/20. *TWENTIETH CENTURY ALARM CLOCK, PROBABLY GERMAN. Movement is anything but devoid of mechanical ingenuity and interest, and the cast case is inspired by older styles. (By courtesy of Bryson-Moore, Portland, Oregon, U.S.A. Photograph by P.H. Kohl)*

No. 55½. Joker Music Alarm Clock, $7 00
Gold gilt front and handle, nickel-plated frame and glass sides. Height, 7 inches; width, 5 inches.
This clock plays two tunes for about ten minutes, instead of ringing of a bell.

Plate XI/21. *GERMAN BLACK FOREST JOKER ALARM. This clock was offered in the Catalogue of the American wholesale house of St. Louis Clock and Silverware Company in 1904. Compare with the American-made Joker by Seth Thomas shown in Plate XII/51. (By courtesy of American Reprints, U.S.A.).*

Plate XI/20 shows an inexpensive eight-day alarm, probably made in the present century, and probably German in origin. It is not without interest because its cast case with birds and love-knot finial recaptures, however crudely, the atmosphere of the earlier Viennese clocks. This clock strikes and repeats hours and half hours. The going and alarm trains are driven from opposite ends of the same spring. Another spring provides power for the strike and alarm. The winding arrangements are particularly neat and ingenious. There is a cylinder escapement which goes very well in spite of the balance lying "on its side". Much mechanical ingenuity is shown in the design generally.

The German Black Forest *Joker* shown in Plate XI/21 was a type of cheap musical alarm made in carriage clock form. The clocks are virtually identical in both appearance and performance to those made in the U.S.A. Compare with Plate XII/51.

The small cube-shaped alarm shown in Plates XI/22

335

XI   Germany

Plate XI/22. GERMAN ONE-DAY ALARM. *This particular model was in competition with the Waterbury* Tourist, *Plate XII/1. (By courtesy of Bryson Moore. Photograph by Phil Kohl).*

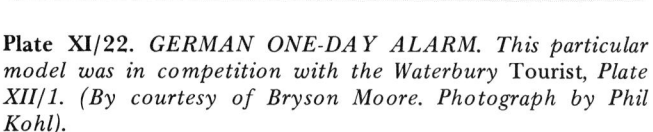

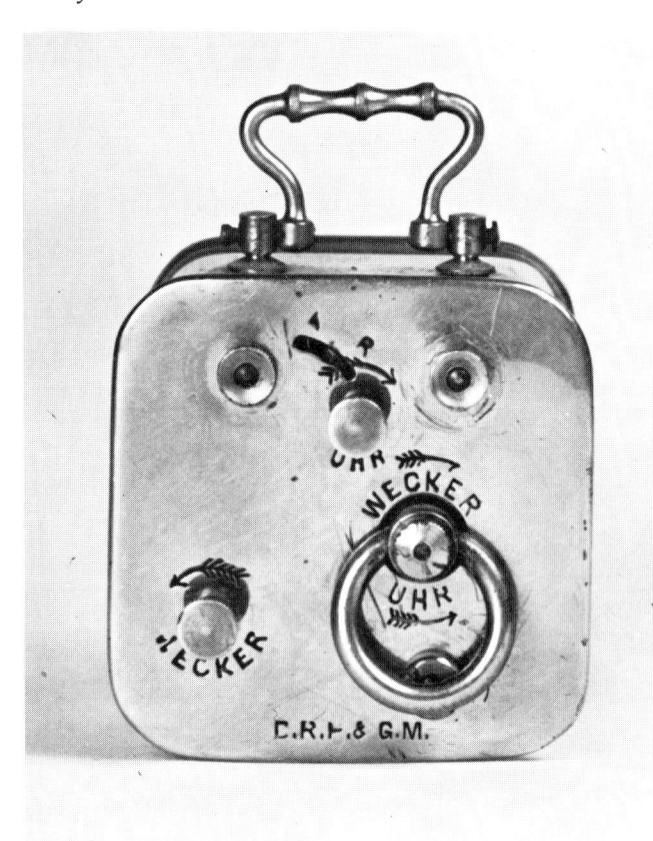

Plate XI/23. GERMAN ONE-DAY ALARM. *The back of the cube-shaped clock shown in Plate XI/22. (By courtesy of Bryson Moore. Photograph by Phil Kohl).*

and XI/23 is of interest because this clock was produced in competition with the Waterbury *Tourist* shown in the top right-hand corner of Plate XII/1. The German clock stands 2½ inches tall with the handle up and the American clock is almost exactly the same size. Both clocks were current circa 1908-1909.

Plates XI/24, 25 and 26 show a small round-section alarm clock with travelling box and which, in all probability, is German. It bears the trade name "ALLROUND" and mechanically its design embodies a number of sensible and useful features in terms of mass production.

The *Allround*, which stands approximately $3\frac{5}{16}$ inches tall with the handle up, has a rather high-class dial, white enamel on copper with blue chapters. The copper

Plate XI/24. ALLROUND ALARM WITH TRAVELLING BOX. *This clock, which stands about $3\frac{5}{16}$ inches high with handle up and which is probably German in origin, goes for 30 hours. (By courtesy of Bryson Moore. Photograph by Phil Kohl).*

XI Germany

Plate XI/25. *ALLROUND ALARM. The side view of the movement of the clock shown in Plates XI/24 and 26. (By courtesy of Bryson Moore. Photograph by Phil Kohl).*

Plate XI/26. *ALLROUND ALARM. The rear view of the clock shown in Plate XI/24. Note the sub-frame and also the single winding square for both going and alarm. (By courtesy of Bryson Moore. Photograph by Phil Kohl).*

plate is soldered to a false plate in the manner of some of the Waterbury clocks. The motion work and the alarm work is accommodated between the false plate and the front movement plate. Solid pinions are used throughout. The clock is wound and set through a back shutter by means of a double-ended key which stows below the base of the clock in a spring-loaded holder. One winding square serves for both going and alarm. There are two clickworks acting in opposite directions. Turning the key clockwise winds the going train, and turning it anti-clockwise winds the alarm. Both mainsprings are contained in going barrels. The part of the going train above the centre pinion is run in a sub-frame contained within the main movement plates. There is a pin-pallet escapement with the balance staff horizontal. Both lever and 'scape wheel are made of brass while the plain balance is of steel.

Although this clock is fairly roughly made with the movement plates stamped out and the edges filed, and although in the example illustrated several arbors are very much out of upright (possibly due to faulty repairing), yet the clock goes well and on the whole compares favourably with mass produced American clocks. A button is provided for silencing the alarm. The alarm silencing mechanism has for some reason been made fairly complicated employing a friction-tight spring-damped pinion moved by a rack.

Chapter XII

American Carriage Clocks

FASOLDT CARRIAGE CLOCK. **See Plates XII/63, 64 and 65.** *(By courtesy of Seth Atwood, Rockford, Illinois, U.S.A.)*

XII
American Carriage Clocks

Introduction. Waterbury Clock Company. Waterbury Watch Company. Ansonia Clock Company. Jerome, New Haven, Welch and Seth Thomas. Joseph Eastman. The Boston, Chelsea and Vermont Clock Companies. Fasoldt.

INTRODUCTION

The Waterbury Clock Company was by far the most prolific maker of mass-produced carriage clocks and alarms in the U.S.A.; while at the other end of the scale Charles Fasoldt (1818-1898) made a fine and unique carriage clock having his special type of escapement. Between the two extremes were made a few notably superior carriage clocks of which the products of Boston, Chelsea and Vermont are typical. At the Waterbury end of the scale Ansonia, Jerome, Seth Thomas and Welch made inexpensive carriage clocks and alarms all in much the same idiom.

Among the most interesting aspects of American clock manufacture (not just carriage clocks and alarms, but all their clocks) was the wonderful ingenuity and resourcefulness evinced between different firms in "beating" each other's patents. The formula was to copy someone else's design or manufacturing "brainwave" without appearing to do so. The secret was to re-style in such a form as would be unlikely to result in dispute or litigation.[1] Clever economies in production techniques and widespread use of standard parts was another feature of American clockmaking. Even when the Americans used fusees, they mounted the fusee on the same arbor as the mainspring and then placed the barrel on the great wheel arbor. This brilliantly simple innovation allowed one design to be used either for a weight-driven clock or for one with spring and fusee. The whole history of American clockmaking is most interesting. It is probable that this industry was one of the cradles of those mass production techniques for which the U.S.A. became famous. The cornerstone was interchangeability of parts, and its first successful application in the U.S.A. was in firearms made on the "American System".[2]

Elsewhere in this book it has been said that at times a hairline breadth of definition divides the cheapest type of carriage clock from the ordinary alarm clock. This fact is particularly apparent in many American carriage clocks, which more often than not go for one day only and which are provided with movements of a quality little different from that found in alarm clocks proper.

In terms of retail selling price it is not easy to compare an eight-day American carriage clock with the French native production. Paradoxically enough, while eight-day American striking and repeating carriage clocks were considerably cheaper than even the most inexpensive French ones, the eight-day American timepieces seem to have been rather expensive even when compared with *Obis*. For instance a Waterbury eight-day timepiece made under the model name of *Sage,* and introduced for the express purpose of competing with *Obis* and *Corniche,* cost $7.15 retail (about £1.16.0d. at the time) in the years 1908-9. In about the same period a French *Obis* timepiece could be retailed at less than a guinea, and a *Corniche* made by one of the lesser makers retailed at exactly a guinea. On the other hand a Waterbury *Sage*, if ordered with

---

(1) Judges were not clockmakers, and it might be fairly difficult to convince a Court that an idea had really been pirated. A good example of this situation was the instance of Hoadley overcoming Terry's Patent wooden "works" by turning the movement layout upside down and also inverting the "verge".

(2) J.W. Roe (see Bib.) quotes the surprising evidence that Thomas Jefferson while in France in 1785 saw early French attempts at making musket parts on an interchangeable basis in small quantities.

XII American Carriage Clocks

their eight-day hour and half-hour striking and repeating movement, cost just $11,50 (£2.17.6d.) in 1908-9. A French *Corniche* of the most ordinary quality, but offering identical facilities, would have retailed at £3.15.0d. soon after 1900. In 1914 an Ansonia timepiece clock sold under the model name of *Bonnibel,* and also marketed in direct competition with the *Obis* and *Corniche,* had a retail price of $7.75 (say £1.18.9d.). The retail prices just quoted may be misleading because they applied to French clocks as retailed in England but to American clocks as sold in their country of manufacture. The figures, however, afford at least some degree of comparison.

One area in which the Americans undoubtedly had a price edge on the French was that the U.S. manufacturers sold their clocks with cheap pin-pallet lever escapements made integral with the movements and not bought elsewhere and added afterwards in the form of platforms. While French carriage clocks at the bottom end of the scale were given cheap platform cylinder escapements[3] virtually all American carriage clocks, even those going for one-day only, had pin-pallet escapements.[4] These may or may not have afforded better timekeeping and longer wearing properties than the French cylinders. Certainly, and despite statements made to the contrary by rival clock-making nationalities, it is not true to say that American clocks were never intended to be repaired. Even the cheapest clocks could be restored. Old manufacturers' catalogues more often than not contain long lists of material for the purpose. The essential difference between the American clocks and those made by the French was that the American pieces had practically no hand-finishing in them. They were assembled quickly and solely from pressings, stampings, etc. That American production of clocks became a worry to the French soon after the middle of the 19th century is apparent from an editorial article on page 42 of the *Revue Chronométrique* of July 1867. This translates to read:— "Numerous makers of small clocks and house clocks have established themselves in America in the last fifteen years, and are now in the position of putting on to the market 700,000 clocks a year.[5] This excessive output gives way under its own weight. Of thirty establishments four were destroyed by fire, nine failed and five closed down bankrupt. In a few years the production had fallen to 100,000 clocks a year. We do not know if this number has increased or diminished since. American clocks sell in that country and are exported to Mexico, to South America, to China and even to Europe; but they have enjoyed little success despite their low price: one dollar (5fr.30) for a simple clock and three dollars for a striker. Hand-work is much more expensive in America than it is here. The comparatively low selling price of an American clock is explicable only in the context of extreme division of labour and above all to the use of machines offering very quick production. This mechanisation makes unaided almost the whole clock; and further when a clock stops they find it cheaper to throw it away and buy a new one". (The author was C. Saunier).

Much mention will be made in this chapter of the old catalogues issued by various American clock manufacturers. These catalogues are important because not only do they reveal the correct names of different models, but also because more often than not they give the retail prices of clocks shown. When studying old catalogues, which are often difficult to understand today, it is necessary to remember that they served a dual purpose and that they were distributed to the trade for two distinct and quite separate reasons. The first object in circulating a new catalogue was to keep retailers abreast with those models which were available. The second purpose was to enable the same retailer to show to his customers a wide range of clocks together with their cost. For this reason the prices quoted beneath illustrations of clocks were always retail prices, or in other words List prices. Trade prices, not intended for the eye of the customer, were usually either printed on a separate sheet or else they were

---

(3) Insistence on a lever escapement raised the price of a cheap French *Corniche* from £1.1s.0d to £2.2s.0d in about the year 1900.
(4) In the mind of the public, or in other words from a salesmanship point of view, any clock which could be said to have a lever escapement (even if it was really only a pin-pallet) sounded far better than one having a cylinder escapement.
(5) Compare with Hiram Camp's figure (see page 375).

XII American Carriage Clocks

listed by themselves in a different part of the book.

Two catalogues of particular interest were not manufacturers' catalogues at all but were those distributed to the retail trade by the two distinquished wholesalers, S.F. Myers and Company of New York, and the St. Louis Clock and Silverware Company of St. Louis, Missouri. Myers, who described themselves as "Manufacturing Jewelers" and who published catalogues annually, prefaced their No. 22 of January 1885 with the statement that it was their "Illustrated Catalogue and Wholesale Price List (For the Trade Only)". A letter of introduction addressed to the "Jewelry Trade" contained the words "We aim to send this catalogue ONLY TO REGULAR DEALERS. Should you know of any instance where they are received by any one not entitled to them we would be obliged for such information". At the front of the Myers Catalogue is a confidential discount sheet on which it is clearly stated that "EVERYTHING QUOTED in Catalogue No. 22, *whether illustrated or not*, is a List Price, subject to the Trade and Cash Discounts; so you can show the Book to anybody, or let it lie on your show case at any time without loss, or your customers knowing your costs. In fact it will often create a demand for goods which you may not have in your stock". The discount sheet mentioned is a table showing List prices and also Net prices running from $0.06 to $810.00. Each clock illustrated has below it its List or retail price.

The St. Louis Clock and Silverware Company also published annual catalogues. The one issued in 1904 says in its preface "We send our Catalogue and printed matter to none but legitimate dealers. We show our name and address on this page only. By covering the same you can use this Catalogue safely to show to your customers... We do not retail, never did, and never will... FOR DISCOUNT SHEET AND INDEX, SEE BACK OF BOOK". Below the clocks are shown their retail prices.

The study of old catalogues, not only for carriage clocks but for all American clocks, is most rewarding. Fortunately today many catalogues are once again available as facsimile reproductions (see Bibliography).

WATERBURY CLOCK COMPANY

The Waterbury Clock Company was established in Waterbury, Connecticut in 1857 and was an offshoot of Benedict and Burnham. In 1922 the Waterbury Clock Company bought Robert H. Ingersoll & Bros.[6] This concern had in 1914 purchased the New England Watch Company, which until 1898 had been the Waterbury Watch Company. In 1944 the U.S. Time Corporation took over the remaining assets.

A Waterbury Catalogue, published for the years 1908-1909 and now available as a facsimile reproduction, provides most interesting reading and shows clearly that the Company then offered a wide choice of designs in both carriage clocks and in alarms. Retail prices are shown.

As will be seen from Plates XII/1, 2 and 3 no less than three pages of the Catalogue were devoted almost entirely to carriage clocks, while two further pages reproduced in Plates XII/4 and 5 offered leather-covered travelling boxes for both carriage clocks and alarms. Several boxes were made to fit more than one model. While the large range of Waterbury clocks included a fine repetoire of differing appearances, most of their models were proudly and wholly American in looks. The various models bore such names as *Spy, Conductor, Companion, Wanderer, Stroller, Sage* and *Convoy*. It is interesting to note that no clock as illustrated in the Catalogue bears either the famous "WCC" monogram or even the words "Waterbury Clock Co. U.S.A.", although in practice this legend seems to have appeared below the chapter ring of every clock sold. The prices ranged from about $5 for a one-day timepiece *(Conductor)* to as much as $13.70 for an eight-day hour and half-hour striker with repeat and alarm in a plain and heavy case *(Comet)*. Also illustrated was a semi-miniature clock, a timepiece called *Speck* and selling at $4.25 with either "gold-plated" or "gun-metal" finish. Most of the clocks went for one day only. Some of them were offered in cases available in several different finishes, while certain models could be ordered with any of up to four different movements of progressively increasing

---

(6) According to *The History of the Ingersoll Company* (see Bib.) from 1922 the business continued as the "Ingersoll Waterbury Company". In London the Ingersoll Watch Company went on trading as before, becoming a public company in 1930. Ingersoll Ltd. are still in business.

XII   *American Carriage Clocks*

DANDY
Celluloid Case—Green, Blue, Tortoise
and Light Onyx Color.
Bright Gold Ornaments.
1 Day Time $3 00
2 inch IVORY Dial.
Beveled Glass. Height, 4½ inches.

HORNET
Rich Roman Gold Plated.
1 Day, Time $3 50
2 inch IVORY Dial.
Beveled Glass. Height, 3½ inches.
Polished Movement, visible through a Glass
Cylinder ⅛ inch thick.
Leather Cases for above, see Pages 16 and 17.

TOURIST
Long Alarm.
1 Day, Time, Alarm. 2 inch Dial.
Beveled Glass. Height, 3¾ inches.
Alarm runs a minute and a half.
Finished in Nickel $3 85
Finished in Rich Gold Plated 4 35
Finished in Gun Metal, with Gold Plated
   Handle, Bezel and Matting 4 35
Leather Cases for above, see Pages 16 and 17.

SPECK
Rich Gold Plated or Gun Metal Finish.
1 Day, Time $4 25
1 inch IVORY Dial.
Beveled Glass at Front and Sides.
Polished Movement, visible through
Beveled Glass. Height, 3 inches.
Leather Cases for above, see Pages 16 and 17.

MIDGE
Rich Gold Plated or Gun Metal Finish.
1 Day, Time $5 25
1½ inch IVORY Dial.
Beveled Glass at Front and Sides.
Polished Movement, visible through
Beveled Glass. Height, 3⅝ inches.
Leather Cases for above, see Pages 16 and 17.

ELEGANS
Rich Gold Plated.
1 Day, Time, Alarm $6 00
1½ inch PORCELAIN Dial.
Beveled Glass. Height, 3½ inches.
Leather Case for above, see Page 17.

BLOSSOM
Rich Roman Gold Plated.
1 Day, Time $4 40
2 inch IVORY Dial.
Beveled Glass. Height, 5½ inches

Plate XII/1. *WATERBURY CLOCK COMPANY CATALOGUE, 1908–1909. Compare the alarm Tourist with its German competitor in Plates XI/22 and 23. All the clocks illustrated went for one day only. (By courtesy of Adams Brown Co., N.H., U.S.A.).*

XII American Carriage Clocks

AID
Rich Roman Gold Plated.
1 Day, Time $5 00
2 inch IVORY Dial.
Beveled Glass. Height, 6¾ inches.

TRUANT
Rich Roman Gold Plated.
Long Alarm with Intermissions.
1 Day, Time, Alarm $5 00
2 inch PORCELAIN Dial. Beveled Glass.
Height, 6¼ inches.
The Alarm will ring a quarter of a minute and
be silent half a minute alternately for
5 minutes.
Can be stopped at pleasure. Attached Keys
and Hand Sets.
1⅞ inch Bell at Back.

SPY
Rich Gold Plated or Gun Metal Finish
with Gold Plated Handle, Bezel
and Matting.
Long Alarm with Intermissions.
1 Day, Time, Alarm $6 00
2 inch Dial.
Beveled Glass. Height, 4¼ inches.
The Alarm will ring a quarter of a minute and
be silent half a minute alternately for
5 minutes.
Can be stopped at pleasure. Attached Keys
and Hand Sets.
1⅞ inch Bell at Back.

CONDUCTOR
Rich Gold Plated.
1 Day, Time (solid back)............ $5 00
1 Day, Time, Alarm 6 60
1 Day, Half-hour Strike, REPEATER... 6 60
2 inch IVORY Dial.
Beveled Glass at Front and Sides.
Height, 4½ inches.
Leather Cases for above, see pages 16 and 17.

COMPANION
Rich Roman Gold Plated or Syrian Bronze.
1 Day, Time, Alarm $6 60
1 Day, Half-hour Strike, REPEATER... 6 60
2 inch IVORY Dial.
Beveled Glass. Height, 5¼ inches.

WANDERER
Rich Gold Plated or Gun Metal Finish with
Gold Plated Handle and Matting.
1 Day, Time, Alarm $6 60
1 Day, Half-hour Strike, REPEATER... 6 60
2 inch PORCELAIN Dial.
Beveled Glass at Front and Sides.
Height, 4½ inches.

Plate XII/2. *WATERBURY CLOCK COMPANY CATALOGUE, 1908–1909. A page of typical American one-day carriage clocks.* Conductor *must have been a very popular and successful model because it was offered with three different alternative movements besides being stocked by such notable wholesalers as the St. Louis Clock and Silverware Company. The hundreds of different model names used by American manufacturers for their various clocks were in fact nothing more than a system of stock numbers. The purpose was to simplify the ordering procedure both for local retailers and their customers. (By courtesy of Adams Brown Co., N.H., U.S.A.).*

XII *American Carriage Clocks*

METEOR
Rich Gold Plated.
1 Day, Time (solid back)............. $6 10
1 Day, Time, Alarm 7 70
1 Day, Half-hour Strike, REPEATER... 7 70
2 inch IVORY Dial.
Beveled Glass at Front and Sides.
Height, 4¼ inches.
Leather Cases for above, see pages 16 and 17.

STROLLER
Rich Gold Plated.
1 Day, Half-hour Strike, REPEATER... $7 70
2 inch IVORY Dial.
Beveled Glass at Front and Sides.
Height, 4½ inches.

SENTINEL
(DESK CLOCK.)
Polished Brass.
1 Day, Time $7 70
2 inch PORCELAIN Dial.
Magnifying Lens.
Measures across the back 3⅝ inches.
Stem Wind and Pendant Set.
Regulates from Outside.

SAGE
Rich Gold Plated.
8 Day, Time (solid back)............. $7 15
8 Day, Half-hour Strike, REPEATER... 11 00
8 Day, Half-hour Strike, GONG,
  REPEATER 11 50
8 Day, Half-hour Strike, REPEATER,
  ALARM 11 50
2 inch PORCELAIN Dial.
Beveled Glass at Top, Front and Sides.
Height, 5⅝ inches.
Leather Case for above, see Page 16.

CONVOY
Rich Gold Plated.
8 Day, Half-hour Strike, REPEATER...$12 65
2 inch PORCELAIN Dial.
Beveled Glass at Top, Front and Sides.
Height, 6¼ inches.

COMET
Rich Gold Plated.
8 Day, Time (solid back)............. $9 35
8 Day, Half-hour Strike, REPEATER... 13 20
8 Day, Half-hour Strike, GONG,
  REPEATER 13 70
8 Day, Half-hour Strike, REPEATER,
  ALARM 13 70
2 inch PORCELAIN Dial.
Beveled Glass at Front and Sides.
Height, 5½ inches.
Leather Case for above, see Page 16.

Plate XII/3. *WATERBURY CLOCK COMPANY CATALOGUE, 1908–1909. A selection of typical American carriage clocks including three models going for eight days. Note particularly the clock Sage made in competition with the French Obis. (By courtesy of Adams Brown Co., N.H., U.S.A.).*

XII   American Carriage Clocks

RED LEATHER CASE
FOR SPECK.
Leather Case (without Clock).......... $1 20
Height, 3¼ inches.

RED LEATHER CASE
FOR CONDUCTOR, WANDERER AND METEOR.
Leather Case (without Clock).......... $2 10
Height, 4½ inches.
When ordering, please state whether Case is wanted for CONDUCTOR, WANDERER or METEOR.

RED LEATHER CASE
FOR WASP, HORNET AND SPIDER.
Leather Case (without Clock) for
    WASP $1 35
Leather Case (without Clock) for
    HORNET 1 45
Leather Case (without Clock) for
    SPIDER 1 90
Height, 3 to 3¾ inches.
When ordering, please state whether Case is wanted for WASP, HORNET or SPIDER.

RED LEATHER CASE
FOR BUGABOO.
Leather Case (without Clock).......... $1 60
Height, 3¼ inches.

RED LEATHER CASE
FOR MIDGE.
Leather Case (without Clock).......... $1 60
Height, 3¾ inches.

RED LEATHER CASE
FOR WASP ALARM.
Leather Case (without Clock).......... $1 65
Height, 4 inches.

RED LEATHER CASE
FOR SAGE AND COMET.
Leather Case (without Clock).......... $3 80
Height, 5½ inches.
When ordering, please state whether Case is wanted for Sage or Comet.

RED LEATHER CASE
FOR TOURIST.
Leather Case (without Clock).......... $1 70
Height, 3½ inches.

Plate XII/4. *WATERBURY CLOCK COMPANY CATALOGUE, 1908–1909. Travelling boxes for both carriage clocks and alarms. Note the* Obis-*like* Sage *in its travelling box. It looks every inch a "French" carriage clock. (By courtesy of Adams Brown Co., N.H., U.S.A.).*

347

XII   *American Carriage Clocks*

BLACK LEATHER CASE
FOR SPECK.
Leather Case (without Clock).......... $1 80
Height, 3⅛ inches.
Spring Doors, see below illustration.

BLACK LEATHER CASE
FOR HORNET.
Leather Case (without Clock).......... $2 25
Height, 3½ inches.
Spring Doors, see below illustration.

BLACK LEATHER CASE
FOR MIDGE.
Leather Case (without Clock).......... $2 40
Height, 3⅝ inches.
Spring Doors, see below illustration.

BLACK LEATHER CASE
FOR METEOR, WANDERER AND CONDUCTOR.
Leather Case (without Clock).......... $3 00
Height, 4½ inches.
Spring Doors.
When ordering, please state whether Case is wanted for METEOR, WANDERER or CONDUCTOR.

BLACK LEATHER CASE
FOR TOURIST.
Leather Case (without Clock).......... $2 60
Height, 3½ inches.
Spring Doors, see above illustration.

BLACK LEATHER CASE
FOR ELEGANS.
Leather Case (without Clock).......... $2 75
Height, 3¾ inches.
Spring Doors, see above illustration.

Plate XII/5. WATERBURY CLOCK COMPANY CATALOGUE, 1908–1909. *Travelling boxes for small one-day carriage clocks, and also for the alarm* Tourist *and for the round timepiece* Hornet. *(By courtesy of Adams Brown Co., N.H., U.S.A.).*

XII American Carriage Clocks

Plate XII/6. WATERBURY EIGHT-DAY CARRIAGE CLOCK, SAGE. A model made to compete with the French *Obis* and shown in the Waterbury Catalogue of 1908–9. Sage was made with four different movements costing from $7.15 to $11.50. See Plate XII/3. (By courtesy of Bryson Moore. Photograph by Phil Kohl).

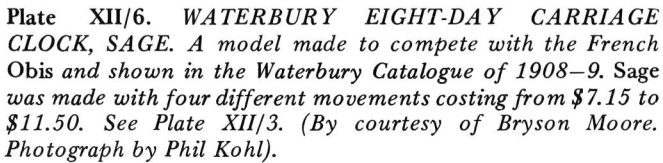

Plate XII/7. WATERBURY EIGHT-DAY CARRIAGE CLOCK, SAGE. The movement of the clock shown in Plate XII/6. This particular clock, which has the most expensive of four alternative available movements, is provided with hour and half-hour strike, repeat and alarm. (By courtesy of Bryson Moore. Photograph by Phil Kohl).

complication. For example, the plain *Obis*-like eight-day clock called *Sage*, of which two views of an actual example are shown in Plates XII/6 and 7, cost from $7.15 to $11.50 depending on the type of movement. No choice of case was offered.

The Waterbury Catalogue is most useful as a means of identifying carriage clocks and also for forming some opinion of their dates. Front and rear views of a fair example of the clock *Conductor* are shown in Plates XII/8 and 9. It must have been a popular model because it was offered with three different alternative movements besides being stocked by such notable wholesalers as the St. Louis Clock and Silverware Company, and being shown in their 1904 Catalogue. *Conductor* stood 4½ inches tall in its brass case which had bevelled glasses front and sides. It had a dial described in the Catalogue as being ivory. No patent numbers were quoted on the back of the movement which bore simply the inscription "Waterbury Clock Co. U.S.A.". In the alarm version at least, solid pinions were used

349

XII American Carriage Clocks

Plate XII/8. *WATERBURY ONE-DAY CARRIAGE CLOCK, CONDUCTOR. A model available in the earlier years of the 20th century if not before. In the Waterbury Trade Catalogue for the years 1908–9, this clock cost $5.00 as a timepiece, $6.60 as a timepiece with alarm and $6.60 striking hours and half hours and repeating hours. Note that the clock has lost its original corner finials which have been replaced by four ordinary nuts. (By courtesy of Bryson Moore. Photograph by Phil Kohl).*

Plate XII/9. *WATERBURY ONE-DAY CARRIAGE CLOCK, CONDUCTOR. The movement of the clock shown in Plate XII/8. This is the timepiece/alarm version of the model. Note the alarm setting dial bottom centre and that the alarm fixed-key is missing. The alarm bell is housed below the base. (By courtesy of Bryson Moore. Photograph by Phil Kohl).*

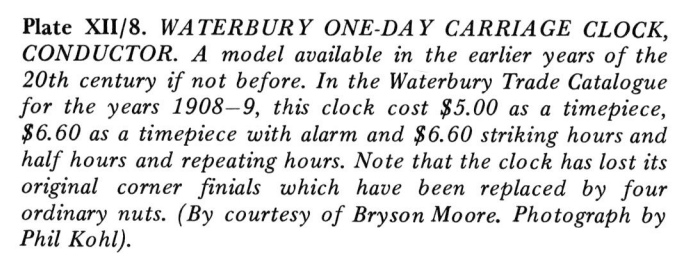

throughout the movement except for the alarm escapement which had a lantern pinion. The pin-pallet lever escapement had cone pivots and flat balance spring and was located between the plates at the top of the movement.

It is probable that the striking and repeating clock shown in Plate XII/10 is a development of the model called *Companion*. It stands just over 5 inches tall with the ornate carrying handle raised. The substantial cast case has no windows except at the front and the solid back door is secured with a latch as is usual with most American carriage clocks. The fold-flat "winders", or more correctly the fixed-keys, are accessible at the back of the movement. The ratchets and clicks are on the back movement plate. The striking train, which employs a rack and also "warning", is on the left hand side of the movement as seen from the back, the opposite from French clock practice. It is also interesting to notice the use of warning and not of "knock-out" striking in a repeating clock. No doubt Waterbury had excellent reasons for this practice and in all probability they were connected with economic production. Compare the Waterbury striking with the notes given under *Hour and Half Hour Strike* in Chapter VIII in connection with standard French carriage clocks. A further interesting design feature of the *Companion*-like clock is the unexpectedly small distance between the movement plates. It is less than five-eighths of an inch. The enamel dial and its false plate are made up as one unit; that is to say they are

XII   American Carriage Clocks

Plate XII/10. *WATERBURY ONE-DAY CARRIAGE CLOCK, PROBABLY COMPANION.* This clock strikes hours and half hours and repeats hours. The clock Companion cost $6.60 as a timepiece with alarm in the 1908–9 Waterbury Catalogue, and the same in an alternative version striking hours, half hours and repeating hours. (By courtesy of Bryson Moore. Photograph by Phil Kohl).

Plate XII/11. *WATERBURY ONE-DAY CARRIAGE CLOCK, PROBABLY STROLLER.* Compare with Stroller as shown in the 1908/9 Waterbury Trade Catalogue (Plate XII/3). This clock strikes hours and half hours and repeats hours. The bell is in the base. Patents quoted on the back door of the case are "Mar 19 1889, May 6 1890, Dec 23 1890, Jan 13 1891". The movement of this clock is very similar to that shown in Plate XII/10 but its case is far lighter. (By courtesy of Bryson Moore. Photograph by Phil Kohl).

soldered together and attached to the movement by feet. Between the dial-and-plate unit and the front movement plate are accommodated not only the motion work but the click work and also the front cock for the pallets of the escapement. The movement bears no patent dates but on the back movement plate in flowing script appear the words "Waterbury Clock Co. U.S.A."

A Waterbury one-day clock called *Stroller* was current in the years 1908-9 and retailed at $7.70 as a clock striking hours and halves and repeating hours. It appears in the Waterbury Catalogue reproduced in Plate XII/3, while a similar clock which may well be a later *Stroller* design appears in Plates XII/11 and XII/14.

Meteor was another one-day model found in the Waterbury 1908-9 Catalogue. An example of an actual clock is shown in Plate XII/12. A peculiarity of *Meteor* is that it has a solid brass case with no doors. The "winders" and hand-setting button are external to the back of the case, while the index may be reached with a pin through a slot. A similar clock is illustrated in Plate XII/13 and its rear view in Plate XII/14.

The small timepiece *Speck* with its travelling box is shown in Plate XII/15, while Plate XII/16 shows on the left a rear view of *Speck* and on the right the back of the

XII   American Carriage Clocks

Plate XII/12. WATERBURY ONE-DAY CARRIAGE CLOCK, METEOR. This model was offered in the Waterbury 1908–9 Catalogue (Plate XII/3). It was available as a timepiece at $6.10, as a timepiece alarm at $7.70, or (as in the case of the example shown) as a strike with repeat at $7.70. The trains employed lantern pinions. (By courtesy of Bryson Moore. Photograph by Phil Kohl).

Plate XII/13. WATERBURY ONE-DAY CARRIAGE CLOCK, RATHER SIMILAR TO METEOR. This clock strikes hours and half hours and repeats hours. It bears no patent dates or numbers. The case has no back door and the clock is wound and set externally. (By courtesy of Bryson Moore. Photograph by Phil Kohl).

Plate XII/14. WATERBURY ONE-DAY CARRIAGE CLOCKS. Rear views of the clocks shown in Plates XII/13 and XII/11. Note two distinct types of Waterbury case, probably current during the early years of the present century, the one having a back door and the other having no doors but external winding and setting. The movements are comparatively similar. (By courtesy of Bryson Moore. Photograph by Phil Kohl).

XII   American Carriage Clocks

Plate XII/15. *WATERBURY ONE-DAY CARRIAGE CLOCK, SPECK. Speck is a timepiece only. Note the surviving travelling box as illustrated in the Waterbury Catalogue. The travelling box has a wooden carcase and is covered in black leather. (By courtesy of Bryson Moore. Photograph by Phil Kohl).*

Plate XII/16. *WATERBURY ONE-DAY CARRIAGE CLOCKS. Left the rear view of Speck. Note the bridge affording easy replacement of the mainspring. Compare with the rather similar clock shown on the right, the front view of which is shown in Plate XII/17. Note that the patents quoted on the backs of the two clocks are quite different. (By courtesy of Bryson Moore. Photograph by Phil Kohl).*

353

XII *American Carriage Clocks*

Plate XII/17. *WATERBURY ONE-DAY CARRIAGE CLOCK. A rear view of this timepiece model is shown in Plate XII/16. It is of small size and is in much the same idiom as* Speck. *Note the incorrect modern hands. (By courtesy of Bryson Moore. Photograph by Phil Kohl).*

Plate XII/18. *WATERBURY ONE-DAY CARRIAGE CLOCK. This timepiece is unidentified. Despite its pediment-topped case it is a close relation to* Speck. *(By courtesy of Bryson Moore. Photograph by Phil Kohl).*

rather similar clock shown in Plate XII/17. *Speck* has a very plain "multi-piece" brass and glass case and stands three inches tall with the handle up. The glasses are bevelled. The back of the enamel dial is soldered to the false plate. The motion work, winding ratchet wheel, click and also the front adjustable cock for the pallet arbor are all accommodated between the dial and the front movement plate. The whole train, including the barrel and the escapement, are planted between the two movement plates; but the barrel arbor runs in a bridge at the back of the clock in order to facilitate the replacement of the mainspring. On the bridge are listed two patents for the year 1890 and one each for the years 1891 and 1897. The barrel great wheel runs directly into the centre pinion. All the pinions are lanterns. The pin-pallet escapement is integral with the clock, having cone pivots and a two-arm brass balance with ornamental "knobs" on its periphery. *Speck* is a well made clock considering the comparatively reasonable price.

Plates XII/17 and XII/18 show two unidentified Waterbury carriage clocks. They are both small one-day timepieces.

A further unidentified Waterbury carriage clock, this

354

XII   *American Carriage Clocks*

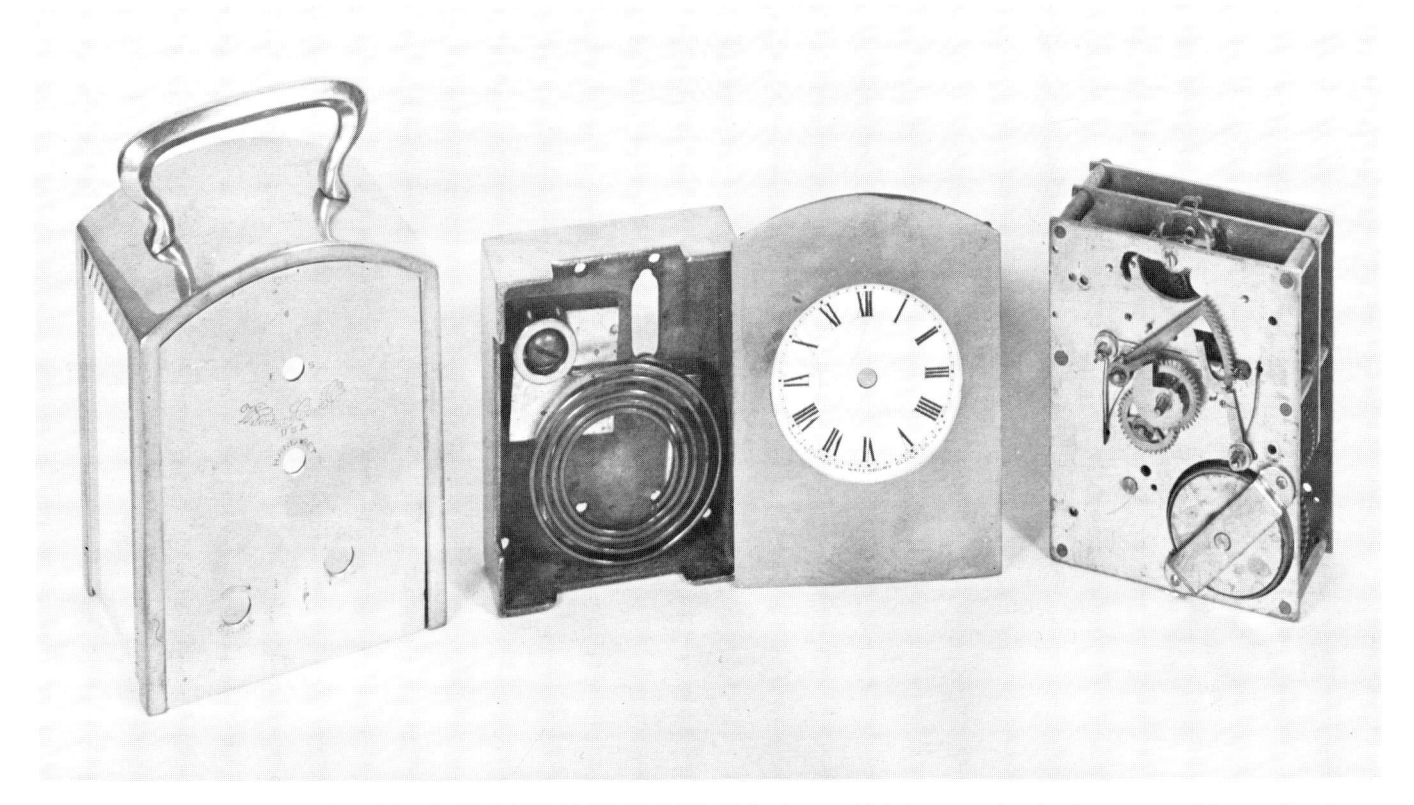

Plate XII/19. *WATERBURY EIGHT-DAY CARRIAGE CLOCK. This piece, which is unusual in having gong striking, strikes hours and half hours but does not repeat. Rack striking is used. Note the strap to allow the use of a deep mainspring for the striking train. The case of this clock is unusually sturdy in construction. (By courtesy of Jim Campbell, Edmunds, Washington State, U.S.A. Photograph by Phil Kohl).*

time going for eight days, is shown in Plate XII/19. This clock strikes hours and half hours on a gong but it does not repeat, although a repeating version may have been made. Gong striking is unusual in Waterbury carriage clocks.[7] In this model the gong "speaks" very well indeed because of the satisfactory "acoustic" arrangements allowed by its situation in the cast brass base. The striking train is on the left of the movement as seen from the back, in reverse to French carriage clock practice, and rack striking is used. A sub-frame between the two movement plates carries the pin-pallet escapement with its "knobbly" balance. The case work of this clock is of particularly good quality by Waterbury standards, the base and handles being solid castings, while bevelled glasses are provided at the sides and front. An enamel dial is sandwiched between a dial surround and a false plate. These are soldered together. The motion work and *cadrature* are accommodated between the dial plate and the front movement plate.

The Waterbury Clock Company, in addition to the various models of carriage clocks which have already been mentioned, did a very large business in cheap one-day alarms and timepieces. Most of these clocks, retailing in 1908-9 from $1.60 to $3.05, were made in true drum-shaped alarm clock style with "tin" cases and paper dials. Plate XII/20 illustrates a number of

---

(7)   Bells were the rule. They were housed in the base of the clock because there was no room in them anywhere else.

355

typical examples taken from the 1908-9 Waterbury Catalogue. The majority of the alarms were topped by large external bells kept in place by carrying rings. One model called *Monitor* incorporated a simple one-day calendar. Two clocks *Patrol* and *Guide*, shown in Plate XII/21 reproduced from the Catalogue, were made in carriage clock form. *Patrol* was available only as a one-day timepiece with alarm and cost $3, while *Guide*, also a one-day clock, cost $3.85 as a timepiece with alarm, $4.10 with hour and half-hour striking but no repeat[8] and $4.35 with hour and half-hour striking and alarm but no repeat.

Just as the Waterbury Catalogue included in the alarm clock pages two "carriage clocks" of strictly alarm-clock quality, so was offered amongst the carriage clocks one comparatively well-made miniature alarm called *Tourist*, besides two drum-shaped timepiece clocks called *Hornet* and *Spider*. *Tourist*, shown in Plate XII/1 and which was made in competition with the German alarm shown in Plates XI/22 and 23, had a cube-shaped case and was capable of sounding its warning for a full minute and a half. It was offered finished in "nickel", "rich gold plated" or "gun metal" and it cost from $3.85 to $4.35 depending on the finish. A leather travelling box was available as an optional extra, thus placing *Tourist* firmly in the travelling clock class. The two drum-shaped clocks *Hornet* and *Spider* were also not without interest. Both employed cases partly made of glass in order to show the movements. Plate XII/22 shows a typical one-day *Hornet* exactly as available circa 1908, while Plates XII/23, 24 and 25 show three views of *Spider* which was an eight-day version of *Hornet*. Plate XII/26 shows rear views of both clocks. It will be seen that they bear the inscription:

"PATENTED
MAY 6. 1890
NOV.11 1890
DEC.23 1890
JAN.13   1891"[9]

Spider is not unnaturally an extension of *Hornet*, the movement of the short duration clock lending itself simply to the addition of an extra frame in order to accommodate a very large mainspring barrel. Although in the photographs the great wheel of the clock appears to be carried on the barrel arbor, it is in fact attached to the barrel itself. The sub-frame carrying the barrel is approximately the same size as the main frame of the clock and the two share a common middle plate. Although lantern pinions are used throughout much of the train, two solid pinions are employed at the beginning. The escapement, as might be expected, is a pin-pallet lever with brass 'scape wheel and steel pallet pins.

WATERBURY WATCH COMPANY

One of the most interesting American carriage clocks ever made was produced by the Waterbury Watch Company.[10] This extraordinary carriage clock was based upon the famous Waterbury Long Wind watch, which in itself was a mechanical curiosity of the first order. For this reason alone it is more than worthwhile taking pains to describe and illustrate the watch thoroughly before passing on to the clock.

The Waterbury Long Wind watches were made for some ten years from 1880. They were un-numbered and were manufactured as Series "A" to "E".[11] The

---

(8) The absence of repeating work, the inclusion of which would have required a different train and mainspring, was an economy and sets *Guide* sharply apart from the repeating carriage clocks shown earlier in the same Waterbury Catalogue.
(9) According to Mr. J.E. Coleman, the noted American horological historian, patent dates on American clocks are apt to be misleading. More often than not they do not apply specifically to the movements on which they appear.
(10) This Company began operations in 1880 in the town of Waterbury, Connecticut. A parent Company, Benedict and Burnham, provided the funds. In 1898 the name of the firm was changed to the New England Watch Company. In 1912 New England failed, and in 1914 it was sold to Robert H. Ingersoll & Brother. In 1922 the latter firm was absorbed by the Waterbury Clock Company.
(11) The Series "A" watches, at least, had skeletonized carriages and dials. The first "long-winders" were signed "Benedict & Burnham". They were made prior to 1880. It would appear that the Long Wind watch was revived briefly some time between 1898 and 1912 because examples exist signed "New England Watch Company"

XII American Carriage Clocks

DOT TIME—Nickel.
1 Day, Time.................$1 60
2¾ inch Dial. Height, 4¾ inches.

TRANSIT—Nickel.
1 Day, Time.................$1 60
4 inch Dial. Height, 5⅞ inches.

CALL—Nickel.
1 Day, Time, Alarm.................$1 70
4 inch Dial. Height, 6 inches.

DOT ALARM—Nickel.
1 Day, Time, Alarm.................$1 75
2¾ inch Dial. Height, 5⅛ inches.

SUNRISE—Nickel.
1 Day, Time, Alarm.................$1 75
4 inch Dial. Height, 6¼ inches.

CALIPH—Nickel.
1 Day, Time, Alarm.................$1 75
4 inch Dial. Height, 6¼ inches.

LUMINOUS-SUNRISE—Nickel.
1 Day, Time, Alarm.................$2 10
4 inch Dial. Height, 6¼ inches.
To obtain the best results, the dial should be exposed during the daytime to a bright light and it will be plainly visible in the dark. At night use at a distance suitable to the eyesight.

MONITOR—Nickel.
1 Day, Time, Alarm, Calendar.........$1 90
4 inch Dial. Height, 6¼ inches.

ALERT—Nickel. Loud Alarm.
1 Day, Time, Alarm.................$2 00
4 inch Dial. Height, 5⅞ inches.
Alarm can be stopped at pleasure.
Attached Keys and Hand Sets.
3¾ inch Bell at Back.
For Side view, see illustration of Spasmodic, Page 63.

Plate XII/20. *WATERBURY CLOCK COMPANY CATALOGUE, 1908–1909. A typical selection of clocks taken from the part of the Catalogue offering one-day alarms. (By courtesy of Adams Brown Company, N.H., U.S.A.).*

XII   American Carriage Clocks

PATROL
Nickel.
1 Day, Time, Alarm $3 00
2¾ inch Dial. Glass Sides. Height, 6 inches.
Gilt Front and Handle.

GUIDE
Nickel.
1 Day, Time, Alarm $3 85
1 Day, Half-hour Strike 4 10
1 Day, Half-hour Strike, Alarm 4 35
2¾ inch Dial. Glass Sides. Height, 7 inches.
Gilt Front and Handle.

Plate XII/21. WATERBURY CLOCK COMPANY CATALOGUE, 1908–1909. These two clocks, appearing in the part of the Waterbury Catalogue offering one-day alarms, were cased in carriage clock form. Patrol stood 6 inches tall and Guide 7 inches. Compare prices and sizes with those carriage clocks shown in Plates XII/2 and XII/3. (By courtesy of Adams Brown Company, N.H., U.S.A.)

Plate XII/22. WATERBURY ONE-DAY TIMEPIECE, HORNET. This clock, shown in the Waterbury Catalogue of 1908-1909, retailed in the U.S.A. at $3.50 (Plate XII/1). In England the Army & Navy Catalogue of 1907 offered Hornet at the incredibly low price of 9s. 3d! How this was possible remains a mystery. (By courtesy of Bryson Moore. Photograph by Phil Kohl).

Plate XII/23. WATERBURY EIGHT-DAY TIMEPIECE, SPIDER. This clock, offered in the carriage clock part of the 1908–9 Catalogue, is a more expensive version of Hornet. Note the glass tube used for the middle of the case. (By courtesy of Bryson Moore. Photograph by Phil Kohl).

XII   American Carriage Clocks

Plate XII/24. *WATERBURY EIGHT-DAY TIMEPIECE, SPIDER. The side view of the movement showing what is basically a* Hornet *movement made to go for eight days by the addition of a large mainspring and barrel. The great wheel is integral with the barrel and drives a steel idler pinion, which in turn drives the going train. The mainspring click is a standard part made to a symmetrical design so that it is capable of working "either-handed". (By courtesy of Bryson Moore. Photograph by Phil Kohl).*

Plate XII/25. *WATERBURY EIGHT-DAY TIMEPIECE, SPIDER. The front view of the movement showing the arrangement of the main and sub frames. The narrow space between the two frames is utilised to accommodate part of the hand setting mechanism and also for the rack and pinion motion for moving the regulation index remotely from the back of the clock. The pin pallet escapement has a brass 'scape wheel. (By courtesy of Bryson Moore. Photograph by Phil Kohl).*

Plate XII/26. *WATERBURY TIMEPIECES, HORNET AND SPIDER. Note the patent numbers and dates. It is worth mentioning that* Hornet *and* Spider *were both available in 1904 as proved by the St. Louis Catalogue (see Bib.). The prices at that time were $3.25 and $5 respectively. (By courtesy of Bryson Moore. Photograph by Phil Kohl).*

359

XII American Carriage Clocks

Waterbury Long Wind system was developed from a patent No. 204,000 taken out in 1878 by D.A. Buck of Worcester, Massachusetts. In a Waterbury Long Wind the train and escapement, which are planted in a circular carriage pivoted at its centre, rotate once an hour together with the minute hand. The "long-winders", which in 1887 sold for as little as $2.50, had very long, weak mainsprings some nine feet in length. The Waterbury "everlasting spring" was housed inside the back of the watch case under a cover secured by screws and carrying the ominous but most necessary words "DON'T REMOVE THIS CAP UNLESS YOU ARE A PRACTICAL WATCH REPAIRER"! Plates XII/27 and 28 are included in order to show that this was no idle warning! The escapement was a duplex, a really good "thumper"[12] having a most ingeniously contrived stamped-out brass 'scape wheel with the two sets of teeth produced in one operation by a press-tool. No "ruby roller" was provided or necessary. Instead a passing-slot was milled in the balance staff itself. Even the impulse "pallet" of the early Waterburys was nothing more than a steel pin.

Not surprisingly, the Waterbury Watches Series "A" to "E", which ran for 28 hours at one winding, took several minutes to wind. People spoke jokingly of running the winding buttons down staircase walls each morning on the way to breakfast. The hands, which were secured friction-tight, were set to time independently of each other by poking them round with a match stick or even with a fingernail, the front bezel having first been removed to allow access to the dial. A similar procedure was necessary in order to regulate the watch. First the bezel had to be removed. Then the index, which rotated with the movement, had to be discovered through one of six slots beside the dial. Once the device was found, then the index could be poked towards "F" or "S" by means of a pin.

The Waterbury Long Wind watches, which were masterpieces of mass production and interchangeability, were enormously successful and vast numbers of them were made. By 1887 production was running

Plate XII/27. *WATERBURY LONG WIND WATCH. "The Warning". (Photograph by Frank Mancktelow, Sevenoaks).*

---

(12) Many were the jokes about this Waterbury watch. As the balance beats, its syncopated "tick" is amplified by the peculiar mechanical arrangements and may be felt through the case. Waterbury anecdotes included that of the girl who remarked how strongly her admirer's heart was beating, only to be told that it was his Waterbury watch that she could feel!

XII   *American Carriage Clocks*

Plate XII/28. *WATERBURY LONG WIND WATCH. "The Warning Ignored". (Photograph by Frank Mancktelow, Sevenoaks).*

at 1,500 a day. In theory "rotating-movement" watches[13] had none of the vertical position errors associated with normal pocket watches; but in fact the purpose behind the peculiar Long Wind system was simply the prosaic necessity of limiting production costs. The clever design has the effect of reducing the total number of parts to as few as fifty-seven, as against about one hundred and sixty in a conventional keyless pocket watch. For instance, it is possible to use a shorter train of wheels and pinions than in normal watch practice. A further economy is effected by the use of identical second and third wheels, each having 50 teeth. A wheel and two pinions are also saved in the motion work. The very weak mainsprings seldom if ever broke, and they also provided more even torques than the short fierce mainsprings of more ordinary cheap

---

(13)   It is worth mentioning here that in the Long Wind Waterburys, and despite the slow rotation of their carriages, the torque to the escapements was transmitted through the carriages themselves as in a tourbillon, and not independently as in a karrusel. See Chapter VIII for further information.

XII American Carriage Clocks

watches. Plate XII/29 depicts a Waterbury Series "E" watch with the mainspring cover removed. No jewels were used anywhere in the Waterbury Series "A" to "E", not even for the balance pivot holes. Even the dials were made of paper. An advertisement booklet, entitled *The Whole Story of The Waterbury* and published in 1887, said that "The cheapest watch in the world" was possible because it was made in vast quantities by means of highly perfected machinery.

In Plate XII/30 are illustrated parts taken from several Series "E" Long Wind watches.[14] The pieces are laid out in order to show both sides of the main components of a single watch, and also to give an idea of the paradox of the sheer simplicity of the units when compared with the subtlety of their operation.

The dial is glued to a plate pressed out of the same material as the body (or "middle") of the case. Slots have been blanked around the edge of the dial to give access to the regulator. Behind the dial is tightly fixed a thick centre wheel of 44 teeth. The pipe or cannon of the hour wheel passes freely through the long central hole in the fixed wheel of 44 and carries the hour hand, friction tight. In the centre of the front plate of the carriage is planted a steel arbor or pivot. This pivot passes through the hour wheel cannon in which it bears. The end of the pivot carries the minute hand set friction tight. The mainspring arbor is attached to the centre of the back plate of the carriage, and pivots in the cover over the mainspring.[15] On the front plate of the carriage is a pinion of 8 leaves. This pinion is carried on the arbor of the second wheel and meshes behind the dial both with the fixed centre wheel and with the hour wheel. The torque exerted by the inner end of the mainspring on the barrel arbor causes the carriage[16] to rotate clockwise. The pinion of 8 "walks" round the fixed wheel of 44, and thus motion is imparted to the carriage train and escapement. This type of gearing arrangement is known as "epicyclic". Whilst the carriage is rotating a second function is performed by

Plate XII/29. *WATERBURY LONG WIND WATCH. The back of the case of this Series "E" has been removed together with the mainspring cover. The mainspring is almost fully wound and the stopwork is about to act. Some collectors consider that long-wind stopwork acts best when the return of the mainspring from its outer hooking on the spring wheel is allowed to rest on the opposite side of the tail of the pivoted stop finger to the last coil of the spring. The method of assembly shown in the photograph also works satisfactorily. It is, however, essential that in all other respects the lay of the mainspring is exactly as illustrated above. (Photograph by Frank Mancktelow, Sevenoaks).*

---

(14) The names used for the parts are the correct Waterbury terminology as used by the makers.

(15) The mainspring is wound from its outer end, which is secured to a post near the edge of the spring wheel. The teeth cut in the periphery of the spring wheel serve the dual purpose of engaging with the winding pinion and also with a click screwed to the body of the case. When the watch is fully wound and the outside coil of the mainspring is drawn towards the centre, it takes with it the tail of a pivoted stop finger. With the watch fully wound, the head of the finger is forced out towards the tips of the spring wheel teeth. It then locks the wheel by engaging in one of eleven detent-steps cut in the inner rim of the case itself.

(16) The design of the carriage is such that it is almost exactly in poise; that is to say that it has no heavy point when supported by its central pivots.

XII American Carriage Clocks

Plate XII/30. *WATERBURY LONG WIND WATCH. This photograph, which does not pretend to offer correct projections, features several Series "E" watches dismantled to show both sides of the principal parts and also their relative positions.*
1st Row – The top row shows the construction of the case with its integral dial, stopwork steps and fixed centre wheel. Note also the mainspring and the spring wheel with its associated parts, including the winding button, winding pinion and clickspring. Clearly visible are the relative positions of the centre wheel, hour wheel and second wheel pinion. On the right the carriage is seen in its working position.
2nd Row – The second row shows the front of the carriage with the centre wheel, hour wheel and second wheel pinion. When the watch is assembled the fan-shaped index scale is accessible through the slots beside the dial. Next on the right in the second row is a carriage with the front plate removed. The second wheel and pinion, third wheel, 'scape wheel and balance are visible. To the right are shown the other side of the spring wheel and the mainspring cover with its solemn warning.
3rd Row – The third row shows the inside of the front carriage plate with the second and third wheels in position. Next in the same row are shown a balance spring pinned to its stud, a balance without its staff, spare hour hands and a long-wind mainspring.
4th Row – In the bottom row are the bezel and back of the watch, separated by two of the boxes of the interchangeable spares at one time freely available.
(Photograph by Frank Mancktelow, Sevenoaks).

XII American Carriage Clocks

the same gearing. This is to advance the hour wheel and hour hand at the correct rate in relation to the minute hand. How all this is done is simple to understand if one can play with the parts of a dismantled watch; but it is far from easy to explain. The mechanical system is as follows:—

The second wheel pinion with its 8 leaves "walks" round the fixed centre wheel of 44 teeth as the carriage rotates. Therefore the pinion must advance by 44 leaves for each turn of the carriage. The hour wheel which has 48 teeth is driven round by the second wheel pinion. The hour wheel would make one turn for each rotation of the carriage were the pinion itself not also rotating by virtue of its engagement with the fixed wheel of 44 teeth. However, during each turn of the carriage, the rotation of the pinion in effect causes the hour wheel of 48 teeth to turn forwards by 48 teeth but at the same time to turn backwards by 44 teeth. The net result is that the hour wheel advances in an hour by only 4 teeth or 4/48, or in other words by 1/12th of a turn. The hour hand therefore moves forward through a one hour division.

Details of the Waterbury Series "E" watch are as follows:—

	Teeth	Leaves	
Fixed centre wheel:	44		
Hour wheel:	48		
Second wheel:	50	Pinion	8
Third wheel:	50	Pinion	6
'scape wheel:	20	Pinion	6

Mainspring length — about 9 feet (30 turns ±)
Mainspring thickness — .006 inch
Mainspring height — .010 inch
The spring wheel has 70 teeth.
The winding pinion has 14 leaves.
5 turns of the winding button represents 1 turn of the spring wheel, which means that the winding button has to be turned about 150 times per day.

The Waterbury Watch Company's carriage clocks, which ran for well over eight days and which incorporated Series "A"-type rotating movements similar to those used in the Long Wind watches, were introduced some time after 1880. Plate XII/31 illustrates such a clock. It bears on the dial plate the inscription "PAT. MAY 21. 1878" (the date of Buck's original patent) and also the monogram "W.W.C.". Plate XII/32 shows

Plate XII/31. *WATERBURY WATCH COMPANY CARRIAGE CLOCK. Date post 1880. This clock incorporates a Series "A" watch-type rotating carriage visible through the dial. The design of the case is ingenious. Top and bottom pressings are separated by the four corner pillars through which pass rods with nuts at either end. There is no back door. Instead the back movement plate is slotted into two pillars in the same way as the bevelled side glasses. A hinged front door provides access to the dial. (By courtesy of Sam Kutner, New Rochelle, N.Y. Photograph by Frank Mancktelow, Sevenoaks).*

a close up view of the dial with part of the carriage clearly visible behind the centre. Front and rear views of the carriage appear in Plates XII/33 and XII/34, while Plate XII/35 shows a side view. The extended period of going between winds is achieved as follows:— on the back of the carriage a pinion of 10 leaves is fixed

XII *American Carriage Clocks*

Plate XII/32. *WATERBURY WATCH COMPANY CARRIAGE CLOCK. A close up view of the front of the movement of the clock shown in Plate XII/31. Note the front of the rotating carriage visible through the dial. (By courtesy of Sam Kutner, New Rochelle, N.Y. Photograph by Frank Mancktelow, Sevenoaks).*

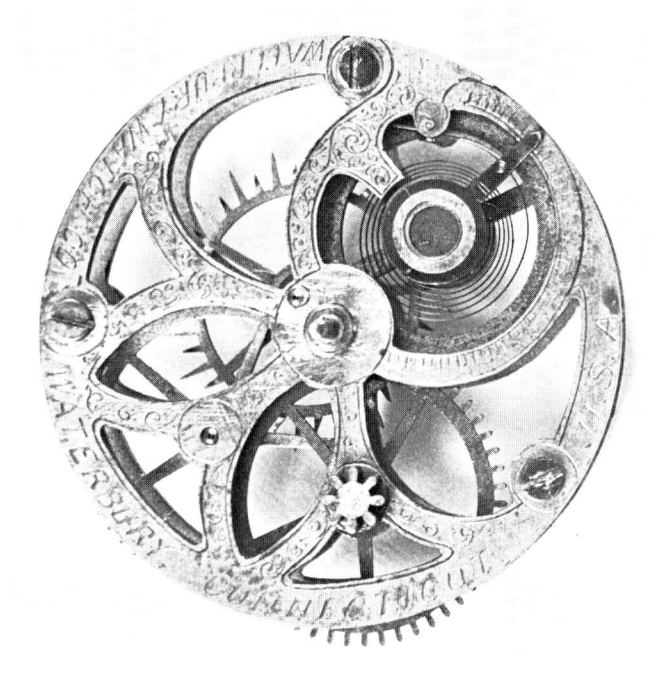

Plate XII/33. *WATERBURY WATCH COMPANY CARRIAGE CLOCK. The front of the carriage. Note the second wheel and its pinion, the third wheel, 'scape wheel, balance, spring and index. The carriage rotates about its central pivot which also carries the minute hand. (By courtesy of Sam Kutner, New Rochelle, N.Y. Photograph by Frank Mancktelow, Sevenoaks).*

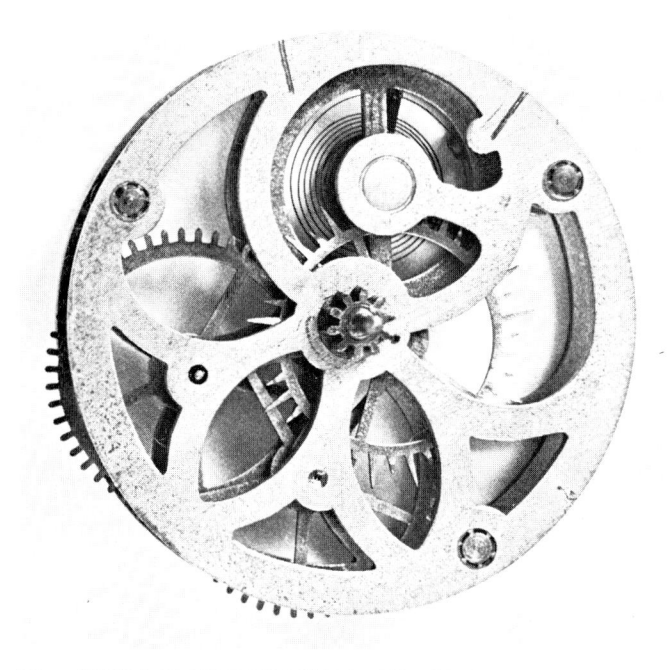

Plate XII/34. *WATERBURY WATCH COMPANY CARRIAGE CLOCK. The rear of the carriage. A pinion of ten leaves replaces the mainspring arbor used in Long Wind watches. Note that the back plate is not engraved because it is not seen. (By courtesy of Sam Kutner, New Rochelle, N.Y. Photograph by Frank Mancktelow, Sevenoaks).*

in the position occupied by the mainspring arbor in a Long Wind watch. This pinion is driven by a great wheel of 120 teeth mounted on an arbor attached to the centre of a mainspring of some 18½ turns. Plate XII/36 well shows the arrangement. There is no stopwork, and thus there are at least theoretically available for rotating the carriage $\frac{120}{10}$ x 18.5, or about 220 turns per full winding, as against a total of some 30 turns in a Series "E" watch. It follows that the clock will go for well over a week.

The method by which the Long Wind watch movement is adapted for use in a clock is very simple indeed. The watch-type carriage is planted behind a skeletonised dial set in a rectangular front movement plate. Mechanically the arrangements of the carriage and motion work are identical to those used in Long

365

XII *American Carriage Clocks*

Wind watches. At the back of the carriage the pivot of the pinion of 10 leaves turns in a hole in a shaped bar or bridge screwed to the front movement plate. The great wheel is attached to the mainspring arbor and thus is driven from the centre of the mainspring. Plate XII/37 shows a side view of the assembled clock movement,

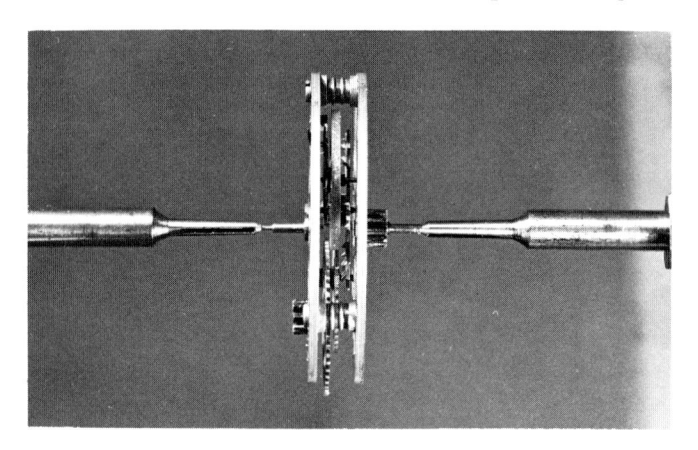

Plate XII/35. *WATERBURY WATCH COMPANY CARRIAGE CLOCK. A side view of the carriage showing the central arbors about which it rotates. Note the second wheel with its external pinion. The carriage is supported between centres for convenience of photography. (By courtesy of Sam Kutner, New Rochelle, N.Y. Photograph by Frank Mancktelow, Sevenoaks).*

Plate XII/36. *WATERBURY WATCH COMPANY CARRIAGE CLOCK. The inside of the movement plates. On the left is seen the front (pillar) plate and above it the bridge in the centre hole of which runs the back pivot of the carriage. The bridge is attached to the inside of the plate by screws and steady pins and is shaped to clear the mainspring barrel. On the right of the illustration is shown the back plate together with the mainspring barrel, clickwork, and above it the external fixed winding key. (By courtesy of Sam Kutner, New Rochelle, N.Y. Photograph by Frank Mancktelow, Sevenoaks).*

Plate XII/37. *WATERBURY WATCH COMPANY CARRIAGE CLOCK. A side view of the movement showing the carriage in relation to the great wheel, barrel and winding arrangements. (By courtesy of Sam Kutner, New Rochelle, N.Y. Photograph by Frank Mancktelow, Sevenoaks).*

XII American Carriage Clocks

and Plate XII/38 shows a top view. The barrel, incorporating the ratchet wheel, must be turned to wind the mainspring. The open end of the barrel faces a recess in the great wheel. The closed end, with the ratchet teeth on the edge, has a short central boss on the outside. This boss projects through the hole in the

Plate XII/39. *WATERBURY WATCH COMPANY CARRIAGE CLOCK. A rear view of the clock shown from Plates XII/31 to 40. Note the fixed winding key which is recessed into the dished part of the back movement plate and which holds the barrel in place. Note also the clever arrangement by which the back movement plate becomes part of the case. (By courtesy of Sam Kutner, New Rochelle, New York. Photograph by Frank Mancktelow, Sevenoaks).*

Plate XII/38. *WATERBURY WATCH COMPANY CARRIAGE CLOCK. A top view of the movement showing the engagement of the carriage pinion with the great wheel. The bar supporting the back pivot of the carriage is plainly visible. (By courtesy of Sam Kutner, New Rochelle, N.Y. Photograph by Frank Mancktelow, Sevenoaks).*

dished part of the back plate of the clock, and carries the winding key. The winding key, as will be seen in Plate XII/39, holds the barrel in place. The hole in the barrel boss provides the rear bearing for the barrel arbor. Plate XII/40 shows an exploded view. The ratchet wheel acts with a click and clickspring planted inside the back movement plate.

367

XII   *American Carriage Clocks*

Plate XII/41. ANSONIA CLOCK COMPANY. *Carriage clock based on* Bee *timepiece. Note the patent date 1878, and also the Ansonia trademark. (By courtesy of Bryson Moore. Photograph by Phil Kohl).*

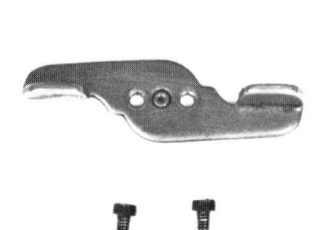

Plate XII/40. *WATERBURY WATCH COMPANY CARRIAGE CLOCK. This photograph is intended to show the mainspring arrangements. Note that the great wheel is mounted on the barrel arbor, and that the ratchet wheel is made integral with the barrel. The clickwork is mounted on the inside of the back movement plate. It is important to understand that in this arrangement, as in the Waterbury Long Wind watches, the piece is driven from the centre of the mainspring and not from its outside. The mainspring is wound by rotating the barrel in reverse to the practice usual in clocks and watches. (By courtesy of Sam Kutner, New Rochelle, N.Y. Photograph by Frank Mancktelow, Sevenoaks).*

XII American Carriage Clocks

ANSONIA CLOCK COMPANY

Plates XII/41 and 42 show two views of a peculiar Ansonia "carriage clock". A dial inscription records a patent for 1878. The clock goes for one day only and it is a timepiece standing just over six inches tall with the handle up. The imitation-repoussé nickel-plated rectangular case is fabricated as one unit. It is made of thin gauge brass sheet, joined with soft solder. The movement is housed in its own drum-shaped container and is detachable as a unit from the main case for the purpose of winding. The winding is accomplished by turning the back of the drum-shaped case, a system known in the U.S.A. as "rim-wind". The same Ansonia movement also appeared in the form of a small drum-shaped timepiece clock sold under the trade name of *Bee*. It was certainly available as late as 1914 and at the time retailed at $1.55 in a one-day version. *Bee*, with porcelain instead of paper dial, retailed at $1.69. There was also an eight-day *Bee;* but it did not have rim-winding. The rectangular repoussé case already mentioned was really simply an accessory by which *Bee,* basically a small and cheap bedroom clock, could instantly be transformed into a carriage clock of sorts. It was just a matter of unscrewing the two feet and the top pendant.

Two Ansonia catalogues have been reproduced in facsimile in recent years (see Bib.) and as a result it is possible to know with a fair degree of accuracy the types of clocks offered by the Company in 1886-1887 and subsequently in 1914. In the earlier catalogue Ansonia[17] showed the eight different "carriage clocks" seen in Plate XII/43. These clocks were classed as "Nickel Novelties", a quality which also included common alarms, a "railway-engine" clock and a number of other highly original pieces including *Cupid's Wreath.* The "carriage clocks" were designed to compete with the French *Obis* and with German and American *Jokers.* Only one "Nickel Novelty" carriage clock went for eight days and that had a duplex escapement. The seven one-day clocks, in the absence of any specific catalogue descriptions, may be presumed to have had some form of pin-pallet lever escapements.

Plate XII/42. *ANSONIA CLOCK COMPANY. Carriage clock based on* Bee *timepiece. In this photograph the drum-shaped case containing the movement has been removed for winding.* Bee *could also be used as a small bedroom clock by attaching two feet and a pendant. (By courtesy of Bryson Moore. Photograph by Phil Kohl).*

---

(17) Ansonia advertised that they made "Solid Cut Pinion Clocks" (as opposed to those with lantern pinions which were cheaper to produce).

XII *American Carriage Clocks*

✴ NICKEL ✴ NOVELTIES. ✴

PEEP-O'-DAY CARRIAGE.
ONE DAY TIME, ALARM.
ONE DAY TIME, ALARM, MUSICAL.
Dial, 3 inches. Height, 7 inches.

PEEP-O'-DAY CARRIAGE.
Fancy Mat, Silver or Gilt.
ONE DAY TIME, ALARM.
Dial, 3 inches. Height, 7 inches.

CARRIAGE PEEP-O'-DAY, STRIKE.
ONE DAY, STRIKE (repeating).
Dial, 3 inches. Height, 6 inches.

CARRIAGE EIGHT DAY.
Duplex Movement.
EIGHT DAY TIME.
Dial, 3 inches. Height, 7 inches.

ORIOLE.
Brass Finish, Enameled in Fancy Colors.
ONE DAY TIME, ALARM.
ONE DAY, STRIKE.
ONE DAY, MUSICAL ALARM.
Dial, 3 inches. Height, 7 inches.

ORNAMENTAL CARRIAGE.
Black Enameled Panels.
Embossed, with Gilt Ornamentation.
ONE DAY TIME, ALARM.
Dial, 3 inches. Height, 7 inches.

ELLIPTICAL CARRIAGE.
ONE DAY TIME, ALARM.
Dial, 2½ inches. Height, 6 inches.

CLIMAX.
ONE DAY TIME, ALARM.
Dial, 2½ inches. Height, 5½ inches.

Plate XII/43. *ANSONIA CLOCK COMPANY TRADE CATALOGUE, 1886-87. A selection of comparatively inexpensive carriage clocks made to compete with French Obis and with German and American Jokers. (By courtesy of American Reprints, Missouri, U.S.A.).*

Between the years 1904-1906, Ansonia manufactured a number of Time Indicators, otherwise called the "Flick" or "Plato" clocks, which were invented by Eugene L. Fitch of New York. These clocks were rather similar to the French example illustrated in Plate VIII/23.[18]

By 1914 Ansonia showed two pages of carriage clocks. These would seem to have been of much the same quality as those offered some twenty-seven years earlier, and it is significant that they are all made to resemble French carriage clocks. Plates XII/44 and XII/45 show ten different models retailing at from

---

(18) In the 1970's reproduction "Plato" clocks were available from Mr. Charles Terwilliger of Bronxville, New York. These clocks were made in Germany.

370

XII American Carriage Clocks

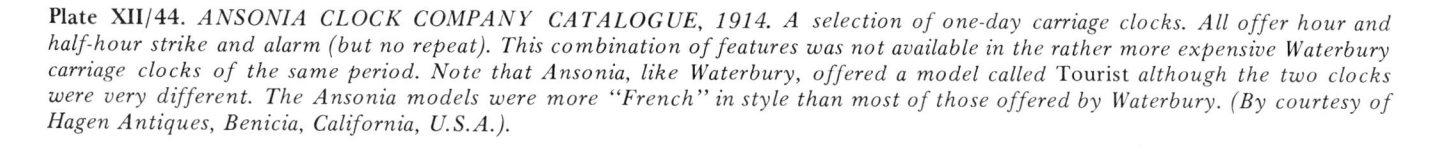

NOVELTIES

Strike and Alarm Clocks.

TOURIST
Roman Dial.
Dial, 2½ Inches.
Height, 7⅛ Inches.
Nickel Plated.
1-Day, Hour and Half-Hour Strike and Alarm.
List, Each, $4.75
Packed 24 in a Box.

TOURIST
Arabic Dial.
Dial, 2½ Inches.
Height, 7⅛ Inches.
Nickel Plated.
1-Day, Hour and Half-Hour Strike and Alarm.
List, Each, $4.75
Packed 24 in a Box.

COMET
Arabic or Roman Dial.
Dial, 2½ Inches.
Height, 7⅞ Inches.
Beveled Glass.
Silver Plated.
1-Day, Hour and Half-Hour Strike and Alarm.
List, Each, $6.60
Packed 24 in a Box.

CARRIAGE EXTRA
Arabic Dial.
Dial, 2½ Inches.
Height, 7½ Inches.
Nickel Plated.
1-Day, Hour and Half-Hour Strike and Alarm.
List, Each, $4.75
Packed 24 in a Box.

CARRIAGE EXTRA
Roman Dial.
Dial, 2½ Inches.
Height, 7½ Inches.
Nickel Plated.
1-Day, Hour and Half-Hour Strike and Alarm.
List, Each, $4.75
Packed 24 in a Box.

Plate XII/44. ANSONIA CLOCK COMPANY CATALOGUE, 1914. *A selection of one-day carriage clocks. All offer hour and half-hour strike and alarm (but no repeat). This combination of features was not available in the rather more expensive Waterbury carriage clocks of the same period. Note that Ansonia, like Waterbury, offered a model called* Tourist *although the two clocks were very different. The Ansonia models were more "French" in style than most of those offered by Waterbury. (By courtesy of Hagen Antiques, Benicia, California, U.S.A.).*

XII American Carriage Clocks

NOVELTIES

PERT
Dial, 1¾ Inches.
Height, 5 Inches.
Beveled Glass, Front and Sides.
Nickel Plated.
1-Day Time and Alarm.
List, Each, $3.40
Packed 24 in a Box.

CLIMAX
Arabic or Roman Dial.
Dial, 2½ Inches.
Height, 5½ Inches.
Beveled Front Glass.
Plain Glass Sides and Back.
Nickel Plated.
1-Day Time and Alarm.
List, Each $3.20
Packed 24 in a Box.

MIDGE
Arabic or Roman Dial.
Dial, 1½ Inches.
Height, 3¼ Inches.
Glass Front and Sides.
Nickel Plated.
1-Day Time.
List, Each, $1.75
Packed 50 in a Box.

BONNIBEL
Arabic or Roman Dial.
Porcelain Dial, 1½ Inches.
Height, 5 Inches.
Beveled Glass.
Front, Sides and Back.
Polished Brass.
8-Day Time.
List, Each, $7.75
Packed 24 in a Box.

BONNIBEL AND LEATHER CASE.
List, Complete, $9.40

SATELLITE
Arabic or Roman.
Porcelain Dial, 1½ Inches.
Height, 5⅝ Inches.
Beveled Glass Front, Sides and Back.
Finished in Rich Gold.
8-Day Time.
List, Each, $10.50
Packed 24 in a Box.

Plate XII/45. *ANSONIA CLOCK COMPANY CATALOGUE, 1914. Note the Obis-like eight-day Bonnibel and the one-day Climax. The one-day Pert should be compared with the Japy clock shown in Plates VI/10 and 11. (By courtesy of Hagen Antiques, Benicia, California, U.S.A.).*

XII *American Carriage Clocks*

Plate XII/46. *ANSONIA CLOCK COMPANY. The one-day model* Pert *shown in 1914 Catalogue.* Pert *stood 5 inches high with the handle up. Note the travelling box with double opening doors. Compare* Pert *with the Japy clock in Plates VI/10 and 11. (Mrs. Charles Allix. Photograph by Frank Mancktelow, Sevenoaks).*

$4.75 to $10.50.[19] Of these clocks *Climax* was also shown in the St. Louis Clock and Watch Company's Catalogue as early as 1904. The retail price was $3.00. Plates XII/46, 47 and 48 show three views of the small and attractive one-day timepiece and alarm called *Pert*, also shown in the 1914 Ansonia Catalogue and then retailing at $3.40. The significance of *Pert* lies in its undoubted connection with the Japy American-style clock described and illustrated in Chapter VI. It remains to be discovered which clock came first. The production of the Japy *type Américain* is said to have begun from about 1880 and it is certain that *réveils Américains* were still in production in 1907. On the other hand *Pert* is not shown in the Ansonia Catalogue of 1886-87 and at present there seems to be no evidence when it was first introduced. *Pert* is shown in the 1914 Ansonia Catalogue; but it may well have appeared far earlier.

JEROME, NEW HAVEN, WELCH AND SETH THOMAS

The fortunes of the Jerome, New Haven and Welch Clock Companies, and even that of Seth Thomas, were to some extent inter-related. None of them seems to have concentrated on carriage clocks to anything like the extent of Waterbury or Ansonia, but it is clear from old catalogues that a number of such clocks were made. Another relevant source of information is *A Sketch of the Clock Making Business 1792-1892* by Hiram Camp.[20]

---

(19) The Ansonia 1914 Catalogue served a dual purpose. The prices shown beneath the clocks are retail (List) prices intended for the customer. The statement that a model is supplied packed so many in a box is information for the retailer.

(20) Camp in his time was not only intimately connected with Jerome and New Haven; but he also, to use his own words, "without doubt had a better opportunity to learn the clock business than any other man living". He rubbed shoulders in one way or another with most of the well-known makers of his day. Camp's historical sketch, written shortly before his death in 1893, was discovered in the files of the New Haven Clock Company 43 years later. It is now available as a facsimile reproduction. Hiram Camp (1811-1893) was a nephew of Chauncey Jerome.

XII American Carriage Clocks

Plate XII/47. ANSONIA CLOCK COMPANY. The back view of the clock Pert shown in Plate XII/46. Note the neat stowage for the winding key. Compare particularly with Plate VI/11. (Mrs. Charles Allix. Photograph by Frank Mancktelow, Sevenoaks).

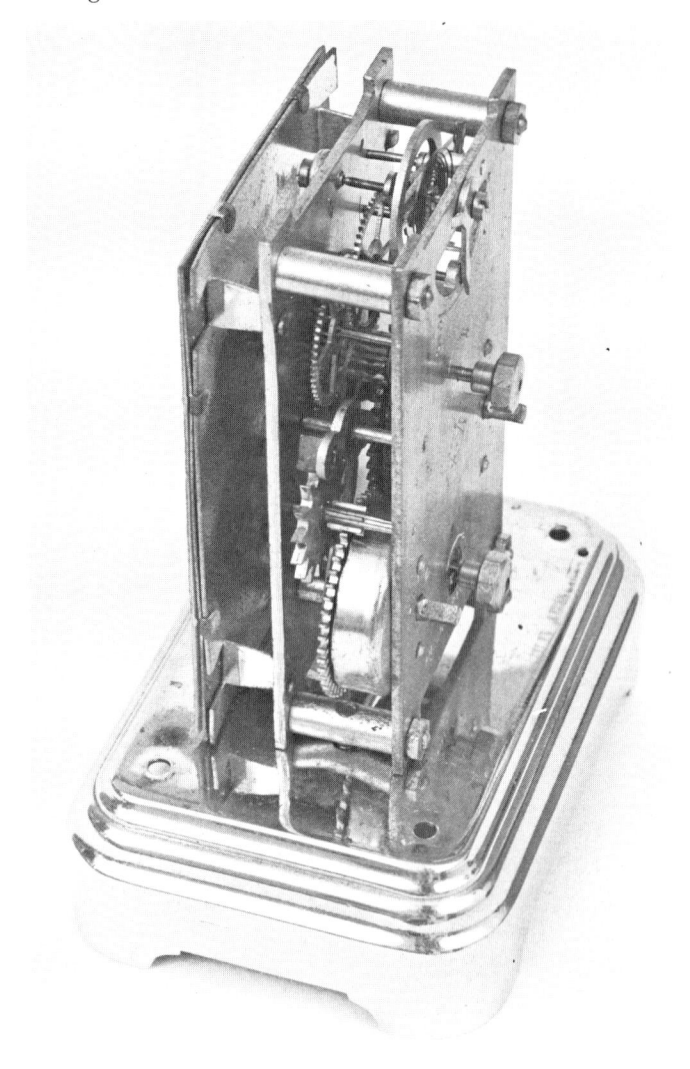

Plate XII/48. ANSONIA CLOCK COMPANY. The movement of the clock shown in Plates XII/46 and 47. Note the use of solid pinions and also the economical methods by which the dial is attached to the movement. (Mrs. Charles Allix. Photograph by Frank Mancktelow, Sevenoaks.).

Chauncey Jerome (1793-1868) began his career as a joiner. He made clock cases from 1811 and came to Bristol, Conn. in 1821. He was partner in various firms from 1826 before forming his own clockmaking company in 1840. Jerome's foreman during the years 1847-50 was none other than Hiram Camp. The firm failed in 1855, being assimilated by New Haven. The New Haven Clock Company was formed in 1853. E.N. Welch, H.M. Welch and also Hiram Camp were subscribers to that new enterprise. Mr. Camp finally retired from New Haven in 1892, when he wrote his memoirs.

Plate XII/49 shows a carriage clock model called *Pilgrim*. Although it bears the Jerome trademark, it was in fact made well after 1855 by the New Haven Clock Company. The photograph is taken from the 1885 Catalogue of the Manufacturing Jewellers, S.F. Myers & Company of New York. *Pilgrim,* which was another clock deliberately made to have the same appearance as

Obis, ran for one day only. It was available as a striker at $4.10, as a timepiece alarm at $3.77, and with both strike and alarm at $4.50. No information was given by Myers about the escapement, but almost certainly it was a pin-pallet lever.

The Catalogue of the E.N. Welch Manufacturing Company *Superior American Clocks and Clock Materials* published in March 1885 illustrates two one-day carriage clocks. They are reproduced in Plate XII/50, from which it is apparent that *Rex* was in the *Obis* idiom, while *Carriage No. 2* was more in the nature of a *Joker*. The two clocks stood 5½ and 7 inches tall respectively and both carried on their dials the Welch

XII American Carriage Clocks

PILGRIM.
Nickel. Height, 6 inches.
Improved Movement. Beveled Edge, Plate Glass Front, Glass Sides and Door.

No. 5104.	1 Day, Strike	$4 10
No. 5105.	1 Day, Time, Alarm	3 77
No. 5106.	1 Day, Strike, Alarm	4 50

Plate XII/49. *JEROME CLOCK COMPANY, PILGRIM. This one-day carriage clock, which appears in Myers' wholesale trade Catalogue of 1885, was made by the New Haven Clock Company. Quite evidently the Jerome trademark was still used at least in certain clocks long after 1855. (By courtesy of American Reprints, St. Louis, Mo., U.S.A.).*

trademark. The E.N. Welch Manufacturing Company, according to their own Catalogue, had origins going back to about 1835. The Welch family bought interests in the New Haven Clock Company when it was first formed in 1853. E.N. Welch died in 1887 and by 1897 the Company was in the hands of the Receiver. The Seth Thomas Clock Company, on the evidence of their own Catalogue of 1879, was established in 1813. It would seem that in the carriage clock sphere Seth Thomas concentrated upon inexpensive one-day carriage clocks of the alarm clock or *Joker* type. Plate XII/51 taken from the 1879 Catalogue shows a one-day *Joker Lever* which stood 7 inches tall. This clock must have enjoyed a very considerable success because it was still in production without apparent modification twenty-five years later in 1904. In this year the same old Seth Thomas *Joker* was still advertised in the Catalogue of the St. Louis Clock and Silverware Company where its retail price was shown as $3.60 or $3.92 as a striker. It was manufactured in direct competition with the German *Jokers*. The St. Louis Company also showed in their 1904 Catalogue the competing German *Joker* musical alarm illustrated in Plate XI/21. This last clock was of almost identical appearance to its American rival; but the German clock, instead of having an alarm on a bell, played two tunes lasting a total of about ten minutes. The retail selling price was $7. Plate XII/52 shows a further inexpensive Seth Thomas clock termed the *Artist Lever*. It is reproduced from the Myers Catalogue of 1885.

Camp comments in his memoirs with reference to the companies Seth Thomas, Waterbury, New Haven, E.N. Welch, Ingraham, Gilbert and Ansonia that the number of pieces made at the time of his writing by the seven companies did not fall much below 250,000 a year, which he says is "... a number far too many either for the trade or profit of the makers. The desire to make and sell great quantities has led the manufacturers to bring out new designs until dealers have become amazed and bewildered to such an extent as to paralyze the trade, the expectation of something new prevents the sale of the old. Those companies that have to the greatest extent been tempted and followed this line have suffered by the accumulation of unsaleable goods and loss on tools for manufacturing the same; while the companies that have been more conservative have found themselves in the better condition". It is interesting to compare

XII *American Carriage Clocks*

REX.
NICKEL AND GLASS.

Height, 5½ inches.   Dial, 2½ inches.
1 Day.   Lever.   Half-hour Repeating Strike and Alarm.

CARRIAGE.
No. 2.

Height, 7 inches.   Dial, 2½ inches.
1 Day.   Lever.   Time.   Alarm.
1 "   "   Strike.
Plain White, Gilt or Fancy Dial.

Plate XII/50. *E.N. WELCH MANUFACTURING CO. The only two carriage clocks shown in Welch's unpriced trade Catalogue of 1885. Clapp & Davies offered Carriage No. 2 in 1886 at $4.80 for a timepiece with alarm, and the same price for a striker. (By courtesy of Adams Brown Co., Exeter, N.H., U.S.A.).*

Camp's remarks made in 1893 with Saunier's comments in 1867 and quoted on page 342 of this chapter.

JOSEPH EASTMAN. THE BOSTON, CHELSEA AND VERMONT CLOCK COMPANIES.

It would seem presumptuous to attempt to unravel here the complicated and often inter-connected histories and mergers of those clock companies associated with Joseph Eastman, who died as recently as December 1931. The manufacturing names at one time and another seem to have included his own enterprise the Eastman Clock Company, besides "Boston", "Chelsea", "Harvard", "Fairhaven" and "Vermont". While some information is given by

376

XII *American Carriage Clocks*

JOKER LEVER.

ARTIST LEVER.
Gold Gilt Front and Handle, Nickel Frame and Glass Sides. 3-inch Dial.

No. 5049.   1 Day, Time, Alarm............$4 10
No. 5049½. 1 "   Strike................... 4 50

3 inch Dial.

1 Day, Time.

1 Day, Time, Alarm.

1 Day, Strike.

Plate XII/51. *SETH THOMAS CLOCK COMPANY, JOKER LEVER. This illustration is taken from the Seth Thomas trade Catalogue of 1879. Compare with the German Joker shown in Plate XI/21. (By courtesy of American Reprints, Missouri, U.S.A.).*

Plate XII/52. *SETH THOMAS CLOCK COMPANY, ARTIST LEVER. This clock was offered in Myers wholesale trade Catalogue of 1885. (By courtesy of American Reprints, Missouri, U.S.A.).*

Brooks Palmer, and while much further research has been made in more recent times by William E. Drost, Edwin B. Burt, George H. Amidon and no doubt by others, the facts (to an uninformed Englishman at least!) remain most confused and far from being fully explained or documented. It does seem clear, however, that Joseph H. Eastman was born in Georgetown, Mass. on September 10th, 1843. He was apprenticed to the E. Howard Watch Company at Roxbury and he could have returned to Howard at a

377

XII American Carriage Clocks

later stage of his life had he so wished.[21] Eastman's most active years were associated with the Boston Clock Company and later with the Chelsea Clock Company which still exists to this day.[22] Eastman's specialities included the "two-way" wind[23] and the "noiseless-tick" as applied to pin-pallet lever escapements. Later in life he became interested in electrical horology.

Plates XII/53 and 54 illustrate two similar Boston

Plate XII/53. *BOSTON CLOCK COMPANY. Clock No. C4586. Date circa 1894. A good quality eight-day timepiece carriage clock. This expensive model was nevertheless manufactured on the interchangeable principle both as regards case and movement. The case is electro-gilded in 18 carat gold. The clock stands 6½ inches tall with the handle up. (By courtesy of Eric Graus, London. Photograph by Frank Mancktelow, Sevenoaks).*

Plate XII/54. *BOSTON CLOCK COMPANY. Another carriage clock similar to that shown in Plate XII/53 but with a circular unsigned dial. (By courtesy of Ernest R. Conover Jr., Aurora, Ohio, U.S.A.).*

---

(21) Edward Howard, the founder, was the apprentice of Aaron Willard, Jr. Aaron was the nephew of Simon Willard, the famous developer of the Banjo clock.

(22) Both Companies had the address 284 Everett Avenue, Chelsea, Mass.

(23) Both striking and going sides were wound through one arbor turned alternately clockwise and anti-clockwise. This system, according to W.E. Drost, was patented by James H. Gerry in 1880.

XII *American Carriage Clocks*

Plate XII/55. BOSTON CLOCK COMPANY. *A rear view of clock No. C4586 shown in Plate XII/53. Note that the back door follows French and not American practice in being made a close fit and having no latch. The case is of very good quality, and is sturdily constructed. (By courtesy of Eric Graus, London. Photograph by Frank Mancktelow, Sevenoaks).*

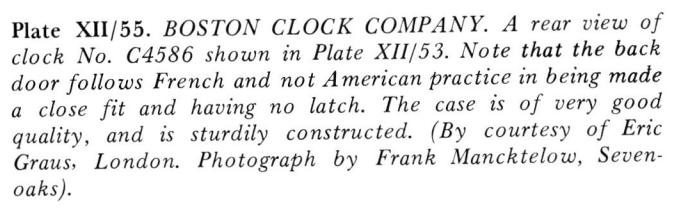

Plate XII/56. BOSTON CLOCK COMPANY. *A rear view of the movement of clock No. C4586. Compare with the movement of the Vermont clock shown in Plate XII/61. (By courtesy of Eric Graus, London. Photograph by Frank Mancktelow, Sevenoaks).*

eight-day carriage clocks, one found in England and one in the U.S.A. Plates XII/55 and 56 show views of the movement of the first clock which has the serial number C.4586.[24] The great interest of these pieces, which are of very good quality and also solidly cased, is that all the case and movement parts were apparently made on an interchangeable system. The Boston clocks, which are timepieces only, are made on the main-frame/sub-frame principle. Plate XII/57 shows a side view of the movement of No. C.4586, and Plate XII/58 a half side view. It will be seen that the front movement plate acts as a false plate and that the enamel-on-copper dial is spaced away from it just far enough to allow room for the motion work between the two. The great wheel on the going barrel is at the front of the movement. There is an intermediate wheel and pinion between the great wheel and the centre pinion in order to achieve an eight-day duration of going. The last two figures of the serial number, i.e. "86",

---

(24) The Boston Clock Company was in existence from 1888 to 1897. There should be no doubt about the date of manufacture of clock No. C4586 because inscribed on the top of the case, and almost certainly engraved when the clock was brand new, are the words "J.A.E. Malone, FROM George Edwardes, Brooklyn N.Y., Xmas, '94".

379

XII *American Carriage Clocks*

Plate XII/57. *BOSTON CLOCK COMPANY. A side view of the movement of No. C4586. Note the main-frame/sub-frame arrangement and also how the motion work is accommodated in the narrow space behind the dial. The escapement is carried on the back of the sub-frame avoiding the use of both contrate wheel and platform. (By courtesy of Eric Graus, London. Photograph by Frank Mancktelow, Sevenoaks).*

Plate XII/58. *BOSTON CLOCK COMPANY. Another view of the movement of clock No. C4586 showing the general layout. (By courtesy of Eric Graus, London. Photograph by Frank Mancktelow, Sevenoaks).*

are repeated on the barrel, barrel cap and front movement plate, while "586" appears on the side of the balance cock foot.[25]

The train count is as follows:—

	Wheel	Pinion
Going Barrel	62	
Intermediate Wheel	44	10
Centre Wheel	64	10
Third Wheel	60	8
Fourth Wheel	70	8
Escape Wheel	15	7

The balance vibrates 18,000 times an hour.
The straight line single-roller lever escapement has jewelled pallets, club-tooth brass 'scape wheel, cut

---

(25) The significance of these numbers is the clue which they provide to production methods.

380

XII American Carriage Clocks

brass and steel balance and flat spring. The holes for the balance pivots are jewelled and have endstones. The pallets and 'scape wheel are under one cock and the banking pins are set in eccentric bushes, slotted to allow adjustment by means of a screwdriver. The main frames are separated by four pillars and the sub-frame plate is spaced from the front movement plate by three pillars. The dial is held to the front movement plate by means of four countersunk screws at the corners and passing through the dial feet into the movement pillars.

Other than the peculiar movement construction, the most fundamental difference between this American carriage clock and a typical French example is that coarser tooth pitches are used throughout the train and also more changes of pitch are required. See Plate XII/59. It is worth emphasising that in this clock the wheel teeth are very well shaped and cut and that the pinions are altogether better formed and finished than those in any but the best French carriage clocks. The train on the whole is more reminiscent of English practices than of French, having sturdy pivots with pronounced back slopes from the shoulders. Note the short arbors used in this American clock. The movement, when dismantled, reveals excellent clean work besides such luxurious details as pinion ends faced and polished and arbors glossed all over. The intermediate pinion, which is planted close to the outside of the movement and which is therefore easily seen, even has its end hollowed and polished in the manner of "best work". The movement plates are electro-gilt, reputedly in 18 carat gold.

The distinctive case of the Boston carriage clock also deserves special notice. It is about the most sensible, sturdy and practical "multi-piece" case that it is possible to imagine. It does not "fall to pieces" during dismantling. No elastic bands, sellotape or other "dodges" are required when assembling it again. Every part lines up correctly without having to be "sprung" into place. The clock movement is secured by means of a shaped strap. This passes over the two bottom movement pillars and is attached to the base plate of the case by means of two screws. As a result the movement may be carefully aligned in its final position with the case fully assembled.

Before leaving the Boston clocks and discussing the mechanically very similar pieces made by the Chelsea Clock Company and by the Vermont Clock Company (and possibly by others) it must be said that it is not fair to attempt to compare these expensive clocks with the frankly-cheap Waterbury, Ansonia, Jerome, New Haven, Welch and Seth Thomas clocks already described. The alarm-clock nature of the latter clocks sets them sharply apart from the high-class productions of the Boston, Chelsea and Vermont Clock Companies. These well stand comparison with good quality French carriage clocks.

The Chelsea Clock Company, which began life as the Boston Clock Company, is still in business. The Chelsea ship's bell clocks (striking nautical hours) and other special pieces made for marine purposes and for the U.S.A. Armed Forces generally, have long been

Plate XII/59. *BOSTON CLOCK COMPANY. Clock No. C4586. A view of the inside of the front movement plate showing the four main pillars, the three pillars for the sub-frame and the planting of the train up to the 4th wheel. Note the comparatively coarse pitches used at the lower end of the train and also the changes in pitch and the proportions of the pivots. The going barrel makes use of 9¾ turns of spring. (By courtesy of Eric Graus, London. Photograph by Frank Mancktelow, Sevenoaks).*

XII American Carriage Clocks

very famous.

The Chelsea Clock Company traces its origins back to Joseph Eastman, whose first independent venture was the Eastman Clock Company started in 1886. In 1888 the name was changed to the Boston Clock Company, and this in turn in 1897 became the Chelsea Clock Company. The new owner was Charles Pearson. In 1928 William Neagle purchased the Chelsea Clock Company from the Pearson estate. In 1945 Walter E. Mutz and George J. King became the new joint owners. In 1970 Automaton Industries Inc. of Los Angeles bought the Chelsea Clock Company. Mr. King retired and Mr. Mutz was retained as

Plate XII/60. *VERMONT CLOCK COMPANY. A good quality eight-day carriage clock made to resemble a French Obis, but having a movement very similar to the Boston clocks. (By courtesy of Bryson Moore. Photograph by Phil Kohl)*

Plate XII/61. *VERMONT CLOCK COMPANY. The movement of the clock shown in Plate XII/60. (By courtesy of Bryson Moore. Photograph by Phil Kohl).*

consultant. His contract expired on 1st January 1973. Meanwhile, on 10th November 1972 the Company was sold to Bunker-Ramo.

Unfortunately it has not proved possible to find a "Chelsea" carriage clock to illustrate in this book; but according to Mr. Mutz the Chelsea Clock Company and some others made carriage clocks of the same type as the Boston. He comments, "This shows me that Mr. Eastman had a hand in the operations". Mr. Mutz also says, in speaking of Boston carriage clocks, "even some of the escapement parts were almost the same as were later manufactured by Chelsea".

Plates XII/60 and 61 are photographs of a clock signed "VERMONT CLOCK CO., FAIR HAVEN, VT, USA".[26] Its height with handle up is approx-

---

(26) The Vermont Clock Company was started in about 1890 and closed in about 1902. The authority for this statement is Mr. Walter E. Mutz already mentioned.

XII American Carriage Clocks

imately six inches. The thick dial acts as its own false-plate and has the motion work behind it. The movement pillars are extended through the front plate and serve also as feet for the dial, which is held in place by four screws.[27] The movement is nickel-plated. The plates have their visible parts decorated with that typically American horological patterning called damascene work. It was sometimes produced free-hand; but Vermont here employed some sort of machine-guided end-mill. The zig-zag waves correspond to "spotting" in English work. The Vermont movement employs a main and sub-frame arrangement exceedingly similar to that found in the Boston clocks.[28] The barrel arbor is run in the main frames with the clickwork inside the back plate. The great wheel is at the front of the movement instead of at the back as in French work. The train is accommodated in the sub-frame. The train has "solid" pinions throughout and employs an intermediate wheel and pinion between the great wheel and the centre pinion in order to achieve an eight-day duration of going. The sub-frame also serves as the escapement platform. The balance, lever, etc. are planted "sideways" under cocks. A straight-line lever is used. There is a cut,

bi-metal balance and a flat balance spring. The "lift" is divided and there is a single roller. The case seems so "French" that it is tempting to believe that it was imported. There is no top window but the escapement is visible through the back door. The door has a latch like most American carriage clocks and unlike the majority of French ones. This is a good quality clock and it is certainly classifiable as a *pendule de voyage* in the true meaning of the term. It was probably made in the present century.

Plate XII/62 shows the Vermont factory at Fairhaven. It is said that Vermont made at least one model of carriage clock in which the eight-day going and striking trains were both wound by a single key, turned first in one direction and then in the other. This system, as already noted, was one of Joseph Eastman's specialities.

One of Eastman's helpers, whose name is still remembered, was Charlie Raines. As a young man Charlie lived in Everett, the next town to Chelsea. He saw the original Chelsea Clock Company factory being built. He began work at the Vermont Clock Company on 12th August 1901. After the Vermont Clock Company closed about a year later, Charlie

Plate XII/62. *VERMONT CLOCK COMPANY. The Vermont clock factory at Fairhaven, Vermont. (By courtesy of Walter E. Mutz, Chelsea, Mass., U.S.A.).*

---

(27) This arrangement is a great improvement on the Boston system which in clock C.4586 has resulted in damage to the dial. Someone has since tried to conceal both screws and damage.
(28) No doubt this shows Eastman's influence.

XII American Carriage Clocks

went to work for the Chelsea Clock Company. He began with the firm on 5th October 1903. Unfortunately, as Mr. Mutz says, all of the "old timers" like Charlie Raines who knew of these matters have passed on.

FASOLDT

Paul Chamberlain, that most indispensable and best-liked of all American horological writers, deals with Fasoldt so thoroughly that it is unnecessary to enter into too much detail here. Chamberlain, how-

Plate XII/64. *FASOLDT. The back of the case and movement of the clock shown in Plate XII/63. Note the very unconventional arrangement of both movement and escapement. (By courtesy of Seth Atwood, Rockford, Illinois, U.S.A.).*

Plate XII/63. *FASOLDT. The doyen of American carriage clocks. This clock, which was made circa 1885, stands 12 inches high and has a gold dial and gold plated case. (By courtesy of Seth Atwood, Rockford, Illinois, U.S.A.).*

ever, explains that Fasoldt came from Germany, arriving in New York in 1849 via Rome and after much political tribulation in his own country. In 1861 he moved to Albany where he started the watch factory which saw most of his major productions.

At the beginning of this chapter is a large photograph of the unique Fasoldt carriage clock with patent double-wheel escapement and which was made circa 1885. It stands twelve inches tall and has a gold dial and gold-plated case. Plates XII/63 and 64 show front and rear views of the clock and movement,

384

XII   American Carriage Clocks

while Plate XII/65 illustrates the escapement. Chamberlain shows several drawings of Fasoldt's arrangement in *It's About Time*. This type of escapement, usually called "Fasoldt's Chronometer", is one of those numerous composite escapements of which so many exist that a book could be written upon the subject alone. Chamberlain says that Fasoldt used his own design for "... some thirty years in clocks and watches". The carriage clock illustrated is so famous that it must be regarded as the veritable doyen of American *pendules de voyage*. It is certainly a carriage clock able to hold its head high in any company, besides perhaps being the only piece of the kind that Fasoldt ever made.

Plate XII/65. *FASOLDT. A top view of the escapement platform of the clock shown in Plates XII/63 and 64. (By courtesy of Seth Atwood, Rockford, Illinois, U.S.A.).*

Chapter XIII

Japan, Italy and The Argentine

XIII
Japan, Italy and The Argentine

Seikosha of Tokyo, Masetti and Maransei, both of Bologna. One Argentine Clock.

JAPAN

The return of American servicemen from Korea has drawn attention to the fact that carriage clocks were at one time manufactured in Japan by Seikosha, now Seiko Time Corporation. These Seikosha carriage clocks, which were probably made between the two World Wars, were mostly found either in Yokohama or in Seoul. Some examples have Korean chapters but others, like the two which are illustrated, have ordinary Arabic numeral dials and also winding and hand-setting instructions written in English. The chief interest in these Japanese carriage clocks, however, lies in the provision of simple calendars showing day of the week and date of the month. Plate XIII/1 shows such a clock. It goes for eight days and bears the inscription "MANUFACTURED BY SEIKOSHA. TOKYO. JAPAN". This signature is under the glazing of the dial as are the numerals.

Plate XIII/2 illustrates another Japanese carriage clock, again having meantime dial and calendar, but this time arranged vertically much more in the manner

Plate XIII/1. *JAPANESE CARRIAGE CLOCK. This example, with calendar for day of the week and date of the month, was made by Seikosha of Tokyo during the present century. (By courtesy of Robert Millspaugh. Photograph by P.H. Kohl)*

XIII Japan, Italy and The Argentine

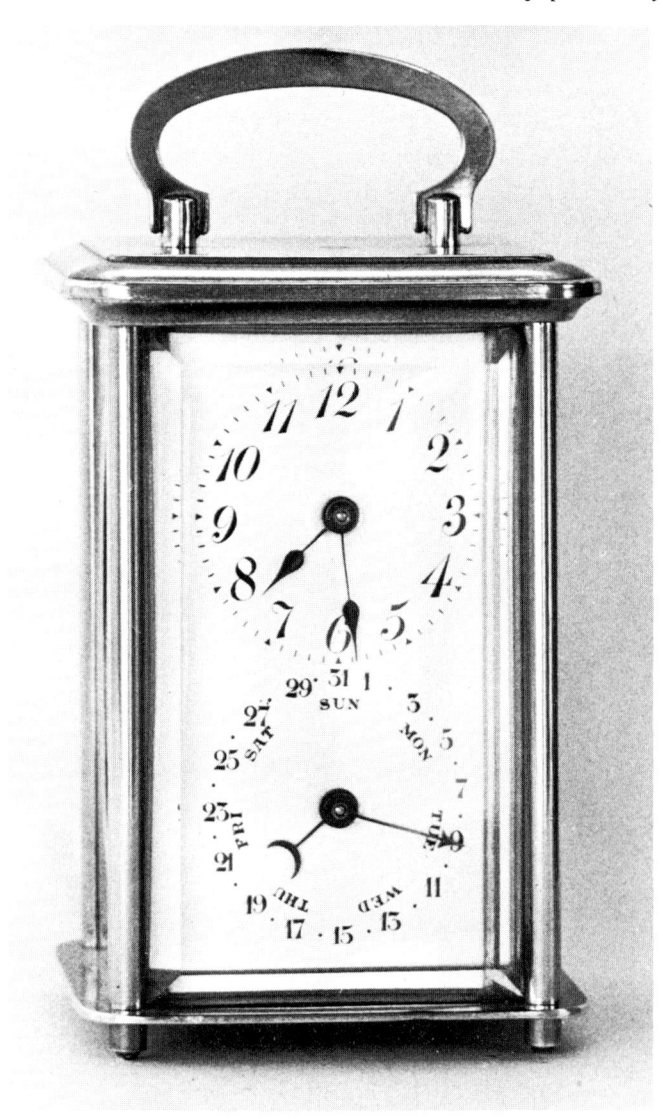

Plate XIII/2. *JAPANESE CARRIAGE CLOCK. This Seikosha travelling clock has the same movement as the example shown in Plate XIII/1. These clocks seem never to have been sold in Europe despite their having English inscriptions and Arabic numerals. Examples also exist having Korean characters on the dials. (By courtesy of Bryson Moore. Photograph by P.H. Kohl)*

of a conventional French carriage clock. The piece is unsigned, but there is no doubt that the maker was Seikosha because the movement is identical to that of the clock already described. The second Seikosha clock stands 6¾ inches tall including the handle which does not fold down. It is worth noting that the bottom of the case of this particular clock is not original. In all probability it consisted of "mouldings" similar to the top, the whole overall effect being not unlike that of a conventional French *Obis* timepiece. The rather plain case of the Seikosha clock is nickel-plated and it has elliptically-shaped side windows. There is no top glass. The construction of the case depends very much upon soft solder and upon bent-over tongues much in the manner of the inexpensive pre-war Japanese "tin toys". The oval side glasses are flat but the rectangular front glass is deeply bevelled. The solid back door is latched at the bottom and at the top is a regulating device for bringing the clock to time. The fixed winding keys and setting-buttons project through the back of the case. The "winders" do not fold flat. The movement is held to the bottom of the case by two screws and washers through slotted holes and at the top by a tongue "turn-buckle". The eight day movement consists of main and sub-frames. The train is run in the main frame. The mainspring and its associated great wheel, made without any barrel, are run in a sub-frame behind the larger plates. The intermediate wheel with its lantern pinion is run between the front plate of the main frame and the sub-frame plate. The centre wheel pinion is also of the lantern type, the remainder being cut from the solid. The enamel dial is mounted on a false plate. Between the false plate and the front plate of the movement lies the motion work and part of the calendar work. The pin-pallet lever escapement has a plain brass balance. The regulating index has a loop instead of two separate index pins. The calendar work is of simple and practical design. It needs correcting every month that does not have thirty-one days. The crescent-shaped hand shows the day of the week while the arrow-hand points to the date. It would appear that the day-of-the-week hand moves every twelve hours, which is not only a means of differentiating between night and day, but also of avoiding the necessity of providing a twenty-four hour wheel for the purpose of moving the mechanism.

U.S.A. readers may have encountered the very attractive copies of American clocks produced by Seikosha during the early years of the present century. One example made circa 1903 is a virtual reproduction of a small steeple-type American eight-day shelf clock (but in fact made without steeples). The name "Seikosha" and also the trademark is proudly displayed on the clock in two places including on the American-type back paper.

"Seiko" are still very much in business today. They make and market clocks and watches on a world-wide basis.

XIII Japan, Italy and The Argentine

ITALY

In the *Liste des Exposants Recompensés* published in connection with the Paris Universal Exhibition of 1867 under *La Classe 23 (Horlogerie)* were Masetti (Barthélemy) and Maransei (Gaëtan) both of Bologna and both of whom received Honourable Mentions for striking clocks and carriage clocks.

THE ARGENTINE

According to Tripplin in *Watch and Clock Making in 1889*, page 93, "The Argentine Republic was a participator in the display...and exhibited two clocks, made evidently by not very skilful hands. One was a skeleton, the other a carriage clock".

Appendices

Appendix (a)

Paul Garnier:

(Biographical Notes and Quotations with references to his carriage clocks, to his escapements and to his career and inventions generally).

1830. An abridged specification of Garnier's two-plane escapement re-published by the *Revue Chronométrique* in 1873.

"30 septembre *Garnier,* Paul (Paris) — Brevet de 5 ans. Echappement à repos, applicable aux pendules, montres, etc. Il se compose de deux roues parallèles fixées sur la même tige. Chaque dent d'une roue, terminée par un plan incliné, répond à un vide de l'autre. L'axe du balancier porte une portion de cercle où se fait le repos par les pointes des dents. C'est sur le devant rectiligne de cette portion de cercle qu'a lieu l'impulsion au passage de chaque plan incliné." This may be translated as:

"30th September *Garnier,* Paul (Paris) — Patent for 5 years. Frictional-rest escapement, applicable to clocks, watches, etc. It consists of two parallel wheels mounted side by side on the same pinion. Each tooth of one wheel, ending in an inclined plane, lies opposite to a space in the other. The balance staff carries a semi-circle which provides dead locking for the points of the teeth. The impulse given by the inclined tooth tips takes place on the front straight edge of the 'disc'."

1849. Peupin on Garnier, with reference to the Paris Exhibition of 1849, translated from *La Tribune Chronométrique,* page 80.

"For some time M. Paul Garnier has held a distinguished position in horology; he is the first who understood the undoubted dividend that accrues from the large-scale production of carriage clocks. If this type of clock has created an ever-expanding industry, in all fairness it should be said that this outcome is due to Garnier's inventive mind and to his grasp of production techniques. M. Paul Garnier, clever mechanic that he is, has figured in a remarkable way at all the exhibitions since 1827, on which occasion he made his debut before the Examiners". Most of the remainder of the article is not relevant in the context of this book; but we are told that Garnier's inventions included multi-dial clock systems, monitoring and timing devices connected with the running of railways, engine-indicators, dynamometers, and counters generally.

1851. Dubois on Garnier's escapement, taken from *La Tribune Chronométrique.*

On pages 231 to 234 of this interesting monthly journal appears an article, edited if not actually written by Dubois, entitled *Echappement d'Enderlin, perfectionné par M. Paul Garnier, Elève de Janvier.* There are six figures, intended to illustrate the escapements attributed to Enderlin, Sully and Garnier respectively, with plan and elevation drawings of each.[1] The elevation drawing of Garnier's escapement is particularly inaccurate. The author of the article says that historically this type of escapement is very old, having been first attempted by Tompion towards the end of the 17th century, but without his having apparently applied it to watches. "Whatever happened" says the *Tribune* "this endeavour came to nothing". Presumably the "Tompion" escapement the writer had in mind is the one illustrated in Gros *Echappements . . . 1913,* fig. 52, and also shown in Rees' *Cyclopaedia* in Plate XXXV on page 217 and also in "Britten". While this design may well have been one of the first dead-beat escapements, it is difficult to see in it much analogy to Garnier's two-plane arrangement. Moreover, if the escapement attributed to Tompion resembles any that later came in to being, then it is either with the "tic-tac" or with the virgule that it might be compared. In any case, there seems to be no

---

(1)   Enderlin's escapement is basically a later variant of De Baufre's.

Appendix (a).

authority for the drawing in Rees, any more than there exists one shred of evidence that an example of "Tompion's" escapement is either known to exist or ever to have existed. However, the writer goes on to say that at the beginning of the 18th century two horologists, namely Enderlin and Sully, tried to put this type of escapement to practical use. In the attempt of the first there is one "disc" and two wheels, and in that of the second one wheel and two "discs". The author quotes Thiout as being very critical of Enderlin's escapement, saying that it shakes about very much, appears to be particularly susceptible to dirt and is also impossible to time in different positions (which he details). The *Tribune* article says that the escapements of both Enderlin and Sully were thoroughly unsatisfactory and had long been forgotten until Garnier decided to use that of Enderlin. The article continues "The first attempt of our clever artist was an instant success, and the *pièces de voyage* or portable clocks in which M. Paul Garnier used his escapement gave results in timekeeping equal to a cylinder escapement or to a well made lever." The writer says that Garnier's new escapement is particularly suitable for carriage clocks because his design not only does away with the undesirable "fluttering" but also largely eliminates the positional errors associated with the earlier versions. He feels that Garnier has achieved this happy improvement by transferring most of the "lift" to the 'scape wheel teeth, allowing the "disc" to be altogether lighter and finer. He claims that this lay-out allows a supplementary arc of about 160 degrees in either direction, which is enough to afford good timing. The writer of the article concludes by saying that he has checked the going of a Garnier *pendule de voyage* for eight days with the following results. The first day the clock gained 20 seconds. The second day it gained 15 seconds. The third, fourth and fifth days time was kept to within a second. During the successive three or four days the clock tended to lose, ending the week a few seconds fast. The report is at pains to point out that the results quoted apply only to Garnier clocks with his own escapement[2] and adds that the observed results left a very favourable impression.

Readers may be left startled by the implication that lever and cylinder escapements could possibly ever have given comparable results. The truth is that they probably did at this date, there being little to choose between an *unworn* well-made and designed cylinder and a badly-designed lever used with an uncompensated balance. In 1971 light-hearted tests were made with a standard early Garnier carriage clock with his escapement. This clock, which was only fairly clean, showed a mean variation of less than a minute a week, the performance scarcely worsening even when the clock was carried about London lying on its side in an attache case.

1851. Garnier's letter concerning his escapement, taken from *La Tribune Chronométrique*, pages 261 and 262.

Following the article on his escapement Garnier wrote on 2nd September 1851 a letter to the editor (Dubois) complaining that in No. 8 of the *Tribune* several errors had crept in to the article dealing with the escapement with two wheels which he said that he had made in 1829 and which he had used in what were called carriage clocks ever since. This letter was published in the *Tribune* together with two drawings of Garnier's escapement which he said were "... the faithful reproduction of the manner in which it is constructed in my workrooms." See Figs. A/1 and A/2. Garnier states that it is not correct to say that his escapement is derived from those of Enderlin and Sully. According to him, the only likeness that exists between his and theirs is the similarity of the relative positions of the 'scape wheels in relation to the axes of the balances. He says that otherwise the actions and the principles of construction are entirely different. He gives a fairly long explanation, which is not easy to follow. Garnier claims that the other escapements wear out rapidly, and that the interaction of the escapement parts also produce bad timekeeping. He feels that none of these faults is present in his own escapement, which on the contrary takes a brisk action due to the "lift" being on the wheel teeth rather than upon the "disc" of the balance staff. He compares the action of his escapement with that of a cylinder, and says that this escapement suggested

---

(2) The significance of this remark is perhaps underlined by Vol. III of the *Revue*, p.60, Oct. 1859, where it is stated that M. Bourlon presented in a watch "one of those many variants of the escapement known as the Paul Garnier escapement ...". The evident proliferation of Garnier-like escapements, following his great success of 1830, is clear from the clocks which will be found having them. One example is a carriage clock signed "BOLVILLER A PARIS" and numbered "26". It does not bear the words "P.G. Breveté", and thus it is possible that the design for the escapement was "pirated".

Appendix (a).

Fig. A1.

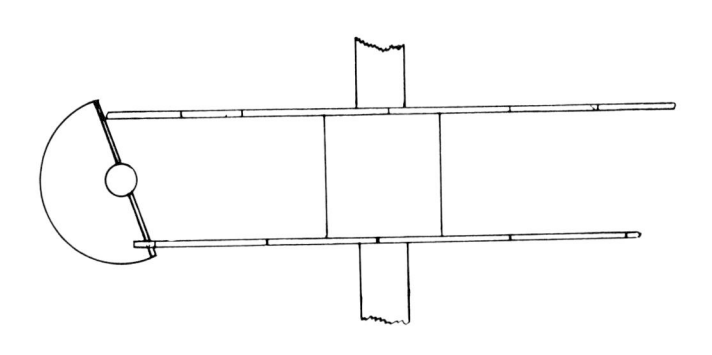

Fig. A2

to him the principles upon which his own is based. Garnier even goes so far as to say that the only difference between a cylinder escapement and his new escapement is that in his when the wheels are "locked" the thrust is downwards on to the bottom balance pivot, while with a cylinder in the same circumstances both the pivots are pressed against the walls of the holes. The letter ends by saying that the writer is obliged to the editor of the *Tribune* for the trouble that he has taken to study the going of one of the very many clocks to which he had applied his escapement, but that he assures him in all humility that his intention in making it was only ever to produce something appropriate to the development of portable clocks, and to be accurate enough for ordinary use rather than suitable for precision pieces.

1853. Moinet on Garnier, taken from *Nouveau Traité Général d'Horlogerie,* second edition, Paris 1853, Vol. II, pages 370-371, footnote 1.

A full page of good biographical notes are introduced as a footnote which translates to read "Apropos escapements normally used in horology, and in order to complete the list, we believe it to be essential to include the description of some of those invented or improved by M. Paul Garnier, which the writer omitted to mention in the first edition of this treatise". The footnote continues, "Nous citerons d'abord l'échappement à repos, vertical, inventé par M. Paul Garnier en 1830, dont les dispositions simples et les moyens d'execution prompts et faciles lui ont permis, en l'appliquant aux pendules de voyage de généraliser à cette epoque, ce genre d'Horlogerie, qui a pris aujourd'hui un si grand dévelopment".

1855. Redier on Garnier, taken from *Revue Chronométrique,* December 1855, Vol. 1, page 79.

"Nous nous attendions à voir décorer M. Paul Garnier, et, si son exposition ne présentait aucune de ces choses qui frappent au premier coup d'oeil, il n'en est pas moins vrai que M. Garnier est le créateur de l'industrie des pendules de voyage à Paris".

1870. E. Flachet on Paul Garnier. Biographical notes based upon his Obituary translated from *La Revue Chronométrique,* February 1870, pages 39 to 50, Vol. VII, reprinted from *Mémoires de La Société des Ingénieurs Civils'*

"Paul Garnier was more than a clever and ingenious artist. Not only was he unusually gifted with his hands but he had the kind of lively and inventive mind that enabled him to appreciate instantly the most varied problems and questions. He has left behind, in addition to some finely executed work, a long series of achievements that have given him a solidly-based reputation." Flachet relates that Garnier was born at Epinal (Vosges) in November 1801 of a musical family who both played and made pipe organs. His father died young, so that Garnier was obliged to start work early. He began with an Epinal printing firm; then tried an apprenticeship with a locksmith, whom in turn he left to start work with a clockmaker.

Appendix (a).

According to Flachet, Garnier had already become a very good workman when he happened to speak to a master-horologist from Luxeuil who enjoyed a very considerable local reputation. Garnier managed to persuade this man to employ him, staying with him until the time came when his master told him that the horizons of Luxeuil were too circumscribed for his future tuition and made him promise to take himself to Paris. In 1820 Garnier made his first appearance in the capital working for the famous firm of Lépine, then at the height of its fame. He immediately became noticed for the excellence of his astronomical clocks. After five years of working in this *atelier* Garnier felt that he was ready to go into business alone. He managed to set up on his own account with the aid of the modest savings which he had been able to set aside from previous earnings. Flachet continues in his article (translated) . . . "In 1826 Garnier submitted to the Académie des Sciences a detached remontoir constant-force escapement which showed seconds from a half-seconds pendulum. This escapement was based upon a new principle. In it the pendulum was detached from the variable driving force, its motion being maintained by a small weight which restored to it at each vibration the impulse necessary to keep its amplitude constant, all this being achieved without any friction. This piece, examined carefully by Messieurs Arago, Molard and Mattieu, was the subject of most complimentary comment." In 1827 Garnier showed at the Paris Exhibition an astronomical regulator which, according to Flachet, made ". . . une grande sensation". This clock is said to have been made throughout by Garnier himself, and the report says that the pinions, although they were cut by hand, showed not only great regularity of tooth form, but also very beautiful division. Garnier received a Silver Medal in recognition of this piece. The next part of the Obituary worth translation in the present context reads "In 1830 the invention of a new escapement, mainly applicable to portable pieces, gave him (Garnier) the means in conjunction with his straightforward designs and quick and easy methods of manufacture of producing simply a new species of clockwork, namely the clocks described as *de voyage ou de voiture*. Up until now these have only been available to rich people because of their high price. This branch of horology was so successful that it increased rapidly. Today it has become such a large scale business that several million carriage clocks have been made in France". In the 1834 Exhibition Garnier presented a range of carriage clocks of quite new types of which several, according to the reporter M. Le Baron Séguier,[3] ". . . offered some remarkable innovations, including one showing on various dials the day, the date of the month, the phases of the moon, etc., etc." In 1839 Garnier showed a further collection of carriage clocks of which he had expanded his manufacture. He also exhibited marine chronometers to which he had applied some sort of special escapement said to have been adapted from the principles which he had conceived in 1826.

The Obituary contains much more information which, although very interesting, is outside the scope of this book. The versatility of Paul Garnier was quite remarkable. The rationalisation of the *pendule de voyage* was but one of his significant industrial contributions. Furthermore, his work in connection with electrical horology during the infancy of that art was very significant and has yet to be fully recognised. The writer concludes "Paul Garnier's life has been, and one can see this, a battle of talent and work against the mass of obstacles confronting the artisan who owes to himself alone the learning without which his intelligence and his courage would be in vain or powerless. This battle began at an early age and only finished with his life. Paul Garnier taught himself while working; he founded an excellent establishment; he gained the regard and nearly always the affection of those for whom and with whom he worked. As a man who had battled against ignorance he brought up a large and distinguished family, dedicating the major part of his income to the education and instruction of his children. Thereby this life had been complete. It has left, with the most honourable memories and unanimous regrets, a great example to be followed."

---

(3) Séguier's Obituary is in *Revue*, July 1876, P.89. He was Vice-President of the Société d'Encouragement. In addition he was an amateur horologist and presented various escapement models to the Conservatoire des Arts et Métiers.

Appendix (b)

Paris Exposition Universelle, 1889

(i) **Clockmaking** by T.D. Wright.
(Partial quotation from Artisans' Report prepared by T.D. Wright who was sent to the Paris Exhibition of 1889 by the Mansion House Committee).

"A comparison between France and England in the manufacture of the ordinary chimney-piece clock, and of the carriage clock, is not possible, because none are made in England;[1] but the more I saw of the system of manufacture in Paris, the more amazed I grew that this too-evident fact should exist. Many of the details of the manufacturing are so much a counterpart of the method of making watches in Clerkenwell, that I have not yet recovered from my surprise, and never shall until some enterprising firm, with money enough, commences a healthy rivalry with our friends on the other side. I had expected to find a number of large factories where the whole manufacture was so complete that the arrangements almost constituted a "corner" in clockmaking, difficult to compete against. I found the reality very different from my expectations. Just as the Clerkenwell watchmaker obtains his rough movement from Prescot, so the Parisian clockmaker obtains his "roulant", for the marble or gilt clock, from the movement-maker of the Jura, or of Saint-Nicolas-d'Aliermont, near Dieppe. The roulant, however, as its name implies, is more finished than the Prescot watch movement; for, although it is incomplete, the whole train is pivoted and run in, and if it is a striking clock, the striking work is all made and planted. The manufacturer, therefore, has not so much of the work to execute as the Clerkenwell watchmaker has, but this only means that he is a little more advanced than *some* of us are, because the growing tendency in England for years past has been to get the work in a more complete state from the movement maker, especially in those branches which, admitting of little variation in method, give to the work no individual character, and lend themselves readily to systems of interchangeability. The cases, the dials, and the hands are obtained from the makers of these specialities — the majority of the last two being made in the French provinces, although a few are made in Paris. The gilt cases are usually made by the Parisian case-makers; nearly all the marble cases *are made in Belgium;* sometimes imported complete, but more often all the finished pieces are separate, and are put together by the clock-maker.

All the separate parts are now given out to the "finisher", who, like the watch finisher of Clerkenwell, works at his own home, and who finishes, polishes, adjusts, and finally completes the clock. Of course all the work in this kind of trade is piece-work, and as the men, like ours, usually work for several employers, and are probably assisted by members of their families, I found it difficult to arrive at a reliable estimate of their average earnings. It is variously stated to be from thirty to fifty francs per week.

The carriage clocks are usually made up by wholesale houses, who confine themselves to this kind of work. M. Margaine, of Rue Beranger, kindly conducted me over his show rooms and workshops. The system is the same as that pursued in the manufacture of the chimney-piece clocks. The roulants are obtained mostly from St. Nicolas D'Aliermont. In the Exhi-

---

(1) What the author meant was that no small mantel or carriage clocks were regularly made in England which in any way resembled French ones. In fact Saunier, writing some twenty years earlier of the Paris Exhibition of 1867, *(Revue,* Dec. 1867, page 138) goes out of his way to praise the beauty of English carriage clocks, pointing out that they were ". . . produced at great expense, and not the result of regular manufacturing".

Appendix (b)

bition, Japy Frères, the well-known makers from the Jura, have a most extensive display of complete clocks of all kinds, and of roulants and separate pieces, but on the whole the work is common and inferior to that exhibited by the St. Nicolas makers. The escapement, cylinder or lever, with platform complete, is obtained from Besançon, or the other watchmaking towns in the Doubs, or from Switzerland. M. Margaine keeps a few workmen on the premises, but most of the men work at their own homes. The specimens shown me here were very well finished and artistic in design.

M. Drocourt, of Rue Debelleyme, who has an extensive English connection, and who manufactures a very excellent class of work, has the most complete factory for this kind of work of any I visited. At his own factory in St. Nicolas, where he employs about twenty hands, the frames, wheels, pinions (which are cut forgings, not pinion wire), rough stampings, etc. are made, and those other processes carried on which suggest themselves as best conducted under the factory roof. He also employs there about sixty families, who work at their own homes — and who, of course, are at liberty to work for other houses — in completing the roulants. At the Paris house — where about the same number of hands are employed as in the factory at St. Nicolas — the cases are made and the clocks are finished.[2]

To the ordinary work Besançon escapements are fitted, but for the higher qualities the escapements are made on the premises. These are lever escapements, English style, with cut compensation balances and Breguet springs, and very good specimens of work they are. About two thousand clocks are turned out annually. The weekly earnings of the workmen in the Paris house vary from forty to sixty francs, according to their abilities and the branch they are engaged in; but in many of the houses the wages are not so good, neither is the work. In this trade, as in the manufacture of the "apartment" or "chimney-piece" clock, the work is always piece-work."

---

(2) The reader may have noticed that the famous firm of Jacot is not mentioned by Wright. Further comment on this omission appears in Chapter V.

Appendix (b)

Paris Exposition Universelle, 1889

(ii) **Watch and Clock Making in 1889** by J. Tripplin.
(Being part of his account and comparison of the exhibits in the horological section of the French International Exhibition).

"Beaucourt and its district was, as manufacturers of clock movements, represented at Paris by the firms of MM. Japy Bros. & Co., S. Marti & Co., Fritz Marti, Mégnin and A. Mougin.

The firm of Japy is an important and old-established one, having been founded as early as the year 1750;[3] its capital is over two millions sterling, and the multiplicity of the works it embraces is enormous; but what we have to consider at the present moment is its establishments of Beaucourt and Badevel (a locality close by), and what, with the help of two thousand hands and engines of 800 horse-power, it produces there in the way of clock movements alone.

From the year 1810 up to 1888 it had manufactured 6,198,639, all sent to Paris clock-finishers, who at the present time it supplies to the extent of 500,000 a year at the value of £120,000, with what is known under the name of "blanc", that is to say, the frame made of top and bottom plates, pillars and barrels, or of the "blanc roulant", which is the same plus the pivoted wheels of the train.

But time goes on and progress also, and with it keen competition, which came originally from America, and subsequently from Germany, and the question arose with patriotic and powerful houses how to resist this invasion. It became evident that Paris, with its parcelled system of manufacturing, a system which has made Clerkenwell and Besançon suffer so much of late, could not offer effective resistance on account of its want of organisation, and hence the resolution come to by MM. Japy and some other important manufacturers to make completely at their factories some cheap and popular form of timepieces. This determination is ominous, as it points to concentration, in other words to the factory automatic system, and also to the displacement of an industry.

The display of MM. Japy occupied one-sixth of the entire French section, namely, 132 square yards, and presented to the eyes of the observer hundreds of differently calibered movements put in a variety of cases of fanciful and tasteful designs, the usual chimney clock shapes, carriage timepieces, dials, long clocks, regulators, different electrical devices to serve for the transmission of time, and in the way of cheapness, small drum clocks of elegant appearance going 30 hours, provided with a lever escapement at a price we scarcely like to mention, but which allows the humblest and poorest to have a friend on their mantelshelf.

Japy Frères by their extensive organisation and wealth wield an enormous power; almost everything in the way of new ideas must be submitted to them for execution and has to pass through the hands of those colossal monopolists, who, receiving the ore direct from the mines, melt and refine it and convert it into bars and sheets of brass, iron, and steel, and subsequently through different automatic processes into the finished work.

The productions of both the Marti do not compare, as far as quantities are concerned, with that of the above mentioned firm, but what they exhibited showed great attention to rather superior work; so do the exhibits of MM. Mégnin and A. Mougin, although in a less degree.

St. Nicolas d'Aliermont near Dieppe, a village of

---

(3) This date is too early. See Chapter VI.

2,500 inhabitants, is according to local traditions the cradle of horology in France. It is from there that, driven by the revocation of the Edict of Nantes, in 1685, some Huguenot families of watchmakers went to the Swiss mountains of the Jura, and implanted there an industry which has since been a source of wealth to that republic.

The yearly production of St. Nicolas now amounts to £100,000 to £120,000, and consisted in alarums, carriage and chimney and turret clocks, ships' chronometers, a great quantity of materials for electric appliances, the entire production of timekeepers alone being between 250,000 and 300,000 per year. The population employed on horology alone amounts to about 1,500, the village is widely spread and besides what is done at the factories everybody works more or less in their own homes at some special part, besides attending to a certain extent to agricultural pursuits in summer.

The exhibit of M. A. Villon, expert to the jury, and Mayor of St. Nicolas, was an important one and represented in a worthy manner the industry of St. Nicolas in all its details and its adaptability to undertake general work. Indeed the appearance of St. Nicolas strangely reminds one of a Swiss village in the canton of Neuchâtel.

The factory of M. Albert Villon is a large, commodious, and elegant structure surrounded by gardens, one storey high and divided into many shops, each devoted to some speciality. A visitor is shown over the designing room, in which new tools, models of cases, fresh designs of dials, etc. are drawn; the machine shop where the tools are made; the rooms where the mountings are moulded, filed, the cornices chiselled by means of circular cutters, then stoned, polished, and gilt; rooms where the plates of the movement are stamped out, the pinions and wheels cut, the screws made, pivoting done; the glass cutting and polishing shop; the assembling, regulating, packing rooms; in fact, a complete factory, as M. Villon professes to be independent of anybody in his system of manufacture. He claims to make 60,000 alarums, 20,000 carriage clocks, and 20,000 chimney clock movements per year.

For the alarums that defective tic-tac escapement is still used; for the better work a cylinder or lever escapement bearer from Montbéliard is employed. We noticed in the way of novelty a relatively cheap clock with a chronometer escapement, and another with a keyless winding and set hands arrangement. In the way of good work we remarked a beautiful regulator whose striking was self-righting, also some remarkable chime clocks.

The cry being here, as it is everywhere else, how to beat foreign competition, the question was to devise some cheap timekeeper, and it was solved by the production of an alarum at a little more than half-a-crown.

The firm of Dessiaux comes next in importance, employing 250 hands. Their exhibit consisted of alarums, chimney time-pieces, and detached parts of the movement.

The show of M. Baveux was, according to competent judges the best of its kind; his movements of carriage clocks, striking repeaters, and other complicated work was unrivalled, and finds its way to the best Paris finishers; unfortunately the production of this maker is limited and we can only regret it.

M. Drocourt, whose name is not unknown in England, besides being a distinguished Paris maker, figures also among the St. Nicolas automatic clock manufacturers, where he constructs, with the help of 40 hands, part of the movements he finishes at Paris. His exhibit showed very creditable products.

M. Guignion exhibited an assortment of alarums and cheap carriage clocks, and, together with a few others showing specialised parts of the movements, completed the display of St. Nicolas clock manufacturers at the Paris Exhibition.

The two American competitors were the firm of A. Kahenn & Co., of St. Louis, who exhibited a cheap drum alarum of a distinct American type, and who claimed to employ 150 hands; and that of Roger & Co., who showed similar articles.

Paris is known all over the world as the great emporium of the clock trade. Nevertheless, the clocks actually made there are few in number, and comprise only the very best astronomical regulators made by hand and at a high price. But the various movements supplied to Paris makers by the manufacturers of St. Nicolas and Beaucourt districts are the cause of about 8,000 hands being employed in that city, directly or indirectly, in the clockmaking and finishing trade, which subsequently gives work to an army of brass smelters, rollers, bronzists, decorators, marble cutters and polishers, glaziers, enamellers, dial painters, engravers and gilders, besides the regular clockmakers engaged in the finishing of the movements. Otherwise

Appendix (b)

Paris, as far as clockmaking is concerned, would long before have ceased to exist were it not for taste ever changing, and buyers constantly demanding something new and pleasing. These incessant changes give the talented designer work, by allowing him scope and play for his imagination. His ideas have been formed in the schools of art which abound in Paris, in its rich museums, containing unequalled treasures of taste and beauty, in the contemplation of its palaces, splendid avenues, and magnificent statues; by inhaling the exhilarating atmosphere which seems peculiar to it, and by the elevated thoughts with which Paris in its grandeur and entirety must impress all thinking minds. All these influences tend to imbue an imaginative designer with refined tastes.

As an illustration of which has just been said let us look at the exhibition of M. Margaine, which is more that of a decorator, so elegant and tasteful are all his works, displaying the greatest refinement. As a matter of course all the movements are good, but the excellence is noticed in the decorations; here every taste is studied, rich, plain, highly decorated; all are pleasing and all remind you that it is only Paris that can create such things. Margaine in the way of clocks is on a par with Champion, the watch engraver.

M. Charles Requier, a member of the jury and one of the officials of the School of Paris, showed some simple and good chimney clocks, travelling clocks, and astronomical regulators, but everything in his exhibit is perfect. M. Requier, although buying the majority of his movements like every other maker, has the science to carefully select what he buys, and the talent to perfectly finish them. M. Requier is considered the first clockmaker of Paris.

The next we come to, that of M. Drocourt, whose name we have noticed when speaking of the products of St. Nicolas d'Aliermont, has been for a long time a Paris firm of high repute; his exhibit combined good taste with sound work..."

Elsewhere in **Watch & Clock Making in 1889** on page 119, Tripplin says under "Platform Escapement Makers":—

"The country round Montbéliard is still the centre for the manufacture of these adjuncts to travelling and other clocks, and the number produced in that district amounts now to not far from 20,000 a year. MM. Gelin, Levy, Dorian, Mégnin, make a great many, as do MM. Coulon & Molitor, whose escapements, as well as the levers of MM. Reuille and Anguenot, appear to us a little better made. M. Besançon-Pillods was the only maker of these escapements from Besançon who exhibited, and his work, like that of M. Yersin, a Swiss (this last especially), was well made. The prices showed much variety. All this work finds its way to Paris or St. Nicolas d'Aliermont. The exhibitors were ten in number."

Appendix (c)

The French Carriage Clock after 1900.
(A personal account by the late Maurice A. Pitcher.)

My own acquaintance with the French carriage clock began in 1915 when, as a boy of seventeen, I joined the family business of E. Pitcher & Co. My father, Ernest S. Pitcher, had entered the wholesale clock trade in 1880 as manager and sole agent for the Paris manufacturer, V. Blanpain. In 1886 he moved to premises at No. 3 Clerkenwell Rd., where the firm was to remain for exactly fifty years before moving to Holborn. Soon after 1900 my father succeeded in obtaining the agency for two important French carriage clock manufacturers: E. Maurice et Cie, who produced some of the best Paris carriage clocks, and Couaillet Frères of St. Nicolas d'Aliermont.

The years between 1860 and 1900 are generally regarded as the heyday of the top quality *(qualité soignée)* French carriage clock, but the production of fine examples certainly continued until the outbreak of the First World War. During the years from 1900 until 1914, carriage clocks must have formed an important part of my father's trade, and he supplied both retailers and other wholesalers. A photograph in my possession, taken about 1910, illustrates a show case filled entirely with carriage clocks. Unfortunately the quality of the photograph is too poor for reproduction, but there must be some 150 carriage clocks on display, of which no two models appear to be alike. A stock book surviving from these years indicates that my father was importing not only from Maurice and Couaillet, but also from other French manufacturers (Chevellier, Corpet, Duverdrey, Hour and Margaine).

The First World War dealt a severe blow to the manufacture of carriage clocks in France. My father had died in 1914, and one of my first tasks on returning from the war was to try to re-establish links with the French carriage clock manufacturers. In 1920 I visited Paris with my manager, Mr. C.E. Wood, who had joined the firm in 1890, and we called on the addresses of some fifty clock manufacturers. Many of these were no longer in existence, and of those who had survived, only a few were still interested in manufacturing carriage clocks.

The production of carriage clocks in Paris did not however entirely cease after 1920. For several years we continued to import carriage clocks from Marcel Corpet (Rue Amelot 84), who was the successor to V. Blanpain, and from Charles Hour (Rue Ste. Anastase 7), whose trademark was "CH.H." on the backplate. This firm had previously been Diette Fils et Hour, and in 1927 it became Hour, Lavigne et

Maurice A. Pitcher

Cie, who are still in existence at the same address. As late as 1935 we were also still importing very fine carriage clocks from the firm of Adolphe Ollier (Rue des Marais 20), probably the last specialist carriage clock manufacturer in Paris. He used no trademark, but his models included the *cariatides*, and clocks with malachite, lapis lazuli, and porcelain panels. Ollier made frequent use of a silvered, engine-turned dial with a plain space above the hour circle for the customer's name.

But the Paris trade as a whole dwindled rapidly after 1920 in the face of intense competition from manufacturers outside Paris. The most important of these manufacturers were Japy Frères (of Beaucourt) — established in 1770 and still in existence today — Duverdrey & Bloquel, and Couaillet Frères (both of St. Nicolas). Duverdrey & Bloquel, originally Duverdrey only, used as their trademark a lion on the backplate until 1939, and their present trademark of "Bayard" on the dial is widely known. But it was Couaillet Frères who produced the widest range of patterns, and who exported the largest quantities to this country. The Couaillet carriage clocks, in particular the "Obis" model (q.v.), were mass-produced, cheap and reliable, and they quickly ousted the Paris products.

Since so many of the French carriage clocks still in circulation in this country were manufactured by Couaillet Frères, it is worth going into a little more detail about this firm. Started in the late 19th century, the firm was built up before the First World War by M. Armand Couaillet. As a small boy, I remember him paying frequent visits to our house. We children were dressed up in our best clothes and presented to him, and I recall that he always looked as if he had come straight from the barber's. We had to speak French to him, since the only English words that he knew could not be used until he was about to leave. These words were "Good night"! By 1920 he had handed over the clock factory at St. Nicolas to his brothers in order to devote himself to his many other business interests: he was the inventor, for example, of a small electric car, which could be operated with one arm and was specially designed for use by invalids of the First World War. The factory continued to produce carriage clocks (as well as many other kinds of clock) until 1939. When my son visited St. Nicolas in the early 1950's, M. Armand Couaillet, who must then have been over eighty, was still alive and living in the village, where the factory continued to operate under his nephew's ownership.

About 1925, not only the Paris trade, but the carriage clock trade as a whole began to decline in importance. Partly the reason for this was that the simple features of the carriage clock seemed old-fashioned among the more daring, modern designs of the 1920's. To some extent, these new tastes were reflected in carriage clocks themselves — for example, in the exaggeratedly elongated numerals which one finds on the squat rectangular models of this period. But more significantly, the carriage clock was "killed" by the advent of the immensely popular, and very much cheaper, travelling alarm in a leather case (the "folder") of Swiss or German manufacture. By 1939 it was an embarrassment to hold large stocks of carriage clocks, and I shudder to recall now some of the fine clocks that were then almost "given away"! Our only regular customers for carriage clocks at this time were Messrs. Goldsmiths & Silversmiths Co. (112 Regent St., London, W.1.) whose catalogue for 1939 includes a page of carriage clocks, all of them manufactured by Couaillet Frères.

During the Second World War, and for some years afterwards, fine carriage clocks continued to be sought after as part of the general trade in second-hand French clocks; but it was not until about 1960 that the revival of interest in carriage clocks began to make itself really felt in this country. This interest has snowballed ever since, and it is intriguing to speculate on the reasons behind it.

One reason of a general nature is that sufficient time has now elapsed for carriage clocks to be regarded as "antiques". Although clocks manufactured between 1860 and 1900 do not strictly speaking always qualify for this description, they can certainly be regarded as antiques of the future, since it is most improbable that clocks of this quality will be manufactured again.

On the technical side, clockmakers have come to realise that carriage clocks, especially those produced by the best Paris makers — Jacot, Drocourt, Margaine, Maurice and so on — are on a very high level of craftsmanship. Within the relatively compact dimensions of the carriage clock, it is possible to find movements of great variety and complexity: Westminster chime (see Plates VIII/4 and 5), *grande sonnerie* and *petite sonnerie*, clocks giving the day and the date, one-minute, five-minute and quarter repeaters, etc. Moreover, providing the clockwork is basically intact, these movements can always be reconditioned, and — as in the case of the "grandfather" clock — given a surprisingly long new lease of life. The cases too, consisting of a large number of small component parts which are simply screwed together, can very often be reconditioned. Even the dirtiest and most tarnished case may be miraculously transformed and made to look almost as new.

As far as questions of taste and fashion are concerned, the basically simple, classic features of the carriage clock have again come into their own. Their compactness, and their lack of bright colour, mean that they can easily be moved about from room to room, and are likely to fit in with almost any style of decor. But the carriage clock is functional as well as decorative. It is a reliable timekeeper, and its bold white dial can be read without difficulty; it may be used as an alarm clock; and it may strike, and at the touch of a knob repeat the hours for you during the night.

To the collector, the carriage clock is not in itself a rare item. It was after all for a number of years the clock most commonly found on the mantelpiece of a middle-class home in England or Scotland (for the Scots have always shown a liking for carriage clocks). But collectors have searched enthusiastically for examples of the more unusual and complex kinds of movement mentioned above; and there has also been great interest in the works of individual manufacturers. The best Paris manufacturers invariably placed their trademarks on the backplate or inside the movement of their clocks, and these "signed pieces", each showing the distinctive style and features of a particular maker, have been much sought after. Finally, there are those carriage clocks whose decoration entitles them to be described as works of art in their own right. I have in mind carriage clocks which are elaborately engraved, or those which display fine enamel or porcelain work, often of the highest quality and the most delicate colouring. To possess a carriage clock with a complex movement, "signed" by a well-known maker, and enclosed in a case of beauty or elegance, is the dream of today's carriage clock collector.

Appendix (d)

The Japy Factory at Badevel

Twelve workbooks and many invoices, etc., relating to the Japy Frères factory at Badevel have survived. The documents all cover the month of January 1907, and four of them contain entries directly connected with carriage clocks.

The relevant workbooks have the titles: *Mouvts Blancs et Finissages Voyages, Voyages, Découpoirs,* and *Réveils Américains.* The other workbooks are less relevant, but they include *Comptoir* (apparently a running record of "Stock in hand"), *Finissage, Finissage Externe,* and *Atelier des Mouvements Blancs.*

The workbooks are all hand-written. They contain many abbreviations as well as horological terms probably no longer in use. Some usages may have been peculiar to the Badevel factory. Many of the entries are also difficult to decipher. However, certain facts emerge clearly, and no doubt much more could be learnt with further careful study.

Here are a few notes compiled in collaboration with Mr. A. Randall:—

1) The names of customers for carriage clocks are given, and in January 1907 they were Margaine (much the largest), Bolviller, Ollier and Lefèbre.

2) Very many operations were involved in the production of a carriage (or other) clock. Because of extreme division of labour each operation had to be noted separately. Usually no one workman carried out more than a very restricted number of operations.

3) A great deal of work was done outside the factory in the village. The work was delivered once or twice a month by wives or daughters. In this connection the name of the Péquignot family appears, including the wife of Edouard Péquignot who is mentioned in Chapter VI.

4) Here and there may be gleaned some small insight into production methods. For instance, the procedure for making train wheels involved first "gashing" the teeth. After this operation each wheel had to be mounted, some on collets and some directly on to their pinions.[1] Last of all the teeth were rounded up and the crossings of the wheels filed and burnished. Each operation was performed by a different person. Even the great wheels, which of course were cut on barrels, had their teeth first "gashed" and afterwards rounded up. Above all, each and every hole which had to be drilled merited an individual entry. It would seem that drilling was a difficult and demanding operation.

The work produced by means of *Découpoirs,* or in other words by presses, included parts for *pendules Paris* and also parts for carriage clocks. Plates, bridges, wheel blanks, bell standards, etc., etc., were produced (and sometimes also had centres marked ready for drilling holes) by means of punches and dies.[2] Afterwards parts were corrected, if necessary, by means of a hammer. Brass and steel sheets were flatted, sometimes by hammering and sometimes by machine.

In the rough movement workshop train planting was done, besides such operations as pinion trueing and click cleaning and polishing after hardening.

The external jobs were usually in connection with smaller parts, such as the finishing of the pallet frames for *pendule Paris,* the finishing of levers, work

---

(1) The soldering of collets on to pinion arbors was done by a woman, and so was the undercutting of the pinion heads ready to receive wheels.

(2) Mr. Jules Lhomme and his daughter made various wheel blanks, cutting-out, crossing and flatting them "with one blow". In January 1907 they booked the staggering total of 72,000 pieces. The removal of burrs after the blanking out of parts was a separate operation.

on hammer arbors, flys, etc., and the finishing, leathering and polishing of strike hammers. Screws were made "by the million" (some 43,000 in January 1907). The fitting of mainsprings and of stopwork was done in the *Atelier des Tours.*

5) It is significant that parts for very many different clock *calibres* were in production at the same time. Well to the fore were *pendules Paris,* with plates of both 80 and 90 mm in diameter and each size made in two types. Other clocks included *régulateurs,* musical and drum alarms, and carriage clocks of various descriptions ranging from *grandes sonneries* to *mignonnettes* to *Réveils Américains.*[3] Also in production in 1907 were speedometer parts and other such non-horological items as electric lamps. Some electrical components made at Badevel were for use in connection with clockwork.

6) From the book *Voyages* it is clear that Coulon *femme* in January 1907 turned 1,312 hammer arbors on a machine, having but 12 rejects! Various arbors were machined square by Julie Cattez. Elsewhere are recorded the cutting of "Maltese crosses" for stopwork, besides the making of stopwork fingers and the slotting of rack arms. On one page is recorded the cutting by Borruat *femme* of 1,499 crown wheels, probably contrate wheels. Another page relates to the drilling by *femme* Joseph Catu of numerous holes. Julie Catu is on record as undercutting pinions. Page 36 records 361 hand-set squares, filed by hand by Henri Bernard.

7) The book *Mouvts Blancs et Finissages Voyages* shows on page one "Frames planted and click work done" by Albert Genez for:

12 8-day Margaine *mignonnettes* ("Small model")
12 ¼ repeat and alarm (Margaine)
12 Alarm/repeat *No. 3* (Margaine)
50 8-day/alarm Size *No. 0.*
24 8-day 49 mm. (round movement?)
12 Simple repeat *No. 3* (Lefèbre)
12 Sinks done for Margaine *mignonnette.*

On page 3 Louis Péquignot *et son ouvrier* did "movement planting and clickwork made" for 259 carriage clocks of various types destined for Margaine. Page 4 relates to the fitting of 162 stopworks, while page 5 shows *repassage;* that is to say the finishing of *blancs-roulants* and making them work (probably between the plates only).

Pages 12–22 contain the names of workers (outworkers?) who presumably also were engaged in finishing carriage clocks, since the clocks are listed and it is not specified exactly what was done to them. In these pages are the names of Charles and Joseph Stouff, Joseph Rousse, Alfred Villers, Eugène Chavanne, Edmond Radiquez and Auguste Krauss.

---

(3) The significance of *Réveils Américains* will be made apparent by reference to Chapter VI and Chapter XII. Vast quantities seem to have been made.

Appendix (e)

(i) **Nail and Cork** by Professor D.S. Torrens

(An article published in the *Horological Journal* in February 1938).

In the report of Commander Gould's lecture, in the December issue of the Journal, page 14, it is stated that the early chronometer makers, "in the main, ... must have worked by trial and error, and by rule of thumb." Similar views were expressed by Mr. Mercer at the same meeting. I hope I may be forgiven if I presume to add some qualification to the statement.

I have several times heard it said that the watchmakers of old time carried out their work almost without special tools, using only turns and graver, file and polisher, "nail and cork." In the very early days of the horological art the workers were, no doubt, limited to such simple tools and methods. By the days of Tompion some advance had been made, and Hatton is not wholly correct when he credits Tompion with being able "to work a piece without engines, and other knack contrivances" (quite apart from the fact that many of Tompion's "pieces" were — even in his own day — worked by hands other than Tompion's).

In the period of Arnold and Earnshaw such a primitive state of things was very far from being the case. No other trade or profession, either then or since, had so many specialised tools and appliances; indeed, the prototypes of most of our modern machine tools are to be found among the appliances of the watchmaker of almost or quite two centuries ago. The first "screw-cutting lathe" in general use was the fusee engine, a simple tool which yet embodied all the essential principles of such lathes. It was used by watchmakers for probably more than a century before Henry Maudslay invented his slide lathe about 1800. The first "milling machine" was the wheel-cutting engine, in use in Tompion's day, and but little changed since. (Its counterpart, the pinion engine, was invented by Joshua Hewitt in Earnshaw's time, and was soon applied to the manufacture of chronometer pinions.) The forerunner of all our modern press tools was the old "stamp", which has been in use for well over two centuries in the production of watch wheels, balances, hands, fusee chains and other small parts. A method equivalent to drop or die forging was early used for the production of certain steel parts, e.g., barrel and fusee arbors for watches, and when, in Earnshaw's time (and probably under his influence), chronometer movements began to be made in Lancashire, the same method was employed in making the blanks for barrel and fusee arbors, the large pinions, detent spring and fusee lob, etc. These were in all cases forged between tools, producing a clean blank that required very little rough turning. The result was, of course, far superior to that of present-day methods, where the blank is turned off in a lathe from the usually over-annealed and decarburised bright drawn bar. (This is one of several reasons why the old chronometers wear on, while most of the new ones wear out.) Repetition work ("mass production") has been the rule and the common practice in the movement-making trades in Lancashire for well over 200 years. Frames were drilled off templates quite early in the 18th century, and frames of the same size and calliper were to that extent interchangeable. The first gauge to come into more than individual use was the Lancashire pinion wire gauge, and this gauge (now getting on for two centuries old) is still widely used, both in its original form and as the "American" twist drill gauge. The latter is the direct descendant of the pinion wire gauge, and differs from its parent by .002 in. or so in most of the sizes. The first catalogue of tools ever published by a manufacturer was, I believe, John Wyke's pattern book of watch and clockmakers' tools (published about 1770, not 1810, as stated by Britten in his "Old Clocks and Watches and their Makers"). With more than sixty plates and some five hundred figures, it indicates how well provided with tools the watchmakers of those days might be.

When chronometer movements began to be made in Lancashire, some time near 1800, the methods that had been in vogue already for watch movement making were extended to suit the larger work. The marine chronometer movement of Earnshaw's time was merely an enlarged copy of the full-plate watch movement, with proportionately greater pillar height to allow for the required extra turns to the fusee. The lay-out and general arrangement of the train remained the same. Except for size and for a few details, there is little difference between the marine chronometer movements of 130 years ago and of to-day. There have been some individual variations between the movements of different makers, such as the reversed arrangement of the fusee, introduced, I think, by Mudge, and now retained by Kullberg alone among English makers.

Whilst the greater part of the work on the early chronometer movements was thus performed with the aid of appropriate tools and machines, there still remained a good deal of hand work on the details of the movement, quite apart from the work of finishing, 'scaping and springing. Furthermore, the small tools (e.g., punches, dies and cutters) were made, as a rule, by the workman who used them, so that a high degree of manual skill was cultivated even in the most restricted branches of the craft. Not so to-day. Space precludes even bare mention of many special tools and devices which have been in use for a century or more in the production of watch and chronometer parts. It is true that the compound slide-rest was not used by the early makers, but they had little need of it, for the lathe was not then used for roughing-out work to anything like the extent that it usually is to-day. Also the turning of plates, cocks and bars, for example, was chiefly face turning or "surfacing," and the frame-maker's lathe had a gallows or swing-rest which performed this function as perfectly as a slide-rest would do, and cost a great deal less to make. Furthermore, a "slide-rest" in fact, if not in name, was an essential part of every wheel-cutting engine, pinion engine, fusee engine, shaping tool, slotting tool and backing-out tool. I have referred chiefly to tools used in the movement-making industry in Lancashire. The same tools are in use to-day, and movements are still made in the old tradition and to the old standards of quality, with tools which have been in daily service for many generations.

In the other branches of chronometer making special tools and appliances also had their early place. Individual ingenuity and enterprise brought forth instruments equal to the demand for accuracy and precision of workmanship. Earnshaw himself tells us how, about the years 1782-3, he "struck out a plan for making the compensation balance with three arms, turning, dividing and cutting them by an engine, in order to have them perfect; by which they were balances indeed in the full sense of the word, equal in every part." On the other hand, he charges Arnold with taking "the worst method of making balances that can be, soldering, filing, bending and hammering, bringing them to their wretched state by mere rule of thumb, without any truth about them."

In the remaining departments of chronometer work — finishing, jewelling, making the escapement, and springing — each workman had his own special tools, often made by himself and usually quite unintelligible to the uninitiated. As examples may be mentioned the many varieties of swing-tool used in polishing such parts as centre-pinion faces, 'scape-wheel teeth, certain portions of the detent, and the like. The jeweller had his mills, diamond drills and cutters, openers and polishers of different kinds, adapted to the working of the different stones he used; the springer had his blocks for hardening and tempering the balance springs, and so on. These tools were, in general, very simple, but none the less important as showing the great amount of thought and effort devoted to the task of producing good workmanship.

At the close of the 18th century and beginning of the 19th the escapement was often made throughout by the same workman, who sometimes also made the balance and sprung and timed the chronometer. Soon the work became further subdivided, and balance-making, pivoting, planting and springing each became the province of a different workman. While there is little doubt that Earnshaw himself was skilled in all of these specialities, this was not the general rule, and even in his time there was some division of labour in chronometer *manufacturing,* as distinct from the movement making.

The chronometer of those days, and until not

many years ago, was a production completely and characteristically English. The mainspring, fusee chain, balance and spring, hands and all other parts were made entirely in the country. The sole material imported was the rough stone for jewel holes, end-stones and pallets. The materials used by the makers of the best movements were selected with the most critical care, and the treatment of them was above reproach. The workmanship in all branches was always excellent, and often superb. Not only so, but all the tools used in every stage of the manufacture were English, too. Even after the decease of watchmaking in England the chronometer still remained the glory of horological art. That glory fast grows dim. Some branches of the manufacture have died out altogether, in others, only one or two workmen are left to recall to us what a splendid school of craftsmen here grew up and flourished for a time. Factory methods have not yet succeeded in producing a complete chronometer approaching in quality or finish the best "hand-made" chronometers, and it is doubtful if they ever will. There are many reasons for this; I shall mention only three. The market for chronometers is far too small to justify the whole-hearted application of modern production methods to chronometer making. The types of brass and steel best adapted for working in modern machine tools are eminently unsuitable for chronometer parts. Lastly, most of the tools and some of the materials and parts must of necessity be imported from abroad.

We move so fast in these days, and do things so badly, that it may be worth while sometimes to consider whether we can still learn anything from the men of times past, who went slowly, and did their work so well. How many of us to-day, with their tools—or with our own—and without their handicaps, can produce work fit to be compared with the best, or even with the average, of theirs?

(Reproduced by courtesy of N.A.G. Press Ltd., and of Messrs. James and Thomas Torrens.)

Appendix (e)

(ii) **Rule of Thumb** by Professor D.S. Torrens
(A letter published in the *Horological Journal*, 1938)

Dear Sir, — I was much interested to read Alderman A.H. Gledhill's lecture, as reported in recent issues of the Journal, and his amusing anecdote of the old Prescot wheel-cutter's "rule of thumb" reminds me that at any rate the epicycloidal tooth form was not unknown in Prescot some 120 years ago.

William Hardy (of chronometer balance, ruby drawplate and other fame) made some time about 1810 a sidereal clock for Greenwich, to Maskelyne's order and "unlimited in price." In this clock "the teeth of the wheels and pinions were epicycloidal, and were all finished in the engine by cutters which I made for that purpose." In his description of the clock (published 1820) he further states: "I have enabled Mr. Thomas Leyland, clock-maker, of Prescot, in Lancashire, to cut wheels and pinions with teeth made truly epicycloidal, from patterns generated by myself."

The excellence of Leyland's work is well known even at this distant day, and his style was freely copied by his contemporaries and by successive generations of wheel-cutters in Prescot and elsewhere. It is true that many or most of these old workmen knew nothing of the theory behind their work — for
"—Knowledge to their eyes her ample page,
Rich with the spoils of time, did ne'er unroll."
—yet they did produce tooth forms often closely approaching the theoretical. Anyone who cares to examine, by the method of optical projection, clock teeth of Leyland's, or watch and chronometer teeth by Preston or the Dokes, will find that almost invariably the profiles approximate to correct geometrical curves much more closely than do the profiles of modern form-milled teeth of similar fine pitches produced with commercial cutters of French or Swiss origin.

The acting surfaces of the Lancashire teeth are, furthermore, very smooth and free from cutter marks, often almost polished, and one has heard of and seen such teeth coming straight off the cutter "with a surface like glass." This is due to the manner of making the cutters and to the care taken in "getting them up". The modern fine pitch wheel tooth, whether generated or form-milled, always has striated surfaces, often quite rough. These surfaces readily "charge" with any abrasive dust that may be abroad (just as does a newly-filed polisher) and the direction of the cutter marks favours such charging. The wheel teeth may then act as grinders constantly attacking the pinion leaves, and so pinion wear takes place comparatively quickly.

One has often found in examining old Lancashire movements, whether watch, clock or chronometer, that the pinions after fifty or even a hundred years of use show very little trace of wear. This is, no doubt, due in part to the good form and pitching of the teeth, but I think another important factor is the smooth, work-hardened surface of the brass wheel teeth, free from cutter marks and less liable to become charged with abrasive dust. In high-class gears of coarser pitch the profiles are corrected, cutter marks removed and the acting surfaces smoothed or polished by special grinding or lapping processes. These processes have not been generally adapted to the production of such fine pitch gears as are used in watches and clocks, nor are they very suitable for brass wheel teeth owing to the risk of particles of the abrasive becoming embedded in the brass.

Hardy's clock still exists, though in much altered form. There is at present in the museum of the Institute a tool for making inserted-tooth cutters after the manner devised by Hardy; it is, however, of later date.

Yours faithfully, DAVID S. TORRENS.

(Reproduced by courtesy of N.A.G. Press Ltd., and of Messrs. James and Thomas Torrens.)

Appendix (f)

La Société Des Horlogers

The Société des Horlogers was founded in Paris on 16th December 1856. The Articles of Association are set out in the first volume of the *Revue Chronométrique* which became the official organ of the Society. The first Presidents were:— Henry Lepaute, Henri Robert, Paul Garnier, Wagner (nephew) and Winnerl. The Secretary was Claudius Saunier and the Treasurer Achille Brocot. Other French founder-members of particular interest in the context of carriage clocks were:— Baron Séguier, Breguet (the son), Edward Brown (of Maison Breguet), Charpentier (successor to Charles Oudin), Couet père (Breguet), Couet (Charles, pupil of Breguet), A. Croutte (St. Aubin-le-Cauf), Desfontaines (Maison Leroy & Fils), Gontard, Japy (Louis), Leroy (Théodore of Maison Breguet), Emile Martin (Saint-Nicolas), Lous Raby and Victor Reclus. The Swiss founder-members were:— Henry Grandjean, Golay-Leresche, A. Houriet and William Dubois, while the only original English member of importance was James Ferguson Cole. Denmark was represented by Urbain Jürgensen and Holland by C. Van Arcken.

Glossary

This glossary, which includes a few French terms, is intended to supplement the text in instances where further explanation seems worth while. Certain definitions, for instance those relating to mainsprings, trains, etc., have been deliberately aligned with their applications to carriage clocks.

AIRY'S BAR. A device invented by Sir George Airy (Astronomer Royal) in 1871 in connection with compensation balances for chronometers.

ALL-OR-NOTHING PIECE. A device invented by Julien Le Roy for use in repeating watches. The purpose was to ensure that a watch would not repeat at all unless the pendant was pushed fully home. Watches having no "all-or-nothing" pieces strike an incorrect number of blows if the pendant is only partially depressed. "All-or-nothing" pieces are found in those types of carriage clocks which are basically timepieces but in which a repeating train is wound and set off at will by the depression and release of a long plunger.

ANTI-FRICTION WHEELS (or Friction Wheels or Friction Rollers). Very occasionally in horology, in an attempt to reduce friction, the train and/or escapement pivots of a piece have been "run" between two or three rollers (themselves supported on pivots) instead of in plain or jewelled holes in the normal way. The best examples of friction wheels are to be seen in the Harrison large Timekeepers at Greenwich. A *pendule portative* by F. Berthoud shown in Plates I/23 and I/24 makes use of friction rollers.

APPARENT SOLAR TIME (True Solar Time or Sundial Time). Time based on the observed intervals (which vary slightly through the year) between two successive transits of the sun's centre at a given place. (Because of the tilt of the earth's axis and because the earth's distance from the sun varies during the year, the length of the solar day varies according to the time of year). See MEAN SOLAR TIME.

APPLIQUE. Applied, pinned-on or overlaid.

ARBOR. A term derived from the Latin *arbor* meaning "tree" and used in horology to denote a shaft or spindle. It is customary when speaking of arbors which embody pinions to refer to them as "pinions".
See further information under TRAIN.

ART NOUVEAU. An art-style first shown to the world at the Paris Universal Exhibition of 1900.

ASSORTIMENT. A complete set of escapement parts which go together to make up one unit. For instance, the *assortiment* of a lever *porte-échappement* would include 'scape wheel, pallet-frame and stones, rollers and impulse pin.
See PORTE-ECHAPPEMENT.

ATELIER. Workroom, especially in horology or allied trades. *Chef-d'atelier* foreman of such.

AUXILIARY COMPENSATION (Secondary Compensation). An extra device attached to a balance to afford compensation for middle temperature error.

BACKSLOPE. On an arbor the part which slopes from the shoulder to the full diameter. On a balance staff the tapered part below the balance seating, and also the conical parts between pivot cones and shoulders. The purpose of the latter backslopes is to prevent the spread of oil away from the pivots.

BALANCE. A balance, together with a balance spring, is used to control the going of a portable clock or watch by performing swings, vibrations or oscillations at a predetermined frequency.
See BALANCIER

BALANCE COCK. See COCK.

BALANCE SPRING (Hairspring). The spring used in conjunction with a balance to control the rate of going of a portable clock or watch. One end of the spring is attached to part of the balance staff, the other to a fixed part of the frame or platform.

BALANCE SPRING STUD. The outer or upper detachable "anchorage" of a balance spring.

BALANCE STAFF. The arbor carrying balance, balance spring, and roller or rollers. In order to reduce friction a balance staff has fine and delicate pivots. See COLLET.

BALANCIER. This word is used in French to denote any device which, by oscillating, controls the going of a clock, watch or chronometer. In other words *balancier* may mean, according to the context in which it is used, either a pendulum or a balance. Sometimes the word is qualified. For example, *balancier rectiligne* means the pendulum of a clock as opposed to *balancier circulaire* which is used to mean the balance of a chronometer, watch or portable clock. *Echappement à balancier* does not mean an escapement used in conjunction with a balance. On the contrary it means an escapement controlled by a pendulum (e.g. Sire, 1870, p.69). On the other hand *pendules compensateurs* are "compensated balances" and not pendulums. See also BALANCE.

BANKING. Banking, when referring to lever or chronometer escapements, is the stop device against which either the lever or the detent is held while the escape wheel is at rest.

BARETTE. A small movement bar of a clock or watch, in which a pivot or pivots is/are run.

BARILLET INDEPENDANT. A mainspring barrel which may be removed without dismantling a movement.

BARREL. See MAINSPRING BARREL.

BARREL, WATERBURY "LONG WIND" SYSTEMS. In a Long Wind watch the carriage is attached to and rotated by the centre of the mainspring. In one Waterbury carriage clock based upon a Long Wind watch the great wheel is carried on the barrel arbor and once again is driven from the centre of the mainspring.

BARREL WITH TWO GREAT WHEELS. Some carriage clocks made in the French Jura, and also some made in Austria, employ a system by which the opposite ends of a single mainspring are used to drive both a going and a striking train. The method of operation is that one great wheel is a normal going barrel-type great wheel made integral with the body of the barrel itself, whilst the other great wheel runs on the barrel arbor and carries the clickwork. The ratchet wheel is made integral with the barrel arbor. In some modern travelling alarms a single mainspring drives both going and alarm trains.

BASCULE ESCAPEMENT. Another name for a pivoted-detent escapement.

BASTARD TRAIN. A train which has such numbers of teeth in its wheels and pinions as to require a balance or pendulum vibrating some unusual number of times an hour. Example:- carriage clock trains giving 16,320 vibrations per hour, or 15,600 vibrations per hour instead of (say) 18,000.

BEAT. The "tick" caused by the locking or coming to rest of a 'scape wheel tooth upon a detent or pallet after the wheel has escaped and impulse has been imparted. An escapement is said to be "out of beat" when, due to incorrect adjustment or to faulty setting-up, the "ticks" occur at the wrong points during the vibrations of the pendulum or balance.

BEZEL. The "ring" round a dial, usually but not necessarily holding a glass.

BI-METALLIC COMPENSATION. A device, usually part of a balance, utilising the difference in co-efficients of linear expansion of two metals (usually brass and steel) to minimise the effect of variations in temperature upon the rate of going of a clock or watch. See COMPENSATION. COMPENSATION BALANCE. COMPENSATION CURB. AUXILIARY COMPENSATION. MIDDLE TEMPERATURE ERROR.

BLACK POLISH. A polish upon steel so perfect that it appears from certain angles to be "black".

BLANC. A partially-completed clock movement, consisting of the two plates separated by their pillars and with mainspring barrel or barrels in place.

BLANC-ROULANT (or simply *ROULANT, tout court*). The frame of an unfinished clock, consisting of the two plates, separated by the pillars, with the barrels and their arbors in place, and with the train or trains pivoted and planted with their depths correct. Such a piece requires many hours of skilled work to see its completion, a series of processes known as "finishing".

BOITE. The case of a clock or watch.

BOW. The ring attached to the pendant of a watch by which it may be secured to a chain or fob.

BREAK-ARCH (Broken arch). In a clock dial the type topped by an arch of a radius less than half the width of the square or rectangular part.

BREVET. In France a *brevet* corresponds more or less to a patent in England. When the word "*Breveté*" is found on a clock it means that the mechanical parts or system are patented in France. Nowadays in France the initials "*S.G.D.G.*" ("*sans garantie du gouvernement*") are used in conjunction with the older word "*Breveté*". The qualification "*S.G.D.G.*" is somewhat paradoxical in its implications because the French Government, while granting a *brevet*, yet "...ne garantit ni la valeur ni la priorité de l'invention".
See MARQUE DEPOSE, MODELE DEPOSE.

CADRAN. Dial.

CADRATURE. That which goes under the *cadran* or dial. The term *cadrature* is used to embrace not only motion work but also certain parts of striking, calendar or alarm work external to the plates.

CAGE. The frame of a clock movement, i.e. two plates separated by pillars. The word *cage* is sometimes used to mean a clock case, though *boîte* or *cabinet* is more usual in this connection. *Cage* is also correctly used to refer to the carriage of a tourbillon.

CALIBRE. The size, or nowadays the shape, of a watch movement (or by extension of a clock movement). *Calibre* may also be used to refer to a particular movement, e.g. *calibre revolver* which was a repeating watch movement made at Le Brassus.

CANNON PINION. The hollow pinion carried friction tight on the centre pinion and driving the motion work. The cannon pinion also carries the minute hand.
See MOTION WORK.

CAPUCINE. A type of French portable clock, the first examples having pendulums and later examples having balances.

CARILLON. In clockwork a set of bells struck mechanically in order to produce a melody.

Glossary

CARRIAGE. In a tourbillon or karrusel the part which carries the escapement and balance and which rotates while the piece is going.

CARRIAGE TRADE. A term used by the tradesmen of Victorian England to refer to those of their customers able to maintain their own horses, carriages, coachmen, strappers, grooms, etc.

CENTRE-SECONDS. A seconds-hand arranged to be concentric with the hour and minute hands instead of at 6 o'clock or elsewhere on a dial (off-set seconds). Today centre-seconds hands are sometimes called "sweep centre-seconds hands", presumably because they "sweep" the whole dial. The term, however, is quite a new one, and in this sense it is not strictly correct when applied to old pieces.

CENTRE WHEEL AND PINION. The wheel and pinion of a going train planted in the centre of the movement. The centre pinion (centre arbor) usually turns in one hour, in which case its extended front pivot carries the cannon pinion and minute hand. In French carriage clocks the centre pinion is driven by an intermediate wheel, and in English carriage clocks by the great wheel on the fusee. Because the centre arbor turns in exactly one hour, train calculations are started from the centre wheel. The numbers of teeth in the wheels and pinions of a carriage clock train from the centre wheel to the escape wheel determine the frequency of oscillation of the balance. The duration of running between winds is fixed by the gearing between the mainspring and the centre pinion. While the design of most carriage clock going trains follows the conventional layout, alternative arrangements are to be found in some rarer clocks. For instance, where a centre-seconds hand is fitted the centre wheel and pinion, while still being required to turn once per hour, are planted excentrically in order to leave room for the centre-seconds arbor. The drive to the minute hand is then made indirectly through additional gearing.

CHAPTERS. The hour numerals on a dial.

CHEF-D'ATELIER. The head man in a workroom; a horological foreman, or manager.

CHINESE-DUPLEX ESCAPEMENT. A duplex escapement in which the long resting teeth have bifurcated tips. Devised by the Swiss circa 1825 for use in watches made in Fleurier for the Chinese market. The object was to enable a centre-seconds hand to "show seconds" from a balance vibrating quarter-seconds (14,400 train). See SINGLE IMPULSE ESCAPEMENT. See DUPLEX ESCAPEMENT.

CHRONOMETER. This term is correctly applied to a marine chronometer or to a high quality watch or clock fitted with a detent escapement and intended for accurate timekeeping. The term "chronometer" was synonymous with excellence in construction, adjustment and performance. This did not prevent the term from being applied to some inferior productions in order to take advantage of the title. Today Swiss watches with lever escapements are called "chronometers" when they have been tested at an official "Bureau d'Observation" (B.O.) and awarded a certificate..
See MARINE CHRONOMETER, See DETENT ESCAPEMENT

CLICKWORK. A horological ratchet and pawl arrangement used in connection with winding work to prevent reversed motion. Clickwork is often used on carriage clock alarm-setting arbors to prevent their being turned in the wrong direction.

CLOCKWATCH. A watch which strikes like a clock of its own accord as it goes. See WATCH CLOCK.

CLUB-TOOTHED LEVER ESCAPEMENT. A lever escapement employing a 'scape wheel having some of the "lift" inherent in the shape of the teeth.

COACH WATCH. A very large watch intended for use on journeys.

COCK. An overhanging bar or bracket screwed at one end to a clock or watch plate and carrying a "hole" in which runs a pivot. Cocks attached at both ends are usually termed "bars" or "bridges".

COLLET. In horology this word has several meanings. It is used to mean the brass ring or collar by which a wheel is mounted on an arbor. A collet is also the means by which the inner end of a balance spring is attached to a balance staff. The "washer" by which a minute hand is retained in place is also called a collet.

COMPENSATION. In horology any device designed to compensate, in so far as is possible, for the adverse effects of temperature variation on the performance of a clock, watch or chronometer. It has been calculated that a change in temperature of 1° will produce a change in rate of about 10 seconds per day in a plain un-compensated balance furnished with a carbon steel balance spring. A pendulum clock having an un-compensated seconds pendulum with iron rod will show a mean difference in rate of about 20 seconds a day between summer and winter temperatures. (See BI-METALLIC COMPENSATION).

COMPENSATION BALANCE. A balance made in such a way that its effective radius of gyration varies with temperature so as to offset the effect of temperature changes on the elasticity of a carbon steel balance spring. In the first quarter of the present century much research was carried out, pioneered by Dr. C.E. Guillaume, to produce a balance spring alloy of which the elastic properties would be unaffected by temperature changes. Modern watches and platform escapements fitted with balance springs made from one of these alloys have plain uncompensated balances.

COMPENSATION CURB. A bi-metallic device similar to the rim of a compensation balance, but acting directly upon the balance spring. A compensation curb is usually carried upon an index and is designed to move one of the index pins towards or away from the other, thus altering the effective length of the balance spring. In this way compensation is made for changes in length and in modulus of elasticity of the balance spring brought about by changes in temperature. (See COMPENSATION BALANCE).

COMPENSATION SCREWS (or Temperature Screws). Screws, usually of gold or platinum, screwed right home in tapped holes in the rim of a balance. There are more holes than screws in the balance rim and the position of the screws is varied in order to adjust the compensation of the balance. The invention is attributed to Robert Pennington, senior (1780-1824). See QUARTER SCREWS.

COMPLICATIONS. Usually any mechanism over and above that required for going and striking is termed a "complication".

COMTOISES *(horloges comtoises).* The generic term for clocks made in the Franche-Comté. Other names are *horloge de Morbier, horloge de Comté, horloge de Morez.*

CONE PIVOT. A cone-shaped pivot, as opposed to one having parallel sides or a slight taper. Cone pivots run in blind "holes" or "cups" of the same shape, and their use is associated mainly with the balance pivots of pin-pallet watches and alarm clocks.

CONSTANT-FORCE DEVICE. Another name for a remontoire (q.v.).

CONTRATE WHEEL. A wheel having teeth cut at right angles to the plane of the wheel in order to transmit motion through 90° to a pinion planted in the opposite plane. In both English and French eight-day carriage clocks the fourth wheel is a contrate wheel. (Contrate wheels are also found in English verge bracket clocks where the contrate wheel is the third wheel, and in verge watches where the contrate wheel is the fourth wheel).

COQUERET. An endpiece usually of steel and retaining the endstone and index in position on a balance cock.

COUP-PERDU ESCAPEMENT *(coup-perdu* means "lost blow" or "lost tick"). The generic name correctly applied to any of a number of pendulum-controlled escapements in which the wheel escapes only once per oscillation. The original reason for the use of *coup-perdu* escapements was to produce clocks capable of showing seconds from half-seconds pendulums (donnant la seconde fixe avec balancier de demi-seconde). The type of *coup-perdu* most often seen is that frictional-rest variety usually called the "deer's foot" and having a divided-pallet; but many *coup-perdu* variants were made at the time of the 1867 Paris Exhibition. Tripplin & Rigg's translation of Saunier's *Treatise on Modern Horology* describes and illustrates a number of examples, as does the *Revue Chronométrique* of 1867. The chronometer (detent) escapement as sometimes applied to a pendulum clock might correctly be described as "à coup perdu". See CHINESE DUPLEX ESCAPEMENT.

CROSSINGS. The arms of a wheel. Crossing-out is the act of cutting out and shaping the crossings of a wheel. Special files called crossing-files are available for this task. The reason for crossing out the wheels is to reduce their mechanical inertia so that the train can stop and start rapidly at each function of the escapement. Very often the apparent quality of a clock or watch is belied by the lack of care and finish imparted to the crossings.

CUT BALANCE. Another name for a compensation balance (q.v.). However, many cheaper carriage clocks fitted with lever platforms will be found to have bi-metallic balances which have never been cut. This non-cutting was an economy measure. The cutting, making true and adjusting of a compensation balance is a lengthy, difficult and expensive business.

CUTTER. A form-tool which is rapidly rotated for cutting wheels, pinions, racks, etc.

CYLINDER ESCAPEMENT. A frictional-rest escapement developed circa 1725 in England by George Graham (1673-1751) to supplant the verge for use in watches. The cylinder in turn was superseded in England by the duplex, the chronometer and later by the lever. On the Continent the cylinder was used extensively in work of modest quality until very recent times.

DEAD LOCKING. The locking or resting of an escapement is said to be "dead" when no recoil is imparted to the 'scape wheel.
See FRICTIONAL-REST ESCAPEMENT. See RECOIL ESCAPEMENT.

DECK WATCH. A precision watch designed for navigational purposes for use in conjunction with marine chronometers (leaving the latter in situ). In small ships deck watches replace chronometers.

DENT'S PATENT BALANCE. A chronometer balance devised by E.J. Dent in 1842 and affording continuous secondary compensation (Pat. Spec. 9302).

DEPOSE. See MODELE DEPOSE.

DEPTH. In horology the degree of engagement of a wheel and pinion working together is said to be their "depth" ("too deep", "too shallow", "correctly depthed"). The term "depth" is also used in connection with the correct planting of escapements. See PENETRATION, PLANTING.

DETACHED ESCAPEMENT. One in which the balance (or pendulum) is free of the 'scape wheel, detent, or lever except during unlocking and impulse.

DETENT. In horology any locking device. Detents are usually associated either with the chronometer escapement or with calendar and striking work. A "knock-out" detent, used in striking work, is an exception in that it is employed to knock out the rack hook as opposed to performing any locking.

DETENT ESCAPEMENT. A delicate, expensive and extremely accurate escapement used in marine chronometers, in chronometer watches and in a few very special clocks. The two main types of detent escapement are the spring-detent and the pivoted-detent. Some cheap and inferior detent escapements were made on the Continent in order to take advantage of the reputation of the "genuine article".

418

Glossary

DIAL PLATE. The plate to which the dial is fitted. The dial plate is usually secured to the front movement plate by means of four pillars.

DISCHARGING PALLET. In a chronometer escapement the discharging pallet is set in the small roller and performs the function of unlocking the detent at the correct moment.

DIVISION. The spacing of the teeth of a wheel or pinion, which spacing should be equal all round. Both wheels and pinions have whole numbers of teeth and therefore equal numbers of spaces. "Modern" wheels tend to have spaces and teeth of about the same width; but in older work the teeth are often narrower than the spaces. Pinions need to have their spaces made wider than their leaves to give plenty of clearance, especially in the presence of dirt or dust.

DOUBLE ROLLER. See IMPULSE ROLLER.

DOUBLE-WHEEL ESCAPEMENT. One in which two 'scape wheels instead of one are used. Usually one wheel is used for locking and the other for impulse.

DRAW. In a lever escapement "draw" is part of the safety action. The locking surfaces of the pallet jewels are set inclined so that when the escapement is locked the pressure of the locked tooth "draws" the pallet deeper into the 'scape wheel and thus holds the lever fork against one or other of the banking pins. In the event of an external shock causing the escapement to try to unlock at the wrong moment, the safety action prevents unlocking and the draw causes the fork to return to the appropriate banking pin. Text books dealing with the design and construction of the lever escapement include W.J. Gazeley's *Escapements*, and J.E. Haswell's *Horology*.

DRAW-BENCH. A device for drawing-out wire to any section through a die (otherwise called a draw-plate or former) of which the internal profile is imparted to the wire.

DRUM CLOCKS. A general term to mean clocks housed in round section cases.

DUPLEX ESCAPEMENT. A single-impulse frictional-rest escapement invented on the Continent reputedly by Dutertre (1715-1742) for use in watches. The duplex escapement, which cost almost as much as a chronometer escapement in a best English watch, was a far better timekeeper than a verge, a cylinder or an inferior single-roller English lever escapement of which very many were made. It was not as good as a chronometer or a best quality lever. The duplex was little used on the Continent but was employed very extensively in England during the 19th century for expensive watches as a slightly cheaper alternative to a chronometer escapement. In the U.S.A. the cheap Waterbury watches had a cleverly contrived form of duplex escapement. A special form of duplex escapement, known as a "Chinese-duplex" was introduced by the Swiss circa 1825 for use in watches made for export to the Orient. In a "Chinese-duplex" escapement the long resting teeth had bifurcated tips. As a result a watch which was provided with centre-seconds hand was able to show jump-seconds from a balance vibrating ¼-seconds (14,400 train).
See SINGLE-IMPULSE ESCAPEMENT.

DYNAMOMETER. An instrument for measuring power developed.

EBAUCHE. An incomplete or "rough" watch movement sometimes called a *blanc*. The latter term is also applied to clock movements.

ECHAPPEMENT NATUREL. A term used by A-L Breguet to describe one of his escapements invented circa 1789, and derived from that of Robin. It is a hybrid employing features of both lever and chronometer escapements. It was said to have "natural lifts". Presumably this esoteric attribution was intended to mean that equal impulse was imparted to the balance in either direction. Many inventors and escapement makers have sought to design escapements having such properties. The advantages, however, would appear to be illusory.

ENDSHAKE. The "end-to-end" movement or "shake" of an arbor, pinion, balance staff, etc. to and fro along the line of its axis. Endshake is kept to the minimum necessary in order to provide working freedom. It is essential for correct functioning.

ENDSTONES. A pair of undrilled jewels serving the purpose of limiting the movement of an arbor, pinion, balance staff, etc. along the line of its axis. The purpose is to reduce friction. The friction of pivot ends against endstones is less than that of the shoulders of arbors against plates or jewelled holes without endstones. Sometimes in clockwork hardened steel end-plates are used to limit endshakes. A second and very important function of an endstone is to retain, by capillary attraction and in conjunction with a pierced jewel, a large reserve of oil in a virtually sealed compartment to lubricate the points of friction of a pivot. For correct oil retention the minimum gap between the jewel hole and its endstone must be 3-5 hundredths of a millimetre.

ENGINE. A term dating from at least the 18th century and applied to any form of machine (particularly horological).

ENGINE-TURNING (Guillochage). A decorative effect produced by machine on the surfaces of metal or of other substances, such as hardwood and ivory, by means of a "rose engine". In this engine a cutting-tool is constrained to move in one direction and is held against the work by hand. The work, or article to be decorated, is either rotated or else it is moved in a straight line past the cutter and at the same time is vibrated in a regular but rapid sequence as determined by a specially-shaped cam called a "rosette". Various patterns may be produced, such as "barley", "basket", "fox-head" and others. Very great skill and experience are needed if the finest results are to be obtained, especially upon non-flat surfaces. Engine-turning was much favoured and also popularised by A-L Breguet, who used it for his gold and silver watch cases and for the dials not only of watches but of some of his finest carriage clocks. See for instance Plates II/1, II/2 and II/10.

Glossary

EQUATION OF TIME. The difference between apparent solar time and mean solar time. The two coincide four times a year.

ESCAPEMENT. The connecting link between the going train and the balance or pendulum. The escapement maintains the vibration of the balance or pendulum by transmitting impulse at the correct moment. Since the going train is controlled by the balance or pendulum it follows that the train and hands will advance at the correct rate.

ESCAPE WHEEL AND PINION. The final and fastest-turning pinion of a going train is called the escape pinion and has the escape wheel mounted on it. The 'scape pinion is customarily driven by the fourth wheel.

FACING. The act of smoothing and polishing the heads (ends) of the leaved parts of pinions.

FALSE PLATE. Another name for the dial plate.

FINISHING. This term had varying meanings at different periods and also according to nationality and context. In general, however, the "finisher" was the worker who took the "rough" movement of a clock, watch or chronometer and made it work.

FIRE-GILDING. A process whereby an amalgam of mercury and gold is applied to a base metal, the mercury thereafter being "driven-off" by heating. Fire-gilding is sometimes termed water gilding or mercurial gilding.

FLICK CLOCK. Otherwise termed "leaf-clock" or "Plato" clock.

FLIRT. A device designed to afford sudden movement to a mechanism. For instance, a jump calendar dial, a knock-out detent in certain types of striking work, and even the controlling device of an independent seconds train.

FLY. A fan-shaped governor used in lieu of an escapement in order to regulate the speed at which an "upwards-geared" train of wheels will run. The most common application of flys is in striking work.

FLY CUTTER. A single-toothed cutter, rotated at very high speed, applicable only to cutting brass wheels in small quantities.

FLY PRESS. A hand-operated press, the centre spindle of which carries a very coarse two-start square-section lead-screw. Used in horology for "blanking-out" shapes in sheet metal by means of dies or press-tools.

FOLIOT. A form of bar balance used in early clockwork in conjunction with a verge escapement.

FONCINE. Another name for a *Capucine*, presumably linking the manufacture of this type of clock with the French Jura, and especially with the villages of Foncine-le-haut and Foncine-le-bas.

FORCE MOTRICE. "Driving-force", i.e. water, steam, electricity, as applied to a mill or factory. Possibly applicable as a generic term to a mainspring or driving weight.

FORK. In a lever escapement that part which interacts with the impulse pin.

FOUR-GLASS. A generic term, apparently only recently introduced, and used to mean those metal-cased French mantel clocks correctly known as *regulateurs* having glass front, back and sides; and in English work those wooden-cased clocks having glasses on four sides and often also on top.

FOURTH WHEEL AND PINION. The next wheel and pinion in a going train after the third wheel. The third wheel drives the fourth pinion. Only clocks having watch-type escapements controlled by balances and springs require fourth wheels. (In a normal pendulum clock the third wheel drives the 'scape pinion) French and English carriage clocks usually have escapement platforms on top of the movements involving the use of contrate fourth wheels.

FRAMES. The movement plates and pillars of a clock, watch or chronometer. The plates are separated by the pillars. In watches the plate to which the pillars are attached is the "pillar plate" and that carrying the potance the "potance plate". In French carriage clocks the *front* plate is the pillar plate and the pillars point backwards. In English carriage clocks the *back* plate is the pillar plate. For this reason the terms "front plate" and "back plate" are used in this book to avoid confusion.

FREE-SPRUNG. A piece having a balance-controlled escapement is said to be free-sprung when it is provided with no regulator or index for making it go faster or slower. Free-sprung pieces are brought to time by means of timing screws on the balance itself.

FRICTIONAL-REST ESCAPEMENT. An escapement in which the 'scape wheel teeth come to rest, when locked between escapes, upon a part which is continually in motion. Good examples of frictional-rest escapements used in watches and also in carriage clocks are the cylinder and the duplex. In the cylinder escapement the 'scape wheel teeth, when locked, rest against the moving shell of the cylinder. In the duplex the long resting teeth are locked upon the moving ruby roller. In pendulum clocks good examples of frictional-rest escapements are the Graham dead-beat and the pin-wheel. See DEAD LOCKING. See RECOIL ESCAPEMENT. See DETACHED ESCAPEMENT.

FRICTION WHEELS, See ANTI-FRICTION WHEELS.

FULL-PLATE. A watch, clock or chronometer is said to be full plate when both balance cock and balance are on the outside of the back plate (as opposed to three-quarter plate, half plate or Lépine-type movements).

Glossary

FUSEE. A special-purpose tapered pulley, usually made of brass, roughly conical in shape and having a spiral groove running continuously from its greatest to its least diameter. The fusee is immovably fixed to a steel arbor upon which is also mounted a great wheel. Fusee and great wheel are connected by clickwork and the fusee is joined to a mainspring barrel by means of a line or chain. The purpose of the fusee is to equalise the progressively decreasing torque delivered by a mainspring between when it is "wound-up" and when it is "run-down". When the mainspring is nearly unwound (not counting the set-up necessary to keep the chain or line in place) then almost all of the line is wound round the mainspring barrel and the fusee end of the line pulls on almost the largest diameter. When the mainspring is fully wound, almost all the line is wrapped round the fusee and the pull is exerted on a very much smaller diameter. A stopwork prevents "over-winding". If the matching of the fusee shape to the mainspring torque is perfect, then the torque delivered by the great wheel will be constant. See notes in Chapter I concerning the invention of this brilliant device. It is also worth noting that the presence of a fusee in a train provides a first "gearing-down" of the torque produced by the mainspring. A mainspring barrel normally provides between 5½ to 7 useful turns of wind. These few turns are transformed by a fusee up to a normal maximum of 16 turns.

In a French carriage clock with going barrel required to go for 8 days between winds, an extra wheel and pinion are needed between the barrel and the centre pinion. In an English 8-day carriage clock with fusee the great wheel drives directly in to the centre pinion.

FUSEE BARREL (or Plain Barrel). A plain barrel used in conjunction with a fusee and connected to it by a line or chain.

FUSEE CHAIN. The chain connecting a mainspring barrel to a fusee.

GADROONING. A type of ornamental edging common in silverware and round clock bezels.

GATHERING PALLET. In rack striking a single-leafed pinion which "gathers" the rack teeth, one at a time, as each blow is struck.

GIMBALS. A contrivance used for keeping instruments such as compasses and chronometers horizontal at sea in spite of the motion of a ship.

GOING BARREL. A mainspring barrel is termed a going barrel when the barrel body incorporates a great wheel. The great wheel teeth are cut on a flange of a larger diameter than the barrel itself and at one end of the body of the barrel. (Breguet and others sometimes employed in watches great wheels set in the centre of going barrels).

GOING TRAIN. A train whose functions are concerned with timekeeping only.

GONG. A length of hardened wire of any section used instead of a bell in striking work, etc.

GRANDE SONNERIE. Striking in which both the hours and the quarters are struck at each quarter.

GREAT WHEEL. The first and slowest-moving wheel in any horological train. In the going train of an English eight-day fusee carriage clock the great wheel is carried on the fusee arbor and drives the centre pinion direct. In the case of a French eight-day carriage clock the great wheel is part of the going barrel and instead of driving the centre pinion direct does so through an intermediate wheel and pinion. For reason see MAINSPRING.

GREENWICH MEAN TIME. The mean solar time for the meridian of Greenwich. See MEAN SOLAR TIME.

GREY ("in the grey"). Parts left from filing, turning or after hardening without being finally smoothed, polished or gilded.

GROS VOLUME. See HORLOGERIE.

GUILLOCHAGE. The French term for engine-turning.

HALF-HOUR SONNERIE. Striking in which full *grande sonnerie* is struck in passing only at the hour and half hour and not at the quarters.

HANGING BARREL. A going barrel of a type in which the barrel arbor is supported at one end only, as in Lépine *calibre* watches.

HELICAL SPRING. A cylindrically-shaped balance spring.

HELIX LEVER ESCAPEMENT. An obscure escapement, usually associated with the MacDowall family, and in which the impulse is imparted by means of a helix on the balance staff. A-L Breguet tried a similar arrangement in his marine chronometer No. 2741, now in the British Museum.

HOLE. In horology the bearing or "hole" in which runs a pivot. ("Holes" may be simple holes in brass; or they may be "jewelled" to lessen friction).

HOOKING. The means of attachment of either end of a mainspring to a barrel arbor or barrel.

HORIZONTAL POSITIONS. In a watch "dial up" and "dial down" are said to be the horizontal positions (as opposed to the vertical positions). The effect of gravity upon the balance and spring system is zero in the horizontal positions. However the rate can be different between "dial up" and "dial down" due to variations in pivot friction and to other factors. For relevance in the context of carriage clocks see Chapter VIII, pages 216-219. See VERTICAL POSITIONS.

HORLOGERIE CIVILE. This term was used in the last century and referred to large public clocks.

HORLOGERIE DE GROS VOLUME. Since about 1900, *Gros Volume* has been applied to mean any horology larger than *de petit volume*, which see. In the nineteenth century the terms *petit* and *gros* were applied differently and also inconsistently. For example the *Revue* (Saunier) and G. Sire interpret the terms variously, the latter classing *pendules de voyage* (one would have thought logically) as *horlogerie de petit volume* on page 68. It seems that the terms must be interpreted in the light of date and context. In the last century all manufacturers in the French Jura (Morbier, Morez) were considered as *Horlogerie de Gros Volume*. To-day alarm clocks by Jaz, Bayard, etc. are included in *Horlogerie de Gros Volume*.

HORLOGERIE DE PETIT VOLUME. Since the beginning of the 20th century this term has been used to mean watches and sometimes small clocks of 18-20 lignes in size. *Gros Volume* is anything larger. See HORLOGERIE DE GROS VOLUME.

HOUR AND HALF-HOUR STRIKE. Striking in which each hour is struck at the hour in passing while one blow is struck at each half hour.

HOUR WHEEL. The wheel in a motion work carrying the hour hand.

HUNTING COG. When gearing was less perfect than it is now the old millwrights avoided whenever possible the use of any wheel with a number of teeth an exact multiple of the pinion with which it engaged. For example, instead of using a wheel of 100 teeth with a pinion of 10, they would use 101 and 10, the extra tooth of the wheel being the "hunting cog". As a result the same wheel tooth did not frequently come into contact with the same leaf of the pinion. Errors due to imperfect tooth form and to inaccurate division were thereby distributed instead of recurring at frequent intervals. In practice the gearing was found to run much more silently and energy was transmitted more evenly. Imagine for instance a 178 toothed wheel working with a pinion of 8.
178 = 2 x 89 (89 is a prime number) and $\frac{178}{8}$ = 22¼
Therefore the pinion must make 4 x 22¼ = 89 turns before a particular tooth meets again a particular pinion leaf. The wheel would make four turns meanwhile. Had the wheel been given 180 teeth the "coincidence" would have taken place after 2 x 22½ = 45 turns of the pinion (2 turns of the wheel). Had the wheel 176 teeth, the "coincidence" would occur after 22 turns of the pinion, or in other words after one turn of the wheel.
"Hunting cogs" found no great application in horology; but the chronometer maker Charles Young used centre pinions of 13 in conjunction with even-numbered great wheels.

IMPULSE FACE. Any part of an escapement which after unlocking has taken place receives the impulse from the escape wheel teeth.

IMPULSE PALLET. See IMPULSE ROLLER.

IMPULSE PIN. See IMPULSE ROLLER.

IMPULSE ROLLER. In chronometer, duplex and in certain hybrid escapements, it is the roller on the balance staff which carries the impulse pallet and which receives impulse direct from the 'scape wheel teeth. Usually the acting surface of an impulse roller is jewelled. Sometimes the "roller" is nothing more than a simple short arm or pallet as found in most duplex escapements. In a lever escapement the impulse roller is the roller in which is fitted the impulse pin (ruby pin). In the simplest form of lever escapement (single roller) one roller alone is employed for both impulse and safety actions. In a double roller escapement the two actions are quite separate, each taking place upon a roller of appropriate diameter. The first English maker to regularly employ a double roller lever escapement was Victor Kullberg. See DISCHARGING PALLET.

INDEX (often called the "regulator"). The device used in a balance-controlled clock or watch for making it go faster or slower by varying the effective length of the balance spring. See FREE-SPRUNG.

INDEX PINS. The two pins carried on an index (or upon any regulating or compensating device external to a balance) and which "embrace" the outer turn of the balance spring. Also called curb pins.

INDEX PLATE. A disc provided with concentric rings of holes or indentations, each with different counts, and used in conjunction with a wheel or pinion engine to divide wheels or pinions.

INTERMEDIATE WHEEL AND PINION. In an eight-day spring-driven clock having a going barrel, such as a French carriage clock, an intermediate wheel and pinion are normally necessary in the going train between the great wheel and the centre pinion in order to achieve a duration of going of eight days. The great wheel drives the intermediate pinion.

ISOCHRONISM. The vibrations of a pendulum or of a balance are said to be isochronous when they are performed in equal time regardless of their amplitude.

JEWELS. See STONES.

JUMPER. See STAR WHEEL.

KARRUSEL. A type of revolving escapement carriage invented in 1894 by Bahne Bonniksen for use in pocket watches. See Chapter VIII. See TOURBILLON.

KNOCK-OUT STRIKING. A type of striking work, common in French repeating carriage clocks, which allows the hour to be repeated right up to the next hour.

LANTERNE D'ECURIE. Another name for a *Capucine*.

LANTERN PINION. A pinion having pins or "trundles" held between "shrouds" to act as leaves; as opposed to a pinion cut from the solid.

LEAVES. The teeth of a pinion.

LEPINE CALIBRE. A movement introduced by J.A. Lépine in which "bars" replace the top plate of the old two-plates-and-tradition of the full-plate watch.

LEVER ESCAPEMENT. A detached escapement invented some time prior to 1769 by Thomas Mudge (1715-1794). The lever escapement in its developed forms is still that universally used today in watches. Because of its robustness and general utility, the lever escapement eventually supplanted all its rivals.
See STRAIGHT-LINE LEVER ESCAPEMENT, See RIGHT-ANGLED LEVER ESCAPEMENT, See CLUB-TOOTHED ESCAPEMENT, See POINTED-TOOTH ESCAPEMENT.

LIFT. In a lever escapement the amount of movement measured in degrees by which the 'scape wheel imparts impulse to the lever. The "lift" may be wholly on the 'scape wheel, wholly in the pallets or be divided between the two.

LOCKING-PLATE (or count-wheel) **STRIKING.** A type of striking work employing a notched wheel to control the number of blows struck.

LONGITUDE. The difference (expressed either in degrees, minutes and seconds of arc, or in hours, minutes and seconds of time) between a standard zero or prime meridian (Greenwich) and any other meridian East or West of that prime meridian. The difference in the local time between two places therefore depends only upon their difference in Longitude. "Longitude East, Greenwich Time least; Longitude West, Greenwich Time best" runs the Royal Navy's "jingle". In other words, a more easterly place will have a later time because the earth rotates from West to East, causing the heavenly bodies to rise in the East and set in the West. In 24 hours the earth rotates completely or turns through 360°. Fifteen degrees of longitude therefore correspond to one hour. A ship at sea is able to obtain the local mean time from celestial observations. If Greenwich Mean Time is also known, by means of a timekeeper (or now by a radio or whatever) the difference between the two times may be used with tables to calculate the longitude of position of the ship.

MAINSPRING. A ribbon-like steel volute spring which is "wound-up" to drive clocks, watches, etc. The spring is a means of storing energy which is then released gradually. To take a specific instance, in an eight-day clock the centre pinion which carries the minute hand is required to rotate 192 times at a single winding (i.e. 24 times for each day run). Most mainsprings are capable of delivering only 5½ to 7 useful turns of development per full winding, and therefore some initial "gearing-up" is necessary before the centre pinion. In conventional eight-day carriage clocks the required duration of going is achieved in one or other of two ways. The first method involves the use of a fusee. This carries a great wheel and drives directly into a centre pinion (as in most English carriage clocks). The second method makes use of a going-barrel. This consists of a barrel and great wheel all in one, and instead of driving directly into the centre pinion, it drives an intermediate pinion and wheel (as in most French carriage clocks). The fusee carriage clock is able to achieve the necessary initial "gearing-up" because, in addition to the upward ratio of the great wheel to the centre pinion, the ratio of the diameter of the mainspring barrel to the average diameter of all the grooves of the fusee is so arranged that the fusee will be rotated 16 times for some 6 turns of the mainspring barrel. If the great wheel has 96 teeth and the centre pinion 8, it follows that 6 turns of the mainspring will suffice to keep the clock going for eight days. The going-barrel carriage clock achieves the same duration as the fusee clock by interposing an extra (intermediate) pinion and wheel between the great wheel and the centre pinion.

MAINSPRING BARREL. A circular box or drum having a removable cover and containing a mainspring.
See FUSEE BARREL, GOING BARREL, STANDING BARREL, HANGING BARREL, WATERBURY "LONG WIND" SYSTEMS, BARREL WITH TWO GREAT WHEELS, BARILLET INDEPENDANT.

MAINTAINING POWER. A device which keeps a clock, watch or chronometer going while being wound. Maintaining power is found in certain weight-driven and fusee clocks, in many watches and in all conventional chronometers. One of the advantages associated with the use of going barrels is that maintaining power as such is not required.

MARINE CHRONOMETER. A machine made specifically to keep accurate time for navigational purposes and normally having a detent escapement. Marine chronometers are usually suspended in gimbals.
See CHRONOMETER, See DETENT ESCAPEMENT.

MARQUE DEPOSE. A trademark registered formally.
See BREVET.

MEAN SOLAR TIME. A time compounded for convenience from the rotation of the earth with reference to an imaginary mean sun and therefore giving days of constant length. In England mean solar time for the meridian of Greenwich is called Greenwich Mean Time and is used as standard time. It is the mean solar time corresponding to the meridian of Greenwich shown by ordinary clocks, and it is based on an imaginary mean (or average) sun travelling at a uniform speed equal to the mean speed of the real sun through the year. See also APPARENT SOLAR TIME and EQUATION OF TIME.

MIDDLE TEMPERATURE ERROR. Middle temperature error is the defect inherent in a normal brass and steel compensation balance (and even to some extent in the vastly superior Guillaume balance) by which the compensation does not match the changes in the elasticity of the balance spring brought about by changes in temperature. Some balances incorporate a device known as auxiliary compensation (or secondary compensation) intended to reduce the effects of middle temperature error. See BI-METALLIC COMPENSATION, AUXILIARY COMPENSATION, COMPENSATION.

MINUTE PINION. See MOTION WORK.

MINUTE WHEEL. See MOTION WORK.

MODELE DEPOSE. When the word *"Deposé"* (short for *"Modèle déposé"*) is found on a clock it means, in theory at least, that the shape and presentation, usually of the case, cannot be copied.
See BREVET.

MONTRE PENDULETTE DE VOYAGE. A small travelling clock, usually Swiss, and furnished with a watch-like movement.
See Chapter X.

MORBIER CLOCKS. See COMTOISES.

MOREZ CLOCKS. See COMTOISES.

MOTION WORK. A train of wheels and pinions driving the hands of a watch or clock and affording a 12:1 reduction ratio (i.e. the pinions drive the wheels) between minute and hour hand. A normal motion train begins with the cannon pinion which is carried friction-tight on the centre arbor. The cannon pinion drives the minute wheel which itself is made integral with a further pinion known as the minute nut (minute pinion). The minute nut drives the hour wheel which is mounted on a pipe sleeved-over the cannon pinion. The cannon pinion carries the minute hand and the hour wheel pipe carries the hour hand. In some clocks the reduction is made in one stage; in which case the cannon pinion has the same number of leaves as the minute wheel has teeth, while the hour wheel has 12 times as many teeth as there are leaves in the minute pinion. In carriage clocks the motion work is usually planted under the dial, forming part of the *cadrature*. In *Obis* clocks the motion work is planted inside the front movement plate.

MOUVEMENT DE PARIS. The standard round-plated movement of a *pendule Paris*. The *blancs* and *blancs-roulants* of *pendules Paris* were made in quantity at both Saint-Nicolas-d'Aliermont and by firms like Japy, Marty and Roux near Montbéliard. The first models, made in the 19th century, had locking plate striking and silk suspensions, while later clocks were given rack striking and "Brocot" suspensions.

NORMANDE (*horloge normande*). A characteristic style of tall clock made in the Seine-Maritime and quite different from a Morez longcase clock.

OFF-SET SECONDS. A seconds hand is said to be "off-set" when it is not concentric with the hour and minute hands.

OGEE. A moulding showing in section an "S"-shaped curve.

ORMOLU. Strictly speaking, bronze decorative castings mercurially gilt.

OSCILLATION. See VIBRATION.

OVERCOILED SPRING (Breguet Overcoil). Invented by A-L Breguet in order to apply a terminal curve to a flat spiral spring. The outermost coil of the spring is raised above the rest, and brought towards the centre in a special curve. It is normally only used in clocks and watches of good or exceptional quality in order to improve their timekeeping abilities.
See also TERMINAL CURVES.

PALLETS. Those parts of an escapement upon which the 'scape wheel acts, whether for locking or for impulse.

PARACHUTE (*Parechute*). A device invented by A-L Breguet for use in pocket watches and which was the first form of what nowadays would be called "shock-protection".

PENDANT. The part of a watch case carrying the bow to which a chain, etc., may be attached. In keyless watches the term "pendant" is used to mean the winding button.

PENDULE DE PARIS (or *Pendule Paris*). A standard French mantel clock of the type having a *mouvement de Paris*.
See MOUVEMENT DE PARIS.

PENDULE DE VOYAGE. Strictly speaking a carriage clock in the established 19th century meaning of the term; that is to say having an escapement controlled by a balance, and usually also a top carrying handle and an outer travelling box. The first true carriage clocks were often called *pendules dit de voyage* showing that the term implied something more than a mere *pendule portative*. Other terms used were *pièce de voyage* and *pendule dit de voiture*.
See also PRE-PENDULE DE VOYAGE and also PENDULE PORTATIVE.

PENDULE PORTATIVE. A small clock, either portable or designed for use on tables etc., in different parts of the house. The generic term *pendule portative* in its widest sense includes

Glossary

not only carriage clocks but also alarms and all manner of small domestic clocks not necessarily having handles, and sometimes (but not usually) having pendulums rather than balances. See Plates II/23-25 and 35-36 for specific instances of what is meant by *pendule portative*.
See also page 322 in Chapter X where it is once again made clear that a *pendule portative* may have a glass shade!

PENDULE SYMPATHIQUE. One of the most remarkable of the inventions of the great A-L Breguet. This type of clock has a companion watch which is placed once a day in a special holder. The watch is then automatically wound by the clock and the hands set to time. In the most complicated version of a *sympathique* the clock even decides whether the watch is gaining or losing and regulates it accordingly.

PENETRATION. Another word for depth. It means the correct (neither too deep nor too shallow) engagement or meshing of wheels and pinions with each other.

PERPETUAL CALENDAR. In clockwork, a calendar which corrects of its own accord for short months and also leap years.

PETITE SONNERIE. Striking in which ting-tang quarters are struck at the quarters, but in which only the hour is struck at the hour.

PETIT VOLUME. See HORLOGERIE.

PIETRA DURA. A form of coloured inlaid work in hard stone (usually marble) and originating in Italy.

PIN BARREL. A barrel or drum carrying radially-planted pins sequentially arranged in order to pluck a music-box comb, to lift strike or chime hammers or to play a mechanical organ.

PINION. A toothed wheel, usually made of steel and cut integrally with its arbor. As a rule pinions have comparatively few teeth. These teeth are known as "leaves". The number of leaves found in pinions used in clockwork varies from about 6 to perhaps 16.
In a going or striking train each pinion is driven by a wheel. The few number of turns available at the great wheel are progressively transformed by the train into the large number of turns required at the escapement, the driving force diminishing in proportion. In other words, a few turns of high torque are concentrated into a large number of turns of low torque. In horology it is usual to use the word "pinion" meaning an arbor complete with its pinion, e.g. "centre pinion" rather than "centre arbor" and "escape pinion" rather than "escape arbor".

PINION WIRE. Brass or steel wire given the approximate section of a pinion by means of "drawing". Pinion wire was first made early in the 18th century; but in English clocks it was little used until late in the 19th century and then only as a cheaper and inferior alternative to pinions cut from the solid. Pinions made from pinion wire require first the removal of the unwanted leafed material and then much finishing of the pinion head.

PIN-PALLET ESCAPEMENT. An inexpensive design of lever escapement, used especially in cheap watches and alarms, in which round-section steel pins replace the conventional pallet-stones (In a pin-pallet escapement all the "lift" is on the tips of the 'scape wheel teeth). Although pin-pallets were used by the Swiss before the middle of the 19th century, the term "pin-pallet" is for ever associated with cheap watch calibres devised by Louis Roskopf circa 1868. By 1972, according to *The Jeweller*, the Swiss for the first time in history exported more "Roskopf models and movements" than jewelled lever watches (nearly 38 million as against 37.2 million). The term pin-pallet is also used to describe certain pendulum-controlled escapements, e.g. Brocot's escapement which employs half-round stone pallets.

PIPE. The part of a detent in which is set the locking stone; or any "pipe" or tube such as the "cannon" part of a cannon pinion.

PITCH. The pitch of a wheel, pinion or other toothed component is the spacing of the teeth. The pitch of a screw thread is the distance it will move along its axis if rotated once. See DIVISION.

PIVOTED-DETENT ESCAPEMENT. See DETENT ESCAPEMENT.

PIVOTS. The two ends of an arbor, pinion, balance staff, etc. which support the arbor by turning in brass or jewelled holes. In order to minimise friction, pivots are always made of a lesser diameter than their arbors. The acting surfaces of pivots are usually burnished or highly polished in order further to reduce friction.

PLANTING. The positioning of holes for train wheel pivots in such relationship to each other as to afford correct depthing of the wheels and pinions to run together and also to be parallel to each other and at right angles to the movement plates or bars.

PLATE. See FRAMES.

PLATFORM. The plate on which an escapement is carried and which may be round, rectangular, square or other shaped..
See PORTE ECHAPPEMENT.

POINCON. A poinçon is a punch-mark such as a maker's mark or a date or town mark.

POINTED-TOOTH LEVER ESCAPEMENT. A lever escapement in which all the "lift" is on the pallets.

POISE. A part (usually a balance complete with screws and balance spring) is said to be "poised" when it is statically-balanced about its pivots.

PORTE ECHAPPEMENT. A platform escapement, i.e. one detachable as a unit as opposed to being an integral or built-in part of a clock.

POSITIONAL ERRORS. Errors in rate of going associated with the positioning of a clock or watch with relation to its balance and spring. Gravitational forces acting on the different

parts of the system produce varying effects depending on the position. Variations in pivot friction, and the amount of freedom both sideways and endways of the escapement parts and other components, also contribute to these effects.
See VERTICAL POSITIONS. See HORIZONTAL POSITIONS.

POTANCE (or Potence). Strictly speaking and originally the cock screwed to the inside surface of the top plate of a full-plate verge watch and carrying the "holes" for the bottom verge pivot and for the inner pivot of the 'scape pinion (then called the "balance-wheel pinion"). By extension, any cock within the frames of a movement of a watch, clock or chronometer especially in connection with the "hole" in which is run the bottom balance pivot.

PRE–PENDULE DE VOYAGE. A term coined by C.A. in the course of writing this book in order to differentiate between a *pendule portative,* which term includes alarm clocks and all manner of small portable domestic clocks not necessarily intended for travel, and those clocks of the type which palpably foreshadow the true carriage clock or *pendule de voyage* in the 19th century meaning of the term.

PULL-WIND. A method of winding a clock or a watch by pulling a cord, chain or pendant part instead of turning a key.

QUADRATURE. The same as CADRATURE.

QUALITE. *Qualité soignée,* when applied to a clock in 19th century literature, usually meant that the quality was good but not best quality, which was *qualité première.* In carriage clocks Margaine offered *qualité supérieure* above *qualité première.* The term *à bon marché* meant frankly cheap.

QUARTER SCREWS (otherwise called Timing Screws). The four screws, or sometimes nuts, left permanently in position at 90° to each other on the rim of a compensation balance. The screws are turned inwards or outwards in order to bring the piece to time. In a two-arm balance two of the quarter screws are in line with the arms.
See COMPENSATION SCREWS.

RACK HOOK. The detent which holds a striking rack both while it is being gathered by a gathering pallet and also in its final locked position.

RACK STRIKING. A type of striking work employing a rack to determine the number of blows struck. See LOCKING PLATE. See KNOCK-OUT STRIKING.

RACK TAIL. The part of a striking rack which falls on to the snail in order to determine the number of blows to be struck.

RAISON SOCIALE. The style of a firm, or in other words the name under which it trades; often shortened to simply *raison.*

RATE. The amount by which a precision clock, watch or chronometer goes fast or slow in the course of 24 hours is said to be its rate. It is the consistency of a rate rather than its amount which is important.

RECOIL ESCAPEMENT. An escapement in which the 'scape wheel is made to recoil by the pallets, as opposed to a "dead" locking.
See FRICTIONAL REST ESCAPEMENT. See DEAD LOCKING.

REEDING. A decorative motif consisting of a row of ornamental semi-circular mouldings.

REGULATEUR. A French glass-sided mantel clock having a *pendule Paris* movement and usually with a twin-jar mercury pendulum. In England today these clocks are often called "four glass" by the antique trade.

REGULATEUR ASTRONOMIQUE. In French usage *regulateur astronomique* almost always means a marine chronometer and nothing else. The term, however, may also mean a precision observatory clock of the type called in English an astronomical regulator.

REGULATOR. A precision clock. Regulator is also another name for the index or device used for making a balance-controlled clock or watch go faster or slower.

REMONTOIRE (Remontoir). The name literally means "re-winder". There are two types of remontoire. The first is a train remontoire and the second an escapement remontoire. A train remontoire is a device incorporated in a going train (or sometimes even in a striking train) by which the train re-winds at regular and usually fairly short intervals a subsidiary spring, far weaker and smaller than the mainspring. The remontoire spring drives the remainder of the train and thence the escapement. The purpose is to produce an improved uniformity of torque to the escapement. Some train remontoires (not suitable for portable clocks) raise small weights which drive the remainder of the train and the escapement by gravity.
In an escapement remontoire, the constant-force principle is incorporated in the escapement itself. The classic examples of train and escapement remontoires are to be found in the work of Harrison and Mudge. Harrison's No. 4 (National Maritime Museum, Greenwich) is provided with a train remontoire, re-winding every 7.5 seconds. Mudge's first timekeeper (British Museum) embodies in its escapement two small volute springs which are alternately wound by the 'scape wheel to a given tension at each vibration of the balance.

REPASSAGE. A movement is said to be *repassé* when it has been made to function, or in other words when the final finish has been applied to all the parts. (The meaning of the term *repassage* appears to have changed in the last 150 years.)

Glossary

REPEAT. To re-strike hours, quarters, etc., as the case may be. See Chapter VIII for various permutations of repeating-work.

REPOUSSE. A style of decoration in relief "raised" by means of hammers and punches. The metal is punched from the back while being held against a block of pitch.

RESSORT MOTEUR. (or simply *ressort*). Mainspring.

RIGHT-ANGLED LEVER ESCAPEMENT. The old English pointed-tooth or "ratchet-tooth" type in which the lever is at 90° to the pallets and all the "lift" is on the pallets.

ROLLER. See IMPULSE ROLLER.

ROULANT. Short for *blanc-roulant* (q.v.).

ROULEAUX (*à rouleaux*). As applied to calendars *à rouleaux* means a type in which the drums bearing the relevant information revolve behind apertures in a dial.

SAFETY ACTION. A generic term for the safety arrangements made in any escapement to prevent accidental unlocking of the 'scape wheel at the wrong moment. See DRAW.

SAINT-NICOLAS. A generic term for clocks made in Normandy especially in or near Saint-Nicolas-d'Aliermont.

'SCAPE WHEEL. See ESCAPE WHEEL.

SECONDARY COMPENSATION. See AUXILIARY COMPENSATION.

SECONDS TRAIN. The going train of a clock, watch or chronometer is said to be a "seconds train" when it has in it any arbor arranged to turn in exactly one minute. Some pieces (usually they are watches) are provided with so-called "independent seconds trains". In most examples there is an extra train for the sole purpose of driving a centre-seconds hand usually showing dead jump-seconds.

SET-UP. The amount by which a mainspring (or for that matter a remontoir spring) is "wound-up" permanently even when "run-down". The most obvious application of set-up is in pieces making use of fusees and barrels or employing going barrels with stopwork.

S.G.D.G. See BREVET.

SIDEREAL TIME. Time as determined by two successive transits of a fixed star across a meridian.

SINGLE-BEAT ESCAPEMENT. Another name for a SINGLE-IMPULSE ESCAPEMENT.
See DUPLEX, CHINESE DUPLEX, DETENT, VIRGULE and TIC-TAC ESCAPEMENTS.

SINGLE-IMPULSE ESCAPEMENT. One in which impulse is given to the pendulum or balance in one direction only. Examples:— chronometer, duplex, single-virgule. In pendulum clocks one form of single-impulse escapement is sometimes termed a *coup-perdu* escapement. See also CHINESE DUPLEX ESCAPEMENT. This last is of interest because it takes the single-impulse principle one stage further and when used in a watch enables seconds to be shown from a balance vibrating quarter seconds. It should perhaps be said that almost all clock and watch escapements are double-beat (double-impulse) although they are not customarily so-called because the double-impulse is taken for granted. Examples:— verge, cylinder, double-virgule or lever in balance-controlled escapements; and in pendulum clocks, verge, anchor, dead-beat and pin-wheel. See COUP-PERDU ESCAPEMENT.

SINGLE-PLANE ESCAPEMENT. One in which the 'scape pinion lies in the same plane as the balance staff or pallet arbor (e.g. Anchor, Graham dead beat, cylinder, duplex, chronometer, lever) as opposed to a two-plane escapement.

SINGLE ROLLER. See IMPULSE ROLLER.

SNAIL. A volute-shaped cam, sometimes stepped. In striking and repeating work snails determine the number of blows struck.

SOLAR TIME. See APPARENT SOLAR TIME.

SPOTTING. A decorative finish, usually associated with best English work, and in which the previously prepared and polished brass plates and associated cocks are covered all over their visible surfaces with rings or circles touching or overlapping each other.

SPRING. See MAINSPRING, BALANCE SPRING.

STACKFREED. A device possibly older than the fusee and intended to equalise torque delivered by a mainspring between "wound-up" and "run-down".

STANDING-BARREL. A mainspring barrel fixed to a movement plate and usually open at one end. The great wheel is attached to the barrel arbor, and is driven from the centre of the spring instead of from the outside as in a going barrel. Standing barrels are often used in continental alarm trains.

STAR WHEEL. A star-shaped wheel held in position by a "V"-shaped sprung-detent termed a "jumper". Once the star wheel has been turned mechanically past the apex of the "jumper" the latter pushes the star wheel forward abruptly, afterwards holding it in its next stable position. Used in calendar work and in striking work.

STEADY-PIN. The horological name for a locating-dowel.

STONES (or "Jewels"). Real or synthetic jewels used to line surfaces subject to friction in order to minimise its effects (e.g. pallet "stones", jewel "holes", etc.).

Glossary

STOPWORK. A device limiting the number of turns of a mainspring which are used. Also the device which prevents a fusee chain from being wound too far on to a fusee.

STRAIGHT-LINE LEVER ESCAPEMENT. In contrast to the old English pointed-tooth or "ratchet-tooth" form of escapement (in which the lever is at 90° to the pallets) a straight-line lever escapement is made with the balance staff, pallet arbor and 'scape wheel all on a line of centres. Straight-line escapements are normally "club-toothed" escapements.
See LEVER ESCAPEMENT.

STUD. See BALANCE SPRING STUD.

SUB-FRAME. A secondary frame (or plate) attached inside or outside a main frame.

SUNDIAL TIME. See APPARENT SOLAR TIME.

SURPRISE-PIECE. A device used in both striking and repeating work to ensure that the correct number of quarter or minute blows is struck. It is brought into action at or near each quarter, or at the hour by its ancillary "V"-shaped jumper in the case of minute-repeating watches and *montres-pendulettes-de-voyage*, or by the star wheel in the case of *grande* or *petite sonnerie* striking clocks.

SYMPATHIQUE. See PENDULE SYMPATHIQUE.

TEMPERATURE SCREWS. See COMPENSATION SCREWS.

TERMINAL CURVES. First applied to helical balance springs by John Arnold, who took out a patent for the idea in 1782. The original purpose was to enable a helical balance spring to "breathe" without wobbling or losing its shape during the motion of the balance. The ends of the helical spring are carried inwards in special curves to their points of attachment at the stud and collet. Arnold later discovered that by altering these curves he could make a spring more or less isochronous to suit the balance and escapement of a given chronometer. The application of the same principle to the outer end of the flat spiral balance spring was discovered by A-L Breguet. Breguet bent the last coil so that it lay above the rest of the spring. He was then able to form a terminal curve on this raised portion, up to the point of attachment at the stud. In more modern times an inner terminal curve to the collet has also been applied to the flat spiral spring for observatory competition. See Haswell, pages 146-172. The correct theoretical explanation of the properties of terminal curves was investigated mathematically by a French engineer named Edouard Phillips, whose results were published in 1861. Phillips' work provided the foundation for modern theory and practice of springing and adjusting. Lossier, Caspari, Billeter, Grossmann, Defossez and others were also important in the field. The best information on springing readily available today is to be found in H. Jendritzki's *Watch Adjustment*.

THIRD WHEEL AND PINION. The next wheel and pinion in a going train after the centre wheel. The centre wheel drives the third pinion.

TIC-TAC ESCAPEMENT. A recoil escapement, usually receiving impulse in one direction only and having pallets embracing a small number of teeth.

TIMEKEEPER. This word is used to describe those early pieces made by Harrison and by Mudge in the 18th century with the specific object of winning the £20,000 reward offered in 1714 by the British Government to anyone who could find any "generally practicable and useful" method of finding longitude at sea.
See LONGITUDE.

TIMEPIECE. A clock having no striking work. (Timepiece clocks may have repeating work wound by pulling cords or by depressing plungers; but they are still basically timepieces).

TIMING SCREWS. See QUARTER SCREWS.

TIPSY KEY. A winding key which "free-wheels" when turned in the wrong direction.

TORQUE. A turning force producing rotation.

TOURBILLON. A type of revolving escapement carriage invented towards the end of the 18th century by A-L Breguet for use in pocket watches. The object was to "mean-out" the errors caused by the effects of gravity upon the balance and spring system in the vertical positions. See Chapter II in which is mentioned a Breguet tourbillon carriage clock. See Chapter VIII in which are mentioned other carriage clocks with revolving escapements, and in which the reasons for the use of tourbillons and karrusels and the essential differences between them are specifically explained.

TRAIN. A progression of wheels and pinions driving from one to another. In a horological going train, for instance, motion is transmitted from one arbor to another by means of wheels driving pinions. Each pinion has a lesser number of teeth than the wheel that drives it, and in consequence the pinion rotates more quickly than the wheel. The pinions are made as an integral part of their arbors, which in consequence are usually spoken of as "pinions". Wheels are mounted either by being riveted directly on to the pinion "heads" (i.e. the actual leaved parts of the pinions) or else the wheels are mounted upon brass collets attached to the arbors away from the pinion heads. The term "train" is also used to denote the number of vibrations performed by a balance or pendulum (usually in an hour) and also the numbers of teeth found in the wheels and pinions of the going train associated with it. Hence horologists speak of the "train-count", and of "fast" or "slow" trains.
See SECONDS TRAIN. See BASTARD TRAIN. See MOTION WORK.

TRANSIT INSTRUMENT. A special purpose telescope used for ascertaining the exact moment of the transit of the sun or a star across the meridian (used for time determination) as well as for making fundamental measurements in Right Ascension.

Glossary

TWO-PLANE ESCAPEMENT. One in which the 'scape pinion lies in a different plane to the "verge", balance staff or pallet arbor (e.g. verge, two-plane anchor, two-plane cylinder, two-plane lever, De Baufre, Sully, Garnier, Samuel, metronomes) as opposed to a single-plane escapement.

"UNDERSLUNG". A term now used for describing an escapement where the parts, with the exception of the balance, are below the platform as opposed to above it.

UP-AND-DOWN DIAL. A subsidiary dial found in certain clocks and watches and in almost all chronometers. Usually it shows how long a piece has run since last wound, how soon it will require re-winding in order to obtain optimum results (for instance a two-day chronometer must be wound after exactly 24 hours), and also how much longer it will run on the same winding.

UPRIGHTING. The parallelism of arbors or pinions in a clock or watch with relation to each other at right angles to the frames. See PLANTING.

USINE. Factory, plant, mill or "works". The term implies machine-production, as opposed to the manual work done in an *atelier*.

VERGE ESCAPEMENT. The first escapement in general use. The verge escapement was in all probability invented in the second half of the 13th century and was subsequently used for both clocks and watches. The verge reigned supreme for about four hundred years, and even then it was only at first supplanted for use in the more expensive pieces. It remained current for the very cheapest English watches until as late as about one hundred years ago. It is a recoil escapement.
(See footnote 6 in Chapter I where is mentioned the possibility that there may be a two-plane recoil escapement even earlier than the verge.)

VERTICAL POSITIONS. In a watch the position "pendant up", "pendant left", "pendant right" as opposed to "dial up" or "dial down" are said to be its vertical positions. The piece will go differently in each vertical position (unless it is a tourbillon or karrusel) because gravity will affect the balance and spring system according to varying balance amplitudes. Changes in timekeeping will naturally result. Gravity produces no effect on timekeeping in the "dial up", "dial down" position. For relevance in the context of carriage clocks see Chapter VIII, pages 216-219. See HORIZONTAL POSITIONS.

VIBRATION. The passage or "swing" of a pendulum or balance from one extreme of motion to the other; as opposed to an OSCILLATION which equals two vibrations, or to a BEAT which implies a "tick" or escape of a 'scape wheel tooth.

VIRGULE (or "Comma") ESCAPEMENT. One of many frictional-rest escapements produced with a view to improving upon the cylinder for use in watches. In the single-virgule escapement the balance receives impulse in one direction only by means of a 'scape wheel of pin-wheel-like form acting upon a steel comma-like impulse pallet. The single virgule was probably invented by J.A. Lepaute (1720-c.1789). The double virgule escapement has two sets of pins, one on either side of the 'scape wheel, and two "commas". Impulse is given in both directions. The double virgule was invented either by J.A. Lepaute or by P.A. Caron (1732-1799). The latter later changed his profession and was the composer of *The Barber of Seville*.

VITRAGE D'HORLOGER. A horologist's large window, especially in France and Switzerland.

VOLUTE BALANCE SPRING. A "flat" spiral balance spring, as opposed to one of some other form such as helical.

WARNING. In a striking clock the transfer of the locking of the striking train from its main detent to a lighter locking ready for its release at the precise moment required.

WATCH-CLOCK. Clock based on a watch movement. See CLOCK-WATCH.

WATERBURY "LONG WIND" BARREL SYSTEMS. See BARREL.

WESTMINSTER QUARTERS (CAMBRIDGE QUARTERS). The quarter sequence of notes struck by the Houses of Parliament clock ("Big Ben"). The Westminster quarters are based on those in Great St. Mary's in Cambridge which are said to have been composed by Dr. Crotch in 1780.

WHEEL. In horology a wheel is usually made from a disc of brass and is mounted on an arbor or pinion. Teeth are cut round the circumference of the wheel for the purpose of transmitting motion to another wheel or pinion.
See TRAIN.

WHITTINGTON QUARTERS. An eight note quarter-striking sequence popular in English bracket clocks.

WINDING STEM. That part of the key-less winding mechanism of a watch to which is attached the winding button. The winding-stem provides the only direct link between the movement and the outside world!

WOLF-TOOTH GEARING. A form of gearing usually employed in winding work and having specially shaped teeth in which the epicycloidal form is only on the front working face of the teeth. This tooth form is designed to give extra strength.

Alphabetical List of Names and Trade Marks in Connection with French and Swiss Carriage Clocks

Most names given have either been seen on actual clocks or else they have been taken from references to *pendules de voyage* found in Exhibition literature, in *Brevets* or in French books and periodicals. Endeavour has been made to omit information which is open to doubt, unduly irrelevant or already well-known. Entries are often deliberately non-committal when the excellent lists of both Baillie and Tardy seem to be more detailed and specific. Omitted from the list are most of the numerous retailers in Paris, London, Hong Kong, Shanghai, Tentsin or Constantinople, whose names are found on the dials of carriage clocks, and sometimes also on the back plates. Among names listed are some of men who did little more than fit movements to cases. Included also are a few Swiss names. This is because the link between France and Switzerland was always so close. The list does not pretend to be exhaustive.

The French say that "... there are no spellings for first names". This means simply that names are often correctly spelt in any of a number of different ways.

ABRAM (père et fils)   Montécheroux
Mentioned by Lebon in 1860 as makers of horological tools from just before 1800. Taken over by Blondeau of Saint-Hippolyte.

ACIER (Emile A.)   Impasse Froissard 9, Paris
Exhibited carriage clocks in the 1889 Paris Exhibition.

ACIER (H)   Paris
Brevet 1867 for carriage clock base. An Acier carriage clock seen in 1974 in a very late "one-piece" case. Interestingly enough, the base bears the words "Breveté S.G.D.G." and the initials "H.A." all in a half roundel.

ALIX   Saint-Nicolas-d'Aliermont
Showed a *pendule de voyage* in the Paris Exhibition of 1889 (*Revue Chronométrique*). According to the Catalogue Général, Alix was a maker of mainsprings.

ALZINGRE   Montbéliard
Maker of horological tools in 1860 (Lebon).

AMSTUTZ   Hérimoncourt
Horological tools.

ANGUENOT   At or near Montbéliard
Tripplin 1890, p.119. Makers of lever platform escapements, "... a little better made" than most from the district.

AUBERT & KLAFTENBERGER   Geneva
See Chapter X for general comment. Probably only a retailer.

AUGUSTE   Paris
One fairly early carriage clock seen. Locking-plate strike on bell, escapement with helical balance spring. Another with rack striking is numbered 357 and has a round sub-frame escapement like those used by Jules and Bolviller. Yet another Auguste seen with revolving escapement. See Chapter VIII. Date of earlier clocks perhaps circa 1840. These early Auguste clocks usually had cases signed "L. Lange". Tardy notes an Auguste working in Paris at Rue Grand Prieure in 1880.

A---------?

(C.A)

"C.A." in a circle. Maker not known. This trademark seen inside front plate of 5-minute repeater of good quality in a *Corniche* case.

BAILLY, COMTE & FILS   Morez du Jura
Great Exhibition, London 1851. Exhibited "A travelling clock, striking the quarters, and going for 30 hours."

BARROIS
See DELEPINE.

BASCHET   Paris
Paris Exhibition, 1855. Honourable Mention for carriage clocks and for mantel clocks.

List of Names and Trade Marks

BAUTTE (J.F. & Co.) Geneva
This name appears upon carriage clocks, in all probability almost entirely French, and made in the second half of the 19th century. Although Bautte died in 1837, it is clear from Tardy's *Dictionnaire* that his firm continued in Rue de la Bourse until at least 1850. See Chapter X. Bautte was succeeded by Rossel & Fils.

BAVEUX frères (Alfred and Louis) Saint-Nicolas-d'Aliermont
Paris Exhibition 1878, Honourable Mention. Paris Exhibition 1889, Silver Medal. Showed carriage clocks and also their *roulants*. Of the 1889 Exhibition Tripplin says that Baveux *mouvements* of carriage clocks, striking, repeaters and other complicated work were unrivalled and found their way to the best Paris finishers. An article in the *Revue* of July 1857, p.32, by A. Croutte, says that a M. Antoine Baveux has found that by using Croutte's newly-invented *fraise dit de secours* he is able to cut twelve to fifteen clock barrels "... sans repasser le crochet", instead of the three or four he had been able to cut before. Croutte's system is ingenious, employing a double-cutter. It would have been applicable to the manufacture of carriage clocks. According to M. Pitou (Chapter IIII) Baveux supplied *blancs-roulants* to Jacot. The old Baveux factory was burned down circa 1914-18. Couaillet, next door, then acquired the site. Today a school stands there. Tardy quotes Baveux at Dieppe as making *pendulettes de voyage* as late as 1922; but I have not checked this. See Tripplin's notes quoted in Appendix (b). In March 1974 Mr. T. White of Isleworth found an eight-day French "Flick" clock in a true carriage clock case (*Anglaise*) and signed "Baveux Frères, Saint-Nicolas-d'Aliermont".

BAYARD Saint-Nicolas-d'Aliermont
Still (1974) making very large numbers of alarm clocks and a few carriage clocks. Evolved from Duverdrey and Bloquel. See Chapter IIII.

BEGUIN Rue Faubourg Saint-Antoine, Paris, 1870
Name on a carriage clock with *grande sonnerie* striking on bells, alarm and calendar. Another seen with duplex escapement, quarter repeat, alarm, day and date, engraved case. Yet another seen in a "one-piece" cast rococo case with a Bolviller/Jules type platform escapement.

BERGER (O.) Paris
Carriage clock No. 6234, strikes hours and halves on a bell and has an unusually old platform escapement with a plain flat balance.

BEROLLA or BERROLLAS family London and Paris
Paris Rue Saint-Martin 1830. Rue de la Tour 1850-60. Rue Oberkampf 1870-80. Showed carriage clocks in Paris *Exposition* of 1839. The report said that the clocks were remarkable for their good and also elegant work and that once again Berolla offered his own peculiar type of escapement. (Bib. Nat., Paris, No. V.38342. *Rapport du Jury Central*). Exhibited carriage clocks London 1851 Exhibition. Firm took out a fifteen year *Brevet* in 1857 for a detent escapement for clocks. Several carriage clocks seen, probably made before 1850. One example has a French locking plate movement striking on a bell, housed in an engraved English case. Clocks signed along top edge of front plate on either side of platform "BEROLLA PARIS".

BERTHOUD family Paris and Argenteuil
See Chapter V where a fine, late *pendule de voyage* with pivoted-detent escapement is described and illustrated.

BERTHOUD (Ferdinand) Paris
Born 1727, died 1807. Pre-*pendule de voyage*, see Chapter I.

BESANÇON-PILLODS Besançon
Lever platform escapements exhibited according to Tripplin (1890, p.119) at Paris in 1889.

BESSON Somewhere on the Doubs
Makers of platform escapements. See Chapter VI. According to M. Lenôtre, of F. L'Epée & Cie, Besson were once the most important makers in the district apart from L'Epée.

BEURNIER Liebvillers
Maker of tools for Montécheroux (Lebon 1860).

BIESSY (Eugène) Monthlery
The Chambre Syndicale de l'Horlogerie awarded a Bronze Medal and *Manuel Roret* to E. Biessy as an apprentice's prize for "good execution of a cylinder escapement for a carriage clock." (*Revue*, July 1880, p.119).

BLANPAIN (V.) Paris
Succeeded by CORPET.

BLONDEAU (*Horloger-mécanicien*) Rue de la Paix 19, Paris
Showed in French *Expositions* of 1827, 1834 and 1839. In 1827 he displayed *une montre de voyage*, with calendar and moon phases, going eight days and with *grande sonnerie* striking. (Bib. Nat., Paris, No. V.38337, pages 96-105).

BLONDEAU frères Saint-Hippolyte (Doubs)
Bronze Medal 1819 *Exposition* report "for horological tools in polished iron and steel". Also mentioned by Lebon in 1860 as being active early in the century. Took over Abraham of Montécheroux according to Lebon.

BLUMBERG & CO. LIMITED Paris and London
Name on a four-dial carriage clock. See Chapter VIII.

BOLVILLER Paris, Rue Saint-Avoye, 1830;
Rue Charlot 1840;
Rue Vendôme 1850-60;
Rue Béranger 1870
An early maker of carriage clocks. A surprising number of his surviving clocks have low serial numbers. Nos. 2 and 67 are illustrated in Chapter II. No. 26 has an escapement very like Garnier's but it is not a Garnier clock and does not bear the inscription "P.G. Breveté". No. 37 is a centre seconds clock. A Bolviller clock with singing bird is illustrated in Chapter VIII. Some carriage clocks will be found signed "Gontard & Bolviller". These two took out a fifteen year *Brevet* in 1848 for a carriage clock escapement. Another *Brevet* in the name of Bolvillers taken out in 1858 was for a dead escapement showing seconds. It was apparently some kind of improved duplex with bi-furcated teeth (a sort of Chinese duplex?). The name Bolviller appears in the Japy workbooks for 1907 in connection with carriage clocks.

List of Names and Trade Marks

BOREL (Henri Justin)   Chaux-de-Fonds
London Great Exhibition Cat., 1851, page 1267, "Two travelling clocks, called imperials, going eight days with great and small chimes and stop, repeater, alarm, days of month, excentric with seconds, chronometer escapement, compensation balance." One of the very few travelling clocks, of any nationality, shown in the Exhibition. The same two clocks are mentioned in the *Tribune Chronométrique*, page 259. See Chapter X.

BOROME   Saint-Nicolas-d'Aliermont
See DELEPINE & CAUCHY.

BOSEET
One truly superb clock mentioned and illustrated in Chapter VIII has *grande sonnerie* striking *(cadrature* on the back plate), special calendar, barometer, thermometer, centre-seconds and minute repeating. The clock also has lunar work. It may be admissible to wonder whether Boseet is another spelling for Bousset (q.v.).

BOUILLON (Emile; de la Maison Japy Frères et Cie)
Mentioned in list of those assistants receiving Medals or Awards in connection with the Paris Exhibition of 1889.

BOURDIN (A.E.)   Paris
Paris *Expositions*. Bronze Medal 1844, Silver Medal 1855, and Bronze Medal 1867 for carriage clocks, etc. Tardy quotes various addresses. The *Tribune Chronométrique* on page 82 says that not only are Bourdin's *pendules de voyage* excellent in all their details but they are really good timekeepers despite all the mechanical complications "hung" on them. The *Rapports du Jury* of the *Exposition Universelle* of 1855 says of Bourdin: "Dans cette riche exposition on remarque une petite pendule portative à grande sonnerie (10,000 francs) avec un échappement à detente à ressort et un balancier compensateur à spiral cylindrique. Le cadran marque l'heure, la minute, la seconde, le jour de la semaine, la quantième, les phases de la lune et enfin la température au moyen d'une lame bimétallique". A clock precisely answering this description was sold at Sotheby's on 19th June 1972, Lot 195. See illustration in Chapter V. This clock is almost certainly partly Swiss. Bourdin certainly sometimes used "H.L." movements. Bourdin No. 3701 is an example.

BOURLON   Maissemy
Revue, Vol. III, 1859, p.60 translates: "M. Bourlon shows one of those numerous varieties of escapement known under the name of the Paul Garnier escapement applied in this instance to a pocket watch. The wheel is single and acts on two platforms, between which it would seem that the oil might remain for a long time".

BOUSSET (V.)   Morez (Jura)
Bronze Medal for *pendule de voyage*, Paris Exhibition 1878 (*Journal Suisse d'Horlogerie*, Jan. 1879, p.112). The *Revue Chronométrique* of June 1880, page 90, speaking of the Exhibition of 1878 specially mentions M. Victorien Bousset as an important Morez maker of very good carriage clocks including complicated ones. See BOSEET.

BRAND (Jonas)   Montécheroux
According to Lebon, writing in 1860, Brand (a Swiss) was a manufacturer of horological tools. Muston, however, mentions that a man called Brand helped F. Japy with factory construction. It is not clear whether this Brand had any connection with the one mentioned above.

BREGUET family   Paris
Abraham-Louis Breguet (1747-1823) was the most inspired and distinguished of all French horologists. Almost certainly he was the first to develop true carriage clocks of discernably exactly the form implied by the term *pendule de voyage*. See Chapter II. Mr. George Daniels has in preparation a definitive book dealing exclusively and exhaustively with the history and work of A–L Breguet. The horological side of the business passed to the Brown family towards the end of the nineteenth century. Mr. Georges Brown sold out to Maison Chaumet in 1970. See CHAUMET. As late as the 1889 Paris *Exposition* T.D. Wright saw ". . . a quarter carriage clock with tourbillon, beautifully finished throughout and costly enough to make one wonder where the customers were obtained." Jeanneret notes in *Biog. Neuchâteloise* 1863, that as early as the 1819 Paris *Exposition* Breguet showed ". . . plusiers pendules de voyage à répétition, réveil, mouvement de la lune et quantième complet;" (p.108, Vol. I).

BROCOT (Achille)   Paris
Born 1817. Died 1878. See Baillie and Tardy for various names, achievements, dates and addresses. An advertisement in *Almanach Artistique* for 1864, p.204, says that the trademark of Achille Brocot was then "A.B." in a five-pointed star and his address Rue de Parc-Royal. It has not proved possible to find even one specific reference to Brocot in connection with carriage clocks amongst the Paris and London Exhibition lists, in Brocot's own advertisements, or in the very numerous references to him in horological literature generally; but the fact remains that carriage clocks are found bearing his trademark. The lowest Brocot carriage clock serial number noted by Mr. Pitcher is No. 51, a *gorge*-cased quarter repeater. The highest is No. 1225, a repeater with Limoges panels. Both carry the Brocot trademark. Brocot's Obituary, written by A. Redier (five pages, but no carriage clocks) will be found in the *Revue Chronométrique*, June 1878, p.93-98. It appears that Brocot's business was continued by his son, and then from circa 1889 by G.E. Gibaudet.

BROWN family
See BREGUET.

BROYOT (de la Maison Japy Frères et Cie)
Listed among those assistants given Medals or Awards at Paris in 1889.

BRUNEAU   Galerie de Valois
                                   150, Palais-Royal, Paris
Mention Honorable at Paris *Exposition* of 1839 for carriage clocks. (". . . pour des pendules de voyage . . ."). Bib. Nat., Paris, No. V.38342, page 231 of report on Section One, Horology.

433

List of Names and Trade Marks

BRUNELOT (Jules)   Rue Oberkampf 10, Paris
Paris *Exposition* 1878, Honourable Mention. Paris *Exposition* 1889, showed carriage clocks.

B------------?
Trademark found on carriage clocks. Origin unknown. Might well be Baveux.

B------------?
This trademark is found on very many carriage clocks of good-ordinary quality. It is not known which firm made these clocks. Brunelot has been suggested; but no evidence has been found.

B------------?
"V.W.B." in a heart. This trademark is found on a superb clock No. 557 having *grande sonnerie* striking, fly-back calendar, alarm, thermometer, and semi-oval case. Could "V.W.B." have been Victorien Bousset?

B------------?
A trademark consisting of an "A" and "B" in an ellipse separated by a star is found on carriage clocks. An example was sold at Sotheby's on 24th April 1972, Lot 8. The dial was signed "Reed & Co., Paris".

CAMPBELL   Place de l'Oratoire 4, Paris
Bronze Medal Paris *Exposition* of 1839. The citation (Bib. Nat. V.38342, page 230) says that he specialised in the working of stones used in horology. It also says that he made many escapements for carriage clocks. Mr. Charles Terwilliger of New York has a fairly early carriage clock in an ornate cast case signed "CAMPBELL A PARIS". This piece strikes hours and halves on a bell, and has repeat, alarm and a duplex escapement.

CAPT (Henry Daniel)   Geneva and Paris
See Chapter X covering Swiss Carriage Clocks. Capt was not a maker. No evidence has been found that Capt sold any *pendules de voyage* other than best-class French work. These bore his name, as was quite customary. Jaquet and Chapuis in their masterpiece *La Montre Suisse*, published in 1945 but since available in English, mention Capt repeatedly and with praise, but only in connection with watches. Neither Baillie nor Tardy say a word to the contrary, but the latter gives Paris addresses of Rue d'Alger for 1850, Rue Scribe 1870 and Rue de la Paix 1880. Tardy further says that Capt was succeeded by Gallopin et Cie, and Baillie links his name with that of Isaac Daniel Piguet of Geneva. In London Henry Capt apparently had a shop at 151 Regent St., being succeeded there by T. Martin & Co. circa 1893.

CARPANO (Louis)   Cluses
Trained by Vissière, Carpano made the cutters used by all or most of the French manufacturers of clock *blanc-roulants*. He received a Gold Medal for his cutters at the 1878 Paris Exposition (vide *Journal Suisse* Jan. 1879 p.112). Louis Carpano, an Italian, was born in 1833 and died in 1919. His firm is still in business as SOMFY.

CARRY (O.)   106, Rue Vieille-du-Temple, Paris
Advertisements in the *Revue* of 1912 were headed "Fabrique d'Horlogerie". Items offered included carriage clocks; but a view No. 106, shown "plastered" with advertisements for the products of many companies, suggests that Carry were nothing but wholesalers.

CAUCHY
See DELEPINE.

CHAMBET (de la Maison Paul Garnier)   Paris
Mentioned in *Revue* 1889 in list of Medals and Awards given to assistants.

CHAMPAGNY (J. B. Le Duc de Cadore)
Born 1756. Died 1834. Minister of the Interior in 1806; was responsible for sending H. Pons to Saint-Nicolas-d'Aliermont to re-organise the clock industry there. Minister of Foreign Affairs under Napoleon I (1807-1811). See Chapter IIII.

CHAMPION (Emile)   23 Rue des Bon-Enfants, Paris
Paris *Exposition* 1889, carriage clocks, watches, etc.

CHANTERET   Paris
According to M. Pitou, made carriage clock cases circa 1900.

CHARPENTIER   Besançon, 1870
Mentioned by G. Sire, *L'Horlogerie à l'Exposition Universelle de 1867*, who says that the carriage clocks displayed by Charpentier were among the most attractive in the Exhibition (p.69). Sire mentions one Charpentier carriage clock with a special escapement. See OUDIN.

CHARTIER   Paris
According to Tardy, there were several men of this name in the Paris horological trade towards 1900. One apparently exhibited a carriage clock. M. Pitou, who is mentioned in Chapter V, was apprenticed to the Chartier of Rue du Pont-aux-Choux and not to the one in Rue St. Gilles who was

List of Names and Trade Marks

his contemporary. One fine carriage clock seen signed "E^D CHARTIER" on the bottom of the case and inside the back movement plate.

CHAUMET (Maison)   Place Vendôme 12, Paris
Took over the Breguet horological interests in 1970. See BREGUET.

CHEVALLIER
See LECHEVALLIER.

CHEVENAL (Joseph; de la Maison Carpano)
Paris Exhibition 1889. Among those assistants listed as receiving Medals or Awards.

CHEVELLIER   Paris
E. Pitcher & Co. in 1913 imported a quarter repeater carriage clock supplied by this firm.

CLEMENT & BOURJOIS   Morez (Jura)
Great Exhibition, London 1851. Official Catalogue page 1191. "An eight-day travelling clock, striking and repeating the hours and quarters. Arnold escapement, jewels in six holes, alarum and calendar, in plain gilt brass frame". One clock seen in Milan in 1972, signed "Léon Clément Beurgeois à Morez". This piece had all the appearances of having been produced partly in the Franche-Comté, and partly in Switzerland. The external striking work was on the back movement plate, while a Fleurier-type of "Chinese-market" platform escapement had ornate decoration and "winged" balance cock. The reference in connection with the clock exhibited in 1851 to "Arnold escapement" would in this context almost certainly have meant the Earnshaw-type of escapement. The 1851 catalogue on page 1200 mentions Clément & Bourjois as makers of kitchen jacks. There exist both these and skylark traps made in the characteristic work style of Morez. According to Tardy's *Dictionnaire*, Clément and Bourgeois made Comtoise clocks circa 1830-50. Note the various spellings of Bourjois. See notes in Chapter X on Swiss carriage clocks.

COQUELLE   Paris
1889 "Specialité pour pièces de voyage". (*Cat. Gen.*)

CORNU (Jean-Louis; de la Maison Carpano)
Paris Exhibition 1889. Among those assistants listed as receiving Medals or Awards.

CORPET (Marcel)   Rue Amelot 84, Paris
Successor to V. Blanpain (shortly before 1914). E. Pitcher & Co. 3 Clerkenwell Road, London E.C.1. were sole agents for both firms from 1880. See Appendix (c).

COUAILLET family   Saint-Nicolas-d'Aliermont
See detailed account in Chapter IIII. Quite a complicated history. E. Pitcher & Co. were the sole Couaillet agents in England. They imported carriage clocks until 1939, supplying the wholesale and retail trade. After the 1939-45 War and up until 1954 Couaillet made a few carriage timepieces with alarm. These were not sold in England. Couaillet in their heyday were the largest exporter of carriage clocks to Britain. While they made and sold completed clocks, their second forte always lay in supplying semi-finished *pendules de voyage* to the French carriage clock industry. Armand Couaillet took over from Delépine-Barrois some time after 1913 and occupied their large house and factory, now No. 14, Rue E. Cannevel. In the Paris Exhibition of 1900, Couaillet was awarded a Silver Medal for carriage clock *blancs* and also for complete carriage clocks. Couaillet used neither trademark nor serial numbers. The various Couaillet firms at all times employed many outworkers. Among these, according to M. Delabarre, were those whose job included the punching of the hand-setting arrows upon the movement back plates. The outworkers used their own punches and they supported the plates upon marble slabs on their benches. This apparently inconsequential piece of information probably means that in the past clocks associated with certain workers could have been identified by their "arrows". One arrow-form found on many Couaillet clocks is:-

COUET (Charles)   Paris
Breguet pupil. Showed carriage clocks at the London Exhibition 1862. (*Revue* May 1862, p.121, Vol. IV) Address: 5 Rue Faubourg Saint-Honoré in 1867. (See *Revue* Mar. 1867, p.280)

COULON & MOLITOR   Hérimoncourt
Exhibited in Paris in 1889. Tripplin writing in 1890 mentions this firm. He says that their platform escapements produced on a large scale were "a little better made" than those of Gelin, Levy, Dorian and Mégnin. According to the 1889 Exhibition *Catalogue Général*, Coulon & Molitor made their own balances. *Brevet* "Coulon" 1860. Means of holding down quickly and "stiffly" *ébauche* pinions in "engines" for either slitting teeth or for finishing their shape. Tardy mentions Coulon as being at Auxonne in 1861.

COURVOISIER family   La Chaux-de-Fonds
Frédéric-Alexandre Courvoisier (1799-1854) was by far the most notable member of this family. He was both the leader and the hero of the Neuchâtel Revolution of 1848. He was also a member of the firm Courvoisier et Cie, which he left to run his own business from circa 1842. The name "Courvoisier" appears both upon fine Swiss carriage clocks and upon some rather nondescript travelling pieces. See notes on Chapters I and X.

CRETIN
Name upon *Capucine* clock.

CROUTTE family   Saint-Nicolas-d'Aliermont
See Chapter IIII. Mathieu Croutte is reputed with Papin to have been prominent in the horological industry at Saint-Nicolas-d'Aliermont circa 1800. Croutte of Saint-Aubin-le-Cauf, probably a descendant, is noted in 1870 by Georges Sire as a maker of *mouvements* of *pendules de voyage* (*L'Horlogerie à L'Exposition Universelle de 1867*). Madame Lejoille-Grard says that his factory was at Blesdal, between Saint-Nicolas and Saint-Aubin. One member of the Croutte

435

family (Auguste according to Tardy) obtained a number of *Brevets*. Included among these are:— striking work, 1846; patent alarm, 1852 (and 1853 with Gouel); perpetual calendar, 1847; speed/distance "chronometer", 1859; alarm clocks, 1864. M. A. Croutte invented a new type of cutter for clock barrel great wheels. See BAVEUX. The *Revue* of Oct. 1859, p.61, says that M. Croutte exhibits at Rouen as many inventions as objects, of which there are more than a few (lists various examples, including tools). Author says Martin, Croutte and also Hollinge are a long way behind the Doubs as regards tools. In this context see HOLLINGE and MARTIN and Chapter VI.

The Croutte family exhibited carriage clocks in Paris in 1889.

CUGNIER, LESCHOT La Chaux-de-Fonds
Very early Swiss *pendule de voyage*. See and compare with Breguet wooden-cased clocks. See also *Pendule d'Officier*, Plate I/28.

C-----------?

= E.T.C.   Maker not known.

Maker's mark punched in a small circle on the left side of the back plate of a "full-sized" *grande sonnerie* clock in a private collection in Washington, D.C., U.S.A. The vendors or wholesalers were Laizon and Deron.

C-----------?

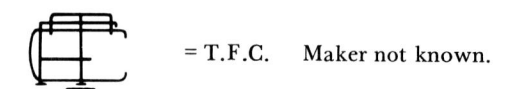

= T.F.C.   Maker not known.

Monogram very finely engraved at the lower left hand corner of the back plate of a miniature *grande sonnerie*, 3 inches tall, in a private collection in Washington, D.C., U.S.A.

DACLIN (J.H.) Lyon
Mentioned in the *Revue* Feb. 1873, p.362, re Lyon 1872 *Exposition*, under *Pendules de Voyage*. Entry translates to read:— "M. J.H. Daclin. Unfortunately the absence of the exhibitor and even of the key to his display window did not allow us to examine his work closely. So far as we could decide from ordinary viewing the appearance of the clocks seemed good and the workmanship thorough. This assessment was confirmed to us by M. Daclin's colleagues, who were able to examine the items exhibited, and which consisted of travelling pieces with simple *sonnerie* and *grande sonnerie*."

DAMIENS-DUVILLIER Rue du Bouloi 10, Paris
Paris *Exposition* 1855, Bronze Medal for carriage clocks. Honourable Mention in 1867 for *régulateurs*. Advertisement for carriage clocks, etc. in *Almanach Artistique*, 1864. M. Emile Damiens, élève de M. Lefranc, was awarded a Bronze Medal for a carriage clock movement in 1880. He may have had some connection with the above.

DEFRANC Saint-Nicolas-d'Aliermont
Exhibited carriage clocks in 1889.

DEHORTER Paris
A very attractive, small-ish *gorge*-cased *petite sonnerie* carriage clock, found in Alpes Maritimes, Italy, in 1969. Striking and alarm on bell. All "instructions" in French; so clock never intended for English market. Particularly fine quality overall. Tardy mentions a Clément Dehorter, Rue Ventadour, Paris 1860.

DEJARDIN (J.) Paris
Trademark "J.D.".
Fine carriage clock seen 1971. Five porcelain panels, the top one measuring 3ins. x 2½ins. On the inside of the back plate "J. DEJARDIN". On the backplate "JD, 728".

Revue Jan. 1880, p.11, Vol XI, Déjardin's escapement for carriage clocks, a type of lever-pirouette. Saunier says "M. Déjardin is one of the youngest but one of the cleverest of our good manufacturers of clocks ...".

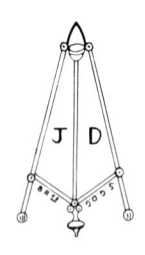

DELABARRE (Alfred) Saint-Nicolas-d'Aliermont
M. Delabarre and his wife both worked for Couaillet. See Chapter IIII. M. Delabarre's father worked for Guignon as well as for Duverdrey and for Couaillet at various times. He was one of those asked to work in St. Petersburg.

DELEPINE (Antoine) 11, Blvd. Bonne-Nouvelle, Paris
 and Saint-Nicolas-d'Aliermont
Mentioned by Sire as having shown top class carriage clock *blancs-roulants* in Paris in the 1867 *Exposition*. The Tribune Chronométrique, pp. 172-175, describes and illustrates a constant-force platform escapement shown by Antoine Delépine in Paris in 1849. Tardy in *Dictionnaire* 1971 illustrates the same escapement, but attributes it to Jean Delépine of Paris. It is clear that some confusion exists concerning the Delépine family. Antoine died during the 1867 Exhibition (*Revue*, Nov. 1867, p.122)

DELEPINE-BARROIS Saint-Nicolas-d'Aliermont
Previously Delépine & Cauchy. Occupied the house and factory illustrated in Chapter IIII. A significant manufacturer of carriage clocks of both ordinary qualities and of *genre Paris*. A Delépine-Barrois Catalogue of circa 1910 records Gold Medals at Brussels 1867, Rouen 1896, and Paris 1900 (not all necessarily for carriage clocks) besides various other awards. See Chapter VIII where is described a Delépine-Barrois carriage clock with a 24-hour alarm kept wound by the going train.

Trademark a lobed shield with the initials "D.L.B." beneath a daisy and above the inscription "B^TE S.G.D.G.". Delépine-Barrois were taken over by Couaillet circa 1913.

"D.L.B." = "Delépine-Barrois"

DELEPINE (Boromé) & CAUCHY Saint-Nicolas-d'Aliermont
According to the *Tribune Chronométrique*, MM. Boromé Delépine et Cauchy succeeded Pons. They certainly made clock movements which included those for *pendules de voyage*. Delépine and Cauchy eventually became "Delépine-Barrois". The *Revue Chronométrique*, speaking in October 1859 of the Rouen Exhibition (Vol. III, p.60) says that Delépine continued to use the name of Pons on all their models. One such clock, bearing the Pons inscription but certainly made in the time of Delépine, is Soldano No. 84. It is a very beautifully-finished *gorge*-cased repeater and is furnished with one of Soldano's special platform escapements.

DELEPINE (Emile) Saint-Nicolas-d'Aliermont
A famous maker of marine chronometers who succeeded O. Dumas of the same town. Sire, writing in 1870 (see Bibliography) listed Delépine as a top chronometer maker. It was to Delépine that the Russian, Bellanovsky, was sent when sponsored by the Empress. (See Chapter IIII). According to Mme. Lejoille-Grard of Saint-Nicolas in 1971, E. Delépine made only chronometers and had no business connection with the Delépines making *pendules de voyage*. The references to the Delépine family found in the *Revue, Tribune* and in the lists of Baillie and Tardy do not always agree either with each other or with the recollections of people still living in Saint-Nicolas. In order not to further complicate this book it has been necessary to largely omit them. They are not of great importance in the present context.

DELMAS Paris
Petite sonnerie and repeat clock in *gorge* case, striking on bell and having a Gontard platform escapement, the latter bearing the P.G. trademark and punched inscription "Gontard BREVETE À PARIS". The whole styling of this clock, besides its steel *cadrature* suggests that it started life as a *blanc roulant* obtained from Jacot, and was then finished and escaped by Gontard. See Chapter VIII. Delmas was a noted maker in his own right of unusual and sophisticated clocks.

DENIS FRERES & CIE (S.U.M.) 68 Rue Edouard-Cannevel, Saint-Nicolas-d'Aliermont
Founded in 1874 by M. Gustave Denis, father of next generation, Earnest and Georges, and grandfather of present partners, Georges and Paul Denis. Gustave Denis specialised in wheel and pinion cutting, supplying makers of *roulants* in the town, including for *pendules de voyage*.

DEPLANCHES-VASSY Saint-Nicolas-d'Aliermont
Paris Exhibition 1889, carriage clocks. Tardy lists Deplanche-Vassy, Rue de Saintonge, Paris, 1870.

DESFONTAINES, MAISON LEROY & SON
13 & 15 Galerie Montpensier, Palais National, Paris
The above style, which readers of Chapter V will know was not strictly correct, was used at the London Great Exhibition of 1851 in connection with a "travelling clock, striking the minutes". Sire, writing about the Paris Exhibition of 1867, singles out as being particularly attractive carriage clocks displayed by MM. Th. Leroy, G. Sandoz, Charpentier, Desfontaines (p.69). The "Th. Leroy" mentioned by Sire was Théodore Marie Leroy of Argenteuil. "Desfontaines" refers to the firm of "Le Roy & Fils" of Galerie Montpensier, 13 & 15. See Chapter V.

DESHAYS Rue Cadet 26, Paris
Gold Medal Paris 1834 for *grands régulateurs*. He also showed a chronometer mantel clock with spring-detent. Report says Deshays also makes *pendules de voyage*. Exhibited at Paris in 1839 a small carriage clock with *grande-sonnerie* and alarm combined with a special dial of his own invention. (Exhib. Report Section 1, Horology, pages 233 and 234, Paris, Bib. Nat. No. V.38342).

DESPEROIS (Pierre; de la Maison A. Baveux)
Saint-Nicolas-d'Aliermont
Paris Exhibition 1889. Among those assistants listed as receiving Medals or Awards.

DESPLANCHES-VASSY Saint-Nicolas-d'Aliermont
Exhibited carriage clocks in Paris in 1889.

DESSIEUX Saint-Nicolas d'Aliermont
Tripplin writing in 1890 says firm then employed two hundred and fifty hands, coming next in importance to Albert Villon, and exhibited in Paris in 1889 "... alarms, chimney time-pieces, and detached parts of the movement." *Réveils Fantasie* were their chief line of business.

DETOUCHE & HOUDIN (or OUDIN) Rue Saint-Martin 228 & 230, Paris
In association with Houdin 1850. Great Exhibition, 1851, "Good travelling clocks". Besançon Exhibition 1860, Gold Medal for *régulateurs*, clocks. London Exhibition 1862, Medallist for carriage clocks, etc. *Brevet* 1852 for grande sonnerie striking by three countwheels of the same diameter, mounted concentrically.

DEVAUX (S.) Paris
A fine striking carriage clock with Garnier's escapement was sold at Christie's on 8th November 1972. The plain gilt case was the same shape and had the same handle as the clock illustrated in Plate II/26.

DEVERTE
Brevet 1863 for "mirror" dials for carriage and other clocks.

DIETTE Paris
According to Tardy, this firm had the address Marché des Enfants Rouges in 1860; but see HOUR, LAVIGNE. Diette with Forin had a *Brevet* in 1866 in connection with *pendules de voyage*; but I have not examined it. Diette was later in partnership with Hour. According to Tardy, their joint trademark from 1891 was "D.H." Diette received Silver Medal

for carriage clocks at the Paris Exhibition of 1878. See Appendix (b). T.D. Wright says that Diette, Fils & Hour showed in the Paris Exhibition of 1889, in addition to a varied selection of chimney-piece clocks, "Those fancy clocks so familiar in London shopwindows during the past few years, such as models of light-houses, windmills, steam-hammers, beam-engines, and other "fantasies" for which this house has a speciality". Diette Fils & Hour were succeeded by Charles Hour, but there is still a fashionable clock shop called Diette in Paris. The proprietress, Mme. Wagowsky, is the French counterpart of Mrs. Oakes in London!

DODILLET (Jean-Henry)   Courtelary
Worked for Japy for three years from 1st May 1786 as a "... founder and forger and whatever other work for the said factory and which will be used for the profit, advancement and good-will of Japy...." See Chapter VI. According to Sahler (see Bib.) Dodillet later established his business at Seppois.

DOLARD (Jean-Baptiste)   Morez
According to Lebon in 1860, Dolard took over various mill-forges along the river Bienne valley, turning them over to production of steel for clockmaking.

DORIAN   at or near Montbéliard
Tripplin in 1890, p. 119. Platform escapement maker on large scale.

DORIUS
Brevet 1869 with Margaine for carriage clock case with decorative panels cut out to show on-background colour. An undated Margaine Catalogue seen offers this as an extra embellishment.

DOUILLON   Saint-Nicolas-d'Aliermont
Madame Lejoille Grard says that Douillon worked near what is today called La Place de la Libération, and that he made carriage clocks before 1900. Tardy mentions him there as late as 1908.

DROCOURT (Pierre and Alfred)   Rue Debelleyme 28, Paris; Rue de Limoges, Paris; also at Saint-Nicolas-d'Aliermont
Superb carriage clocks, predominantly decorative and always attractive in appearance. Besançon Exhibition 1860, Bronze Medal for carriage clocks. London Exhibition, 1862, Honourable Mention. Paris 1867, Bronze Medal. Paris 1878, Silver Medal. Paris 1889, Gold Medal. The son, Alfred Drocourt, succeeded Pierre his father *(Revue,* May 1880, p.69, which says that Alfred had workshops at Paris and in Saint-Nicolas-d'Aliermont). Pierre Drocourt was exhibiting in the 1860's and Alfred was exhibiting in 1880 and in 1889. The 1889 Exhibition Catalogue lists "A. Drocourt" of Saint-Nicolas-d'Aliermont and "A. Drocourt" of Paris as separate exhibitors — but they are presumably one and the same person. Trademark: the letters "D" and "C" with a carriage clock between them. Drocourt clocks sometimes had the name and address on the dial, and sometimes "DROCOURT PARIS" in an oval on the backplate. Known serial numbers range from 1248 to 35,233. The serial numbers of Drocourt's carriage clocks were usually repeated on the winding keys. Drocourt's manager at Saint-Nicolas was Lechavallier. According to Madame Lejoille Grard, E. Delépine owned the building. It was almost opposite to that of O. Dumas.

DRUGEON   Paris
Saunier, speaking of "Les Pendules Portatives dites Pièces de Voyage" shown at Paris in 1878 *(Revue Chron.* June 1880 p.89, Vol. XI), says that M. Drugeon, the long-standing and clever *chef d'atelier* of M. T.Leroy has taken his place as a manufacturer of carriage clocks and is showing them under his own name for the first time. What Saunier fails to make clear is whether Drugeon worked for Théodore-Marie Leroy of Argenteuil and Paris, or for Théodore Leroy of Galerie Valois, Palais Royal, Paris. A good carriage clock with porcelain panels and signed "DRUGEON F^{NT} PARIS" was sold at Sotheby's on 24th April 1972, Lot 35. It bore a trademark consisting of the letters "M.D." over a bugle in an ellipse. Tardy shows a further "M.D." punchmark.

DUBOIS (Emile)   Paris
Revue, July 1880, p.121. Chambre Syndicale De L'Horlogerie. Année 1879-1880. Prizes awarded to workmen and to *Patrons Horlogers.* "To M. Emile Dubois, working with M. Margaine, in Paris, beautifully executed finishing of a carriage clock movement, striking the quarters."

DUBOIS frères   Montbéliard
Noted by Lebon in 1860 as a maker of mainsprings for watches and clocks.

DUCOMMUN   Paris
Mainspring maker for large and small clocks, circa 1870. Ref. G. Sire, p.83.

DUCOMMUN (Paul et Cie)   Travers (Neuchâtel)
Mentioned in the *Catalogue Général* of the 1889 Exhibition in connection with *pendules de voyage (? montres pendulettes de voyage).*

DUMAS (A.)   44, Rue Bonaparte, Paris (and probably Saint-Nicolas-d'Aliermont)
A few carriage clocks seen, all having porcelain dials and panels. These clocks have low serial numbers (for instance No. 568, which also carries the inscription "Breguet 2641"). Dumas' trademark was "A.D." Another trademark was "A.D." in a diamond. This must not be confused with Alfred Drocourt whose trademark was "D.C." The *Revue Chronométrique* June 1880, p.89, Vol. XI, says that Dumas specialises in *mignonettes* which he makes very well, ("ses pendules de voyage ont un cachet de bonne exécution qui témoigne des soins que cet horloger apporte dans ses travaux"). Saunier says that Dumas even makes under his own resources such items as cases and gongs.

List of Names and Trade Marks

DUMAS (O.)   Saint-Nicolas-d'Aliermont,
Rue de Four Saint-Louis, Paris

Onésime Dumas was a chronometer maker of repute. The report *Visite à L'Exposition Universelle de Paris 1855* says that he was a worthy successor to his father-in-law Motel. References to O. Dumas in both the *Tribune* and the *Revue* say that Motel was his uncle. Eventually Dumas took over Gannery's chronometer manufacturing interests, becoming the most important maker in France. The *Revue Chronométrique*, July 1862, p.194, mentions Dumas manufacturing *roulants* of chronometers. See Chapter IIII. Tardy's *Dictionnaire* shows Dumas' portrait and says that he was born in 1824 and died in 1889 or 1890.

DUTERTRE (J.B.)   Paris

Progenitor-type 18th century pre-*pendule de voyage* sold by Kugel of Rue Saint Honoré of Paris in recent times, very much like the Hessen clock in Plate I/29. This J.B. Dutertre was most likely the grandson of the famous Jean-Baptiste and born in 1743. Another J.B. Dutertre travelling clock with two trains and alarm is illustrated by Monreal, plate LXII (see Bib.).

DUVERDRY & BLOQUEL   Saint-Nicolas-d'Aliermont

Founded by Albert Villon from about 1867. For full history see Chapter III. A most interesting firm; still going strongly as Bayard alarms, and making a few carriage clocks under the old name of Duverdrey & Bloquel. Trademark a "Lion", often seen on carriage clocks of fairly modest quality. One such "Lion" seen on a miniature carriage clock, sold in Hong Kong between 1909 and 1918 and bearing the name of the German retailer "CHS. J. GAUPP & CO. HONG KONG". Duverdrey & Bloquel's main English agent was Landenberger & Co.

In the early days from 1867 and before D. & B. became a self-contained factory, most of the work was done in the homes of outworkers. It is probable that those whose job included the punching-on of hand-setting arrows did so using their own punches. This reasoning suggests that at one time the "arrow" may have identified the worker. One arrow found with the "Lion" trademark is like this:—

DUVERNOY   Montbéliard

Maker of files (Dubois, 1860).

D----------?

"A.D." in an oval seen on gong blocks of French carriage clocks.

D----------?

"F.D." in an oval is found on the supporting brass blocks of many gongs. Presumably F.D. was a gongmaker.

D----------?

Calendar clock with moon seen signed "J.D." *Cadrature* on back plate suggests Franche-Comté origins.

D----------? ("R.E.D."?)

This monogram is found upon clocks with circular plates, fusees and two-plane lever escapements described in Chapter VIII.

ECALLE (Auguste)   Palais Royal 93, Paris

Bronze Medals Paris Exhibitions 1878 and 1889 for carriage clocks. Agent for Dent of Cockspur Street, London. Tardy also cites addresses Galerie Beaujolais 1880-1890, Boulevarde de la Madeleine 1900.

ELFFROTH (D.)   Switzerland

Journal Suisse d'Horlogerie, 1876-77, pp.134-135, and plate VII. Report on "... une petite pendule portative...". This piece was 61 mm. high, offering *grande sonnerie* striking and repeat, besides calendar for day, date and month, and phases of the moon. The article says that Elffroth made almost all of the clock himself, including its complete movement, dial and case, and that it was most unusual for one workman to produce all these parts. Elffroth's clock clearly attracted widespread approbation. Adrien Philippe mentions it with enthusiasm in *Etudes ... 1878*. The clock is described charmingly and wisely as being the product of the *verte vieillesse* of M. Elffroth. See Chapter X.

ELYOR   Paris

An anagram of Le Roy used by Basile Charles during the Revolution.
(Documentation: *notice advertissement* by L. Leroy & Cie for the St. Louis Exhibition in 1904).

ERBEAU (L.)   100, Boulevard Sébastopol, Paris

Exhibition 1889, Paris. Escapements for carriage clocks.

EXPOSITION COLLECTIVE DE SAINT-NICOLAS-D'ALIERMONT

Silver Medal, Paris Exhibition 1889.

FAIVRE   Trévillers near Montbéliard;
or Pontarlier

Mentioned by Lebon in 1860 as a noted maker of finished clocks including those with *carillons* and *à musique*. There may be some link here with those carriage clocks mentioned in Chapter VIII and having "birds" and/or music.

List of Names and Trade Marks

FARRET   Rue Chapon, 23, Paris
Mentioned in *Tribune Chronométrique*, p.85, as having shown at the *Exposition des Produits de L'Industrie Nationale*, 1849 "... un petit mouvement de pendule de voyage dont la cadrature est en acier". The reporter was M. Peupin.

FENON (Auguste)   Paris
Exposition 1878, carriage clocks (*Revue Chronométrique*, 1880, p.69). T.D. Wright mentions Fenon's excellent astronomical regulators exhibited in Paris in 1889. Wright goes out of his way to say that Fenon's regulators are made throughout, rather than from unfinished movements as obtained by most "makers" from Saint-Nicolas-d'Aliermont. Tripplin, p.44, says that Fenon is watch and clock maker to the Paris Observatory. It is probable that by astronomical regulators both mean "marine chronometers". The first term is often used in France to mean the second. Tardy says that A-V Fenon (1843-1913) was a pupil of Winnerl and became head of the Besançon horological school in 1892.

FERNIER (Louis et Frère)   Besançon and Paris
A maker of carriage clocks. Trademark: "L.F. Paris" punched on back plate.

FORIN
See DIETTE.

FUMEY   Franche-Comté
Evidently a maker of platform escapements. One, a cylinder, seen on a *pendule de voyage* made by Michoudet, Foncine-le-bas, Jura. A J-M. Fumey, Foncine-le-Haut, received a Bronze Medal at Paris in 1855.

FUTVOYE   Paris
One carriage clock seen, striking on bell, probably made before 1850.

GANTOIS (Amédé; de la Maison Drocourt).
Paris Exhibition 1889. Among those assistants listed as receiving Medals or Awards.

GARNIER (Paul)   Addresses variously:
Paris, 8 Bis Rue Taitbout,
25 Rue Taitbout.
6 and 14 Rue Taitbout.
Born 1801. Died 1869. An associate of Janvier and a founder member of the Société des Horlogers. Not by any means the inventor of the French carriage clock, but beyond question the man who first standardised and rationalised the type leading to the development of a very large industry. See Chapter II and Appendix (a). Paul Garnier received Silver Medals in the Paris Exhibitions of 1827, 1834 and 1839 for exhibits which included carriage clocks, and Gold Medals in 1844 and 1849, besides awards in provincial exhibitions. He was awarded a Medal of Honour in 1855. In 1860, in recognition of his many public services, Paul Garnier was named *Chevalier de la Legion d'Honneur*. Paul Garnier's son succeeded him and died in 1917. Tripplin said (in 1889, p.75) that Garnier had supplied railway station clocks throughout France "... ever since the beginning of the railway enterprise". Paul Garnier signed himself variously: "Eléve de Janvier", "Horloger du Roi", "Horloger de la Marine" and "Ing^{eur} M^{cien}". Paul Garnier, the son, was still exhibiting carriage clocks in the Paris Exhibition of 1889.

GARNIER (Thomas).
See LEROY, Galerie Valois.

GASTELLIER   Paris
Chef d'atelier to M. George Brown of Breguet for very many years until the closing of his shop in 28 Place Vendôme in 1970. See Chapter V. Gastellier was associated with the finishing of the last Breguet "humpback" clock completed in 1931.

GELIN   Montbéliard
Exhibited platform escapements in Paris Exhibition of 1889, where the *Catalogue Général* said plainly that he made them for carriage clocks and that he made both lever and cylinder escapements. Tripplin, writing in 1890, p.119, mentions Gelin as manufacturing platforms on a large scale.

GINDRAUX   Paris
Exhibited in the London Exhibition of 1862 "Stones for Clockmaking".

GIRARD (Peter)   Canton of Neuchâtel
London 1851 Exhibition Catalogue, p.1269, says that Girard showed "A travelling clock, eight-day movement possessing an alarm, and showing days of the month, excentric great chimes during the night and small during the day, repeats at will, enamel dials, anchor escapement, visible pallets, compensation balance, 17 holes in rubies, case engraved, the movement electro-gilt". The *Tribune* describes the clock "... à carillon bruyant pendant la nuit, modéré pendant le jour, etc...", page 258. See Chapter X. There were Girards at Besançon in the watch business in 1870 (Sire, p.134).

GONTARD (C.P.)   Paris
Paris Exhibition 1855, Bronze Medal. Exhibited small carriage clocks. Showed carriage clocks in London 1851. Also see Tardy's *Dictionnaire*. Gontard obtained a *Brevet* with Bolviller in 1848 for a lever-chronometer escapement intended for pocket chronometers or for carriage clocks. Later Gontard alone patented two further escapements of a rather similar nature. See Chapter VIII. The *Revue Chronométrique* of August 1872 said that the Gontard escapement gave excellent results when placed by the inventor and maker in carriage clock *roulants* obtained from Jacot (an advantage not open to everyone!). Gontard's trademark consists of the initials "P.G." with a representation of a star and a pair of dividers inset in a triangular punching. Underneath his platforms is found the inscription "GONTARD BREVETÉ À PARIS." See BOLVILLER.

                                                                     P★G

GOUEL
See CROUTTE.

List of Names and Trade Marks

GOUNOUILHOU & FRANÇOIS   Geneva
Sotheby's auction catalogue 21st May 1965. Clock described as *pendule d'officier*, ormolu, in original fruitwood travelling case. Sold with documents saying that it belonged to Napoleon at Longwood, St. Helena, and was bought by Sir Hudson Lowe on Napoleon's death.

GRARD (Bénédict)   Saint-Nicolas-d'Aliermont
Worked for Emile Delépine. Went to St. Petersburg (Leningrad) in about 1900-01 to teach horology. His daughter, Madame Jeanne Lejoille-Grard, who went with him to Russia, still lives in Saint-Nicolas. See Chapter IIII.

GRENON (Paul)   39, Rue des Trois-Bornes, Paris
Advertisements *Revue* 1912. *Pendules de voyage;* but evidently wholesalers.

GRIGNON-MEUSNIER   Paris
"Giant" carriage clock, striking on a bell and with chronometer escapement (Sotheby's 3 April 74, Lot 79).

GUIBAUDET (Gustave E.)   Rue du Parc-Royal, Paris
Succeeded Brocot circa 1889.

GUIGNON (Julien Louis)   Saint-Nicolas-d'Aliermont and Paris
Paris Exhibition 1889. Exhibited an assortment of alarums and cheap carriage clocks (Tripplin, Appendix [b]). The father of Alfred Delabarre worked for Guignon as well as for Duverdrey and for Couaillet. Also mentioned by Madame Lejoille-Grard in 1971 as a maker of *pendules de voyage*.

GUILMET   Paris
Brevet No. 119,862 in 1878 for a four-dial carriage clock. Beillard says in *Recherches* that in about 1867 Guilmet made mystery clocks of the statue-lady-holding-pendulum type. At the same period, says Beillard, Robert Houdin and Henri Robert were making *pendules à marche mystérieuse* of the multi-glass dial types.

GUYERDET (l'aîné)   Paris
Name on Paul Garnier clock with his escapement. Tardy gives Guyerdet address Rue Meslay in 1830.

G----------? ("J.G.")
Initials found on top of back plate of "V.A.P." timepiece carriage clock, described in Chapter VIII.

G. & B. (Gontard & Bolviller?)   Paris
Probable name on some carriage clocks finished by Jacot. See GONTARD and also Chapter VIII.

HARRIS & HARRINGTON   New York
Importers of French *pendules de voyage*. Even before the U.S. Customs required that names of manufacturers or importers be stamped on the movement back plates, some American importers identified themselves with their names or trademarks. Harris and Harrington sold clocks identified by the initials "H & H" within an outlined box.

HENRY   Paris
This name is on the back plate of a very typical "Bolviller-type" early carriage clock.

HESSEN (André)   Paris
Pre-*Pendule de voyage*, see Chapter I.

HOLDT (R)   Paris
Name on carriage clock with top sundial. See Chapter VIII.

HOLINGUE (or Hollingue or Hollinge) family   Saint-Nicolas-d'Aliermont
Revue Oct. 1859, Vol. 3, p.61. Showed at Rouen Exhibition some very pretty "quadratures de pendules de voyage". The *Revue* says that M. Holingue, without having either the workshops or the tools of Martin and Croutte, yet manages to compete with them. "The making of all types of movements come easily to him . . ." According to Tardy, at least some of the Holingue and Martin interests were eventually merged. Albert Hollinge is said to have been bought out by Drocourt c.1875-1876. (See Chapter IIII). The *Tribune*, p.92, refers to "M. Holingue fils" as receiving a Bronze Medal for *roulants* (type not specified) shown at the *Exposition des Produits de l'Industrie Nationale* in 1849. The *Paris Normandie* of 26th Aug. 1932 (see Bib.) mentions particularly a M. Alexandre Holingue, whose establishment was set up in 1869. He is shown with his wife, but the article does not state specifically what he made, although "une horloge à poids, une vraie normandie..." and "... de délicats mouvements de pendulettes..." are mentioned. The first clock was probably a "Saint-Nicolas" as mentioned in Chapter IIII.

HOUR (Charles)   7 Rue Sainte Anastase, Paris
Founded in 1852 under the name DIETTE. According to Tardy, Hour's punchmark was "C.H." from 1900. Mr. Pitcher says that Hour also used "CH.H" and that after 1927 the concern became "Hour, Lavigne et Cie". See Appendix (c). According to M. Oliver Meyer, the present director, carriage clocks were made up to the First World War. *Roulants* were obtained from Montbéliard, or alternatively from Couaillet at Saint-Nicolas-d'Aliermont. Escapements came from L'Epée. All the finishing was done in Paris. "Ch. Hour" were still advertising carriage clocks in the *Revue Chronométrique* as late as 1912, using the address 7 Rue Saint-Anastase. See DIETTE.

HOUR (Edouard; de la Maison Diette)   Paris
Paris Exhibition 1889. Mentioned in list of Medals and Awards to assistants.

HUBER (W.J.)   London
In 1972 produced an admirable reproduction French timepiece *pendule de voyage* in *Anglaise* variant case, with *Obis*-type movement. This clock has a lever platform escapement with excellent timekeeping capabilities. At approximately serial No. 1000, the trade mark changed to "A.C.G." in an ellipse. Miniature timepiece carriage clocks were produced from 1973.

(WJH)

List of Names and Trade Marks

HULMANN. At or near Montbéliard
Platform escapement makers on a fairly modest scale. See Chapter VI.

JACOB (Aimé) Saint-Nicolas-d'Aliermont and Rue J.-J. Rousseau 3, Paris
A chronometer maker mentioned in 1855 Paris Exhibition report (*Visite à L'Exposition Universelle de Paris 1855*) as being equal to Dumas. Saunier in 1872 mentions Aimé Jacob as making duplex escapements for carriage clocks. A very well-made semi-finished regulator movement, complete with original packing paper and wood box, with dead beat escapement on the back plate, appeared at Christie's on 7th June 1972. The maker was "Jacob à Paris" (but in fact it is Saint-Nicolas work). According to his obituary in the *Revue*, Dec. 1889, Jean-Aimé Jacob (1793-1871) was a pupil of Berthoud.

JACOT (Henri) 31 Rue de Montmorency, Paris
Also said to have had a factory at Saint-Nicolas-d'Aliermont
Paris *Exposition* 1855 and London Exhibition 1862, Bronze Medals. Paris 1867, Silver Medal for carriage clocks and movements. Paris 1878 Silver Medal for carriage clocks. Paris 1889, Gold Medal for carriage clocks. Paris 1900, Gold Medal for carriage clocks. The first Henri Jacot died in 1868. He was succeeded by a nephew of the same name. Responsible for the production of superb carriage clocks. Trademark a parrot and initials "H.J.". A *Jacot* is a French parrot, or "Pretty Polly". According to Saunier and Sire, Jacot made all parts of his own clocks. According to M. Pitou, who worked for Jacot, the firm used *blancs-roulants* (at least latterly) obtained from Baveux of Saint-Nicolas-d'Aliermont. Moreover, we ourselves saw in 1970 in M. Pitou's ex-Jacot stock of materials, *roulants* made by Japy of Beaucourt. Known Jacot serial numbers range from No. 387 to over 19,000. These Jacot serial numbers were repeated on the winding keys of his clocks. According to M. Pitou, the Jacot concern finally ceased operations in about 1920.

JANVIER (Joseph) Saint-Claude
A *Capucine* clock No. 62 signed "JANVIER Cadet" is most important as suggesting the French Jura origins of this type of clock. Antide Janvier is known to have been born in Saint-Claude in 1751. He returned there briefly circa 1791-92. Janvier Cadet was Antide's younger brother. He was born in 1754 in Saint-Claude, remained there for most of his life, and died there in 1820.

JAPY (Albert) Beaucourt and Paris
32 Rue Michel-Ange. Now Giens
Director Général from after the First World War up to about 1943. Great, great, great grandson of Frédéric Japy, founder. See Bib.

JAPY frères Beaucourt, Badevel, Paris, etc.
For full history see Chapter VI. Founder Frédéric Japy (1749-1812). He was a pioneer of mass-production. He began in 1772 with a small factory at Badevel turning out *ébauches* of watches. He first made clock *blancs-roulants* in 1777, but soon became not only the largest but commercially by far the most important clockmaking concern ever in France. Jaquet & Chapuis in *The Swiss Watch*, p.167, mention that in 1836 Japy of Beaucourt supplied watch *ébauches* to Charles Montandon and Girard of La Chaux-de-Fonds, Switzerland! Tripplin records that in 1889 Japy alone occupied about one-sixth of the entire French section of horology in the Paris Exhibition. *Brevets* include (*Revue* Mar. 1878, p.40) Japy et Millot, carriage clock hour and quarter repeat mechanism. Carriage clock at a greatly reduced price (Ibid. p.42). Calendar work for clock or watch (Ibid. p.43). Brevet 1862, *pendule de voyage* alarm, lever escapement.
We saw in 1970 Japy *blancs-roulants* still in their original paper wrappings. Attached labels read "JAPY FRERES & CIE. A Beaucourt et à Paris Rue de Chateau-d'Eau 7, 3 VOYAGES REPETITION SIMPLE NO 3, DEPOT A PARIS Japy, Marti, Roux, Mougin & Cie, 75 Rue de Turenne 75". Awards received by Japy Frères included medals in 1819, 23, 27, 34, 39, 44, 49, to say nothing of the Grande Medaille d'Honneur (en Or) received in 1855.

JAPY (Louis) à Berne, commune de Seloncourt (Département du Doubs)
M. Albert Japy mentions a "fabrique des mouvements de pendules... à Berne près de Seloncourt". The *Tribune*, p.90-91, refers to this establishment in 1849 as showing (Exhib. Prod. Nat.) not only "movements of clocks of all kinds" but, most interesting to note, rough and finished movements of watches of the Lépine calibre for supply to Switzerland. Louis Japy, the proprietor and founder was the son of Louis of Beaucourt and grandson of Frédéric (the illustrious). Muston records (Vol. 1, p.93) that the Berne-Seloncourt factory was "... une fabrique de mouvements de pendules, d'ébauches de montres et de finissages". M. Albert Japy says that brass sheet, etc. was also produced there.

JAPY, MARTI, ROUX. Rue de Turenne, 75, Paris
The large factories of the Montbéliard area were in direct competition with each other for many years. In 1863 they formed one *maison commerciale* in Paris. In this association, initially undertaken for 20 years, each House was responsible for the production of a special kind of movement.
(Sahler/Galliot, 1954. See Bib.)
See MOUGIN.

List of Names and Trade Marks

JEAN de Paris   Fifteenth century
Said to have made a travelling clock for Louis XI of France. See Chapter I.

JEANNERET-GRIS (J.J.)   Le Locle
A clever early maker of automatic lathes and of machine-tools generally. Undoubtedly made some contribution to the early success of Frédéric Japy (see Chapter VI).

JOLY   Bayeux
Name found on the back plate of the movement of an 18th century pre-*pendule de voyage* which also has the name Lhoest on the enamel dial. This clock has a verge escapement with "rosette" bridge cock on the back plate. The *cadrature* is also on the back plate. The case bears the town mark of Rennes. The movement has fusee and chain for the going train but a going barrel for the striking.

JOROSAY   Palais Royal
Name on a (? replacement) dial of an early French *pendule de voyage* having *fusee* and maintain on the going side, but with barrel for the *grande sonnerie* striking. The *cadrature* is on the back plate. A Franche-Comté clock. Perhaps the real name was Jarossay? Old writing is often difficult to read and very likely a mistake was made if the dial was replaced.

JOSEPH (Charles P.H.)   Rue Amelot 114, Paris
Paris Exhibition 1889 – showed carriage clocks. Paris Exhibition 1900 – Gold Medal for exhibits which apparently included carriage clocks. Tardy shows Joseph's photograph and quotes for him the dates 1852-1935.

JULES   Paris
One of the earlier carriage clock names. See Chapter II. Several examples seen with locking plate striking, helical balance spring and signed on backplate "JULES PARIS". The date of these clocks is circa 1840. Many Jules cases will be found signed "L. Lange". No. 133 is of this kind but it has rack striking and is a repeater. This clock has an engraved "one-piece" case and a flat-spring lever escapement decorated with straight-line engine turning.

KLENTSCHI (Charles Frédéric)   La Chaux-de-Fonds
Born 1774, died 1854. A notable clockmaker. According to Chapuis in *Histoire de La Pendulerie Neuchâteloise*, Klentschi ran a horological school with Maillardet from about 1814. One of their pupils was "Fritz" Courvoisier. Klentschi also worked for the Courvoisier family. See Chapter X.

KREMER (Adam)   Paris
A maker of carriage clock cases for which he received a Silver Medal in the 1889 Paris Exhibition. According to M. Pitou, Kremer was still in business as late as about 1900.

LACHAUX   Montbéliard
Maker of files in 1860 (Dubois).

LAMAILLE (E.G.)   London and Paris
Established in Farringdon Road, London, before 1900. Taken over by H. Williamson Ltd. Trademarks "E.G.L." and "G.L." (sometimes in an oval). The *Horological Journal* of March 1880 on page 94 mentions Messrs. Gay, Lamaille and Co. of Paris and London as securing the patent for an improved star-wheel invented by Moritz Immisch applicable to carriage clock strike and calendar work.

LAMAZIERE & BUNZLI   Saint-Nicolas-d'Aliermont
Succeeded by Ateliers Vaucanson, q.v.

LAMBERT (Enregistreurs)   Saint-Nicolas-d'Aliermont
Still advertising "Distribution automatique de l'Heure, Controle de Traveil, Horloges-mères, Horloges Monumentales" in 1969.

LANDENBERGER & CO.   91 Aldersgate St., London
The English agents for Duverdrey & Bloquel.

LANGE (L.)   Location not known
An early maker of carriage clock cases. His name is found stamped on the base castings. Lange numbered his cases. For instance, Jules clock No. 133 has a Lange case No. 108 which is a "one-piece". Another, Auguste No. 337, has Lange case No. 79, a "multi-piece". No. 26886 made by Japy Frères and sold by Henry Marc has an engraved "one-piece" Lange case No. 968. (See Chapters II and VI). In the Lange "multi-piece" cases each glass will be found to be set in its own frame which in turn is slotted in the frame pillars.

LAVIGNE
See HOUR.

LEBREJAL   Saint-Nicolas-d'Aliermont
Manager of Exacta c.1930-1935.

LECHEVALLIER (Auguste)   Saint-Nicolas-d'Aliermont
In charge of the workrooms of Drocourt in Saint-Nicolas. Authority: Madame Gamard quoted by Dr. Dickie in 1963 ("Le Chevallier") and Madame Lejoille Grard writing to C. Allix in 1971. This information has since been confirmed by the *Revue* 1889, where Auguste Lechevallier "de la Maison Drocourt" is mentioned in the list of Medals and Awards given to assistants.

LECOULTRE (Samuel)   Le Sentier, Switzerland
Produced the "rough" pinions for the last Breguet "humpback" carriage clock, completed in 1931 and numbered 759.

LEFEBRE
Customer of Japy Frères for carriage clocks in 1907.

LEFRANC   Rue Vielle-du-Temple, Paris
Revue June 1880 under the heading "Pendules Portatives dites Pieces de Voyage" says that Lefranc showed for the first time in 1878, and made an excellent *début*. The writer says that the appearance of the clocks ". . . .à effets multiples, dont quelques dispositions nous ont paru heureuses..." rather defy description without sketches!! (Vol. XI).

List of Names and Trade Marks

LEGASTELOIS (J-P and J-M; de la Maison Diette fils et Hour)
Among those assistants listed as receiving Medals or Awards in the Paris Exhibition of 1889.

LEGRAND Paris
Represented Japy in Paris (ref. M. Pitou).

LEMAIGNEN, LECHEVALLIER & MERCER
Saint-Nicolas-d'Aliermont
A firm with horological traditions. Still making "Mouvement d'horlogerie pour enregisteur à disques", fishing reels, radio, etc. See LECHAVALLIER.

LENOTRE Sainte-Suzanne
Manager to Madame L'Epée, producing platform escapements.

LEPAUTE Rue Saint-Honoré 27, Paris
Described as *horloger-mécanicien* in the official account of the Paris *Exposition* of 1827 (Bib. Nat. No. V.38337, pp. 96–105). Showed a *petite pendule portative*. This clock struck *grande sonnerie*, had repeat, alarm, calendar showing through a dial aperture, and what must have been a platform lever escapement ("... échappement libre, trous en pierres fines, montée sur un plateau qui permet de l'élever séparément, contenue dans une boîte de bronze à jour, bien ciselée et dorée au mat;"). This Lepaute was presumably Jean-Joseph (1768-1846) who according to Baillie was a grandnephew of the famous J.A. Some fine carriage clocks exist bearing the name "Lepaute"; but by 1889 the firm is listed only as exhibiting tower clocks and precision clocks.

L'EPEE (Frédéric & Cie) Sainte-Suzanne (Doubs)
This firm founded their reputation as makers of excellent music boxes. Still very much in business, with a fine, clean factory making *porte-échappements* on a large scale. L'Epée platform escapements are exported all over the world, including to England. They have very many uses in both science and industry. The present head of the firm is Madame Henry L'Epée. See history of the firm in Chapter VI. L'Epée succeeded Marti and also Jacques Viotte (Authority: M. Albert Japy, Giens, 1971).

Trademark, crossed épées.

```
       ALC
        ×
       ×·×
      DEPOSE
```

LEPINE family Paris
Involved early with carriage clocks. See Chapter II. Alarm carriage clock seen signed "LEPINE Place des Victoires 2", and also "Palais Royal" and "Rue de Valois 124".

LEROY (Théodore Marie) Argenteuil and Paris
Born 1827, died 1899. Son of Marie-Balthazar Leroy. Exhibited carriage clocks in 1867; but his main interests and accomplishments were centered upon the improvement of marine chronometers and regulators. Théodore Marie was associated with Philipps in the development of terminal curves for balance springs. Although his work is mainly outside the scope of this book, he was in fact by far the most important of the 19th century Leroys. He was created *Chevalier de la Légion d'Honneur*. Saunier was quite scandalised in 1867 that Théodore Marie Leroy should have been awarded a Silver Medal, not for his excellent chronometers but for his carriage clocks! A biographical note on Théodore Marie Leroy (*Rapport du Jury ... Exposition Univ. 1900*) freely translates to say "... the making of carriage clocks kept him occupied for a few years but ultimately his business developed along such lines as to result in his devoting himself exclusively to the construction of marine chronometers ....".

LE ROY (later L. LEROY & CIE; now LEROY)
Paris, Galerie Montpensier, 13 & 15 Palais Royal, and also London. Now at 4 Faubourg Saint-Honoré
Business founded by Basile Charles Le Roy, circa 1785. Soon after the Revolution he occupied Galerie Montpensier, 13 & 15 Palais Royal. After his death in 1828 the business was carried on by his son Charles-Louis. Sold 1845 to Desfontaines who kept the name "Le Roy & Fils". In 1889 Louis Leroy (son of Théodore Marie Leroy, q.v.) became a partner in the firm. The name then became "Ancienne Maison Le Roy & Fils, L. Leroy & Cie Successeurs". (In due course, and certainly by 1900, the reference to Ancienne Maison Le Roy & Fils was dropped). In 1895 Leon Leroy (brother of Louis) entered the firm. In 1899 the business moved to 7 Boulevard de la Madeleine. In 1938 L. Leroy & Cie moved to 4 Faubourg Saint-Honoré. Today the concern is run by Pierre and Philippe, the sons of Leon; but they no longer make carriage clocks. The trading name was recently changed to Leroy. See Chapters II & V for further history of the firm. Clocks by L. Leroy & Cie are usually signed in block capital letters on the edge of the right side of the back movement plate where the number of the clock also appears. Watches sold by L. Leroy & Cie were sometimes signed "L. Leroy & Cie" and sometimes "L. LeRoy & Cie". Examples of both were sold at Christie's on 26th June 1973 on which occasion there were also sold a number of watches by the other firm Leroy & Fils.

LEROY Paris
Galerie Valois, 114 & 115 Palais Royal; latterly Avenue Opéra.
Founded by Théodore Leroy (apparently no relation to Théodore Marie Leroy q.v.). Moved to the Palais Royal from the Rue Saint-Martin in 1813. In 1827 partnership with Auguste-Pierre Lepaute. On 22nd June 1827 Théodore obtained the *Brevet d'Horlogers du Roi* for himself and for Lepaute. Used trading name "Leroy et Fils" from 1839. Sold out in 1843 to M. Fraigneau who continued to use the name of "Leroy et Fils". Subsequent ownerships were Schaeffer in 1871, Clericetti in 1883, and Thomas Garnier in 1924, all using the trading name "Leroy et Fils", which however will often be found written on watches thus "LeRoy & Fils". The firm moved to Avenue Opéra sometime after 1883. They finally closed down in 1960 following litigation instigated by L. Leroy & Cie. See further history in Chapters II & V. Most interestingly, and since this book was written, a collection of watches sold at Christie's on 26th June 1973 included examples produced both by the firm originally in the Galerie Montpensier and also by the firm originally in the Galerie Valois.

List of Names and Trade Marks

LETOREY (R.)   13 Rue des Arquebusiers, Paris
Couaillet's Paris agent. (1914 *Catalogue*).

LEUBA (Henry, sen.)   Bâle
London Exhibition 1851. "Two travelling clocks, furnished with alarm, etc."

LEVY   at or near Montbéliard
Tripplin 1890, p.119. Platform escapement maker on a large scale; but not so large as either L'Epée or Besson. See Chapter VI.

LEVY HERMANOS   Hong Kong
A retailer whose name appears upon a French carriage clock with a digital dial. There are two trains. The one is the going train. The other is provided to move the digital system without its imposing an unnecessary load on the going train. See Chapter VIII.

LEZE   Paris
The maker of a carriage clock platform escapement (spring-detent) both signed and numbered. Lézé succeeded Blondeau in Rue de la Paix.

LIORET
Brevet No 122,278 (25th Jan. 1878). "Application of an horological mechanism, called a winder for a carriage clock, and the peculiar location of this mechanism". See Chapter VIII. Lioret is credited in the *Revue* Vol. IX with what today is called a "Jacot tool".

LUCIEN   Paris
Several carriage clocks seen. All appeared to have been made pre-1850 and struck on bells. Name "LUCIEN PARIS" on backplates.

L--------?
E.G.L. in an ellipse. Mark seen on carriage clock backplate of circa 1885 (a *gorge*-cased repeater).

L----------?
"G.L." in an ellipse seen on a *Corniche*-cased strike/repeat clock.

(G.L)

L---------?
"H.L." trademark found on the inside of plates of the movements of very many French carriage clocks. Origin unknown. Probably a maker of *blancs, blancs roulants*, or possibly of finished clocks or of finished *mouvements*. Lepine has been suggested. No evidence. "H.L.", who ever he was, must have been well known. A feature of these movements is that often the pillars are attached to the front plate (pillar plate) by large steel countersunk screws. Henry Lepaute is a further possibility. Some Paul Garnier clocks with Garnier escapement are marked "H.L." in addition to Garnier's signature. Lepaute certainly exhibited a *pendule de voyage* in 1827, while Garnier produced his only two years later; so the dates would agree.

MAILLARDET family   La Chaux-de-Fonds
According to Chapuis, writing in 1917, one member of the family ran an horological school with Klentschi in La Chaux-de-Fonds from about 1814. "Fritz" Courvoisier was one of their pupils. The same author also says that Henri-Louis Maillardet (1790-1842) worked for the Courvoisier family. He also links the name of Maillardet with Jaquet-Droz and Leschot. There are a *Magicien* automat and also a two-singing-bird cages by Maillardet in the Chaux-de-Fonds Museum.

MAILLARD-SALINS   Hérimoncourt
Maker of pinions of all sizes. Lebon, writing in 1860, says that the steel came from England and that almost all the pinions were exported, especially to Switzerland. Active circa 1821; but by 1826 the business had diminished.

MALLET (Louis)   Paris
One travelling clock seen signed on the dial "L. Mallet, Horloger de S.A.R. Mgr le Duc d'Orléan, A PARIS". This piece has a round-plated movement, wound from the front, going eight days with *grande sonnerie* striking on two bells. There is also a pull-wound alarm attachment and a pull-repeat cord. The escapement is a cylinder mounted on its side at the top of the backplate. The case and dial appeared to have undergone an unspecified amount of restoration and/or alteration.

MARC (Henry)   Paris
One of the names found most commonly upon sound but quite unremarkable 19th century French pendulum clocks *(pendules de Paris)* often in wooden cases. Carriage clocks exist signed on their dials "H^{RY} MARC PARIS". One example has four dials and bears Blumberg's name. There seems to be no evidence whether Marc was a maker or simply a retailer (probably the latter). One fairly early clock, No. 26886, was supplied to Marc by Japy Frères, who probably made many "Marc" clocks.

MARCHAND
Brevet No. 127,663, 1878, for carriage clock playing music at the hours and for the alarm. See Chapter VIII.

MARCOT   At or near Montbéliard
Platform escapement makers in a small way. See Chapter VI.

MARGAINE (François-Arsène)   Rue Beranger 22, and Rue Bondy 54, Paris
A very famous maker of carriage clocks. Trademark "A.M." with a clock in the form of a beehive or skep between the "A" and the "M". Pitcher & Co. imported Margaine clocks up to 1914. *Revue Chronométrique*, May 1880, p.70, says that Margaine used *roulants* both from Franche-Comté and from St. Nicolas. *Brevet* Margaine 1869, alarm system for *mignonnettes*. Same year with Dorius, *Brevet* carriage clock cases with decorative panels. Serial numbers of clocks seen range from 2355 to 18303. Margaine was advertising regularly in the *Revue Chronométrique* as late as 1912. Japy's workbook "Mois de Janvier 1907, Mouvements blancs et finissages voyages" shows Margaine as the main customer of the period. See DUBOIS and also DORIUS. Some Margaine clocks, in addition to the trademark, will be found punched "MARGAINE PARIS" inside their back plates. Margaine was

445

awarded a Silver Medal at the Paris Exhibition of 1889, and a Gold Medal in 1900.

MARTI (S. & Cie)   Le Pays de Montbéliard
Mentioned by Sire in 1867 "in same breath" as Japy Frères and Roux. *Revue* March 1867, p.276, "Japy, Marti, Roux, à Paris, Boulevard du Prince-Eugène, 3." Vinter's history says, P.182, Japy-Marti-Roux formed in 1863 to market for all three in Paris their clock *mouvements*. Marti were makers of *roulants* and customers of L'Epée for *porte-échappements*. Sire, writing in 1867, p.82 specifically mentions Marti in connection with carriage clock *roulants*. The firm are said still to be in business making parts for motor cars (Peugeot). Their *porte-échappement* business was absorbed by L'Epée according to Albert Japy. Mantel clock seen 1972 with compound pendulum and with typical "Doubs" *coup-perdu* gravity escapement in the dial. This clock is in the form of a *régulateur*, having a *mouvement de Paris*, suitably re-designed where necessary. Tardy gives a Paris address Rue Vieille-du-Temple in 1870-1890.

MARTIN (Emile)   Saint-Nicolas-d'Aliermont
Georges Sire, writing of the 1867 Paris *Exposition*, lists Emile Martin with A. Delépine and Sauteur frères, all of Saint-Nicolas-d'Aliermont, as top-class makers of *mouvements* of carriage clocks. *Revue* Oct. 1859, p.60, says that M. Martin is the largest manufacturer in the village. Indeed, his *atelier* would not seem out of place in the Doubs or Haut-Rhin. He has 200 workers.

MATHEY-TISSOT   Les Ponts-de-Martel, Switzerland
Makers of *montres pendulettes de voyage* from just before 1900 until about the Second World War. See Chapter X.

MAURANNE   Saint-Nicolas-d'Aliermont
This family at one time fitted platform escapements to carriage clocks for Couaillet.

MAURICE (E. & CO.)   Rue Charlot 75, Paris
Showed carriage clocks in Paris Exhibition, 1889. A very fine maker, specialising in *cloisonné* enamel, in unusual dials and in small clock sets *(Garnitures de Cheminée)*. One of the very few makers ever to use black dials. Sole agent in Britain, E. Pitcher & Co., 3 Clerkenwell Road, London, E.C.1. Succeeded by Blanpain. See Appendix (c).

MAYER & SCHATZ   Paris
Name found on carriage clock with unusual composite escapement, being in effect a duplex without the frictional rest. See Chapter VIII.

MAYET (frères)   Morbier
Mentioned by Lebon in 1860 as having been the founders in the 17th century of the horological industry in the French Jura. See Chapter VI.

MEGNIN   At or near Montbéliard
Tripplin 1890, p.119. Platform escapement maker on large scale.

MICHOUDET (Jean-Marc)   Foncine-le-bas, Jura
Early carriage clock showing signs of having been based upon a style of movement derived from *Capucine* practices, and having a Franche-Comté platform cylinder escapement. See Chapter VI.

MILLOT   Paris
Brevet with Japy for carriage clock hour and quarter hour repeating work 1877.

MIROY (frères)   Rue d'Angoulême du Temple, Paris
Showed carriage clocks London Exhibition 1851. Several examples seen, date probably circa 1860, with "Miroy Frères Prize Medal" on dial, and name and address engraved on the backplate.

MOLITOR
See COULON.

MONNIN   Paris
Relation of Fritz Japy. In charge of Paris warehouse at 108, Rue du Temple from about 1848. He was certainly there in 1851. Later the offices were moved to Monnin's *appartement*, and the stores, etc., to Quai Jammapes (Muston). Georges Monnin et Cie. S.A. are still manufacturing watches at Charquemont. Lebon in 1860 mentions Monnin in connection with ruby "stones" and Pierre Monnin as a balance maker in 1860.

MONTANDON (frères)   Rambouillet and Paris
Mainspring makers on a very large scale for both clocks and watches. Beillard mentions them as Lefèvre et Montandon, 1851-57. *Recherches* p.122. They were awarded a Silver Medal at Paris in 1855. Sire records them as active in 1867 and mentions that their factory at Rambouillet had a *force motrice* of thirty horsepower. *Revue* March 1867, p.377, "Montandon Frères à Paris, Rue Culture-Sainte-Catherine, 28". Tardy's newly published List gives the Paris address Rue Lions St-Paul for the year 1825 for Montandon frères. Montandon exhibited springs in the London Exhibitions of 1862.

MOSER   Boulevard du Temple 15, Paris
London Exhibition 1851, "... travelling clocks of all descriptions", *Tribune Chronométrique* p.257. One clock seen dating from the first half of the 19th century. Strike on bell. Name in full "MOSER A PARIS" on backplate. See Tardy's newly published List for addresses both before and after 1851. A wooden-cased alarm *pendule portative* exists signed "MOSER et Cie". According to Tardy's new *Dictionnaire* this signature should date the clock pre-1860. It has a Garnier-type escapement. Moser was apparently connected with Marti of Montbéliard. Moser is also mentioned by Lebon in 1860 as having in 1834 started in Montbéliard a factory making clock movements.

List of Names and Trade Marks

MOUGIN & CIE   Rue de Turenne 75, Paris
Name appearing on wrapper of Japy *blancs-roulants* of carriage clocks as "Dépôt à Paris" with Japy, Marti, Roux. Almost certainly connected with the maker of *roulants* at Hérimoncourt (Doubs).

MOULINIE   Geneva
See Chapter X.

NUMA-HINFRAY   Saint-Nicolas-d'Aliermont
Silver Medal for "... un mouvement de pendule de voyage, petite dimension, grande sonnerie et réveil" at Concours Annuel de la Chambre Syndicale de l'Horlogerie de Paris 1887-88.

OLLIER (Adolphe)   20 Rue des Marais, Paris
The last regular manufacturer in Paris of really fine carriage clocks with porcelain panels. Styles offered included cases with malachite and lapis lazuli panels and cariatides. No trademark. Not manufacturing after 1930. Dials usually silvered and engine-turned with a plain space above the hour circle for the customer's name. See Ollier Catalogue, mentioned in Bib. *Brevet* No. 116,868 of 1877 (*Revue* Mar. 1878, p.40) is for a carriage clock with musical alarm, with music after the alarm or at any other time desired. This Ollier may have been the father of Adolphe. Japy's *pendule de voyage* workbooks for Jan. 1907 show Ollier to have been a customer for quarter repeaters with alarm. E. Pitcher & Co. were the sole importers of Ollier clocks in Great Britain.

OUDIN   Paris
E. Bruton in *Clocks & Watches*, 1968, illustrates on p.76 a really fine *petite sonnerie* carriage clock with calendar, alarm and seconds dial. Mr. Pitcher had a repeater with porcelain dial (hunting scene). It was signed "Ch. Oudin Paris" on dial, backplate and base plate. These clocks were certainly made long after the death of the famous Charles Oudin, who was active c.1807-25, and they were probably sold by his son. See DETOUCHE. See OUDIN-CHARPENTIER.

OUDIN-CHARPENTIER   Paris, London, New York,
                                          St. Petersburg, Moscow,
                                               Madrid and Geneva
A. Charpentier was successor to Oudin *fils*, son of Charles (See *Revue Chron.* Vol IV, Nov. 1862, p.291; misprinted as "1852" at foot of page). To accompany the display at the 1862 Exhibition, Oudin-Charpentier produced a book, which may be seen at the British Horological Institute Library, entitled *Catalogue of Chief Exhibits by Oudin-Charpentier, principal clockmaker to their Majesties The Queen and King of Spain and to the Imperial Navy*. Of particular interest are items 10, 11 and 13. No. 10 showed "a new system to facilitate the regulating of carriage clocks". Exhibit No. 11 consisted of two very fine carriage clocks, with artistic cases and chased silver mounts. Both were *grande sonnerie* with alarm, day and date. One of the dials was in crystal and the motion work made in tempered steel was visible. No. 13 was a clock made for his Majesty the King of Spain. Description:- "Small carriage clock, quarter repeat, lever escapement, chronometer regulation; the plain case surmounted by a small crown acting as a repeat button; on the glasses round the clock the coat of arms and the monogram of the King are engraved and inlaid with silver". There are photographs of two very fine carriage clocks in Oudin-Charpentier's *Catalogue* (Universal Exhibition London, 1862, Class 15, No. 1590 French Section).

PAPIN   Saint-Nicolas-d'Aliermont
Tradition associates the name of Papin with that of CROUTTE, which see.

PARODI   55 Via Abbruzzi, Milan
Couaillet's Italian agent. (1914 Catalogue.)

PATAY   Paris, Rue Vielle du Temple, 1860
                              Rue des Filles-du-Calvaire, 1870
London Exhibition, 1862 (*Revue* May 1862, p.121, Vol. IV). Hon. Mention for carriage clocks according to *Almanach Artistique* 1862. Patay filed a *Brevet* in 1859 for improvements relating to carriage clocks (*Revue Chron.*, list of *Brevets*, usually Vol. VIII).

PEQUIGNOT (Edouard)   Badevel
Aged 93 in 1971. He worked for Japy Frères from 1890 to 1940 including on *pendules de voyage*.

PERRENOT & MONGE   Pontarlier (Doubs).
Makers of round and square section steel for horology. Established 1826. Noticed by Lebon in 1860.

PEUGEOT (frères)   Hérimoncourt (Doubs)
Paris Exhibition official report, 1819, "Makers of excellent steel for springs of clocks and watches". One of the *familles bourgeoises* who started the Industrial Revolution in the Franche-Comté. Also mentioned by Lebon writing in 1860. The Peugeot motor car factory today is at Montbéliard.

PEUGEOT (Jacques)   Hérimoncourt
Made 4,800 files a year in 1859 (Dubois, 1860).

PEUGEOT (Pierre)   Hérimoncourt
Made 3,000 files a year in 1859 (Lebon, 1860).

PHILIPPE (Adrien)   Geneva
Born 1815. Died 1894. Very famous in the history of Swiss watchmaking. Mentioned here because of his suggestion made in about 1878 that the workman of Geneva should take up the manufacture of carriage clocks. See Chapter X. Biography and portrait Beillard *Recherches*, p.186.

List of Names and Trade Marks

PIAGET (frères) La Côte-aux-Fées, Switzerland
Worked in connection with the lever escapement of the last Breguet "humpback" clock numbered 759, and completed in 1931. The Piaget frères were the father and uncle of M. Gérald Piaget, head of the firm of the same name.

PICARD (?) Besançon
Platform escapement makers on a fairly modest scale.

PIERRET (Victor-Athanase) Paris
Revue, May 1862, p.121, Vol. VI. "Pendules Portatives". *Brevet* 1853 calendar work for day of week, the date and the month, visible through three apertures in the dial. Described *Revue* Oct. 1855, page 54-55 with illustration. That Pierret used duplex escapements in his clocks is clear from a description in the *Revue* of Feb. 1856, p. 105, of his device to prevent accidental occasional over-large balance amplitudes resulting in double-escapes for a few beats.

PINCHON Paris
A very well made and fairly early carriage clock has both the going and the striking trains driven by one mainspring, using opposite ends in a manner similar to some carriage clocks sold by Breguet and to Jura work.

PITCHER (E. & CO.) 6 Dyer's Buildings, Holborn, London, E.C.1.
French carriage clock specialists. Ernest S. Pitcher agent for V. Blanpain (Corpet) from 1880. In 1886 moved to 3 Clerkenwell Road where firm remained for fifty years. Agent after 1900 for E. Maurice et Cie of Paris and for Couaillet Frères of Saint-Nicolas-d'Aliermont. Also imported from Chevellier, Duverdrey & Bloquel, Hour and Margaine. See Appendix (c) Mr. Maurice Albert Pitcher, the last proprietor, retired in 1973 after 58 years in the carriage clock trade. He died on 7th May 1973 in his 76th year.

PITOU Rue Turenne 85, Paris
The last craftsman in Paris to finish *blancs roulants* of carriage clocks, still using in 1970 materials bought from Jacot some fifty years previously. M. Pitou at one time worked for Jacot, the nephew. Subsequently, and for very many years, M. Pitou worked for himself, producing clocks sold almost exclusively for the House of Breguet. See Chapters II & V.

PONS (Honoré; otherwise Pons-de-Paul)
Paris & Saint-Nicolas-d'Aliermont
Early 19th century. Pons was brought to Saint-Nicolas-d'Aliermont in 1806 by the Minister of the Interior in order to re-organise the flagging industry there (See Chapter IIII). The *Tribune Chronométrique*, page 91, mentions the arrival of Pons at St. Nicolas in 1806, saying that until then small clocks were made there "badly enough", but that Pons changed all this, and that it was to him in the first place that France largely owed her important clock export trade. Since 1830, says the *Tribune*, the *blancs-roulants* produced under the control of Pons have been constantly improved, the tooth-forms which he introduced being particularly exact and even ordinary clocks made under his direction well ahead, as usual, of those made by others. The *Rapport du Jury Central*, on the occasion of the 1823 Paris Exhibition, makes it plain that even by then Pons was already specialising in machine-made movements "pour pendules portatives" made in conjunction with various escapements (". . . échappements mixtes combinés par M. Pons lui même"). Pons won Silver Medals in 1819 and 1823, and a Gold Medal in 1834. He received the Légion d'Honneur in recognition of his services to French horology. Pons abdicated in favour of Delépine sometime after 1844 (probably in 1847). The *Revue Chronométrique* (Vol. III, Oct. 1859, p.60) says that Delépine continued to use the name of Pons on all his models. See DELEPINE & CAUCHY.

POTONIE (Léon) 5 Rue Neuve Saint-Francois, Paris
London Exhibition 1851. Exhibited many kinds of clocks, including *pendules de voyage*. Depôt, 20 Red Lion Square, London.

POUGEOIS (Ch.) Paris
Probable name on some carriage clocks finished by Jacot as evidenced by punches once belonging to Jacot and now in the possession of M. Pitou.

POULIGNOT Montécheroux (Forge de Soulce)
Mentioned by Lebon in 1860. In 1834 there were 50 workers engaged in production of *fournitures d'horlogerie*.

P---------? ("V.A.P.")
These letters in an oval are found on clocks of modest quality, of late nineteenth or early twentieth century origins, including carriage clocks.

QUIBEL (Eugène) Saint-Nicolas-d'Aliermont
Exhibited carriage clocks in Paris in 1889 (*Revue Chronométrique*). Quibel is also on record as having exhibited in 1889 screws of all descriptions (*Catalogue Général*).

RABY (Louis) Paris, and Versailles
Breguet pupil. Made a *pendule Sympathique* to go with watch Breguet No. 722 circa 1822. See Chapter II. An exceedingly fine gold pocket chronometer exists. It is small and slim, having pivoted-detent escapement. It is signed "Louis Raby. Horloger de l'Empereur" and is No. 1116. The barred movement is made in aluminium. Raby showed this watch, or one like it, at the Paris *Exposition Universelle* in 1855.

RAINGO (frères) Paris
Best known for Orrery clocks, etc; but at Sothebys on 31st October 1966 (Lot 213) appeared a highly unusual carriage clock signed "Raingo Frères". It had *grande sonnerie* striking on two bells and cylinder escapement on the backplate. The front and back glasses were removable for winding and setting. The silvered dial had a seconds circle at the top. Mr. Pitcher saw in 1971 a simple but early carriage clock, strike on

List of Names and Trade Marks

bell, with "RAINGO FRERES, PARIS" on the back plate in a circle. Another clock signed "Raingo Frères à Paris" appeared at Christies on 26th October 1971. It had outside locking plate and struck on a bell. The (?Doubs) skeletonised lever platform escapement was in the English style. Raingo exhibited a carriage clock in the *Exposition des Produits de l'Industrie Francaise* at Paris as early as 1834. The *Notice* gives his address as "8 Rue de Touraine, à Paris au Marais". It is made clear that his *pendule de voyage* had glass panels, balance escapement, *grande sonnerie* striking, repeat and alarm. (Ref: Bib. Nat., Paris, No. V.38339, page 74). Dents bought carriage clocks from Raingo in the 1860's (shown in old stockbooks).

RAIT (D.C.) Paris
A name found on a mid 19th century clock having an early "multi-piece" case, perhaps foreshadowing the later standard *Corniche*.

RECLUS (V.) Rue du Temple 14, Paris
Showed "...a beautiful collection of carriage clocks..." at the Paris Exhibition of 1867. Silver Medal for carriage clocks 1878. Trademark, the initials "V.R." with a "sunburst" above. Reclus did a large business in alarm clocks. Some "V.R." movements will sometimes also be found to bear Japy's trademarks.

REDIER (Antoine) Rue de Châtelet, 2, Paris
Saint-Nicolas-d'Aliermont
Born 1817. Died 1892. First in Paris 1832. A friend of Arago. Great Exhibition 1851, London, "Travelling repeating clock, new invention". Biography with portrait in Beillard *Recherches*. Exhibition report 1855 mentions that Redier makes cheap alarms or portable clocks almost all destined for the English market, which absorbs annually 35,000 or 40,000 pieces. *Brevet* 1857 for enamel dials with no false plates.

RENESON
Brevet 1868. Striking system for full-sized carriage clocks and for the small type called *mignonnettes*. Réneson's arrangement is described and illustrated in the *Revue Chronométrique* of October 1872. One such miniature clock seen, 3½ inches tall with handle up. It strikes hours and halves on a bell in the base. The *cadrature* is on the back plate, which is inscribed "R. B^TE S.G.D.G." in an ellipse. There are great wheels at either end of a single barrel.

REQUIER (Charles L.M.) Rue Debelleyme 5, Paris
Showed carriage clocks Paris Exhibitions 1878 (Gold Medal) and 1889. Carriage clocks comprised only part of his trade. See Tripplin's remarks quoted in Appendix (b). He says that Requier, although buying the majority of his movements like every other maker, has the science to carefully select what he buys and the talent to perfectly finish them. He is considered the first clockmaker of Paris.

REUILLE Morteau (Doubs)
Tripplin 1890, p.119. Platform escapement makers on large scale. Their levers said to be "a little better made" than most from the area.

RICHARD (C.A. et Cie) Rue de Bondy 32, Paris
According to Mr. Pitcher this concern was founded in Paris as Lemaître & Bergmann in 1848. A branch was opened in London 1857. In 1867 the branch became Richard & Co. of 24 Cannon Street, London. This firm is still in business trading as French Clock House Ltd., 8/9 Greville Street, London, E.C.1. In the Paris *Exposition* of 1889, Richard exhibited carriage clocks receiving an Honourable Mention. Trademarks:- Small "R" and "C" with snake's head between. It is probable that the inscription "R & Co." without the trademark denotes clocks handled by Richard, but not made by them. Richard & Co.'s advertisement in Kelly's *Directory of the Watch and Clock Trades*, 1887, offering "French clocks and carriage clocks (own make)" goes out of its way to say "All Own Made Goods are of Guaranteed Quality and bear our Trade Mark".
Quinet & Richard 1880. Quinet & Richard awarded Bronze Medal at the International Exhibition, Sydney, N.S.W. 1879. The *Horological Journal* of Nov. 1884 describes on pages 31-32 Richard's patent carriage clock striking work in which the hours are sounded before the quarters at each quarter as in a normal French *grande sonnerie* clock; the sequence being reversed at the hours (i.e. the quarters struck first).

ROBERT (Henri, fils)
Revue May 1880, p.69, Vol. XI. Improvements in carriage clock *roulants* said to result in better timekeeping, and first thought of by his father.

ROBERT
See COURVOISIER.

ROBIN Paris
Brevet No. 115,438 (9th Nov. 1876). Relates to improvements in calendars and hand-setting for carriage clocks and *mignonnettes*.

RODANET (Julien Hilaire) Paris and Rochefort
Born 1810, died 1884 or 1885. Paris Exhibition, 1878 Silver Medal for carriage clocks. A very distinguished horologist, famous alike for watches, chronometers and carriage clocks. He received many awards and took particular trouble with his *élèves pensionnaires*. One punch found in M. Pitou's workroom in Paris, and which punch he acquired with Jacot's old materials, proves beyond reasonable doubt that Jacot supplied at least some of the carriage clocks sold by Rodanet fils. See Chapter V. A biography and a portrait are to be found in Beillard. Adrien Philippe in *Etudes... 1878* goes out of his way to praise the finish and superior timekeeping qualities of Rodanet's escapements for portable clocks.

449

List of Names and Trade Marks

ROLLIN   Paris
A few carriage clocks seen, each with strike on bell and probably made before 1850. Named in full "ROLLIN A PARIS" on backplates. One example had a cylinder escapement. Not necessarily a maker.

ROUSAILLE (Denis Jaques) Rue de la République 58, Lyon
Paris Exhibition 1889, showed carriage clocks and alarms striking the hours. President of the Trade Union Committee of Clockmakers in Lyon. Account in book form of the 1889 Exhibition written by him gives no further information on the carriage clock section than that of Tripplin.

ROUX ET CIE   Montbéliard
According to Sire, this firm was in 1867 engaged in manufacture of *blancs* and of *roulants* for clocks including for *pendules de voyage*. "Japy, Marti, Roux et Cie" was the combined marketing arrangement of three firms selling similar products. The Paris address was Boulevard du Prince-Eugène, 3. See *Revue* Mar. 1867, p.276, and Sire p.82. The *Tribune Chronométrique* tells us that Roux succeeded Vincente et Cie and made "... mouvements roulants de pendules, des mouvements de lampes de divers pièces d'horlogerie de toutes grandeurs et de différent calibres". They received various awards for *roulants*. *Pendules de voyage* are not specifically mentioned by Sire in this connection.

SAMUEL   (?) Paris
Made extravagant claims for his carriage clocks in which he used an escapement derived from that of Sully. See Chapter VIII.

SANDOZ (Gustave)   Paris, Galerie de Valois 1870
                                     10 Rue Royal, 1900
Mentioned by G. Sire, p.69, who says that the most attractive *pendules de voyage* to be seen at the 1867 Paris Exhibition were in the displays of Th. Leroy, G. Sandoz, Charpentier and Desfontaines. Sotheby & Co. sold on 24th April 1972 a miniature carriage clock signed "Gve Sandoz", described in the catalogue as "... the dial centre chased as a sawn tree trunk and the surround as three planks... the case well modelled as a thatched woodland hut decorated with owls, a cockerel, a lizard, ivy, bulrushes and woodland flowers...". Sandoz' address was misprinted as "30 Rue Royale".

SAULET   Paris
Tardy gives Saulet's addresses firstly as Galerie de Pierre, Palais Royal, then Galerie de Valois in 1820, and finally Rue Richelieu in 1850. See clock going a year described in Chapter VIII.

SAUTEUR frères   Saint-Nicolas-d'Aliermont
Mentioned in 1870 by Sire (p.82) as first class makers of carriage clock *mouvements*.

SAVOIE ROLLIN (Savoye de Rollin)
Préfét of Seine-Inférieure in 1806. It was at his request that Pons was first sent to Saint-Nicolas-d'Aliermont.

SCHELTZ (Adolphe; de la Maison Japy frères et Cie)
Mentioned in *Revue* as among assistants given Medals and Awards in connection with the Paris Exhibition of 1889.

SILVANI   Paris
Name inscribed on Garnier "Series I" *pendule de voyage*.

SIMON (Henri)   Paris
Received a Bronze medal from the Chambre Syndicale de L'Horlogerie for a carriage clock movement with a perpetual alarm (*Revue*, July 1880, p.121). See Chapter VIII for a similar mechanism.

SOLDANO   Paris & Geneva
Paris Exhibitions, Bronze Medals 1855 and 1878 for carriage clocks. "The Exhibition of this establishment showed substantial progress in the art of case decoration. On show were a group of carriage clocks ornamented with great taste and originality. The finish of these movements was conventional, but the escapements appeared to us to have been treated with particular care. Several items merited the highest praise." (*Revue Chronométrique*, 1880, p.70, report on 1878 Exhib.) The same *Revue* says that Soldano used both Franche-Comté and Saint-Nicolas *roulants*. Trademark:-

                       ⬡ JS PARIS ⬡

Repeating carriage clock seen with trademark and "J. SOLDANO" on the escapement platform.
Brevet No. 107,046 (3rd March 1875) is for "Veuve de Soldano fils. Mouvements de pendule de voyage à quarts, à grande sonnerie et à répétition à minutes". Dent's of Cockspur St. were regular customers of Soldano circa 1860-1865. Soldano's gongs were often made by "M.V." The beautiful "J.S." escapements usually found in Soldano clocks were in all probability produced for him in Switzerland.

STAUFFER (Onésime)   Les Ponts-de-Martel, Switzerland
Maker of *montres pendulettes de voyage*, circa 1930.

STEVENARD   Boulogne
Carriage clock in fairly early "multi-piece" case, signed on dial "STEVENARD HORLOGER MECANICIEN A BOULOGNE" (King & Chasemore, Pulborough, 31st. Oct. 1973).

STOFFEL (Fernand A.)
                      Rue Faubourg Saint Honoré 24, Paris
Showed carriage clocks in the Paris Exhibition of 1889.

List of Names and Trade Marks

THEVENIN (?) Paris
An escapement maker. See Saunier 1872 in Bib. The actual passage translates to read, with reference to Thévenin and the duplex escapement for portable clocks, "People have tried for some years past to democratize this escapement, in order to make it available to the everyday clock trade and at the same time reasonably profitable. We believe that we remember that those who made this attempt had begun by causing the late Thévenin to make two trial pieces, and taking advantage of the poverty in which he lived had promised him orders for hundreds in order to obtain their trial pieces cheaply. After that he never saw them again". (*Revue* Aug. 1882, p.276). See THEVNON (??)

THEVNON Enghien near Paris
According to Tardy, a maker of platform escapements circa 1910-1914.

THIEBLE-VIEVILLE 1 Rue Commines, Paris
Advertisements *Revue* 1912. Offers hands, pendulums, suspensions and all materials for clocks, carriage clocks and alarms, and speaks of factories at Ruyaulcourt and at Bapaume (*Fabrique de Fournitures d'Horlogerie et de Décolletage*).

THOUROT frères Montbéliard
Maker of files (Lebon in 1860).

TRONCHET (Julien; de la Maison Carpano)
Paris Exhibition 1889. Among those assistants listed as receiving Medals or Awards.

TURLIN (E.) Paris
Mr. Pitcher saw a carriage clock No. 2592. Repeat and alarm. Very heavy case. Turlin was probably a retailer.

VARSOVIE 38, Swietzojerska, Pfefferberg
Couaillet's agent in St. Petersburg (Leningrad) (1914 Catalogue).

VAUCANSON (Ateliers) Saint-Nicolas-d'Aliermont
Succeeded Lamazière and Bunzli and took present name from 12th February 1914. Firm made early speedometers for locomotives, cinema projectors, calculating machines and magnetos. One factory was in Paris, one in Blesdal between Saint-Nicolas and Saint-Aubin-le-Cauf.

VENOT FRERES ET CIE 134 Rue du Temple, Paris
Wholesalers of clocks, watches and tools, carrying the products of Japy. (Advertising *Revue* 1912).

VEYRIN Paris
Octagonal-cased portable clock, also signed "LE ROY & FILS, HR DU ROI, Palais Royal".

VILLON (Albert) Saint-Nicolas-d'Aliermont from 1867
Showed carriage clocks at Paris Exhibition, 1889. Tripplin writes (p.85) "The exhibit of M. A. Villon, expert to the jury and Mayor of St. Nicolas, was an important one and represented in a worthy manner the industry of St. Nicolas in all its details and its adaptability to undertake general work. Indeed, the appearance of St. Nicolas strangely reminds one of a Swiss village in the canton of Neuchâtel. The factory of M. Albert Villon is a large, commodious, and elegant structure surrounded by gardens, one story high and divided into many shops, each devoted to some speciality. A visitor is shown over the designing room, in which new tools, models of cases, fresh designs of dials, etc. are drawn; the machine shop where the tools are made; the rooms where the mountings are moulded, filed, the cornices chiselled by means of circular cutters, then stoned, polished, and gilt; rooms where the plates of the movement are stamped out, the pinions and wheels cut, the screws made, pivoting done; the glass cutting and polishing shop; the assembling, regulating and packing rooms; in fact, a complete factory, as M. Villon professes to be independent of anybody in his system of manufacture. He claims to make 60,000 alarms, 20,000 carriage clocks, and 20,000 chimney clock movements per year". See Chapter IIII for history of firm which became Duverdrey and Bloquel, and today is Bayard.

VINCENTI & CIE Montbéliard
Succeeded by Roux (*Tribune*, p.91). According to Galliot, Vincenti was a Corsican who set up circa 1823 in Montbéliard, making rough movements of clocks and watches using machines of his own invention. Vincenti went bankrupt in 1824, the factory from then onwards being managed by Roux. There were 30 to 40 workers in 1825, and "M. Vincenti et compagnie" at Montbéliard (Doubs) were still showing in Paris in 1824, receiving a Silver Medal for clock movements.

VIOTTE (Jacques) (?) Besançon
Mentioned by Albert Japy (letter 11th Jan. 1971) as being absorbed by L'Epée.

VISSIERE (Simon) Argenteuil and later at Le Havre.
Born 1822. Died 1887. A famous maker of marine chronometers. He was taught by Louis and Charles Berthoud for whom he long worked. He showed marine and pocket chronometers for the first time at the *Exposition des Produits de l'Industrie Nationale* in 1849, being awarded a Silver Medal (Vide *Tribune Chronométrique*, p.75). One of Vissière's pupils was Louis Carpano who in about 1846 opened at Cluses the first factory in France making wheel and pinion cutters. Another Vissière pupil was Théodore-Marie Leroy.

VOISIN (Charles) Paris
Retailer supplied by Drocourt, and probably by others. Trademark "C.V." on backplate of movement.

VOLOGNE Paris
Shown in Dent stockbooks of 1860-65 as supplying carriage clocks. May only have been a wholesaler.

V---------?
Evidently a gong-maker, for the initials "M.V." are found on gong-blocks.

451

YERSIN (a Swiss) (?) Somewhere in the Doubs
Tripplin 1890, p.119. Very well made platform escapements (presumably levers) exhibited at Paris, 1889. The *Catalogue Général* of the 1889 Exhibition also mentions a "petite pendule de voyage, mouvement terminé". (This is likely to have been a *montre pendulette de voyage*).

Bibliography

The works listed below are either mentioned in the text or else they contain at least some material having direct relevance to the present book *Carriage Clocks*.

ABBOTT (Henry G.) *The Watch Factories of America.* Chicago, 1888 and reprint c.1972.

ABBOTT (Henry G.) *A Pioneer. A History of The Waltham Watch Company of Waltham, Mass.* Chicago, 1905 and reprint 1968.

ABBOTT (Henry G.) *The Watch Factories of America. Past and Present.* Chicago 1888. Now available as an Adams Brown Company reprint.
Abbott mentions (p.101) the retail price of $3.50 in connection with the original Waterbury Long Wind watch, but the booklet *The Whole Story of the Waterbury* published in 1887 says clearly in two places that the Long Wind retail price was then $2.50.

ALLIX (Charles) *Some Notes on Thomas Reid and on the Astronomical Regulator Clocks made by Reid and Auld, Edinburgh, with an Appendix upon similar Clocks by William Hardy, London.* Written 1948-49, but not published to date 1974.

ALMANUS MANUSCRIPT. See LEOPOLD.

AMERICAN HOROLOGIST AND JEWELER, Sept. 1971. Contains reprint of W.J. Gazeley's article on the famous Bridgman and Brindle watch.

ANSONIA CLOCK COMPANY. *Catalogue of the Ansonia Clock Company,* 1886/7. This Catalogue was supplied by the manufacturer to the retailer. It was intended to be shown to the customer; but no prices are given beneath the illustrations of the clocks. Presumably separate lists were issued to the retailer showing List prices and discounts.
Re-issued by American Reprints, St. Louis, Missouri, 1967.

ANSONIA CLOCK COMPANY. *Ansonia Clock Catalogue 1914.* This Catalogue was supplied by the manufacturer to the retailer, and was intended to be shown to the customer. Beneath the illustrations of each clock appears its List price, or in other words the retail price. Presumably the trade prices and discounts were shown on a detachable page or on a separate sheet. At the back of the Catalogue is a very comprehensive list of clock materials intended for repair work.
Re-issued by Hagen's Antiques, Benicia, California, 1969.

ANTIQUARIAN HOROLOGICAL SOCIETY (The). *Antiquarian Horology,* published quarterly since December 1953.

ANTIQUARIAN HORLOGICAL SOCIETY (The). *Tercentenary of Christiaan Huygens' Pendulum Clock, Dec. 1656 – Dec. 1956.* Catalogue of Exhibition 4th Dec. 1956 to 24th Feb. 1957.

ARCHIVES NATIONALES. The Central reference archives containing virtually all surviving documents relating to the history of France. Opened in 1789. Now housed in the Hôtel de Soubise in the Marais in Paris.

ARETINO (Pietro) *Carte Parlanti . . . ,* Venice, 1543.

ARMY & NAVY STORES. *Catalogue* of 1907, reproduced in facsimile 1969 under the title *Yesterday's Shopping.* This enormous Catalogue, showing retail prices, was distributed to the general public. A variety of French eight-day carriage clocks were illustrated ranging from the expensive to the mediocre. Various one-day American carriage clocks and alarm clocks were also shown including the Ansonia one-day *Climax* at 7s.6d., the Ansonia one-day *Bee* at 2s.11d., a Seth Thomas *Joker* at 10s.9d., and a Waterbury *Hornet* at 9s.3d.

ASSOCIATION NATIONALE DES COLLECTIONNEURS ET AMATEURS D'HORLOGERIE ANCIENNE. *Bulletins* published at irregular intervals from June 1968. Ten issues to November 1973.

AURICOSTE. See THOMAS.

BACHELIN, (A.) *L'Horlogerie Neuchâteloise.* Neuchâtel 1888 (not seen).

BAILEY (F.A.) *An Old Watchmaker's Workshop.* Published in the transactions of the Ancient Monuments Society, 1953. Deals with Joseph Preston & Sons of 19 Eccleston St., Prescot in the days of Harry Pybus.

BAILLIE (G.H.) *Watches.* London, 1929.

BAILLIE (G.H.) *Watchmakers and Clockmakers of The World.* London, 1947.

Bibliography

BAILLIE (G.H.) *Clocks and Watches. An Historical Bibliography.* London, 1951.

BAINES (Edward) *History, Directory and Gazetteer of the County Palatine of Lancaster,* Liverpool, 1825.

BALL (R.E.) and **GILBEY** (T.). *The Essex Foxhounds,* 1896.

BEAUREPAIRE (town of) *Notes Historiques.* Deals with Saint Nicolas-d'Aliermont.

BECKETT DENISON (E., later Lord Grimthorpe). *A Rudimentary Treatise on Clock and Watch Making,* 1850, and last edition 1903.

BEESON (C.F.C.) *English Church Clocks, 1280-1850.* London, 1971.

BEILLARD (Alfred). *Recherches sur l'Horlogerie, ses Inventions et ses Célébrités. Notice Historique et Biographique d'après divers Documents de la Collection de l'Ecole de Horlogerie d'Anet.* Paris, 1895. (printed by Achard, 10 Rue de Flandres, Dreux). Contains short biography of Paul Garnier with portrait.
(Alfred Beillard must not be confused with François le jeune Beliard)

BELIARD (**François le jeune**) *Reflexions sur l'Horlogerie en général et sur les Horlogers du Roi en particulier.* The Hague, 1767, 22 pp. A most interesting, and also exceedingly rare, pamphlet giving an informed contemporary opinion upon many horological matters. Unfortunately it would require much specific research to discover the identities of the various makers who are mentioned, but not specifically named.

BERLIOZ (M.J.). *L'Horlogerie, Rapports sur L'Exposition Universelle de 1878.* About 60 pages, three folding plates and other illustrations. Paris, 1878.

BERNER (G.A.). *Dictionnaire Professionnel Illustré de l'Horlogerie.* Chaux-de-Fonds, 1961.

BERTELE (H. von). *Book of Old Clocks and Watches.* London 1964.

BERTHOUD (F.). *Histoire de la Mesure du Temps,* 1802.

BESANCON (Imprimerie de Sainte-Agathe Ainé). *Mémoire présenté par les Fabricants et Ouvriers en Horlogerie de Besançon.* Nov. 1848.

BESANCON. (Imprimerie Dodivers) *Notice sur la Fabrication de l'Horlogerie à Besançon et dans la Département du Doubs, présentée à l'Appui de son Exposition Collective.*
Besançon 1867, with reference to the Paris Exhibition.

BIBLIOTHEQUE NATIONALE. Founded in Paris in 1926 to centralise a certain number of French libraries. Much in the nature of the British Museum Reading Room.

BLOCH-PIMENTEL COLLECTION. *Auction Catalogue.* Hôtel Drouot, Paris, 5th May 1961. Many fine items, mainly watches, early coach watches, and early clocks.

BONNIKSEN (B.). *The Karrusel Watch by Its Inventor.* London, 1905, and reprint 1967.

BORSENDORF (L.). *Un Coup de Loupe à L'Exposition Universelle de 1855.* Paris 1855. Sold by author and by principal bookshops. This author published regularly horological pamphlets. He was author/publisher of *La Loupe de L'Horloger,* 1850-1861. The importance of *Coup de Loup* lies in its descriptions of the actual *roulant* manufacturing methods used respectively in Saint-Nicolas and at Montbéliard.

BOVILL (E.W.). *The England of Nimrod and Surtees, 1815-1854.* London 1959. (Oxford University Press)

BREGUET (2 Rue Edouard VII, Paris). *Centenaire d' A.-L. Breguet (1823-1923).* A printed circular in booklet form asking owners of Breguet pieces to loan them for the Musée Galliera *Exposition* (Mrs. Mackenzie's copy).

BREGUET (C.). *A-L Breguet Horloger.* French typescript (Roneo) 1961 and subsequent translation into English.

BREGUET (**Maison Breguet, Neveu et Cie**). *Exposition de 1834. Produits de la Maison Breguet, Neveu et Cie,* Paris 1834. *(Notice advertissement.)*

BREGUET (**Maison Breguet, Neveu et Cie**). *Exposition de 1844. Produits de la Maison Breguet, Neveu et Cie,* Paris 1844. *(Notice advertissement).*

BREGUET. see **GALLIERA.**

BREVETS. See Glossary for notes. The *Revue Chronométrique* in 1873 re-published certain *Brevet* abridgements for "patents" relating to clocks, watches, etc. during most of the period 1792-1869. Some of these *Brevets,* where applicable, are mentioned in the Alphabetical List of Names. Subsequently further lists were published.

BRITISH HOROLOGICAL INSTITUTE (The). *Correspondence Course in Technical Horology.* Currently available. A most useful means of gaining a good general knowledge of the working of clocks and watches.

BRITTEN (F.J.). *Old English Clocks. The Wetherfield Collection.* London, 1907.

BRITTEN (F.J.). *Old Clocks and Watches and Their Makers.* 6th edition, London 1933, and photo-litho reprint 1971.
See also **CLUTTON.**

BRUTON (Eric). *Clocks & Watches, 1400-1900.* London, 1967. (Arthur Barker). Pages 67-69 contain a few notes on carriage clocks.

BRUTON (Eric). *Clocks & Watches.* London, 1968 (Paul Hamlyn). Fine carriage clock named "Oudin, Paris", illustrated p.76. It is a quarter repeater on gongs and has alarm and calendar. The case is engraved.

BUCKNEY (Daniel Patrick). Correspondence concerning the Dent family, written from Spain to C. Allix in 1972.

Bibliography

BUGGINS (Geoffrey). *The Maker Behind the Clock.* This work, written soon after 1960, has apparently not so far been published. It concerns "Thwaites" clocks sold under other names.

BUGGINS (Geoffrey) and TURNER (A.J.). *The Context of Production, Identification and Dating of Clocks by A. and J. Thwaites.*
Article in *Antiquarian Horology*, Sept. 1973.

BURT (Edwin B). *Derry Manufacturing Company.* N.A.W.C.C. *Bulletin*, April 1951.

CAMP (Hiram). *A Sketch of the Clock Making Business 1792-1892.* Written 1893. Unpublished for twenty-three years. Now available in facsimile from the Adams Brown Company. Camp began in the clock business in 1829 at the age of 18 with Jerome, who was his uncle. Afterwards Camp was associated with the New Haven Clock Company.

CARPANO (Louis). *Prix-Courant des Fraises pour l'Horlogerie et Petite Méchanique de Précision.* Cluses (Haute Savoie) post 1920. 27 pages, illustrations, portraits. Very wide range illustrated and tabulated, including cutters for wheels and pinions in brass and steel, constant-profile, hobs, fine-tooth cutters, convex, concave, pointed for screw-cutting, etc. Also offered are knurling-tools, 'scape wheels, racks, worms, etc. The outer cover of this Catalogue is over-stamped "Etablis Carpano & Pons". Louis Carpano (1833-1919). Charles Carpano in charge from 1902. Louis was trained by Vissière, the famous chronometer-maker at Le Havre. I have not traced the Pons connection; but probably it has some bearing upon the famous Pons family.

CARPANO & PONS. *Catalogue Général des Fraises*... Cluses and Paris (Fabrique d'Outillage de Saint-Etienne, 113 Boulevard Richard-Lenoir). Undated, but photograph of works in 1920. In addition to the expected cutters, the Catalogue offers topping-tools for watches, complete with two sorts of cutters *(fraises à arrondir)*.

CHAMBERLAIN (Paul). *Its About Time.* 1941 and 1964.

CHAPUIS (Alfred). *Histoire de la Pendulerie Neuchâteloise.* Published in Paris and Neuchâtel in 1917. A very important book, and Chapuis' first. It established him immediately as a horological author.

CHAPUIS (Alfred). *La Montre Chinoise.* Neuchâtel, 1922.

CHAPUIS (Alfred). *Pendules Neuchâteloises. Documents Nouveaux*, Neuchâtel 1931.

CHAPUIS (Alfred). *Fritz Courvoisier, 1799-1854, Chef de la Révolution Neuchâteloise.* Neuchâtel, 1947.

CHAPUIS (Alfred). *A-L Breguet et La Suisse durant la Terreur (1793).* Article in the *Revue International de l'Horlogerie*, Chaux-de-Fonds, 15th March 1951.

CHAPUIS (Alfred). *A-L Breguet Pendant La Révolution Française, à Paris, en Angleterre et en Suisse.* Neuchâtel 1953.

CHAPUIS (Alfred). *Grands Artisans de La Chronométrie.* Neuchâtel, 1958 (Editions du Griffon). Contains most valuable biographies and history.

CHAPUIS (Alfred) and JAQUET (Eugène). *La Montre Suisse* Switzerland 1945. Translated into English in 1953 under the title *The Swiss Watch*.

CHELSEA CLOCK COMPANY. *Company History.* Typescript, undated, U.S.A.

CHRISTIE, MANSON & WOODS, London. Auction *Catalogues*.

CHURCHILL (Winston S.). *A History of the English-speaking Peoples.* Vols. III and IV.
London 1957 and 1958. (Cassell & Company Ltd.)

CIPOLLA (C.M.). *Clocks and Culture, 1300-1700*, London 1967.

CLAPP & DAVIES. *Illustrated Trade Catalogue and Price List.* Chicago 1886. Robert Spence re-issue, 1964. The only carriage clock offered is the Welch *Carriage No. 2*, shown in Welch's own wholesale trade Catalogue of 1885.

CLUTTON (Cecil) and DANIELS (George). *Watches.* London, 1965.

CLUTTON (Cecil, with C.A. Ilbert and G.H. Baillie). *Britten's Old Clocks and Watches and Their Makers.* Eighth edition. London, 1973.

COLE (J.F.) and CUSS (T.P.). *A Watchmaking Centenary. Usher & Cole, 1861-1961.* London, 1961.

COUAILLET FRERES. *Catalogue. Tarif 1914.* This trade catalogue and price list has 32 pages but no illustrations. The cover states "Maison A. Couaillet, fondée en 1892. Acquéreurs et Successeurs de Maisons PONS et DELEPINE-BARROIS". Agents:— M.P. Letorey, Paris; M. Pitcher, London; Varsovie, St. Petersbourg; Parodi, Milan. A huge selection of possible permutations of movements and cases offered in column tabulation, ordered by quoting numbers. *Qualité soignée* had cylinder platforms. *Qualité supérieure* had levers. *Obis*, of course, had cylinders. *Mignonnettes* were offered in a wide range.

COUAILLET FRERES. *Catalogue, 1931-1932.* Saint-Nicolas-d'Aliermont.

COURVOISIER (Paul-Frédéric). *Fritz Courvoisier.* Paul-Frédéric Courvoisier was the son of the famous "Fritz". This biography of his father was first published by Jeanneret in 1863. Then later it was used again by Professor Emile Farny to add to his notes on "Fritz" published in *La Chaux-de-Fonds*.

Bibliography

Son Passé and Son Present, 1894. Finally Chapuis, in 1917, incorporated much of the same material in his *Histoire de la Pendulerie Neuchâteloise*.

CROMMELIN (C.A.). *Descriptive Catalogue of the Huygens Collection.* Leiden 1949.

CROUTTE et CIE (Auguste). *Etablissement d'une fabrique d'horlogerie à Saint-Nicolas-d'Aliermont.* 1847, 15pp. (Archives départementales de La Seine-Maritime, cote BHS.M.364/32).

DANIELS (George). *English and American Watches.* London, 1967. Very useful for history of progressive watch development, and for readily understandable drawings and explanations of escapements.

DANIELS (George). *The Art of Breguet.* London, 1974.

DAVIES (W.O.). *Gears for Small Mechanisms.* London 1953.

DAWSON (Percy G.). *The English Carriage Clock.* Article in *The Antique Collector*, Dec. 1957.

DE CARLE (Donald). *Practical Clock Repairing.* London, 1952.

DE CARLE (Donald). *Watch and Clock Encyclopaedia.* London, 1959/65.

DEDHAM TRANSCRIPT. *Joseph H. Eastman Perhaps Solves Perpetual Motion.* Newspaper article Dedham, Mass., 1st Jan. 1932.

DEFOSSEZ (L.). *Théorie Générale de L'Horlogerie.* La Chaux-de-Fonds, 1952.

DELEPINE-BARROIS. *Manufacture D'Horlogerie. Pendules de Voyage soignées (Genre Paris). Pendules Economiques. Carriage Clocks.* An undated Catalogue, mentioning on the cover the Brussels Exhibition of 1910 and with the "D.L.B." trademark. Intended for the English market. Illustrates 79 carriage clocks, identified by names and numbers. No prices. Some clocks, including several *mignonnettes*, are stated to be "GENRE PARIS" to distinguish them from "PENDULES ZERO bis" and from "PENDULES DE VOYAGE ECONOMIQUES (CHEAP CARRIAGE CLOCKS)". The Catalogue cover announces "SPECIALITE PENDULES DE VOYAGES, GRANDES SONNERIES, Délais, Répétitions, Mignonnettes, GENRE PARIS". Addresses of *Maisons de Vente* are given as 5 Rue St. Claude, Paris and 69 Hatton Garden, London. The factory at Saint-Nicolas-d'Aliermont is illustrated, almost exactly as it remains (empty) today except that now there is no tall chimney.
Printer: Roy & Valade, Amiens.
A facsimile reproduction will be produced in 1974 by Antique Collectors' Club, Clopton, Woodbridge, Suffolk, in collaboration with Charles Allix.

DENT (E. & Co.) *Advertisement of the Adoption of a Trade Mark by E. Dent & Co., Watch, Chronometer, Astronomical and Turret Clock Manufacturers, in Pursuance of the Trades Marks' Act, 1875.* London, 1879. The advertisement says that "Messrs. E. Dent & Co. beg to notify their friends and the public, that for the purpose of preventing fraudulent imitation of the horological and other instruments of their manufacture, they have found it necessary to protect them by a distinctive 'Mark', and have, therefore, under the above Act, registered the device. Every instrument constructed by them, after and including No. 38,000, has this 'Mark' and without it any instrument will be spurious." The "device" was the word "DENT" in a triangle and the pamphlet was produced mainly with a view to differentiating between "E. Dent & Co." and "M.F. Dent".

DESSIAUX (E.). *Catalogue*, circa 1880-1885. Shows mainly *Alarms Fantaisie*. See Chapter IIII. Not dated; but M. Forest of Bayard knows that the first steam engine at the Dessiaux factory (illustrated) was installed in 1880, and that certain lines shown had been discontinued by 1885. A trade Catalogue.

DEVELLE (E.). *Les Horlogers Blésois aux XVI et XVII siècles.* Blois 1913 and 1917.

DICKIE (A.B.). *French Carriage Clocks.* Article in *Antiquarian Horology*, June 1963.

DIDEROT AND D'ALEMBERT. *Encyclopédie ou Dictionnaire raisonné des Sciences, des Arts, et des Métiers.....* Paris, 1751-1780.

DIEPPE CHAMBER OF COMMERCE. *Enquête industrielle de 1809. Seine-Inférieure.* (Archives Nationales, Paris, sous la cote F.12.1569). Report on the horologists of Saint-Nicolas-d'Aliermont.

DITISHEIM (Paul). *A Famous French Horologist. The late Mr. Louis Leroy.* Article in *The Horological Journal*, London, March 1935. Portrait.

DREPPARD (Carl. W.). *American Clocks.* U.S.A. 1947 and 1958.

DROST (Wm.E.). *Boston Directory Record of Joseph Eastman.* Typescript dated 19th September 1955.

DROST (Wm. E.). *Clocks and Watches of New Jersey.* U.S.A. 1966.

DROST (Wm.E.). *The Various Boston Clock Companies.* Typescript, undated.

DROVER (C.B.). *The Brussels Miniature. An Early Fusee And A Monastic Alarm.* Article published in *Antiquarian Horology*, September 1962.

DRUMMOND ROBERTSON (J.). *The Evolution of Clockwork*, 1931 and re-edition 1972.

DUBOIS (Pierre). *Lettres sur les Fabriques d'Horlogerie de la Suisse et de la France.* Paris 1853. Lectures first published in *La Patrie*.

Bibliography

DUBOIS (Pierre). On Paul Garnier, see Appendix (a).

DUVERDIER (Antoine). Source quoted by Portal and de Graffigny for *anecdote* concerning the supposedly early travelling clock of Louis XI of France. See Chapter I.

DUVERDREY & BLOQUEL. *Pendules de Voyage. Carriage Clocks*. A large illustrated Catalogue, intended for the trade only and published circa 1910. Some 270 different models and shapes of carriage clocks are illustrated. This is the most important surviving carriage clock catalogue so far discovered. A facsimile reproduction will be produced in 1974 by Antique Collectors' Club, Clopton, Woodbridge, Suffolk, in collaboration with Charles Allix and by courtesy of Réveils Bayard, Saint-Nicolas-d'Aliermont.

ECKHARDT (G.H.). *Pennsylvania Clocks and Clockmakers*. U.S.A. current printing 1955.

EDEY (Winthrop). *French Clocks*. London, 1967.

EMPLOYMENT (Dept. of). *British Labour Statistics. Historical Abstract 1886-1968*. H.M. Stationery Office, 1971.

EXACTA. *Catalogue*, illustrated, with prices. Saint-Nicolas-d'Aliermont, 1933. Paper, 22 pp. History, illustrations of *Pendules de Paris* and platform drum timepieces and strikers, but no carriage clocks. Trade.
(Loaned by M. et Mme. Jules Delabarre, Saint-Nicolas-d'Aliermont).

EXHIBITION *Catalogues*. Paris (various), Besançon 1860, Nantes 1861, Vienna 1873, Sydney N.S.W. 1879, Melbourne 1881, Amsterdam 1883, etc. (See Chapter II).

EXPOSITIONS DES PRODUITS DE L'INDUSTRIE FRANCAISE. *Catalogues*. Paris 1819, 1823, 1827, 1834, 1839, 1844, 1849. See Chapter II.

EXPOSITION UNIVERSELLE. *Catalogue Général, Groupe III, Classe 26 (Horlogerie)*, Paris 1889.

EXPOSITION UNIVERSELLE INTERNATIONALE DE PARIS, 1878. *Catalogue Suisse*, Zurich 1878.

EXPOSITION UNIVERSELLE INTERNATIONALE DE 1900. *Rapport du Jury International. Group XV. Industries Diverses. Premier Partie. Classes 92 à 97*. Paris 1900.

FARNY (Dr. Emile). See LA CHAUX-DE-FONDS.

FEBVRE (Lucien). *Philippe II et la Franche-Comté*. Paris, 1911.

FEBVRE (Lucien). *Histoire de Franche-Comté*. Paris, 1912.

FENNELL (Geraldine). *A List of Irish Watch and Clock Makers*. Dublin, 1963.

FERRIDAY (Peter). *Lord Grimthorpe 1816-1905*. London, 1957.

FLACHET (E.). on Paul Garnier. *(Revue Chronométrique)*. See Appendix (a).

FOULKES (R.K.). *A Tompion Travelling Clock*. Article in *Antiquarian Horology*, Mar. 1958.

FRANCOEUR. *Extrait d'un Rapport.... sur les Ateliers d'Horlogerie de M. le Roy...* 30th Dec. 1840, Paris.

FRANKLIN (A.), *La Vie Privée d'Autrefois. Arts et Métiers. Modes, Mœurs, Usages des Parisiens du XIIe au XVIIIe Siècle, d'apres des Documents Originaux ou Inédits*. Paris 1888. Mentioned by Chamberlain in connection with a list of *Horlogers du Roi* during the different reigns. Fortunately Beillard reproduces this list.

FRIED (Henry B.). *Cavalcade of Time*. Dallas, 1968.

FRODSHAM (Charles & Co. 84 Strand). *Catalogue of English Clocks and Timepieces... and of French Clocks and Timepieces, Made for, Examined and Guaranteed by them*. Not dated, but published soon after 1891 (The Admiralty Chronometer Trials of that year are mentioned). At this period a Frodsham English carriage clock with chronometer escapement cost retail from £52.10.0d. A French *Corniche* of good quality and with lever escapement compensation balance, and striking hours and halves on a gong, cost £9.0.0d. A *Corniche* timepiece cost £6.6.0d. A *gorge* hour and half hour striker, repeating hours and with alarm, lever escapement and compensation balance cost £15.0.0d. An engraved *gorge* was 15/- more. The clocks are illustrated: but the models are not named. This Catalogue was issued to customers.

FRODSHAM (Charles & Co. 173 Brompton Rd.). *Charles Frodsham & Co. Ltd*. A Catalogue published in December 1950 containing a brief history of the firm. This Catalogue illustrates the "humpbacked" perpetual calendar carriage clock mentioned in Chapter IX. It was styled "The Cole". This Catalogue was available to the public.

FRODSHAM (G.E.). *Illustrated Catalogue of Watches, Clocks, Chronometers, and Gold Chains, manufactured by G.E. Frodsham, Ltd., 31, Gracechurch Street, London, Established 1796*. London, 1889. Illustrates on p.19 four English carriage clocks. Of these two are identical and one identical except for the dial with examples shown in Moore's pattern book dated 1849. Since these cases would have been available *before 1849* and *after 1889*, the reader will more readily believe the observation in the chapter upon English carriage clocks that their appearance remained largely constant from start to finish. It is also disconcerting to have to report that the 1849 and 1889 catalogues show the very same four-glass cases,

457

bracket clocks, mantel clocks, "Gothic-cased" clocks, and regulators.

FRODSHAM. See MOON.

GALLET (Michèl). *Paris Domestic Architecture of the 18th Century.* London, 1972. The author is Assistant Curator of the Musée Carnavalet in the Marais; and the translator (James C. Palmes) was formerly librarian of the Royal Institute of British Architects.

GALLIERA (Musée). *Centenaire de A-L Breguet 1747-1823. Exposition de son Oeuvre d'Horlogerie et de Chronométrie.* Paris, October, 1923. Illustrated Catalogue of Exhibition. See BREGUET, REPUBLIQUE FRANCAISE, CHAPUIS.

GALLIOT (Hélène). *Le Métier d'Horloger en Franche-Comté.* Part of a thesis written in 1954. The Galliot thesis is based on SAHLER, q.v.

GARNIER (Paul). *Brevet* for his escapement in two-planes. See Appendix (a).

GARNIER (Paul). *Plates.* Four *pendules de voyage* are illustrated, including two with calendars. Also shown are two ornate clocks, one on a stand, one on a bracket — not strictly carriage clocks despite handles. Both have calendars. "Lith. de Cattier, Rue Tiquetonne 18", at foot of one sheet. No date.

GARNIER (Paul, fils). *Notice sur Les Travaux de M. Paul Garnier.* Circa 1889.

GAZELEY (W.J.). An article on Bridgman and Brindle published in *The Watchmaker, Jeweller and Silversmith,* Aug. 1953.

GAZELEY (W.J.). *Clock & Watch Escapements* London 1956 and re-issue 1973.

GAZELEY (W.J.). *Watch and Clock Making and Repairing,* 1965 and re-issue 1971.

GAZIER (G.). *La Franche-Comté.* Paris, 1914.

GELIS (E.). *L'Horlogerie Ancienne.* Toulouse, 1949.

GIBBS-SMITH (C.H.). *A History of Flying,* London, 1953.

GIBBS-SMITH (C.H.). *Leonardo da Vinci's Aeronautics,* 1967.

GOLDSMITHS AND SILVERSMITHS CO. LTD., London. *Catalogue for 1939.* Illustrates carriage clocks which were made by Couaillet.

GOOD (Richard). *The First Lever Watch... by Thomas Mudge.* Published in the *Horological Journal* Dec. 1956. Reprinted with notes by Col. Quill in *Pioneers of Precision Timekeeping* (q.v.).

GOOD (Richard). *Two Masterpieces by Nicole Nielsen.* Article in the *Horological Journal,* 1969.

GOODRICH (Ward L.). *The Modern Clock,* 1950 and reprints.

GOULD (R.T.). *The Marine Chronometer,* London 1923, and re-edition 1960.

GOULD (R.T.). *John Harrison and his Timekeepers..* First published in 1935 in *The Mariner's Mirror,* but currently in print.

GREAT EXHIBITION 1851. *Official Descriptive and Illustrated Catalogue,* by authority of the Royal Commission. Three volumes. Vol. 1, section II, class 10. Horological instruments. London, 1851. (Spicer Brothers, wholesale stationers; W. Clowes & Sons, Printers). See INTERNATIONAL EXHIBITION.

GREAT EXHIBITION 1851. *The Crystal Palace Exhibition.* Illustrated catalogue published by *The Art Journal,* 1851. Facsimile reprint New York 1970.

GROS (Charles). *Les Echappements des Horloges et des Montres.* Paris, 1913.

GROSSMANN (Jules and Hermann). *Lessons in Horology* Philadelphia ("Keystone") 1905, and Bienne/Paris 1911-1912.

HALL (John James). *Charles MacDowall, 1790-1872.* Article in the *Horological Journal,* Vol. XVI, Sept. 1873, pp.5-8.

HAMILTON (Ronald). *A Holiday History of France.* London 1971.

HART (I.B.). *The World of Leonardo da Vinci.* London.

HASWELL. *Horology.* London 1928 and 1947.

HAWKINS (John). *Thomas Cole, A Nineteenth Century Clockmaker and Precision Metalworker.* Due to be published in 1974 by The Antique Collectors' Club, Woodbridge, Suffolk.

HIGGINBOTHAM (Charles T.). *Incidents in the American Watchmaking Industry.* Published in *National Jeweler and Optician* on 1st Jan. 1912. This article, not seen, is said to contain interesting notes on Japy Frères.

HOGG (W. & Co.). *The Goldsmiths', Jewellers', Silversmiths, Watchmakers, Opticians, and Cutlers. Directory for London, Birmingham, Liverpool, Manchester and Sheffield; The Watch Trade at Coventry; The Movement Manufacturers, File Makers and Tool Makers of Prescot.* London 1863.

HOROLOGICAL JOURNAL. The official journal of the British Horological Institute, London. Published from September 1858 continuously to date.

Bibliography

INFORMATIONS DIEPPOISES, 13th May 1947. Article on Saint-Nicolas-d'Aliermont. (Bibliothèque Municipale de Rouen).

INGERSOLL LTD. *The History of The Ingersoll Company*, London, 1966.

INTERNATIONAL EXHIBITION, 1862. Illustrated Catalogue of the Industrial Department, British Division, Vol. I. Class XV Horological Instruments. See GREAT EXHIBITION.

INVENTION AND INVENTOR'S MART. Issue of October 1885, pages 1083-1084, contains an obituary with portrait of Victor Kullberg.

JAGGER (Cedric). *Paul Philip Barraud*. London, (Antiquarian Horological Society) 1968.

JAPY FRERES. *Notes sur les Ouvriers de Beaucourt et sur les Institutions créés en leur Faveur*. The second paragraph is entitled "Vulgarisation de la montre" and is important as showing the impact that Japy alone had in placing watches within the reach of all. Undated. Published at Beaucourt.

JAPY FRERES. *Catalogue*, 1923. Seen at Badevel in 1971 in the home of M. Péquignot. Illustrates alarms, carriage clocks, clock sets, régulateurs, etc., etc. According to M. Péquignot, Japy supplied "Vienna regulators" to Vienna until 1914. A trade Catalogue.

JAPY FRERES. *Plaquette*. A commemorative brochure entitled *1749-1949. Japy*. Published at Beaucourt on 22nd May, 1949, the bi-centenary of the birth of Frédéric Japy. The Japy Frères address in Paris in 1949 was 6, Rue Marignan, Paris. See MUSTON. See VINTER.

JAPY FACTORY AT BADEVEL. WORKBOOK. *Comptoir*. Badevel, Jan. 1907.

JAPY FACTORY AT BADEVEL. WORKBOOK. *Découpoirs*. Badevel, Jan. 1907.

JAPY FACTORY AT BADEVEL. WORKBOOK. *Atelier des Tours*. Badevel, Jan. 1907.

JAPY FACTORY AT BADEVEL. WORKBOOK. *Atelier des Mouvements Blancs*. Badevel, Jan. 1907.

JAPY FACTORY AT BADEVEL. WORKBOOK. *Mouvts Blancs et Finissages Voyages*. Badevel, Jan. 1907.

JAPY FACTORY AT BADEVEL. WORKBOOK. *Voyages*. Badevel, Jan. 1907.

JAPY FACTORY AT BADEVEL. WORKBOOK. *Finissage*. Badevel, January 1907.

JAPY FACTORY AT BADEVEL. WORKBOOK. *Finissage Externe*. Badevel, Jan. 1907.

JAPY FACTORY AT BADEVEL. WORKBOOK. *Réveils Américains*. Badevel, Jan. 1907.

JAPY FACTORY AT BADEVEL. WORKBOOK. *Atelier des Réveils Américains*. Badevel, Jan. 1907.

JAPY FACTORY AT BADEVEL. WORKBOOK. *Atelier des Mouvts = Dots*. Badevel, Jan. 1907.

JAPY FACTORY AT BADEVEL. WORKBOOK. *Atelier des Mouvts = Dots. Externs*. Badevel, Jan. 1907.

JAPY (Albert) Giens, 1971. *Letter* re family history.

JAQUET (Eugène) & CHAPUIS (Alfred). *Technique and History of The Swiss Watch from its Beginnings to The Present Day*. English translation 1953.

JAQUET (Eugène) & GIBERTINI (Dante). *La Réparation des Pendules*. 3rd edition. Bienne 1958. Contains excellent chapter on Comtoise clocks, and particularly striking work.

JEANNERET (F.A.M.). *Etrennes Neuchâteloises*. Contains much valuable local history. Locle, 1862.

JEANNERET (F.A.M.). *Biographie Neuchâteloise*. Two vols. Locle, 1863. The Biography of A-L Breguet includes a valuable list of the sources from which it was compiled. These include many scarcely known titles. Jeanneret also quotes an important biography of "Fritz" Courvoisier written by his son.

JENDRITZKI (H.). *Watch Adjustment* Lausanne 1963.

JEROME. *Chauncey Jerome Manufacturer of Brass Clocks*. New Haven, Conn., probably 1853. An unpriced trade Catalogue with illustrations (no carriage clocks). Edited reprint, 1971, Kenneth D. Roberts, Bristol, Conn.

JEWELLER (The). The official journal of the British Watch and Clockmakers' Guild. Published in London on the 15th of the month.

JOURNAL DE ROUEN, 25th Sept. 1932. Article on Saint-Nicolas-d'Aliermont. (Bibliothèque Municipale de Rouen).

JOURNAL SUISSE D'HORLOGERIE. Published from July 1876 to date. Contains very little information upon carriage clocks.

KELLY'S. *Directory of the Watch & Clock Trades*. London 1887.

Bibliography

LA CHAUX-DE-FONDS (Imprimerie National Suisse). *La Chaux-de-Fonds. Son Passé et son Présent. Notes et Souvenirs Historiques.* Published 1894 in La Chaux-de-Fonds on the centenary of the fire of 5th May 1794.

LANG (?Léopold). An *Account* from Lang, Horloger (presumably Léopold Lang who, according to Baillie, was Master in Paris in 1776) to Monsieur Le Marquis de Montesquiou covering the period 1st October 1776 to 1st March 1780. About 25 repair jobs are listed totalling 371 livres 4s. Lang received 100 livres on 7th May 1780 and a further 271 livres on 3rd July 1791. (No wonder there was a French Revolution!)
(Documentation: Collection J-C Sabrier, Acquigny).

LA PATRIE (the Paris newspaper) 1851. See Chapter II, footnote 27.

LAROUSSE (Libraire). *Nouveau Petit Larousse en Couleurs.* Paris, latest edition.

LEBON (E.). *Etudes Historiques, Morales et Statistiques sur l'Horlogerie en Franche-Comté.* Besançon, 1860 (and also 1862, according to Tardy).

LEE (R.A.). *The First Twelve Years of the English Pendulum Clock, or The Fromanteel Family and their Contemporaries, 1658-1670.* London, 1969.

LEECH (John). *Pictures of Life and Character.* From the collection of "Mr. Punch" 1842-1864, re-published in book form, circa 1886/1887.

LEOPOLD (J.H.). *The Almanus Manuscript.* This is an intensive study of a MS. found thirty years ago in Augsburg City Library. It deals with the practical side of mediaeval clock work as recorded in Rome circa 1475-1485 by the German, Paulus Almanus. See Chapter I. Published London 1972 (Hutchinson).

L'EPEE (Frédéric & Cie). A brochure issued in 1964, entitled *L'Epee livre dans le Monde entier depuis 1839.* See Chapter VI.

L'EPEE (Henry). *Causerie sur la Fabrication des Boîtes à Musique à Sainte-Suzanne.* Lecture given to the Société d'Emulation de Montbéliard, 31st Oct. 1942. Montbéliard, 1945.

LE ROY. *La Maison Le Roy de 1764 à 1900.* Paris 1900.

LE ROY (150ᵉ Anniversaire de Pierre le Createur de la Chronometrie). *Invitation à l'Exposition de Chronomètres de Marine du 7 au 15 Décembre 1935 chez L. Leroy & Cie, Horlogers de la Marine, 7 Boulevarde de la Madeleine, Paris.*

LE ROY (Charles Louis, Palais-Royal 13 et 15). *Notice sur L'Horlogerie de Paris.* This gives details of watch sales and production by the firm of which he became head in 1828, showing the proportion of French to Swiss work at various periods from 1800 to 1840. In the years close to 1800, work sold was mostly French, but by 1840 the position was reversed. (Not dated.)

LEROY (L. & Cie, 1785-1904). *High Class Watch and Clock Manufacturers L. Leroy & Cie. 7 Boulevard de la Madeleine, Paris. St. Louis International Exhibition, 1904.* (Notice advertissement).

L. LEROY & CIE versus **THOMAS GARNIER & CIE ET AUTRES.** (See TRIBUNAL).

LIMAN (of Besançon). *Compte-Rendu de L'Exposition de Nantes.* Besançon 1861. No special mention of carriage clocks.

LLOYD (H. Alan). *Some Outstanding Clocks Over Seven Hundred Years, 1250-1950.* London, 1958.

LLOYD (H. Alan). *The Collector's Dictionary of Clocks.* London 1964 and 1969.

LOUVRE (Musée de). *Collection Paul Garnier. Horloges et Montres, Ivoire et Plaquettes.* Paris 1917.

MAINE STATE CORPORATION DIVISION. Letter dated 4th Jan. 1951 re dates of the Boston Clock Company. (the dates given seem not to be correct)

MARGAINE (F-A). *Catalogue of the Best Paris Made Clocks.* No printer's name. Size 6ins. x 4¾ins. approx., paper covers, undated but certainly after 1869. This Catalogue was intended for the English trade; but unfortunately the clocks are identified only by Catalogue Numbers and not by names of models or shapes. Good plates and price list, showing thirty-two clocks of differing appearances, each available with nine different movements. Most of the clocks illustrated are variants of either *Corniche, cannelée, Anglaise* or *gorge.* The two best movements are each offered in two qualities — "1st Qty", the normal high standard of the firm, and "Superior Qty", more expensive. Three versions of the cheapest clock, in *Corniche* case and with lever escapement, are offered as an afterthought in "2nd Quality". A miniature is offered, "*Corniche No. 0*". It is available both as a timepiece and with alarm. The presence of the alarm in a *mignonnette* serves to establish the date of the *Catalogue* as after 1869, as in this year Margaine patented such a clock. At the end of 1875 Margaine took out *Brevet* No. 110,280 for a balance remontoire escapement for portable clocks and "especially for carriage clocks".
A facsimile reproduction will be produced in 1974 by Antique Collectors' Club, Clopton, Woodbridge, Suffolk, in collaboration with Charles Allix.

MAROUF. Auction *Catalogue.* Düsseldorf, October 1971. Plate 65 of this Catalogue illustrates a pre-*pendule de voyage* by Robert and Courvoisier, which is mentioned in Chapter I. Also pictured is the clock signed "Weisse, Dresden" discussed in Chapter X. The same Catalogue is notable for an illustration of a square 17th century German table clock complete with its travelling box.

Bibliography

MATHEY-TISSOT. *Album Spécial de Pendulettes-Mignonettes, 8 Jours à Répétition Quarts, 5 Minutes et Minutes.* Undated, but certified by M. Etienne Ch. Mathey as 1905. Paper covers, 6¾ins. x 9¾ins. approx. Illustrates 23 *mignonnettes*, 1 repeating watch in folding "snakeskin" case, 3 cases without movements and one page of interchangeable parts with identification numbers. All descriptions are in French, German, English and Italian. In addition to *Corniche* and *Indienne* cases, are offered others in plain polished silver or niello, hammered finish, sun-ray pattern, *moiré*, art nouveau, enamel and *guillochée* enamel, engraved, high-relief, etc. The Catalogue was printed by Couvoisier, La Chaux-de-Fonds. (Loaned by M. Etienne Ch. Mathey, Les Ponts-de-Martel, Switzerland). See Chapter X, Swiss Carriage Clocks.

MEGNIN (Georges). *Naissance, Développement et Situation Actuelle de l'Industrie Horlogère à Besançon.* Besançon, 1909. Gives the production of *porte-échappements* in the Montbéliard region as 18,000 per year, c.1900.

MERCER (Dr. V.). *Andre Hessen's Verge Lever Escapement.* Article published in *Antiquarian Horology*, Mar. 1964.

MERCER (Dr. V.). *John Arnold & Son, Chronometer Makers, 1762-1843.* London (A.H.S.) 1972.

MILLER (L.F.). *Carriage Clock Repairs.* Article published in *The Horological Journal* 1969.

MOINET (Louis). Nouveau Traité Général d'Horlogerie. Paris, 1848 and 1953. Pages 370-371 of the two-volume second edition contain a long biographical footnote on Garnier, whom the writer says that he omitted to mention in the first edition. See Chapter II and Appendix (a).

MONGRUEL (A.). *150 Modèles d'Aiguilles de Pendule.* Paris 1936 (not seen).

MONGRUEL (A.). *L'Aiguille de Pendule, Son Evolution Décorative.* A manuscript album containing photographs of clock hands, mainly French, mostly in pairs and also mostly identified. The photographs include fleur-de-lys hands mutilated during the Revolution. The book *150 Modèles d'Aiguilles de Pendule* was clearly based on these photographs, but it appears that Mongruel added others afterwards.

MONGRUEL (A.). *French Clock Making during the 1789 Revolution.* An incomplete and apparently unpublished typed manuscript in the possession of C.R.P. Allix. (Although the manuscript is unsigned various references in the text make it virtually certain that the author is A. Mongruel. The translator was probably the late Malcolm Gardner.)

MONREAL Y TEJADA (Luis). *Relojes Antiguos (1500-1850).* Barcelona 1955.

MONTANES (Luis). *Relojes Españoles.* Madrid 1968.

MOON (Mr.). *Typescript;* untitled, but being an un-completed and unpublished history of the Frodsham business and their connections. Despite corrections and additions in the hands of P. Clowes, Malcolm Gardner and C.A. Ilbert, much further work, and above all checking of facts, remains to be done.

MOORE (B.R. & J.). *Designs for Clocks 1849.* Folio volume illustrating some 116 clock cases and brackets from original pencil drawings. The *Horological Journal*, Vol. 1, Sept. 1858, refers to Moore's pattern book *Designs for Clock Cases 1848* as being sought for the then non-existent B.H.I. Library. In the same issue Moore themselves advertise "TO CLOCK MAKERS. ONE HUNDRED and TWENTY DESIGNS for Clock Cases and Brackets, in one large folio volume, by B.R. & J. Moore, 38 and 39 Clerkenwell Close, London". These Moore catalogues were intended for the use of the clock trade and perhaps also for the public.

MOREAU (Gabriel). *Cours de Rhabillage de L'Horlogerie Comtoise.* Anet. 3rd edition 1973.

MORPURGO (Enrico). *L'Orologio et il Pendolo.* Rome 1957.

MORPURGO (Enrico). *Gli Orologi,* Milan 1966.

MUSTON (M. le Docteur). *Essai de Statistique Industrielle, 1859/1860.* Mentioned in Saunier *Almanach* 1861.

MUSTON (M. le Docteur). *De la Fabrication des Pièces à Musique Méchaniques dans Le Jura.* Article in *Revue Chronométrique*, Oct. 1862, pp. 262-267, dealing mainly with L'Épée.

MUSTON (M. le Docteur). *Histoire d'un Village.* Three volumes, Montbéliard (printed by Barbier Frères) 1882. Vol. I, 355 pp., index, 2 plates, map. Vol. II, 447 pp. index, acknol. Vol. III, 361 pp., 6 plates. Very rare. Contains the history of *Le Pays de Montbéliard*. Also covers the local industries and gives histories of Japy, L'Épée, etc. The Japy family is covered in encyclopaedic detail, and the work is as much a history of Japy as of Beaucourt. Muston was not only the local historian but also a relation by marriage and the family doctor. "Pendant le second Empire j'ai vu mourir les trois frères Japy dont j'etais le médecin". (Vol. II, p.110)

MUSTON (M. le Docteur). *L'Horlogerie dans les Montagnes du Jura.* Published in *Revue Franche-Comtoise*, April-May, 1885.

S.F. MYERS & CO. *Illustrated Catalogue and Wholesale Price List (For the trade only),* New York 1885. Myers were not manufacturers but wholesalers. This Catalogue was intended for the use of retailers. It is so arranged that beneath the illustrations of each clock is shown its List or in other words the retail price. At the beginning of the Catalogue is a discount sheet showing both Net and List prices. This page was supposed to be cut out and pasted on a sheet of cardboard for the eyes of the retailer only. It was then safe to leave the Catalogue lying about on the shop counter. Travelling clocks or alarms made by Seth Thomas and E.N. Welch are featured. Re-issued by American Reprints, St. Louis, Missouri, 1964.

McLEAN (Bruce). *The Webster Family of Clockmakers.* Article in *Antiquarian Horology*, June and Sept. 1956.

NAUTICAL JOURNAL (The) 1842. Illustrations of E.J. Dent's patent balance, showing more than one arrangement.

NELTHROPP (Rev. H.L.). *A Treatise on Watch-work, Past and Present*, London 1873. This oft-maligned book nevertheless contains fascinating, and also very accurate, descriptions of late 19th century trade malpractices.

OLLIER (Adolphe). An illustrated *Catalogue* in the form of an album with actual photographs of clocks pasted in place. Date c.1900. Facsimile reproduction proposed for 1974 by Antique Collectors' Club, Clopton, Woodbridge, Suffolk, in collaboration with Charles Allix and by courtesy of Harvey Pitcher.

OUDIN-CHARPENTIER. *Catalogue of Chief Exhibits by Oudin-Charpentier, principal clockmaker to Their Majesties the Emperor and the Empress, to the Pope, to Their Majesties the Queen and King of Spain and to the Imperial Navy.* London, 1862.

PAIGET (Arthur). *L'Histoire de La Révolution Neuchâteloise.* (Not seen. Do not know date.)

PALMER (Brooks). *A Treasury of American Clocks.* New York/London, 1967.

PALMER (Brooks). *The Book of American Clocks.* U.S.A. 1928 and new printing 1967.

PARIS NORMANDIE. *A Saint-Nicolas-d'Aliermont. Les Confidences d'un Reveille-matin.* Newspaper article 25th, 26th & 27th Sept. 1932.

PERREGAUX (Charles) and PERROT (F-Louis). *Les Jaquet-Droz et Leschot.* Neuchâtel 1916.

PERRON. *Histoire de l'Horlogerie en Franche-Comté.* Besançon 1860. Mentioned by Saunier in *Almanach Artistique*, 1861.

PEUPIN. On Paul Garnier. See Appendix (a).

PHILIPPE (Adrien). *Etudes sur L'Horlogerie à L'Exposition de Paris 1878*, published by *Journal de Genève* in 1878 (or 1879) "... se trouvé chez l'auteur Maison Patek, Philippe & Ce et chez Les Principaux Libraries à Genève, en Suisse et à l'Etranger..." This article is of great interest because in it Philippe, writing during thoroughly bad times for the watch business, said that carriage clocks were comparatively easy to make and could provide a new source of industry for Geneva. Had such pieces been already in current production, such advice would scarcely have been necessary!!! Philippe is best remembered as author of *Les Montres sans Clefs... 1863*.

PHILLIPS (M). *Mémoire sur le Spiral Réglant des Chronomètres et des Montres*, 1859 and other issues.

PHILLIPS (M.). *Manuel Pratique sur le Spiral Réglant des Chronomètres et des Montres*, 1865.

PIGOT (& Co.). *Commercial Directory, Lancashire, 1822.* (T. Hyde's extract, circa 1960)

PLANCHON (Mathieu). *La Pendule de Paris*, Paris 1921. (not seen)

PONS (Honoré). *Exposition de 1823. Produits nouveaux, perfectionnés, de la manufacture d'horlogerie de M. H.Pons.* Paris, 1823.

PORTAL (Camille) and DE GRAFFIGNY (Henri). *Les Merveilles de l'Horlogerie.* Paris, 1888. 296 pp., 112 illustrations. A book corresponding approximately to the English Wood's *Curiosities of Clocks and Watches*, 1866.

POST OFFICE LONDON DIRECTORY. Various years from 1800 (latterly published by Kelly).

PRAGER (Frank D.). *Brunelleschi's Clock?* Published in *Physics*, No. 3, 1968, Florence. Contains a uniquely valuable bibliography of early clockwork including an illustration of a helical mainspring proposed by Filippo Brunelleschi the Florentine architect (1377-1446) for use with fusee.

PRECLIN (E.). *Histoire de la Franche-Comté.* Paris, 1947.

QUILL (Col. H.). *John Harrison, The Man who found Longitude*, Dublin 1966.

RAWLINGS, A.L. *The Science of Clocks and Watches.* Second edition, New York 1948 and re-issue 1974.

REDIER (Antoine). On Paul Garnier. *Revue Chronométrique*, Dec. 1855, p.79. See Appendix (a).

REDIER (Antoine). *Les Récompenses de la Classe 23.* Brochure published by author in 1867 with reference to the Paris Exhibition of that year.

REDIER (Antoine). Obituary of Achille Brocot in *Revue Chronométrique*, June 1878, pp. 93-98 (usually found in Vol. X).

REID (Thomas). *A Treatise on Clock and Watch Making.* Edinburgh 1826, and subsequent editions including one in Philadelphia.

REPUBLIQUE FRANCAISE (La Municipalité de Paris). *Invitation pour Deux Personnes... assister à l'inauguration de l'Exposition... du Centenaire de Breguet qui aura lieu au Musée Galliéra... le Jeudi 25 Octobre 1929, à deux heures et demie.*

REVERCHON (Léopold). *Petite Histoire de L'Horlogerie.* Besançon, c.1930. Contains brief biography of Frédéric Japy.

REVUE CHRONOMETRIQUE. JOURNAL DES HORLOGERS. This important periodical, which became the official organ of the Société des Horlogers founded in 1856, began publication in 1855 with Claudius Saunier its first Editor.

ROBERT (Henri, fils). *Exposition Universelle Internationale de 1878. Les Récompenses de la Classe 26.*

ROE (Joseph Wickham). *English and American Tool Builders.* Oxford 1916. According to Roe, the first really successful interchangeability system was developed by gun makers in America, although attempts were made in France as early as 1717 and 1785.

ROGERE (Claude). *Les Horloges de Saint-Nicolas-d'Aliermont.* Rouen 1973.

ROSE (G.A.) & WESTON (R. McV.). *A Marine Chronometer with a Very Unusual Form of Escapement.* Article in *Antiquarian Horology* December 1964. (The escapement gives impulse through a helix, and bears the name of J. Eden MacDowall).

ROSENBERG (Charles). *American Pocket Watches.* U.S.A. 1965.

ROUSAILLE (D.). *Quelques Études sur l'Horlogerie à l'Exposition Universelle de Paris 1889,* Lyon, 1890. Pages 34 and 35 mention carriage clocks, but really add nothing to the information given by Tripplin and by T.D. Wright. Rousaille himself showed carriage clocks at Lyon in 1889. Rousaille's name will sometimes be found spelt Roussaille and sometimes Roussiale.

ROYER COLLARD (F.B.). *Skeleton Clocks,* London 1969.

SAHLER (Pierre-Louis). *Ma Vie en Deux Mots.* Written in 1835 at the age of 73 years. Contains Japy history used in 1954 by Hélène Galliot.

SAHLER (Pierre-Louis). *Notes sur Montbéliard.* Written in 1835 at the age of 73 years. Contains valuable Japy history, used afterwards by Hélène Galliot.

ST. LOUIS CLOCK AND SILVERWARE COMPANY. *Twelfth Annual Catalogue,* St. Louis, Mo., 1904. St. L.C. & SW Co. were not manufacturers but wholesalers. This Catalogue was intended for the use of retailers. It is so arranged that beneath the illustration of each clock is shown its List, or in other words its retail price. Both Waterbury and Ansonia and German Black Forest clocks and alarms are featured.
Re-issued by American Reprints, St. Louis, Missouri, 1962.

SAINT-NICOLAS-D'ALIERMONT. *Bulletin Officiel.* No. 1, 1966; No. 2, 1969; No. 3, 1969.

SALOMONS (Sir David). *Breguet (1747-1823).* London, 1921 and 1923. The 1921 edition is in English. Both a supplement and appendix to that supplement were published subsequently. The 1923 edition is in French.

SAUNIER (Claudius). *Revue Chronométrique. Journal des Horlogers, Scientifique et Pratique.* The *Revue* was a periodical of great importance published in Paris from June 1855 to March 1914. Claudius Saunier (1816-1896), author of *Traité d'Horlogerie Moderne* and of many other works, was the founder and also the editor of the *Revue* from its first issue until its 42nd. year in 1896. He was a very distinguished horologist and was Secrétaire Général de la Société des Horlogers, Paris. It is more than a pity that his *Treatise* and its companion *Handbook* are apparently all that were ever translated from the French. Other than the short-lived *Tribune Chronométrique*, the *Revue* was the first French periodical devoted to horology. It incorporated in 1883 the *Journal d'Horlogerie*, which was previously published in 1881 and 1882 by a rival concern. The dating system and pagination of the *Revue* is very unsatisfactory and also confusing. Saunier distributed his review each month. Later every year or two he assembled the unsold runs for sale in one volume which he then dated at the time. For example, the volume for the first and second years shows on the title page the date 1857; but if the reader looks at the following page here is the title of the first monthly issue with the correct date of distribution, June 1855. *Table Générale des matièrs* confirms this.

SAUNIER (Claudius). *Échappements Nouveaux pour Pendules Portatives.* Article published in the *Revue Chronométrique* Aug. 1872 and June 1873 (usually found bound in Vols. VII and VIII). Saunier says that Gontard's escapement has given excellent results when made in the establishment of the inventor and making use of *roulants* coming from the hands of Henry Jacot, "... two great aids to success far from being available to all". The duplex escapement, says Saunier, is equally successful, but only if superbly made and allied to a perfectly-finished *roulant*. The duplex carriage clocks of Aimé Jacob left nothing to be desired, but they cost the price of a chronometer. Several attempts to cheapen the escapement, for instance based upon trial pieces made by Thévenin had come to nothing (see a note about him in List of Names). Lever escapements, says Saunier, are not bad in well-made, expensive carriage clocks. As for the cylinder, that is still less satisfactory; and he does not feel that he takes much risk in saying that it never really keeps time. Saunier then goes on to describe and illustrate a whole series of 2-plane escapements, more or less of his own imagining, which he feels will do for *pendules de voyage*, and for *pendules portatives* generally.

Bibliography

SAUNIER (Claudius). *Almanack Artistique et Historique des Horlogers, Orfrèvres, Bijoutiers, Opticiens.* Published each year in Paris from 1859, the *Almanack* was a booklet containing current horological news. Apparently it was distributed mainly from the offices of the *Revue Chronométrique*.

SAUNIER (Claudius). *Exposition Universelle de 1867. Compte Rendu De L'Horlogerie...* Included in the *Revue* from July 1867 to December 1868 (Vol. VI) with Index, plans of stands, and illustrations; but afterwards offered as a separate volume. Contains list of Exhibitors, awards and many of the items exhibited.

SAUNIER (Claudius). *Treatise on Modern Horology in Theory and Practice.* Foyles 1952 re-issue of the translation into English by Tripplin and Rigg.

SAUNIER (Claudius). *The Watchmaker's Hand-Book.* The Technical Press 1945 re-issue of the translation into English by Tripplin and Rigg.

SCHMITH (Thomas). *Examining French Carriage Clocks.* An article published in the *Horological Journal* between October 1894 and February 1895.

SEGUIER (The Baron). A well-known amateur horologist, the author of a number of reports upon *Expositions*. Gave a number of fine escapement models, made by Pons, to the Conservatoire National des Arts et Métiers in 1852, where they may still be seen.

SETH THOMAS CLOCK COMPANY. *Seth Thomas Clock Co's Price List,* January 1st, 1879. New York and Chicago. This Catalogue was issued by the manufactuer to the retailer. The prices of clocks are not shown beneath their illustrations. At the front of the Catalogue is a list of Net prices, or in other words of prices to the trade. This list was certainly either meant to be torn out or at least kept from the eyes of the customer.
Re-issued by American Reprints, St. Louis, Missouri, 1973.

SHEARS (P.J.). *Huguenot Connections with The Clockmaking Trade in England,* London 1960.

SIMONI (Antonio). *Orologi Italiani dal '500 all' '800.* Milan 1965.

SIRE (Georges). *L'Horlogerie à l'Exposition Universelle de 1867.* Besançon, 1870. (Imprimerie de Dodivers, Grande-Rue 87). Sire was the Director of the École Municipale d'Horlogerie at Besançon.

SMITH (John). Old Scottish Clockmakers from 1453 to 1850. Edinburgh and London, 1921 (second ed.).

SMITH (S. & Son Ltd.). *Guide to The Purchase of A Clock.* Second edition, undated but post 1900. Facsimile reprint, U.S.A. 1971. Many carriage clocks are shown. Most of them were Duverdrey & Bloquel's models. This Catalogue was available to the general public.

SOCIETE D'EMULATION (at Montbéliard). A literary and scientific body founded by Dr. Muston. Their "transactions" contain a certain amount of local horological history.

SOCIETE D'EMULATION DU DOUBS. *Mémoires,* 1869. Page 170 lists a few makers of *assortisments* whose names apparently have not been re-published until now.

SOCIETE D'ENCOURAGEMENT POUR L'INDUSTRIE NATIONALE, Paris. *Bulletins.* These are quoted from time to time in this book.

SOCIETE D'ENCOURAGEMENT POUR L'INDUSTRIE NATIONALE. *Bulletin.* Vol. 8, 1809, pp. 326-327 contains "Note sur la fabrique d'horlogerie de St. Nicolas-d'Aliermont". A factory founded by Croutte made clock *mouvements* by slow methods. Pons was sent to Saint-Nicolas by the French Government to re-organise the industry there. He started with the factory of Croutte. See Chapter III.

SOCIETY FOR THE PROMOTION OF SCIENTIFIC INDUSTRY. *Artisans' Reports upon the Vienna Exhibition 1873.* Manchester, 1873.

SOTHEBY & CO., London. Auction *Catalogues.*

STENDALL (i.e. Henri Beyle). *Le Rouge et Le Noir. Chronique du XIXe Siècle.* Paris, 2 volumes, 1830.

STEWARD (W. Augustus). *Gold, Silversmithing and Horology at the Paris Exhibition, 1900.* Sixty-five pages, illustrated throughout. London, 1900. (Heywood & Co. Ltd., 150 Holborn, E.C.). Excellent illustrations of the art nouveau cult. The section on horology mentions Drocourt's carriage clocks. Two clocks are illustrated as exhibited by "Ch. Hour". They are in the form of a naval vessel and of a locomotive respectively. The actual maker was in all probability E. Dessiaux of Saint-Nicolas-d'Aliermont. (See Chapter IIII)

STONE (Orra. L.). *Excerpts Taken From History of Massachusetts Industries.* Vol. 2. Published 1930.

STRAUS. *Carriages and Coaches,* 1912.

SYMONDS (R.W.). *Masterpieces of English Furniture and Clocks.* London, 1940.

SYMONDS (R.W.). *Thomas Tompion, His Life and Work,* 1951 and new facsimile edition 1969.

TAIT (Hugh). *Clocks in The British Museum.* London, 1968.

"TARDY" (Henri Lengellé). *Bibliographie Générale de La Mesure du Temps.* Paris, 1947.

"TARDY" (Henri Lengellé). *Les Échappements de Montres.* Paris 1952.

"TARDY" (Henri Lengellé). *La Pendule Française.* Three volumes, latest editions, Paris.

"TARDY" (Henri Lengellé). *Dictionnaire des Horlogers Français.* Two parts; A-K and L-Z. Paris, 1971-73.

TEJADA. See MONREAL.

THOMAS. See SETH THOMAS.

THOMAS (E.). *Horlogerie d'Art et de Précision.* An illustrated Catalogue or *Notice d'advertissement* of some 54 pages. Undated, but published some time after 1906, as evident from a reference to that year. Offered are precision clocks, deck watches, pocket watches; besides many domestic clocks including *régulateurs*, reproduction period clocks, *garnitures de cheminée*, and a wide range of *pendules de voyage*, almost certainly obtained from many different sources.

THOMPSON (Flora). *Lark Rise to Candleford.* London 1945, and subsequent reprints.

THOMSON (Richard). *Antique American Clocks and Watches.* U.S.A. 1968.

THURY (Professeur M.). *Notice Historique sur l'Horlogerie Suisse. Paris Exposition Universelle 1878.* Neuchâtel n.d.

TORRENS (D.S.). *Nail and Cork.* Published in the *Horological Journal,* Feb. 1938. A superbly written reply to a report of a lecture by Commander Gould "... in which it is stated that early chronometer makers... in the main... must have worked by trial and error, and by rule of thumb..." See Appendix (e).

TORRENS (D.S.). *Rule of Thumb.* Published in the *Horological Journal,* July 1938. See Appendix (e).

TORRENS (D.S.). *Carriage Clocks and Chronometers.* Published in the *Horological Journal,* August 1946.

TOWNSEND (George E.). *Almost Everything You Wanted to Know About American Watches and Didn't Know Who to Ask.* U.S.A. 1970.

TREMAYNE (Arthur). *London's Great Family of Horologists.* Article in the *Horological Journal,* Dec. 1947. (The Dents).

TREVELYAN (G.M.). *English Social History.* London, 1942. (Longmans, Green & Co. Ltd.)

TRIBUNAL DE COMMERCE DE LA SEINE. *Judgement du 5 Avril 1937. Affaire Société L. Leroy & Cie contre Société Thomas Garnier & Cie et Autres.* Paris, 1937.

TRIBUNE CHRONOMETRIQUE. This important horological periodical began publication in 1850, although in 1852 it ran into administrative difficulties and suppressions occasioned by Louis Napoleon Bonaparte's coup d'état which ended the Second Republic. The Tribune, published monthly by Pierre Dubois, unfortunately had a short life, the final numbers not in fact appearing until 1853.

TRIPPLIN (J.). *Watch and Clock Making in 1889, being an Account and Comparison of the Exhibits in the Horological Section of the French International Exhibition. With a View of the British Watch and Clock Making Section.* London, 1890. (Crosby Lockwood and Son). See Appendix (b).

ULLYETT (Kenneth). *Carriage Clocks. A Study in Depth.* An article published in *Antique Dealer and Collectors Guide,* London, March 1971.

USHER, (James Ward). *An Art Collector's Treasures illustrated and described by himself,* London 1916.

VIDALENC. *La petite métallurgie rurale en Haute-Normandie sous l'Ancien Régime.*

VINTER (Ernest). *Histoire des Établissements Japy Frères depuis leur Création jusqu'à nos jours, 1777-1943.* Paper covers, 220pp. including Index. No illustrations. Beaucourt, 1944. A very complete and exhaustive history. No French firm ever offered such a wide range of clocks, watches and ironmongery as the various *usines Japy.* Vinter had 40 years service with Japy, including 25 years as *Chef d'Atelier.* He was apprenticed to the firm, and spent the whole of his working life at Beaucourt.

VUAGNEUX, (Isaac). (Notary of Locle). *Letter of Apprenticeship.* Vol. VI, 1770-1776. Document preserved in the archives of the State of Neuchâtel and relating to the apprenticeship of Frédéric Japy.

VULLIAMY (B.L.). In the Public Record Office are some papers dealing almost exclusively with Vulliamy's ornamental work, such as chimney pieces, candelabra and inkstands, made in the period circa 1800-1820.

WATCH AND CLOCKMAKER. Published monthly by R.J. Donovan, 260a Whitechapel Road, London, for the British Horological Institute. Vol. II, No. 8, October 1883 (one penny).

WATERBURY CLOCK COMPANY. *Catalogue No. 154,* 1908-1909. This Catalogue was supplied by the manufacturer to the retailer. It was intended to be shown to the customer and beneath each illustration appears its List price, or in other words the retail price. At the end of the Catalogue is a comprehensive list of clock materials intended to assist the retailer with his repairs. Presumably trade prices and discounts

were shown on a separate sheet.
Re-issued by Adams Brown Company, Exeter, New Hampshire, U.S.A.

WATERBURY WATCH COMPANY. *The Whole Story of the Waterbury. Why the Waterbury can be sold for two dollars and a half and yet be a reliable time-keeping watch.* New York 1887 (Copyright S.C. Patterson, 177 Broadway). This is an advertisement booklet for the Waterbury Long Wind watch. It contains nine illustrations, three of them reproduced by Henry Abbott in *The Watch Factories of America*. It is useful because it gives the correct names of Long Wind parts as used by Waterbury themselves.

WEBSTER COLLECTION. Sotheby's sale *Catalogue*, 27 May, 1954 and 19 Oct. 1954.

WELCH MANUFACTURING COMPANY, (E.N.). *Superior American Clocks and Clock Materials*, March 1885. This illustrated Catalogue was intended for the use of retailers but no prices are shown either List or Net. Probably these were printed on a separate sheet. An introductory letter is addressed to the trade.
Re-issued by Adams Brown Company, Exeter, New Hampshire, 1970.

WINS (A.). *L'Horloge à travers les Ages*. Paris/Mons. 1924. A very thorough history, always to be taken seriously.

WRIGHT (T.D.). *Paris Exhibition 1889. Reports of Artisans delegated by the Mansion House Committee to visit and note improvements in their respective trades. Clockmaking.* London, 1889. See Appendix (b). Thomas D. Wright was born on 13th June 1847 and died in 1933. He worked for the well-known London firm of Bracebridge and Co. from 1866, becoming foreman. In 1891 he took over the business under the style of "Wright and Craighead", continuing in the same names on the death of his partner about a year later. Wright taught theoretical horology at both the British Horological Institute and at the Northampton Institute. His lectures were published under the title *"Notes on Technical Horology"*.

YOUNG (Arthur). *Voyages en France.*

ZINNER (Ernest). *Aus der Frühzeit der Räderuhr.* Munchen 1954. (Deutsches Museum, Heft 3).

General Index

This Index is intended for use in conjunction with the Index of Plates, the List of Names, the Bibliography, and the Glossary of Terms. It has seemed both impracticable and unnecessary to pick up every reference to such frequently recurring words as "alarm", "hour-and-half-hour strike", "*grande sonnerie*", "mainspring", "calendar" etc. Restraint has also been used concerning names of makers, authors and places. References to footnotes are generally indicated by the letter 'F'; except where the same word or allusion is also present in the text above.

ABBOTT (George)
    Maker of superb chronometer detents 235,283F
ACADEMIE DES SCIENCES 398
ALARM WORK
    Detachable for table clocks of 2nd half of 16th cent. 8-9
    Common in coach watches 12
    In small English travelling clocks 16
    In French pre-pendules de voyage 23F
    In Capucines 27
    In Swiss travelling clocks 27
    In French "sedan-like" wall clocks 31
    In early Breguet carriage clocks 40,42,44
    Alarm work described 193-195
    Alarm clocks closely related to cheapest carriage clocks 329
    German alarm clocks 335-337
    Cheap alarms replaced carriage clocks 405
    45,55,62,119,123,144,167,168,174,185,186,189,238,308,
    310,326,328,332,333,334,349
ALLAINE, RIVER 138,151
ALMANUS MANUSCRIPT
    Shows Italian spring-driven clocks current c.1475-1485 7
    Probability that Almanus saw fusees before going to Rome 7F,8
ALUMINIUM
    Gorge carriage clock case 121F
    First developed in Paris c.1827 187
AMANT ESCAPEMENT 108
AMERICA
    Copies of French clocks 139,140
    Shelf clocks 190
    Carriage clocks and alarms 341-385
    Clockmaking, general 341
    Competition for Obis and Corniche 341,342,370,374
    "Interchangeable System" 341F,379
    Value of old catalogues 342
    Japanese reproductions of American clocks 390
    140,208,319,401,409
AMIDON (George H.) 377
AMSTERDAM 68
ANGLAISE SHAPE
    Described 168-169
    Price 188
    186,187,298,320
ANSONIA CLOCK COMPANY
    Carriage clocks and alarms 369-373
    Pert and Japy type Américains 140,373
    140,341,342,375,381
ANTI-FRICTION WHEELS
    Use by Berthoud in astronomer's clock made 1795 20
ANZIN. Japy 133
ARAGO 398
ARCKEN (C. Van Arcken) 413
ARGENTINE. Carriage clock 391
ARMY AND NAVY STORES
    Prices for French clocks 188
    107F,160,189

ARNOLD (J.R.)
    234,235,246-247,250,267,268,269,270,409,410
ARROWS. Possible significance 125F,435,439
ART NOUVEAU STYLE
    Described 176,178
    234F,320
ASHBOURNE
    English clockmaking centre in Derbyshire 98
ASHBURTON (the Lord)
    His Jump Clock 290
ASHMOLEAN MUSEUM
    Breguet tourbillon carriage clock 40,49
ASPREY & CO., LTD. 210
ASTLEY (T.). Eight-day "sedan" clock 28F
ASTROLABE. Incorporated in table clock 8
AUDEMARS 211
AUDINCOURT. Japy 133F
AUGSBERG. Early cradle of horology 8
AUGUSTE
    Revolving escapement 218
    69F,73,145,158
AURICOSTE 102
AUSTRIA AND AUSTRO-HUNGARIAN EMPIRE
    Pendules d'officier 22
    Quarters struck before hours 196
    Link with Franche-Comté 130
    Pre-pendules de voyage and carriage clocks 325-337
    Influenced by other clockmaking traditions 326
AUTOMATON INDUSTRIES INC. 382
AUTOMATONS. Coach watches with moving scenes and figures 14
BADEVEL. Japy factory run by Charles and Fidot,
    sons of Frédéric 135,141
    Workbooks for 1907. Factory 407-408
    84,85,131,133,134,135,137,139,140,141,143,144,148,149,188
BAILLIE (G.H.)
    Highly important horological historian in recent times 5,6,9,14
BAKER with Frodsham 266,268
BALANCES
    Bar-balance in Zech's clock of 1525 6
    Used long before pendulums 7
    Evidence of use by year 1283 or before(?) 7F
    German table clock c.1550 8
    Circular balance in Roweau clock c.1580 10
    Portable clocks dependent upon balances 13,154
    Breguet, compensated 43
    Dent's patent 252,253,264
    Earnshaw 265
    Kullberg's 283
    Bridgman and Brindle 288
BALANCE SPRINGS
    General adoption c.1676 7
    No balance spring 10
    First development in Holland and England c.1658 . . . . 10F,14
    Transformed timekeeping 13
    Added to existing clocks and watches 13
    Overcoiled 27,112
    Use of overcoil by Breguet 43
    In gold (Breguet) 48
    Bridgman and Brindle 288
BAMBOO STYLE 170,175-176
BARILLET INDEPENDANT 146F,147
BARNET. "Sedan" clock dated 1839 . . . . 29F
BARNSDALE (William). Did carriage clock work for the Jumps . . 290
BAROMETER. Boseet clock 111,204

467

General Index

BARRAUD & LUND 250,276-278
BARTHOLDI. Le Lion de Belfort 131
BARWISE 233,273
BAS-LES-FONDS 133,135
BAUDIN (Madame). Breguet clock 44,53
BAUTTE (J.F.) & Cie. 196,307,314F,318,320
BAVEUX 92,97,107,113F,116,402
BAYARD 89,93,94,98,405
BEAUCOURT
    Japy factory run by Fritz, Louis and Pierre,
    sons of Frédéric 135
    70F,85,114,131,133,134,136,137,142,143,144,148,188,401
BEAUVAIS 90
BEESON (C.F.C.). English church clocks from 1280. Research 7F
BEGUIN 73
BEILLARD (A.)
    Comments on Horlogers du Roi 76,78
    French fear of American competition 140F
BELFORT 130,131,132
BELGIUM. Marble clock cases 399
BELIARD (François)
    Informed observer of French horology in 1767 ..... 76,77,78F
BELLANOVSKY (A.)
    Russian clockmaker trained in Saint-Nicolas 101
BELLEFONTAINE 25
BELLS 43,62,199,200,250,355F
BELMONT (Henry)
    Research in Franche-Comté on Japy 134,137F
BENCH. Chronometer work 286
BENEDICT AND BURNHAM 343,356F
BENNETT (Sir J.) 278,301
BERLIN 325
BERNE 133,135
BEROLLA 67
BERTELE (H. von)
    Expert on Viennese clocks 330
BERTHOUD FAMILY
    Ferdinand's portable astronomer's clock in 1795 19
    F.B. exhibited in Paris in 1802, but not carriage clocks 65
    Family history 111-112
    Chronometer carriage clock 112,210
BESANÇON 67,68,129,130,131,133,283,400,401
BESANÇON-PILLODS
    Only exhibitor of Besançon-made platform escapements
    in 1889 403
BIENNE VALLEY 154
BIRD. English chronometer work 286
BLACK FOREST 98
BLACK, STAR AND FROST 322
BLANCS ROULANTS
    42,47,56F,84,85,90,91,95,100,105,107,116,133,134F,
    137,144,146,147,195,213,215,401,407,408
    See also Rough movements
BLANPAIN (V.) 404
BLAVET. Parisian maker of glass shades 322
BLOCH-PIMENTEL COLLECTION 15
BLOCKLEY (H.) and LUND & BLOCKLEY 276-277
BLONDEAU. Exhibited montre de voyage in 1827 55,67,70
BLUMBERG & CO. Four-dial clocks 219
BLUNDELL 294
BOIS-D'AMONT 130
BOLOGNA. Carriage clocks 391
BOLVILLER 71,72,73,74,107,145,158,203,210,221,222,396F,407
BOND AND SON (BOSTON). Kullberg carriage clock 284
BONNIBEL, Competition for Obis and Corniche 342
BONNIKSEN (B.). Karrusel 216,218
BORDEAUX 67
BOREL (P.A.). Japy 134
BOREL. Swiss carriage clock 317,322
BORNE 160,175
BORNE SHAPE .... See Humpbacked
BOQUET (David). H.M. The Queen's coach watch of c.1650 12-13

BOSEET 111,113,204,205
BOSTON CLOCK COMPANY 341,376,378,379,381,382
BOTTOM-WINDING 123,219-221
BOUCHOT FRERES 143
BOURDIN (A.E.) 109,110,113
BOURGEOIS FAMILIES IN THE FRANCHE-COMTE 132
BOVILL (E.W.). Expert on stage coaches 32
BRACKET CLOCKS (small travelling) 16-17
BRAMAH LOCKS 244,263
BRAND 136
BREGUET FAMILY AND SUCCESSORS
    Coach watch in 1810 11
    Berthoud's astronomer's clock of 1795 19-20
    A-L Breguet portrait 36
    First French carriage clocks 37
    Daniels. Important new book on A-L Breguet 37
    Son, grandson, nephew 37
    Three main types of carriage clock 38
    No. 179 sold 1810. Perhaps earliest unaltered clock 38,39,40
    Serial numbers 38,39,40,48-54
    Books of Firm and certificates 39
    Repeating work 39,40,42-45,48-54
    Tourbillon 40,48,49,54,216,217
    Moon 39,42,43,44,48-52
    Echappement naturel 41,49
    Salomon's collection 41,50,51
    Carriage clocks exhibited in 1819, 1834, 1844 42,43,45,65
    Alarm 42,45-54
    Calendar work 39,42,43,44,45,48-54
    Robin escapement 41,42,49,50
    Perpetual calendar 42,54
    Lever escapement 43,48-54
    Compensated balance 43,45
    Parachute 43,45
    Overcoil 43,45
    Use of fusee with Harrison maintain 43
    Chronometer escapement 43,45,49,51,52,53
    Up and Down dial 43,49
    Draw 43,45
    Thermometer 44
    Swiss background 45
    Double roller 43,45
    Last "humpbacked" clock delivered 1931 45
    Gastellier 47
    Pitou finishing carriage clocks until 1970 47
    Pendules sympathiques 47,48,51,52,65F
    Table of carriage clocks 48-54
    Balance spring in gold 48
    Clock prices 48-54
    Striking work 41,44,48-54
    Equation of time 49,51,52
    Use of minute repeating 53
    Duplex 53
    Multi-piece cases 60
    Importance in context of carriage clocks 65
    Exhibited in Paris in 1798, but not carriage clocks 65
    Constant Force escapement 65F
    Early references in Exhibition documents 67
    More recent history 105
    Daniels, Pitou 105
    59,169,175,187,195,203,204,209,211,234,236,237,238,
    250,309,310,331,413
BREVET ABRIDGEMENTS 136
BRIDGMAN & BRINDLE
    Sidereal time 288
    287-289
BRITISH MUSEUM 10F,19,28,42,53
BRITTEN (F.J.) 11F,16,17,409
BROCK. "Humpbacked" clocks 234
BROCOT 91,413
BROWN FAMILY. Breguet firm 37,48,105,413
BRUNEAU 67,70

468

General Index

BRUNELLESCHI (Fillippo)
   Early drawings of clocks with helical mainsprings
   and fusees 6F
   Possibly earliest illustrator of fusees, pre-1446 .... 7F
BRUNELOT 106F
BRUSA (G.). Leonardo's escapement 7F
BRUSSELS MANUSCRIPT. Early spring-driven clocks. Fusee 6
BUBBLE CLOCK 224
BUCHANAN (A.). "Sedan" clock probably pre-1800.
   Goes 3 days 28,29
BUCK (D.A.). Patentee of American long wind watch 360,364
BUCKNEY FAMILY 250,251,254,262
BUDAPEST 325
BUGATTI (Ettore). Last Breguet "humpbacked" clock 46,53
BUNKER-RAMO 382
BURT (Edwin B.) 377
BUTCHER (K.A.G.). E. Dent & Co. 252
CADRATURE
   Defined 90
   26,74,84,199,202F,203,204,309,355
CAILLY 67
CALENDAR WORK
   Early use in coach watches 12F,13
   In early Breguet carriage clocks 39,43,44,202
   Perpetual 42,51,54,204,205,237,238,254,262,263,291
   Semi-perpetual 204
   Various types described 202-206
   Fly-back 203,204
   Islamic 314
   Japanese 389
   45,48,49,50,51,52,53,54,55,59,60,91,109,111,118,
   121,144,186,205,207,234,239,308,310,322,389,390
CAMEMBERT CHEESE BOXES 130
CAMERER-CUSS & CO. 287
CAMERINI. Enigmatic pendulum clock dated 1656 .... 7F
CAMP (Hiram) 342F,373F,374,376
CAMPBELL 67
CANNELEE SHAPE 160,166,185,187
CAPT (Henry) 307,318
CAPUCINE CLOCKS 21,23-26,30,152-154
CAPUCINS MONASTERY AT SAINT-CLAUDE 154
CARIATIDES
   Described 172
   114-115,118,179,185,189,405
CARNAVALET (Musée) 105
CARNER (Dr. Mosco). His understanding of the Viennese
   spirit 325
CARPANO (Louis) 85
CASES OF CARRIAGE CLOCKS
   Margaine praised 118-119
   Aluminium 121F,187
   Lange 145,158
   French, various standard patterns described 157-187
   French case names seldom used in England 160
   Proliferation of designs 160
   Confusion of names 190
   English shared sources of supply 233
   English styles 233-234
   English Kullberg silver case weighed 93½oz. 284
   Engraved 241,279,295
   Holmden for Vulliamy 241
   85,86,95,243,381,383,385
CATALOGUES, FRENCH
   Margaine 119,161,162,189
   Duverdrey-Bloquel 161,162,185,189
   Couaillet 161,162,189
   Delépine-Barrois 190
CATALOGUES, ENGLISH
   Frodsham 188
   Army & Navy 188,189
   S. Smith & Son 189

Grimshaw, Baxter 190
Moore's pattern book 234F,280
CATALOGUES, SWISS
   Mathey-Tissot 320
CATALOGUES, AMERICAN
   General comment 342
   S.F. Myers 343,374
   St. Louis 343,349
   Waterbury 343-349,351,356
   Ansonia 369-373
   Welch 374,375
CENTRE SECONDS
   Coup-perdu clocks; "Chinese-duplex" watches 70F
   Duverdrey and Bloquel 92
   Boseet 111,204
   Japy/Reclus 145
   Michoudet 152
CHAMBERLAIN (Paul) 211F,236F,241,247F,281F,384,385
CHAMPAGNY (Le Duc de Cadore) 65F,90
CHAPUIS 308,312,315,322
CHARPENTIER 413
CHARQUEMONT 130
CHARTIER 107,111,113
CHAUMET. Acquired Breguet business in 1970 37,105
CHAUX-DE-FONDS
   37F,45,113,130,135,137,307,308,310-315
CHELSEA CLOCK COMPANY 341,376,378,381-384
CHEVAUX-DE-MARLY 182
CHEVELLIER 404
CHINA
   Sacking of Summer Palace 14
   American clocks in China 342
CHINESE DUPLEX 210
CHINOISERIE
   Boîte Chinoise described 179
   Panels on copper described 184
   139F,176
CHRONOMETER CARRIAGE CLOCKS
   F. Berthoud's portable astronomer's clock 20
   By Breguet 43,45,49,51-53
   Vogue in England 210,232-235,252-254,270,280,
   284-287,293
   French examples 73,109,112,145,210-212,226
   The famous Fasoldt clock 384-385
   Swiss examples 210,310,311,313-315
CLERICETTI. Leroy et Fils 123
CLERKENWELL 30,84,98,105,106,188,190,399,401
CLOCKMAKER'S COMPANY 12,13,31
CLOCK-WATCHES. Coach watches usually clock-watches 12
CLOISONNE ENAMEL 166,182,186
CLOWES (P.) 266F,267
CLUB-TOOTH LEVER ESCAPEMENT 43,45,72,74,202
CLUTTON (C.) 6F,14,49,233F
COACH WATCHES 11-15
COAL 130,132
COCCLAEUS (J.). Reference to Henlein in 1511 5
COCKERILL (Sir Charles). Breguet clock 51
COLE (J.F. and Thomas) 38,232,234,236-241,245,278,413
COLEMAN (J.E.)
   Noted contemporary American horological historian 356
COLLMAN (Johann)
   Austrian travelling clock late 18th cent. 326
COMPASSES
   On tops of carriage clocks 223-224
COMPENSATION
   Mudge. Polworth clock 18
   Berthoud. Astronomer's portable clock 20
   Breguet 43,45
   Garnier 60
   Bolviller 72
   Bourdin 109

General Index

Bi-metal curb 243
Now see Index of Plates under "Compensation"
COMPOSITE ESCAPEMENTS
   Hessen 22F
   Tompion 116
   Robin, etc. 211,212
COMTOISE CLOCKS 25,154,199
CONNELL 294
CONSERVATOIRE DES ARTS ET METIERS 398F
CONSTANT-FORCE ESCAPEMENT 65F,215,216,398
CONVERSIONS
   Changing old platforms for new 185F
   Bell to gong 199
   Petite sonnerie to grande impossible 202
COOLE (P.). British Museum 19
COOPER. Duplex watch movement in "sedan" clock 29F
CORINTHIAN ..... See Pillared Cases
CORNICHE CAREE 168,186
CORNICHE SHAPE
   Described 167-168
   Prices 188-190
62,75,160,184-187,320,341-342
CORPET (M.) 106F,404
COSTER (Salomon)
   First satisfactory pendulum clock with Huygens in 1656 7
COTE-AUX-FEES 46
COTTAGE CLOCKS 29,195
COTTAGE INDUSTRY
85,86,94,98,130,134,234-236,286,317,320,322,399,402,
407,408,410
COUAILLET
   History 94-97
84,85F,92,98,100,101,107,113F,125F,147,150,160,162,
174,185,189,190,298,404,405
COUET 42,413
COULON & MOLITOR 84,403
COUP-PERDU ESCAPEMENT 70,210F
COURTELARY. Dodillet 136
COURVOISIER FAMILY
   History and family tree 308,310F,315,316
   Carriage clocks 307-315
   Cugnier, Leschot 310
26,317,322
COUSTOU (Guillaume père)
   Les Chevaux de Marly 182
COUTTES (J). Name on Vulliamy clock 245
COVENTRY 94F,98,286
COWAN (James) 27
CROTCH (Dr.) Cambridge Quarters 196
CROUTTE. Saint-Nicolas 90,97,413
CUBIQUE SHAPE 174
CUGNIER, LESCHOT
   Swiss pendule d'officier 22
   Early carriage clock 310
CUTTERS 85,96
CYLINDER ESCAPEMENT
   Two-plane 15
   In Hessen travel clock 23
   "Sedan" clocks 29
   Carriage clock platform 75
   Manufacture in Franche-Comté 129
   Backward-turning 150
26,154,185,190,193,209,221,238,326,334,335,342,396,
397,402
DAMPIERRE-LES-BOIS 143
DANIELS (G.)
   Important new book on A-L Breguet 37
   Recent research 39
6,14,105,238
DANUBE 325
DAWSON (P.G.) 272,277

DE BAUFRE. Escapement 57,209,395F
DECORATIVE PANELS
   Described 182-184
   Modern 183
117,119,166,176,405
DEJARDIN 209F,436
DELABARRE
   Living link with Saint-Nicolas in the past 95,96,99-101
DELANDER (Daniel)
   Small English travel clock with balance 17
DELEPINE FAMILY
67,90,92,94,95F,97,99,100,101,190,193,195F
   Escapement — See List of Names
DELFT 183
DELMAS 213
DELORME (Etienne). Engraved coach watch dial (?) 12
DEMIDOFF (Count). Breguet clock 44,52,53
DENFERT-ROCHEREAU (Col.). Le Lion de Belfort 131
DENIS FRERES 100,102
DENT FAMILY
   Dates and addresses 250-252
   Trade mark 251
   List of carriage clocks 252-254
   Patent balance 252,253,264
   Giant clocks 254,262,263
61,233,234,235,246,247-264,278,280,301
DESBOIS 232,273-275
DESFONTAINES FAMILY 121,413
DESHAYS 55F,67,70
DESOUTTER (L.) 234
DESSIAUX 91F,402
DETOUCHE & HOUDIN 106F
DIALS
   Garnier 57
   Erasure of Horloger du Roi 79
   Jacot 114,115
   Digital 206-207
   Willis 293
   Japanese 389
86,119,168,185,186,187,214,233,337,350,355,362,383
DIAMOND CUTTING AND JEWELLERY 105,130
DICKIE (Dr.) 117
DIEPPE 89,90,181
DIETTE 404
DIGITAL DIALS 206-207
DIJON 67
DODILLET (Jean-Henry). Machine tools for Japy 136,137
DOKE (Richard) 235F,264,412
DOLARD (Jean-Baptiste) 154
DORIAN 403
DORIUS 119F
DOUBLE-ROLLER
   Explained 237F
45,202
DOUBLE-STRIKING 153,199,237
DOUBS
   Forges and Foundries 129-130
67,75,86,400
DOUCINE SHAPE
   Description and Prices 160,171,188,189
DOUILLON 67,97
DRAW. Breguet 43,45,51
DRESDEN 325
DROCOURT
   History and general discussion 117-118
   Unsigned clocks (?) 124-125
   Grande sonnerie striking train count 201
91,92,97,106F,109,113F,117,125,176,179,184,186,187,
188F,190,264,400,402,403,405
DROST (William E.) 377,378F
DRUM-SHAPED TABLE CLOCKS 5-9

General Index

DRURY. Gong-maker 108
DUBLIN 28,29,232
DUBOIS
   On Japy and Pons 154
   On Garnier 395,396
107,119,413
DUCKER. English chronometer work 286
DUCLAIR (Claude). Coach watch 12
DUMAS (O.) 99,118
DUMMY-BLOWS IN CARRIAGE CLOCK STRIKING 202F
DUNN THE ASTRONOMER
   His carriage clock 288
   The Bridgman & Brindle watch 288F
DUPLEX ESCAPEMENT 29,53,70,71,209,211,326,360,369
DURER (Albrecht). Knight, Death and The Devil 5
DU TERTRE (J.B.). Travel clock c.1780 22
DUVERDIER (A.). Anecdote re Louis XI's clock 6
DUVERDREY & BLOQUEL
   History 92-94
   No grande or petite sonnerie 92
84,101,107,125F,160,162,163,166,170,171,175,179,184,
185,187F,189,190,404
EARNSHAW 72,265,266,409,410
EAST (Edward). Coach watch 12
EASTMAN (Joseph)
   His clock companies 376-378,382-383
   Two-way wind 378
EBAUCHES FOR WATCHES 131,133,135,136,148
ECHAPPEMENT NATUREL 41,49
EDINBURGH 27
EFFENBERGER (F.) 333F
ELECTRICITY AND ELECTRICAL EQUIPMENT
   Electricity death-knell of repeaters 307F
   Paul Garnier early in electrical horology 398
65,102,133,139,408
ELFFROTH (D.). Swiss miniature carriage clock 322
ELYOR. Anagram for Le Roy used by Basile Charles
   during the Revolution 121,439
EMPIRE STYLE 38,47,48,108,169,203,238,331
ENDERLIN ESCAPEMENT 57F,395,396
ENGLAND
   Early travelling clocks and coach watches 12-19
   Sedan clocks 29,30
   Mail-guard's watches 31,32
   Principal market for French carriage clocks 83
   Carriage clocks not mass-produced 114,399F
   Carriage clocks described 231-301
   Carriage clocks large, expensive 231,232
   Chronometer carriage clocks 210,232,234,
235,252-254,263,270,280,284-287,293
   Lever carriage clocks 232,234,235,237,239,
245,252-254
   Difficulty in establishing dates 233
   Well-known workmen 233-236
   J.F. Cole very early in carriage clocks 236
   English clock in French case 295
   Torrens' article 409-412
129,196,327
ENGLISH-STYLE LEVER ESCAPEMENTS
   Franche-Comté 75,76,150,151,154,400
EPICYCLIC GEARING 362
EQUATION OF TIME 39,43,49,51,52,254
ERBEAU (L.) 106F
ESCAPEMENT MODELS 398F
ESCAPEMENTS
   See: Verge, Leonardo, Richard of Wallingford, Cylinder, Duplex,
   Chinese Duplex, Lever, Robin, Chronometer Carriage Clocks,
   Hessen verge-lever, Goutard, Saunier, Samuel, Earnshaw, Garnier,
   L'Epee, Pons, Raby, Pin-pallet, Rubisola-duplex, Fasoldt, Thiout,
   Virgule, Tic-Tac, Constant-Force, Remontoire, MacDowall, Randall-
   Theuvillat, Breguet, Echappement Naturel, De Baufre, Sully, Rochat,
Savage, Japy, Coup-perdu, Dejardin, Tompion, Frictional Rest,
Revolving, Amant, Soldano, R.E.D., V.A.P., Escapement Models,
English-style in France, Delépine, Enderlin
   In addition see pages 209-219
ETUPES. Japy 133,135
EVERETT 383
EXACTA 95
EXHIBITIONS
   Paris. History and dates from 1798 65-68
   Value of catalogues in tracing makers 65
   Paris, 1819 .... 86F,90
   Paris, 1823 90
   Paris, 1827 55,169,398
   Paris, 1834 43,55,398
   Paris, 1839 55,56,59,398
   Paris, 1844 45
   Paris, 1849 92,395
   London, 1851 67,83,84,264,278,281,315
   Paris, 1855 54,67,137
   Rouen, 1859 97
   London, 1862 262
   Paris, 1867. Plan of exhibitors 66
   Paris, 1867 .... 67,92,333F,391,399F
   Paris, 1878 319,322
   Philadelphia, 1878 .... 169F
   Paris, 1889 54,83,116,159,399-403
   Paris, 1899 .... 130F
   Paris, 1900 .... 121F,122,148,176,287
   Brussels, 1910 97
   Turin, 1911 97
   "Exhibition work" 232F
FACTORIES AND PRODUCTION GENERALLY
   Breguet 37-54
   Garnier 54-65
   Other early makers 69-76
   Strange organisation of manufacture 83-86
   Saint-Nicolas-d'Aliermont 89-102
   Villon (Duverdrey & Bloquel) 92-94,402
   Couaillet 96,100,101
   Delépine-Barrois 100,101
   Along the Doubs 129
   Japy at Beaucourt, Badevel, etc. 133-149,401,407-408
   America 341-342
   General 399-403
FAIRHAVEN 383
FAIRHAVEN CLOCK COMPANY 376
FARMS AND AGRICULTURE
   Links with Horology 89,98,130,132,182
FASOLDT 341,384-385
FAUCHERRE. Engraver 295
FENNELL (G.) 28
FERNAND NUNEZ (Duke of) 50
FERTBAUER (Philipp). Viennese travelling clock 329
FESCHOTTE 133,135,137-139,141,143
FIREARMS
   Early attempts at interchangeable parts 341F
   Made on the Doubs 129
FITCH (E.). "Flick" or "Plato" clock patentee 208F
FIVE MINUTE REPEATERS
   Described 197-198
320-321
FLACHET (E.). On Garnier 397-398
FLEURIER 210
FLICK CLOCKS 207-209,370
FLORENCE 184
FLYS
   Escapement used instead of fly 45
200,334
FONCINE CLOCKS 21,23-26,152
FONTAINE-MELON 320

471

General Index

FORD, WHITMORE & BRUNTON
    Makers of horological tools 236
FOREST (E.). Bayard 93
FORGES AND FOUNDRIES 129,130,154
FOSTER (T.) 235
FOULKES (R.K.) 16
FOUR DIAL CLOCKS 219
FOURNERY 108
FOURNET 130
FRAIGNEAU. Leroy et Fils 123
FRANCE
    Early clockwork 5,8
    Louis XI's clock 6
    Berthoud's portable astronomer's clock of 1795 19
    Proliferation from c.1775 of small travelling clocks of
    pre-pendule de voyage type 21-23
    Portable hanging clocks 30-31
    Breguet A-L — First French carriage clocks 37
    Garnier, Paul — creator of Parisian carriage clock industry 55
    Decline of watch production particularly from 1750 .....77F,195F
    England soon principal market for French carriage clocks 83
    Heyday of carriage clock production 1889 83
    French clocks in English cases 86F,321
    Strange organisation of carriage clock manufacture 84-86,399
    French clockmakers in Russia 101
    Horology for Switzerland 129,135
    Import of American clocks 140F
    Fear of English competition 295-297
    Further links with Switzerland 307,320
    Fear of American competition 140F,342
    131,325,381,383,390
FRANCHE-COMTE
    History of area and industries 129-154
    23,75,84,85,118,167,199,204,222
FRANÇOIS DE BOURBON Breguet clock 39,48
FRANCO-PRUSSIAN WAR
    Japy factories ran normally 143F
    130,131,132
FRANKFURT-AM-MAIN. Amman and Sachs 9
FRANKLIN (S.). Comments on Horlogers du Roi 76
FRENCH JURA 25,84,85,86,98,129-154,199,309,399
FRICTIONAL-REST ESCAPEMENT ..55,56,57,151,212,395,396,397
FRODSHAM FAMILY
    Addresses 267
    Family tree 268
    125,188,189,233,234F,241F,266-272,278,280,289,
    293,295
FULL-SIZED CARRIAGE CLOCKS 184,185,186
FUMEY. Platform escapements 152
FUSEE
    Early use 5-10
    Used by Mudge in Polwarth clock 18
    Used with Harrison maintainer in Berthoud's astronomer's
    portable clock 20
    Used by Hessen late 18th cent. 23
    Not a feature of standard French carriage clocks 23
    In at least one Breguet carriage clock 43,51
    English pre-occupation 200
    Late use in French travelling clock 214
    Kullberg's keyless work 283
    Scarcely used in America 341
    Reversed by Mudge and Kullberg 410
GALERIE MONTPENSIER (13 & 15 Palais Royal) 67F,69
GALERIE VALOIS (114 & 115 Palais Royal) 64F,69
GALILEO
    Considered pendulums as a means of controlling clocks 7
GALLIERA (Musée) 48,49,50,51,52,54
GALLIOT (Hélène). Franche-Comté history 143F
GALOWSKIN (Count). Breguet clock 49
GAMMARD (Madame) 117
GAMET (G.). Coach watch 12
GANNERY 99

GARDNER (Malcolm) 265
GARNIER (Paul)
    First semi-mass-produced carriage clocks from 1830 54
    Evidence Garnier creator of Parisian carriage clock industry 55
    New range of carriage clocks for 1834 Exhibition 55
    His two-plane escapement 55,57,395-397
    Sale of carriage clocks in England 55
    Other escapements 55,398
    Predilection for rack striking work 56
    Hands of first clocks set to time with fingers 57
    Made pendules portatives as well as carriage clocks
    of various types 58-65
    Lever escapement 59,62,63
    Paul Garnier's son 65
    Other achievements 65
    Importance in context of carriage clocks 65
    Frequent mentions in Exhibition literature 67
    Biographical notes and quotations 395-398
    106F,107,108-109,158,187,209,212,318,322,413
GARNIER (Thomas). Leroy et Fils 123
GARRARDS (Goldsmiths & Silversmiths Co.) 241,405
GASTELLIER. Recent work for Breguet firm 47
GAUTRIN. Complicated coach watch 15
GAZELEY (W.) 288F
GELIN 84,403
GELIS (E.)
    Comment on Louis XI's portrait 6F
    Horlogers du Roi 77,78
GENEVA 134,307,318,319,320
GENEVA CLOCK COMPANY 320F
GEOMETRIC CHAIR 20
GERMANY
    Early spring-driven clocks 5-8
    With Austro-Hungarian Empire 325-335
    17th cent. decline in horology 326
    Alarm clocks 336-337
    129,131,132,208,329,369,401,405
GERRY (J.H.). American 2-way wind 378
GIANT CLOCKS
    Described 185-186
    Dent 254,262,263
    69,218,264
GIBBS-SMITH (C.H.). Work on Leonardo 8
GILBERT 375
GIRARD (P.). Swiss carriage clocks 315,317
GLASGOW (D.). Worked for Losada 295F
GLASS SHADES FOR PENDULES PORTATIVES 322
GLOVER 264
GODMAN (J.S.). Clever modern clockmaker 234,290,291
GOING BARRELS 26,28,200
GOLAY-LERESCHE 413
GOLIATH WATCHES 32
GONGS
    Bells superseded by gongs 199
    27,43,108,118,189,200,237,250,315,355
GONTARD. Escapement 209F,212-213
GONTARD AND BOLVILLER 107
GOOD (R.) 18,293
GORGE SHAPE
    Described 165
    Prices 188-190
    English (?) 277
    108,114,115,117,160,166,184,185,186,187
GOTHIC 233,234F
GOUGE (Lamy). Horloger to Louis XVI 78F
GOULD (R.T.) 20,21,409
GOULLONS. Coach watch 12
GOUNOUILHOU. Possible Napoleon link 22F
GOWER (Lord). Breguet clock 51
GOWLAND (J.) 294
GRAHAM (George). Lantern clock with travel box 16

General Index

GRANDE SONNERIE
   Described 195-196
   Striking train count 201
   Dubious conversions 202
   French convention 334F
   English convention 334F
   Austrian convention 334F
   Swiss convention 196,309,310,312,314,315
   Other references include 26,42,43,44,45,92,108,110,
   111,114,118,121F,144,168,176,185,186,189,197,198,
   199,202F,203,204,205,237,238,245,246,263,293,308,
   309,310,313,321,322,326,332,333,334,408
GRANDJEAN (Henry) 413
GRARD (Madame Lejoille) 90F,99,100,101
GREEN LANES (the General of the) 286
GREENWICH TRIALS 216F,283
GRENOBLE 90
GRIMALDE 267
GRIMSHAW, BAXTER AND J.J. ELLIOTT 190
GRIMTHORPE (E. Beckett Denison, Lord Grimthorpe) 196,232F,281
GROS (C.). Escapements 213
GROS-PRE. Japy 143
GROUPE SYNDICAL DE L'HORLOGERIE 116
GUIGNON 92,97,101,402
GUILLOCHE. Enamel 182
GUILMET. Four-dial clock 219
GUINARD & GOLAY 320F
GUTKAES (Franz). Vienna
   Quaint travelling clock, conventional carriage clock,
   both same date 331
HAHN 320
HALF HOUR SONNERIE 124,196
HALF-QUARTER REPEATING. Breguet 42,51
HANCOCK 238,239
HANDLES
   Margaine specialises in handles 119,187
   Obis handles 186
   Typical Drocourt 187
   Jacot 187F
HANDS
   Single-handed pieces 9,12
   Single hands the rule until advent of balance-springs 13
   Set to time with fingers 57,152,153,199,360
   Mongruel collection 79
   Vulliamy 269
   Pendleton 239,280,293
   62,71,86,119,187,233
HAPPACHER (Philipp). Austrian carriage clock 332
HARDY (William). Tooth forms 412
HARRIS. Small English travel clock with balance 17
HARRISON
   Maintaining 43,51
   Connection with Frodsham's improbable 266F
HARVARD CLOCK COMPANY 376
HASWELL (J.E.). Important technical book 219
HATTON (Thomas). Remarks on methods used by Tompion 409
HAUTE-SAONE 129,130
HAUTE-SAVOIE 130F
HAWKINS (John). Research on Thomas Cole 241
HENLEIN (Peter)
   Portable striking clocks by 1511 5-6
   Use of stackfreeds (?) 6
HESSEN
   Travel clock late 18th cent. 22,23
   Verge-lever escapement 22F
HEWITT (Joshua). Pinion engine 235F,409
HIGGINBOTHAM (C.T.). Possible additional information on Japy 137F
HINTON (C.R.) 290
H.L. MOVEMENT 318F
HOADLEY (S.) 341F

HOFMANN (Carolus)
   Typical Viennese travelling clock of c.1790-1800 329
HOLDT (R.). Sundial clock 223
HOLINGUE 97,118
HOLLAND
   First satisfactory pendulum clocks in 1656 7
   Influence 327
HOLMDEN. Wooden cases for Vulliamy 241
HONG KONG 184
HOOKE (Robert)
   Probably first to conceive idea of balance spring 14
HORLOGE DE COMTE 25
HORLOGE DE SAPIENCE. Fusees 6F
HORLOGER DE LA MARINE 61,62,63
HORLOGER DU ROI
   Title in dis-use between 1848 and 1870 60,62
   History discussed 76-79
   Erasure of H du Roi on dials 79
HOTELS PARTICULIERS IN THE MARAIS IN PARIS 105
HOUR (Charles) 106F,120,404
HOUR-AND-HALF STRIKE
   Described 195
   53,71,108,123,167,183
HOUR-GLASSES 5,9
HOURIET (A.) 413
HOWARD (Tom) 236
HOWARD WATCH CO. 377,378F
HUBER (W.J.). Reproduction "French" carriage clocks 298-299
HUGHES (William). Coach watch with automata 14
HUGUENOTS 89,402
HUMPBACKED CASES (BORNE)
   38,41,42,45,48,175,234,236,238,289,290
   But see also Borne
HUNT & ROSKELL 238
HUTCHINSON (B.). British Museum 19
HUYGENS (Christiaan)
   First satisfactory pendulum clock with Coster in 1656 7
   Development of balance spring in Holland from c.1658 14
HUYTON 266
ILBERT COLLECTION 9,42,44
IMMISCH (Moritz) 207
INDIA
   McCabe clocks 281
   Sundial clocks 223
INDUSTRIAL REVOLUTION 83
INGERSOLL (Robert H. & Bros.) 343,356F
INGRAHAM 375
INKSTAND CLOCK. T. Cole 239
INTERCHANGEABLE SYSTEM 341F,379
IONIC . . . . See Pillared Cases
IRON MOVEMENTS
   Use in early travelling clocks 5-6
   Use in German table clock of 1550 8
IRONWORK 129,132
ISLE-SUR-LE-DOUBS. Japy 133,143
ITALY
   Early spring-driven clocks 5-8
   Gradual eclipse of Italian horology 5
   Double-striking 153
   Carriage clocks 391
IVORY
   Ivory-carvers in Dieppe 90
   Case described 181
JACOB 42
JACOT
   History and general discussion 113-117
   Not mentioned by Tripplin and Wright 116,400
   Unsigned clocks (?) 124-125
   Going train count 201
   47,85F,91,92F,105,107,109,118,125,165,184,186,187,
   188F,190,196,198,199,205,213,264,405

473

General Index

JANVIER (Antide)	26,65,68F,154
JANVIER CADET. Possible link with Capucines	25,26
JAPAN	93,184,389-390

JAPY
    Detailed history 133-149
    Frédéric, apprenticeship and career 134-138
    First workroom opened in 1770 or 1772 135
    Japy frères. Names and raisons sociales 135,135F
    Watch production 135,136
    Machine tools 136,137
    Clock blancs-roulants from soon after 1800 137
    The mill clock factory at Badevel 137-142,407-408
    Pendules de voyage 139-141,143-149
    Type Américain and Ansonia Pert 140,373
    Charles Japy monument at Badevel .141
    Good work relations during trying times 142,143
    Busts of Frédéric Japy at Beaucourt 142
    Ingénu Japy in charge at Badevel 143
    Last travelling clocks said to have been made 144
    Un-used blancs roulants in original wrappers 144
    Clock made and cased in two hours in 1882 144
    Work books for 1907 120,149,407-408
    Names of certain workers at Badevel in 1907 408
    70F,84,85,95F,107,113F,116,119,131,132,150,154,158,
    188,209,210,213,222,320,400,401,405,413

JAUNCEY (J.). Collection	20
JAZ ALARMS	133
JEAMBRUN	130

JEAN DE PARIS
    Said to have made travelling clock in 1480 6

JEANNERET (F.A.M.). Notes on Neuchâtel	41,134F

JEANNERET-GRIS (J.J.)
    Early machine tools. Japy 134F,136

JEAN-RICHARD (Daniel). Mass production of watches	134,136

JEFFERSON (Thomas)
    French interchangeable parts for muskets in 1785 . . . . 341F

JEROME CLOCK COMPANY	341,373-376,381
JOHNSON (W.)	267,295
JOKERS. German and American	335,369,374,375
JONES COLLECTION IN V. & A. MUSEUM	17
JOSEPH (Charles)	106F
JULES	73,74,145,158

JUMP FAMILY
    Lord Ashburton's clock 290
    38,234,289-291

JURA	See French Jura and Swiss Jura
JURGENSEN	283,413
KAHENN (A.) & CO. American alarms shown in 1889	402

KARRUSELS
    Described 216-219
    Fundamental difference between tourbillons and Karrusels 217,361F

KESSELS. Work for Breguet	42
KEW TRIALS	216F
KING (George, J.)	382
KLAFTENBERGER	124,184

KLENTSCHI (Charles-Frédéric)
    Link with Courvoisier 310F,311,313,317,322

KNOCK-OUT STRIKING	195,196,309,350
KNOTTESFORD (William). Coach watch	12-13
KOREA. Japanese carriage clocks	389
KREMER	85

KULLBERG (V.)
    Workbooks 233,283-285
    Double-roller 237F
    History and achievements 282-287,302
    List of carriage clocks 284
    232,234,410

LA CASSERIE. Japy	143
LACEMAKERS IN DIEPPE	90
LA CHAUX-DE-FONDS	11
LAMBERT. Saint-Nicolas-d'Aliermont	102
LAMBERTI (The Count). Has early fusee clock	6F
LANCASHIRE	409,410,412
LANDENBERGER	92
LANGE. Carriage clock cases	145,158
LANTERNE D'ECURIE CLOCKS	21,23
LAROCHE. Japy	133
LAVIGNE	404

LEBON (E.)
    Franche-Comté history 25
    Controversial remarks on Geneva watch industry 320

LE BRASSUS	130,320
LE CHEVALLIER	118
LECHEVALLIER & MERCIER	102
LE COULTRE	46,320
LEE (R.A.). English pendulum clocks from 1658. Research	7F
LE FAUCHEUR FILS	78
LEFEBRE	407,408
LEFRANC	111,113,205
LEGION D'HONNEUR	68,90F,137
LE HAVRE	283
LE LANDERON	320
LE LOCLE	130,134,135,330
LEMAIGNEN	102
LEMAILLE	207

LEONARDO DA VINCI
    Considered pendulums as a means of controlling clocks 7
    Escapement perhaps older than verge 7F
    Sketches of fusees 7F,8
    Work scarcely known until 1797 8

LEOPOLD (J.H.). Work on Almanus	7F

LEPAUTE
    Travelling clock similar to "Marie Antoinette" 17
    Exhibited pendule portative in 1827 .55
    69,78,90,114,123,169,413,444

L'EPEE
    Musical boxes and platform escapements 149-151
    Peculiar platform escapement with twin scape wheels 151
    85,95,130,131,132,133F,149,154,209F

LEPINE	70,133,158,195,209

LE ROY (Basile Charles)
    Travelling clock of c.1780 21
    120,121

LE ROY (Julien and Pierre)	45,68,77,78,111

LEROY/LE ROY
    History and general discussion. Litigation 120-124
    Bottom-winding 123,219-221
    30,31,62,63,64,67,68,69,107,113,114,176,185F,187,
    199,203,204,210F,318

LEROY (Theodore)	69,218
LEROY (Théodore-Marie)	121,413
LE SENTIER	46,130,320
LES PONTS-DE-MARTEL	307,320,322

LEVER ESCAPEMENT
    Mudge's lever in 1769 or before 18
    Hessen verge-lever 22F
    In Swiss portable clocks 18th/19th cent. 26,27
    "Sedan" clocks 29
    In mail-guard's watches 31
    Use by Breguet 43,45,48,50,51,52,53,54
    Probable early use by Lepaute 55
    Garnier 59,62,63
    Manufactures on the Doubs 129
    Use by McCabe 213,214
    Two-plane 213
    41,72,74,75,76,84,93,114,115,146,150,154,190,202,
    209,211,214,219,232,234,235,237,239,245,252,253,
    254,280,309,313,318,342F,380,383,396,400,402
    See also Pin-Pallet Lever Escapement

LEVY	403
LEYLAND (Thomas). Used Hardy's tooth forms	412
LIBERTY & CO.	182
LIMOGES ENAMEL	182
LINARD. Last in charge of Japy factory at Beaucourt	137F

General Index

LINDQUIST. Kullberg assistant, often confused with Lundquist . . .286
LIORET (Henri). Patent winding work 221
LITTLEWORT (G.). Mail-guard's watch 31
LIVERPOOL 98,232
LLOYD (H. Alan). Horological historian .16,17
LOCKING PLATE STRIKING 56,75,197,199
LOCKS
    On early travelling boxes . . . . .8(I/5),10,11(I/11 and I/12)
    On mail-guard's watch 31
    Frédéric Japy's father 134
    On English carriage clock boxes .231,244,263
    Bramah 244,263
LONDON
    Great Exhibition of 1851 67,83,84,264,278
    Exhibition of 1862 67
    Best clocks 232
94F,98
LONG WIND WATCHES 356-364
L'ORIENT DE L'ORBE 320
LOSADA 294,295
LOUIS XI OF FRANCE
    Owned travelling clock by 1480 (?) 6
    Portraits 6F
LOUIS-PHILIPPE (King). Breguet clock 52
LOUPPE (E.). Capt's agent in New York 319
LOUVRE 65
LUGRIN 320
LUNDQUIST (S.)
    "Victor Kullberg's" last proprietor 234,283,286
LYCETT. English chron. work 286
LYNCH. English chron. work 286
McCABE (James)
    India 281
232F,233,235,266,279-281
MacDOWALL FAMILY 281-282
MACHINERY . . . . .See Tools
MADELEINE 169,170
MAICHE. Escapement parts 129,130
MAIL-CARTS. Introduced in 1784 32
MAIL COACH 31
MAIL-GUARD'S WATCHES 31,32
MAILLARDET. Link with Courvoisier 310F,311F,317
MAINSPRING
    In use by c.1480, if not before (?) 5-8
    Brunelleschi, perhaps the first 6F
    Manufacture 7
    Portable clocks dependent upon mainsprings 5,7,13
    Two trains driven from one 50,53,75,137,152,309,335
    Manufacturers and sources of supply 85,86,95
    Waterbury. Peculiar arrangements 361,362F,366
MAINTAINING WORK 20,31,43
MALLET COLLECTION 49
MANDEURE. Roman amphitheatre on the Doubs 130
MANN (Henry). "The great pivoter" 235
MAPPIN & WEBB 182
MARAIS. History 105,106
MARANESI (G.). Italian carriage clock 391
MARBLE PANELS 184
MARC (Henry) 58,145,219
MARCHAND 194
MARGAINE
    History and general discussion 118-120
    Unsigned clocks(?) 124-125
    Clock prices 119
    Specialised in dials 119
    Unusual handles 119
106F,109,113,125,160,167F,170,182,185F,187,189,190,
399,400,403,405,407,408
MARIE ANTOINETTE. Her alleged travelling clock 17
MARINE CHRONOMETERS (Box chronometers)
20,21,55,91,99,102,111,112,120F,232,234,235,250,254,
263,264,269,270,283,286,294,328,398,409,410,411,412
MARSEILLES 67

MARTI 85,95F,150,401
MARTIN (E.) 92,97,413
MASETTI (B.). Italian carriage clock 391
MASSEY 309
MATHEY-TISSOT
    Montres pendulettes de voyage 307,320,321,322
MATHIEU 67,398
MAUDSLAY (Henry). Screw-cutting lathe .409
MAURANNE 96,100
MAURICE (E. & Cie.) 106F,404,405
MAYER. Austrian carriage clock 333
MAYET. Possible link with Capucines 25,446
MEGNIN 401,403
MEINER (Louis) 143
MELBORNE 68
MELZI (F.). Leonardo's notebooks and sketches 8
MERCER (Thomas & Son) 234,235,293-294,409
MERCER (Vaudrey)
    Hessen verge lever 22F
    Arnold research 246,247,269
MERCIER 320
METAL REFINING 130
METZ 67
MEXICO. Imported American clocks 342
MEYER & SCHATZ 209F,211,212
MICHOUDET (Jn. Marc) 152,154
MIGNONNETTES
    Described 184-185
    Price 188
    Japy 144,408
119,160,166,168,179,185,239,320,322
MILLS 129,132,133,137,138,139,143,154
MINIATURE CLOCKS . . . . .See Mignonnettes
MINUTE REPEATERS
    Use by Breguet 53
    Described 198-199
    Jacot 198
    Boseet 204
67,121,122,205,320,321
MOINET
    Nouveau Traité 58F
    On Garnier 297
MOLARD 398
MONGRUEL COLLECTION OF CLOCK HANDS 79F
MONNIN. Japy 85,132,143
MONTBELIARD
67,84,95F,114,116,130,131,134,135,144,148,151,403
MONTBELIARD-WURTENBERG (Prince of) 134
MONTRES DE CARROSSE 11,12
MONTRES PENDULETTES DE VOYAGE
182F,199,307,320-322
MOON-WORK
    Early use in coach watches 12F
    Breguet 39,42,43,44,48-52
    Bourdin 109
    Margaine 118
    Moulinie 318
    Japy 144
    Franche-Comté(?) 203
    Boseet 204
MOORE'S PATTERN BOOK OF CLOCK CASES 234F,280
MORBIER 24,25,199
MOREZ 25,67,97,130F,131,153,154,199
MORPURGO (Enrico)
    Research in early clockwork 6F
    Camerini clock 7F
MORTEAU 84,129,130F,131
MOSER 106F
MOTEL 99,112,288
MOUGIN 401
MOULINIE 145,307,318
MOUVEMENT DE PARIS 63,95,407,408
MUDGE (Thomas)
    Gen. Clerk's clock 17
    Polwarth clock pre-1769(?) 18,19
    "Queen's" watch of 1769 18
    Reversed fusee 410
MULTI-PIECE CASE
    Described 158-160
60,71,74,75,186,381

475

General Index

MUSICAL WORK
   26,73,85,147,149-150,154,194,210,221-222,335,408
MUSTON (Dr.). Japy historian
   85F,133,134,135,136,141,142,143,148,151
MUTZ (Walter, E.). Chelsea Clock Co. 382,384
MYERS (S.F.) 343,374
NAME-PUNCHES USED BY JACOT 107
NANTES
   Proposed exhibition 1861 67
   Edict of, and Revocation 89,402
NAPOLEON 22F,129,137,142,441
NATIONAL MARITIME MUSEUM 19
NEAGLE (W.) 382
NEUCHATEL
   Revolution, 1848 .... 310F
   26,134,307,310,317,322
NEW ENGLAND WATCH COMPANY 343,356F
NEW HAVEN CLOCK COMPANY 373-376,381
NEWSUM (B.). Two-tier clock movement 10F
NEW YORK
   Henry Capt's advertisement c.1900 318
   Louppe. American agent 319
NICOLE NIELSEN 234,267,291-293
NORMANDY 89
NORMANDY CLOCKS 97
NORRY (N.). Coach watch 12
NORTH (R.B.) 267
NUREMBERG. Cradle of horology in Central Germany 5
OBIS
   Described 162-165
   Train count 201
   95,160,167,168,171,178,179,182,184,185,186,187,298,
   341,369,374,390,405
OBLONG SHAPE 174
OLLIER (Adolphe) 175,194,405,407
ONE-HANDED CLOCKS 13
ONE-PIECE CASES
   Described 157-158
   56,57,60,74,75,212
ORLEANS 67
OUDIN (Charles) 111
OUDIN-CHARPENTIER 110,190
OVAL SHAPE 166
PALMER (Brooks) 377
PAPE. English chron. work 286
PAPIN. Saint-Nicolas history 90
PARACHUTE 43
PARIS
   Exhibitions. History and dates 65-68
   Today and yesterday 105-125
   The Marais 105
   Wide cultural influence 325
   La Société des Horlogers 413
   37,47,67,84,85,86,90,91,93,95,114,130F,143,144,
   149,151,165,167,172,186,204,283,297,318,399,401,
   402,403,404,405,406
PARKINSON & FRODSHAM 266,267,268,278
PATEK (Philippe) 319,320
PAULET. Small metal-cased travelling clocks with balance 17
PAYNE 238,278
PEARSON (Charles). Chelsea Clock Co. 382
PECKOVER. Small portable clock 16
PENDLETON (Peter). Prescot handmaker 239,280,293
PENDULE DE CHEMINEE 170
PENDULE PARIS 17,58,195,407,408
PENDULE PORTATIVE
   Defined 58
   19,56,322
PENDULE SYMPATHIQUES
   Raby 47
   Breguet 48,51,52,65F
PENDULES D'OFFICIER 21,22,23,322
PENDULES DE VOYAGE
   Mainspring and balances first step 5,7,13
   Final French shape and form foreshadowed 22,23
   An evolution, not an invention 37
   Breguet carriage clocks "fully-fledged" by 1810 37
   Garnier. First semi-mass-produced clocks 54,56
   England soon principal market for French carriage clocks ... 83
   Heyday of French production 83
   Saint-Nicolas-d'Aliermont 91-102
   Japy and Franche-Comté generally 139-154
PENDULUMS
   Ante-dated by balance 7
   Controversial Camerini clock dated 1656 .... 7F
   Early balance-controlled clocks converted to pendulums 13F
   213,214
PEQUIGNOT (Edouard)
   Living link with Japy Badevel factory 144
   140,407
PERRELET (Jean-Jacques). F. Japy's apprenticeship 134,135
PERRET. Not Japy's master 134
PERT. Connection with French type Américain 140,373
PETITE SONNERIE
   Described 195-196
   Striking train count 201
   Dubious conversions 202
   Other references include 26,42,43,44,45,55,92,119,
   168,178,188,197,204,252,279,309,310,321,326,327
PEUGEOT 130,132,133,140,447
PEUPIN ON GARNIER 395
PHILIPPE (Adrien). Advice in 1878 to Geneva workmen 319
PHILIPPS (Edouard). Balance spring theory 121F,428
PIAGET BROTHERS 47
PICARDY 89
PILGRIM. Jerome Obis-appearance clock 374
PILLARED CASES
   Described 169-170
   Corinthian 169,170,185,187
   Ionic 170
PINCHON 75
PIN-PALLET LEVER ESCAPEMENT
   27,89,143,337,342,350,354,356,369,374,378
PIPES FOR SMOKING 130
PITCHER
   Personal reminiscences 404-406
   94,147
PITOU
   Carriage clock finisher until 1970 47,105
   54,106,107,108,113F,115,157,188F
PLATFORM ESCAPEMENT
   Early examples 23,26,27
   Lepaute pendule portative in 1827 55
   Non-use by Paul Garnier 57
   L'Epeé, an important maker 150
   Annual production c.1900 150
   Discarding old ones reprehensible 185F
   Not used in cheap American carriage clocks 342
   72,74,84,85,100,114,116,129,145,148,151,153,188,190,
   202,215,216,219,223,224,234,235,237,263,309,310,311,
   313,318,322,334,383
POINTED TOOTH LEVER ESCAPEMENT
   English type in French work 75,150,154,400
POLWARTH MUDGE TRAVELLING CLOCK 18-19
PONDICHERRY. McCabe clocks 281
PONS (H.)
   Revives flagging clock industry in Saint-Nicolas from 1806 90
   42,56F,64,65F,86,92,98,154
PONTEVICO (Comino da). Spring-driven fusee clock in 1482 5
PORCELAIN PANELS 119,182,405
PORTABLE CLOCKS
   First examples 5-8
   Dependence upon mainsprings 6,13
   Dependence upon balances 7,13
   Uncertainty where first made 7
PORTAL AND DE GRAFFIGNY. Louis XI's clock 6
POST-CHAISE WATCHES 32
POST OFFICE. Mails 31,32
POUGEOIS (Charles) 107
POURTALES (Count). Breguet clock 46,50
PRAGER (Frank D.). Research on early clockwork 6F
PRAGUE. Zech's clock in 1525 6
PRE-PENDULES DE VOYAGE
   Early history 20-27
   Austrian examples 325-330
PRESCOT
   84,94F,98,102,235,238,266,280,281,285,293,399,412
PRESSBURG. Viennese clocks 325
PRESTIGE (Sir John and also S.E.)
   Collections 16,44,48,51,53

General Index

PRESTON (Joseph & Sons)
    Lancashire movement-makers 235F,264,283,285,412
PRICES OF CARRIAGE CLOCKS
    French 136,147,187-190,342
    English 289
    American 341,342,355,356,360,373,374,375
PRODUCTION FIGURES, ACTUAL OR IMPLIED
    French 54,55,56,65-74,83-86,90-102,105-125,150,
    401-403,407-408
    American 342,375
PROGENITORS OF CARRIAGE CLOCKS 5-32
PUNCHES 107,125F,435,439
PYBUS (H.). Prescot movement maker 283
QUALITIES 119,187-190,320,404
QUARE (Daniel)
    Small portable clocks 16
    Watch movement in "Sedan" clock(?) 29
QUARTER REPEATING 39,45,59,320
QUARTER STRIKING . . . . .See Petite Sonnerie
QUEEN CHARLOTTE'S WATCH 18
QUESNEL 96
QUILL (Col. H.) 18,20,21,234,266,289,290,291
RABY (Louis) 47,51,187,211,413
RAILWAYS
    Death knell of the stage coach 32
    Sleeping-cars. Watch-holders 32
RAINES (Charlie). Eastman's helper 383,384
RAINGO 55F,64,67,264
RAIT (D.C.) 75
RAMSEY (General). Breguet clock 50
RANDALL (A.)
    Escapement 215
    Japy research 407
RAWLINGS (A.L.) 234
RECLUS (Victor) 146,210,413
R.E.D. Escapement 213
REDIER (Antoine)
    On Garnier 397
    84F,129,163
REGULATEURS 91,96,139,141,408
REGULATORS 55F,232,250,402
REID & AULD
    Reid's Treatise 17
    Sedan clock made post 1806 27,28
REMONTOIRE 51,215,216,225,398
RENNESON. Alarm work for mignonnettes 185F
REPEATING WORK
    In coach watches 12
    In small travelling bracket clocks . . . . 16,17(I/20)
    In pre-pendules de voyage 22 (I/29)
    In Capucines 26
    In Swiss pre-pendules de voyage . . . . 26,27(I/37)
    In French "sedan-like" wall clocks 31
    In early Breguet carriage clock 39,40,42,44
    Use by Paul Garnier 55,59
    90,144,195,196,197,198,199,201,202,205,237,243,244,
    253,254,270,274,279,280,289,308,309,310,312,318,
    320,321,322,326,332,333,342,343,350,351,355,356
REQUIER (Charles) 106F,403
RETTICH. Austrian carriage clock 332
REUILLE & ANGUENOT 403
REVEILS AMERICAINS 139,140,149,373,407,408
REVERCHON (L.). Japy 137,147
REVOLUTION
    Effect on clock dials and signatures 60,62,79
    37F,78,90,105,106,120F,121,124,129,318
REVOLVING ESCAPEMENTS 40,48,216-219,293,356-367
REX. Welch Obis-appearance clock 374
RHONE 131
RHINE 85,89,131
RICHARD & CO. 196F
RICHARD OF WALLINGFORD'S ESCAPEMENT
    Possibly older than verge 7F
RIM-WIND. American clocks 369
RIPPON (R.). Dent history 247,250,251
ROBERT 26,413
    & Courvoisier 26,308,314F,322
ROBIN (Robert)
    Marie Antoinette, her alleged travelling clock 17

Escapement 41,42,49,50
    211,212
ROCHAT. Escapement 211
ROCOCO STYLE
    Described 179
    In Austrian clock 327
RODANET (A.H.) 107,117
ROGER & CO. American alarms shown in 1889 402
ROLLIN (Savoie). Saint-Nicolas history 90
ROMANS 130,139
RONDELOT. Japy 137,143
ROSKELL. Liverpool-type work 232,233
ROSSEL ET FILS 196,318
ROUEN 89,90
ROUGH MOVEMENTS
    233,264,283,285,409,410,411,412
    See also Blancs Roulants
ROUMANIA. Japy 133
ROUSAILLE (D.) 84,118,159F
ROUSSES (Franche-Comté) 25
ROUX ET CIE. 85,95,150
ROWEAU (I.). Travelling table clock late 16th cent. 9-10
ROXBURY 377
ROYER-COLLARD (F.B.)
    Important book on skeleton clocks 281
RUBISOLA—DUPLEX 211
RUETSCHMANN (Joseph). Austrian travel clock late 18th cent. . .327
RUSSEY. Escapement parts 129
RUSSIA
    Royal family and aristocracy. Breguet clocks 48,49,50
    Overtures to French horologists .101,147
    Napoleon's campaign 142
SACHS (Hans). Plate and verse with Amman, 1568 9
SAGE
    Competition for Obis and Corniche 341
    343,349
SAHLER 135F,136
SAINT-BARTHELEMY MASSACRE 89
SAINT-CLAUDE 24,26,154
SAINT-HIPPOLYTE 129,130,135
SAINT-NICOLAS-D'ALIERMONT
    Clock industry re-organised by Pons from 1806 65
    Past and present 88-102
    56,57,84,85,86,107,114,116,117,118,130F,137F,147,
    148,151,162,163,167,168,174,181,182,183,185,188,
    219,222,224,297,298,399,400,401,402,403,404,405
SAINTE-SUZANNE 85,95,131,149,150,151
SALOMONS. Breguet Collection 41,50
SAMUEL (J.). Two-plane escapement 57F,210,211
SAULET. Year clock 225,226
SAUNIER. Noted French horological writer
    On escapements 209F,211F,213
    85,86,113,115,117,148,342,376,399,413
SAUTEUR (Frères) 92,97
SAVAGE TWO-PIN ESCAPEMENT 239F
SAVOYE. A-L Breguet's nephew 37F
SCHAEFFER. Leroy & Fils 123
SCHLOSS COLLECTION 11F
SCOTLAND 29,406
SEDAN CLOCKS
    General comments on origins and dates 27-30
    French counterpart 195
SEGUIER (The Baron). Noted amateur horologist 67,398,413
SEIKOSHA. Japanese carriage clocks 389,390
SEIKO TIME CORPORATION 389,390
SELONCOURT. Japy 133,135
SEOUL. Japanese carriage clocks 389
SERRIER (Citoyen). Altered dials inscribed H. du Roi 79
SETH THOMAS . . . . See Thomas
SHEFFIELD
    Steel for French mainsprings, which then sold in England 85F
SHIBYAMA PANELS 184
SHIP'S BELL STRIKE
    Described 199
    Chelsea 381
SIDEREAL TIME. Bridgman & Brindle 288
SILKE. English chronometer work 286
SILLS (William No. 1 & No. 2)
    Noted chronometer workmen 235,286
    The General of the Green Lanes 286

General Index

SILVANI 57F
SIMONI (Col. A.)
    Early reference to mainspring 5
    Research in early clockwork 6F
SINGING BIRD CLOCKS 221-222
SINGLE-ROLLERS 75,237F,239,243,275,313,380
SIRE 95F,115,129,147,148
SIZES OF CARRIAGE CLOCKS 184-186
SMITH (S.) & SON
    Sold D.-B. clocks 160
    French clock prices 189
    English clock prices 289
SNOW FACTOR IN HOROLOGY 130
SOCIETE BELFORTAINE DE
    MECHANOGRAPHIE (S.B.M.) 135F,136,142
SOCIETE SUISSE HERMES-PAILLARD (Bolex) 135F
SOCIETE DES HORLOGERS. List of founder members 413
SOLDANO (J.) 118,264,320,450
SOUTH AMERICA. Market of N. American clocks 340
SPAIN
    Astrolabe for Toledo in German table clock 8
    Link with Franche Comté 130
    Export by Japy 143
SPECTACLE-MAKING 105,130
SPENCER (Lord). Breguet clock 42,50
SPRING-DRIVEN CLOCKS
    Early examples 5-11
    Late 15th cent. origins(?) 6
SQUARE CASES
    Described 181-182
    Price 188
ST. LOUIS CLOCK AND SILVERWARE COMPANY 320,349,375
STACKFREEDS. Supposed use by Henlein 6
STAGE COACHES (The "Wonder", etc.) 31,32
STANDING BARRELS 26
STAUFFER (Onésime). Montres-pendulettes de voyage 322
STEAM, COMING OF 32,129,133,136,139
STOPWORK
    Use in first Garnier carriage clocks 56
    Current practice of discarding 202F
STRASBOURG. Franche-Comté. Invasion corridor 132
STRIKING WORK
    Used in portable clocks by 1493 5
    Henlein's portable clocks by 1511 6
    Louis XI's clock of 1480(?) 6
    Roweau clock 10
    Travelling table clocks 11
    Coach watches 12
    Gautrin 15
    Paulet .... 17(I/20)
    Pre-pendules de voyage 23F
    Capucines 26
    Swiss portable clocks 26,27
    Early Breguet carriage clocks 41
    Early Garnier carriage clocks 56
    Various types described 195-201
    Striking trains, French 201
    Abortive conversions – petite sonnerie to grande 202
    Now see Hour and Half Hour Strike, Petite Sonnerie,
    Grande Sonnerie, Half Hour Sonnerie, Westminster Quarters,
    Five Minute Repeaters, Minute Repeaters, Double Striking,
    Ship's Bell Strike
STRUT CLOCKS 238-239
STUBS (Peter). Lancashire tool-maker 236
SULLY. Escapement 57F,211,395,396
SUNDIAL CLOCK 223
SUNRISE, SUNSET 144,238
SWEDEN 23,85F,95,283
SWISS JURA 89,134,309,402
SWITZERLAND
    Pendules d'officier 22
    Pre-pendules de voyage 26
    Early carriage clocks 37F
    Some dependence upon France 129
    Many French watches in fact Swiss 195F
    Swiss striking 196,309,310,312,314,315
    Carriage clocks and montres pendulettes de voyage 307-322
    Further links with France 307,320
    Philippe's advice to workmen of Geneva 319

Lebon's remarks on Geneva watchmaking 320
9,10,27,44,84,86,98,130,132,134,138,150,152,153,182,
199,210,215,216,222,325,400,405
SYDNEY. Exhibition 1879 68
SYMONDS (R.W.). Tompion 16
TABERNACLE CLOCKS
    Vertical table clocks 10-11
    Many examples converted from balance to pendulum 11,13
TABLE CLOCKS. Various types, shapes and nationalities 8-11
TAILLADE (Jacques). Coach watch 12
TALLEYRAND (Prince) 40,41,48
TARDY 40
TAYLOR BROS. Prescot Manufacturers 235F
TAYLOR. Tourbillon 217
TEL AVIV. Breguet clock 50
TERRY (Eli) 341F
TEXTILES 130,133F
THERMOMETERS
    Use by Breguet 44,49,52
    Bourdin 109
    Boseet 111,204
    V.N.B. 203
    On sides and tops of carriage clocks 223-224
THEUVILLAT. Constant-force escapement 215
THEVRAY (Madame) 101
THIOUT. Escapement 213
THIRKELL (F.L.). Frodsham 267
THOMAS (Seth) 341,373-376,381
THORNHILL (W.) & CO. 176
THORNTON (Philip). Gifted modern clockmaker 290,291
THWAITES & REED 275-276
TIC-TAC ESCAPEMENT 395,402
TIEDE (Fridrich)
    German travelling clock mid to late 18th cent. 327,328
TIMBER INDUSTRY 130,132
TIPSY-KEY 15
TOMPION
    Coach watches 12
    Small travelling clocks with composite escapements
    and metal cases 16
    Miniature bracket clocks with pendulums 16
    Watch movements re-used in "Sedan" clocks(?) 29
    Watch escapement 395,396
    Had good tools 409
TOOLS AND MACHINERY
93,94,96,97,102,114,134F,136,137,143,151,236,342,
362,407-408,409-412
TORRENS (D.S.)
    Nail and Cork and Rule of Thumb 409-412
231,232,235,236,239F,283
TOSCANE (Duchess of). Breguet clock 50
TOUCH-STUDS. To tell time in dark 9
TOURBILLONS
    Breguet 40,48,49,54,216
    Described 216-219
    Taylor 217
    Nicole, Nielsen 217,293
    Fundamental difference between tourbillons
    and Karrusels 217,361F
    Auguste 218
    Waterbury long wind watches 219,356-364
TRAINS
    French carriage clock trains 200-202
    Bastard 202
    Waterbury Series "E" watch 364
    Boston carriage clock 380
TRAVELLING BOXES
    In use certainly by mid-16th cent. 8
    In table clock with detachable alarm .... 9(I/6)
    Table clocks .... 10,11(I/11 and I/12)
    Upright table clocks .... 10,11(I/10)
    Graham lantern clock 16

General Index

With royal cypher 17
Chronometers 21
Capucine 23
Breguet 44,52
Garnier 58,63
231,232,238,263,272,277,280,284,293,310,312,315,
343,347,348,405
TRAVERS (John). Noted chronometer workman 234
TRIPPLIN (J.)
   English observer Paris Exhibition 1889 84
   Never mentions Jacot in 1889 116
   Description of Paris 1889 Exhibition 401-403
67,84,85F,92,94,116,117,118,133,144,391
TURKEY 49,314,315
TWO-DIAL CLOCKS 219
TWO-PLANE ESCAPEMENT
   Garnier's 57,395-397
   Samuel's vertical 57F
TYPE AMERICAIN .... See Réveils Américains
UP-AND-DOWN-DIALS
   (Développement de ressorts) 43,49,52,112,270,271
USHER & COLE 287-289
VACHERON & CONSTANTIN 320
VALLEE DE JOUX 322
VALLIER (Jean Baptiste). Coach watch 12
VALLIN. Table clock c.1590 11
V.A.P. Escapement 214
VAUCANSON. Saint-Nicolas 102
VENTURI (J.B.)
   Publication in 1797 of much of Leonardo's work 8
VERGE ESCAPEMENT
   Current by 2nd half of 13th cent. 7F
   Miniature Watson bracket clock .... 16(I/19)
   Paulet travelling clock .... 17(I/20)
   Hessen verge-lever 22F
   Capucines 26
   "Sedan" clocks 27,29
VERMONT CLOCK COMPANY
341,376-381,382,383
VESOUL. Metal refining. Franche-Comté 130
VEYRIN. French "Sedan-like" clock 31
VICTORIA AND ALBERT MUSEUM 17
VICTORIAN AND EDWARDIAN PERIODS
176,231,232,239,266,283
VIENNA
   Exhibition, 1873 68
   Cultural influence 325,326
   Influenced by Paris 325F
   Pre-pendules de voyage and carriage clocks 325-335
   Influenced by England 326F
VILLON 91F,92,94,95,97,189,402
VINTER (Ernest). Japy historian 133,140,141,142,143,144
VIRGULE ESCAPEMENT 395
VISCONTI (Gaspare)
   Spring-driven portable striking clocks in 1493 5
VISSIERE 112,121F
VITRAGES D'HORLOGERS 98,100,130
VOLANT. Table clock late 16th cent. 11
VOLOGNE. Supplied clocks to Dent 264
VOUJEAUCOURT. Japy 133
VUITAL (Louis) 42
VULLIAMY
   Workbooks 233,241F,242,243,245,246
   List of carriage clocks 284
232,235,241-246,250,266,267,269,270,289,301,334F
V.W.B. Trademark 203

WAGES AND SALARIES 190,399
WAGNER (Nephew) 413
WAR OF THE ENGLISH SUCCESSION 89F
WARNING IN STRIKE WORK 195,350
WARSAW. "Viennese" clocks 325
WATCH-CLOCKS 28F,29,195
WATCH PRODUCTION
   Japy 133,135,136,137
   Swiss upsurge 134
   Lebon on Geneva 320
WATERBURY (Clock Co. and also Watch Co.)
   Clock Company, history 343
   Carriage clocks and alarms 343-367
   Watch Company, history 343,356F
   Long wind watches 356-364
   Carriage clocks based on long wind watches 364-367
140,219,336,337,341,375,381
WATERLOO, BATTLE OF, 1815 83,264
WATSON (Samuel)
   Miniature English travelling bracket clock of c.1710 .. 15-16
WEBSTER COLLECTION 8,10F,12,14
WEBSTER (I. & C.). Movement maker 235F
WEBSTER (R.W. & Son) 272
WEEKS. Decorative metalwork 241
WEISSE. German travelling clock dated 1787 328
WELCH CLOCK COMPANY 341,373-376,381
WELSBY (Jonathan). Pinion maker 235F
WENNERSTROM (George and Peter). Kullberg's nephews 286
WESTMINSTER QUARTERS
   Described 196-197
263
WEST'S (of Dublin) 232
WESTMINSTER CLOCK ("Big Ben") 196,250
WHITELAW (James) 233,273
WHITTINGTON QUARTERS 239F
WICKS. Chronometer work 286
WILLARD (Aaron/Simon) 378F
WILLIS. Noted English dial-maker 293
WINNERL 413
WOLF-TEETH IN WINDING WORK 43
WORKBOOKS
   Japy 120,149,407-408
   Vulliamy 233,241F,243,245,246,283-285
   Kullberg 233,283-285
WRIGHT (T.D.)
   English observer at Paris in 1889 84
   His astonishment at strange organisation of French
   carriage clock industry 84,98,399
   Never mentions Jacot in 1889 116,400
   Report on Paris Exhibition of 1889 399-400
67,83,117,144
WYKE. Horological tools 236,409
YEAR-GOING CLOCKS 55F,224-226
YERMOLOFF (General). Breguet clock 51
YERSIN. Swiss platforms 403
YOKOHAMA. Japanese carriage clocks 389
YONGE (W.) 294
YOUNG (H.C.). Living link with Victor Kullberg 286
ZECH (Jacob). Fusee clock in 1525 6
ZERO
   Size 184,189,199
   Petit 184
   M 185,189
   S 185
   Bis .... See Obis
ZINNER (Dr.). Mentions fusee clock dated 1509 .... 6F,8

Index of Plates

All plates are listed; but in addition many features of specific interest are indicated under various headings.

ABBOTT (James) IX/95
AFTER-PAINTED DIAL VII/41
ALARM CLOCKS AS SUCH
    IIII/2, V/26,27, XI/20,21,22,23,24,25,26, XII/1,20
ALARM WORK OF UNUSUAL INTEREST
    Twenty-four hour VIII/1
    Miniature VIII/2,3
    Singing bird VIII/44
AMMAN (Jost)
    "I make the clocks for travellers", 1568 .... I/7
ANGLAISE VII/16,21,43, IX/94
ANGLAISE RICHE VII/C2,C3,C4,C5
ANONYMOUS ENGLISH CLOCK IX/86,87,88
ANSONIA CLOCK COMPANY XII/41-48
ARNOLD (J.R.) IX/16,17,18,19,45
ART NOUVEAU VII/33,34,35,36
AUGUSTE II/46, VIII/36,37,38
AUSTRIAN TRAVELLING CLOCKS
    Unsigned XI/9,10,11,13
    Page 324
AUXILIARY COMPENSATIONS
    Dent's IX/24,31
    Kullberg's IX/71,72
BADEVEL
    Japy factory as in 1971 .... VI/5,6,7,8,9,12
    Jean-Charles Japy's memorial as in 1971 .... VI/13
    E. Péquignot (Japy 1890-1940) in 1971 .... VI/16
BAMBOO VII/19,32,33
BAROMETER VIII/14
BARRAUD & LUND IX/59
BARWISE IX/53,54,95
BAUTTE & CO. (J.F.) X/15,16
BAYARD. Factory, 1970 .... IIII/6
BEAUCOURT
    Japy's clock factory mid 19th cent. VI/4
    Frédéric Japy. Memorial bust VI/14
    Japy. Dis-used factory as in 1971 .... VI/15
BEGUIN (J.B.) VII/38
BELFORT. Bartholdi's Lion de Belfort .... Page 128
BENNETT (Sir John) .... Page 302
BERTHOUD
    Astronomer's portable clock made 1795 .... I/23,24
    Chronometer carriage clock, 19th cent. V/12,13
BLANCS-ROULANTS V/3
BOITE CHINOISE VII/37
BOITE CLASSIC VII/6
BOITE JONC VII/5
BOLVILLER II/43,44,45, VII/3, VIII/43
BOQUET (David). Coach watch c.1650 .... I/14,15
BORNE VII/31
    See also Humpbacked Clocks
BORRELL VIII/11
BOSEET VIII/14,15
BOSTON CLOCK COMPANY XII/53-59
BOTTOM-WIND VIII/40,41
BOURDIN .... Page 104, V/7,8
BREGUET FAMILY AND SUCCESSORS
    A-L Breguet portrait .... Page 36
    Carriage clocks II/1-II/18
BRIDGMAN & BRINDLE IX/73,74
BUBBLE CLOCK VIII/47,48

BUCHANAN. Sedan clock, late 18th cent. I/40
CALENDAR WORK
    Simple II/26,29, V/7,18,24, VII/3,38,43, VIII/2,11,12,14,
    IX/4,18, X/5,8,13, XI/12, XII/20, XIII/1,2
    Complicated or Perpetual II/1,2,10,12,14,15,16, V/23,
    VIII/13,14,15,16,17,18, IX/2,8,35,58,76,78,79,80
    Fly-back VIII/13,14,15, IX/2
    Islamic X/11
CAMPBELL II/40
CANNELEE VII/C9, VII/10
CAPT (Henry) VIII/4
CAPUCINE CLOCKS I/32,33,34,35
CARIATIDES VII/24,25,26,27
CENTRE SECONDS
    V/18,24, VI/17,18,19,20,22,29,30,31, VII/27, VIII/14,15,42,44,
    IX/46
CHARTIER (Edouard) V/11
CHEVAUX DE MARLY VII/C9
CHINESE INFLUENCE VII/26,32,33,37,42
CHINOISERIE PANELS VII/42
CHRONOMETER CARRIAGE CLOCKS
    II/2,3,45, Page 104, V/7,8,12,13, VIII/27,28,32,33,34,
    IX/21-37,39,40,47,68-72,84,85,95, X/5,6,7,11,12
CLOISONNE ENAMEL VII/C2,C3
COACH WATCHES
    French, c.1620 .... I/13
    French, 18th/19th cent. I/17
    English, c.1650, Boquet I/14,15
    English, c.1780, Hughes I/16
COLE (James Ferguson) IX/2,3,4,5
COLE (Thomas) IX/6,7,8
COLLMAN (Johann) XI/1,2
COMPASS VIII/45,46
CORINTHIAN COLUMNS VII/17, VII/C2,C3,C5
CORNICHE .... II/32, Page 156, VII/14, VII/41,44, VII/C11,C12
CORNICHE CARREE VII/15
COTTAGE INDUSTRY
    Vitrage d'horloger IIII/14
    Dis-used small factories IIII/17,18
COUAILLET
    Factory, c.1892 .... IIII/7
    Later dis-used factory as in 1970 .... IIII/8,9,12
    Clocks IIII/10,11, VII/6, VII/16, VII/28,29,30
    Wheel engine, 19th cent. IIII/13
COURVOISIER
    Portrait of Fritz .... Page 306
    X/1,2,5,6,7,8,9,10,11,12
COUSTOU'S CHEVAUX-DE-MARLY VII/C9
CUBIQUE VII/30
CUGNIER, LESCHOT
    Pendule d'officier, c.1780 .... I/28
    Carriage clock X/3,4
DELABARRE (M. and Mme.)
    Living links with old Couaillet enterprises IIII/19
DELEPINE-BARROIS
    Former factory. Dis-used, as seen in 1970 .... IIII/8,9,12
    Clock movement VIII/1
DELEPINE (Emile). Former home IIII/16
DELFT PANELS VII/C11
DENT FAMILY
    Clocks IX/18-IX/40,95
    Patent balance IX/24,25
    E.J. Dent portrait .... Page 301
DESBOIS IX/55,56,57,95

Index of Plates

DESK CLOCK	IX/6
DESSIAUX (E.). Réveil Fantaisie	IIII/2
DIGITAL DIALS	Page 192, VIII/19,20,21,22,23
DOUBLE-STRIKING	VI/33,34
DOUCINE	V/25, VII/22,23
DRESSING-CASE	IX/1
DROCOURT (Alfred)	V/18-21, VII/14,33, VII/42, VII/C1,C13
DRUM-SHAPED CONTINENTAL TABLE CLOCKS	
16th century	I/2,3,5,6
DUMAS (O.). Former home	IIII/16
DURER (Albrecht)	
"Knight, Death and the Devil", 1513	I/1
DU TERTRE (J.B.). French travelling clock, c.1780	I/26
DUVERDREY & BLOQUEL	
Factory, c.1910	IIII/5
Clocks	IIII/3,4, VII/37
EARNSHAW II (Thomas)	IX/41,42
EMPIRE-STYLE	II/1,2,3,4,5,6,7,17,18, V/24, IX/4,18,19, X/3,4, XI/12
ENGLISH CLOCKS IN FRENCH IDIOM	IX/91,92,93,94
EQUATION OF TIME	II/2, IX/32,35,36
ESCAPEMENTS, WELL SHOWN	
Lever. In English clocks	I/22,IX/14,17,49,50,54,57,62,74,88
Lever. In French clocks	II/11,13,16,27,43,47,51,57, V/6,10, 14,17,21, VI/23, VIII/30,31
Lever. In Swiss clocks	VIII/35, X/9
Lever. In American clocks	XII/56,61
Chronometer. Pivoted-detent	V/13, VIII/28, Fig. VIII/1,2, X/7, X/12
Chronometer. Spring-detent	II/3(?), VIII/32,33,34, IX/24,29, 31,71,72
Constant-Force	VIII/32,33,34,35
Cylinder	I/31,33, II/52, VI/32, XI/19
Duplex	II/42
Chinese Duplex	VIII/24
Two-plane. Garnier	II/20
Two-plane. Japy	VI/23
Two-plane. R.E.D.	VIII/30
Two-plane. Samuel	VIII/25,26
Revolving. Auguste "tourbillon"	VIII/38
Revolving. Waterbury rotating movement	XII/33,34
L'Epée	VI/28
Samuel's Vertical	VIII/25,26
Duplex Pivoted-Detent variant	VIII/27
MacDowall's	IX/66,67
Fasoldt's "chronometer"	XII/65
In Dial	VIII/45,46, XII/31,32
In Bubble	VIII/47,48
E.W.S.	IX/91,92,93
EXHIBITION, PARIS, 1867	
Plan of stands with names of exhibitors	Page 66
FASOLDT	Page 340, XII/63,64,65
FAUCHERRE. Engraving on clock	IX/89,90
FERTBAUER (Philipp)	XI/7,8
FIVE MINUTE REPEATER	VIII/6,7, X/19
FLICK CLOCKS	VIII/21,22,23
FOUR-DIAL CLOCK	VIII/39
FRANCHE-COMTE	
General	Page 128, VI/1-VI/34
Characteristic local clocks	VI/29,30,31,32,33,34
FRENCH CASE FOR ENGLISH CLOCK	IX/89
FRODSHAM (Charles)	IX/43,44,45,46,47,48,49,50
FUMEY. Cylinder platform escapement	VI/32
GARNIER (Paul)	
Carriage clocks	II/19-II/22, II/26-II/35, VII/1,4
Pendules portatives	II/23-II/25, II/36
GIANT CARRIAGE CLOCKS	II/37,38, VII/43, VIII/36, IX/32,33,34,35, VII/C1
GIRARD (P.)	X/13,14
GODMAN (J.S.)	IX/79,80
GOLD-PLATED CASE	Page 340, XII/53,54,55,63,64
GONTARD	Fig. VIII/1,2, Plate VIII/28
GORGE	V/4,19, VII/8,9,42,43, IX/59, VII/C7,C8
GRANDE AND/OR PETITE SONNERIE	II/2-9,14,15,16,37,38, V/3,5,7,8,9,15,18,19,23, VI/29,30,31, VII/35, VIII/4,5,8,9,13,14,15,16,17,18, IX/2,3,4,13,15 39,40,81,82,83, X/1,2,3,4,8,10,11,13,14, Page 324, XI/3,6,12-18
GUILLAUME BALANCE	VIII/33
GUILLOCHE ENAMEL	VII/C6
GUTKAES (Franz)	XI/12
HANCOCK	IX/8
HANGING CLOCKS	
Sedans	I/38,39,40
French	I/42,43,44
English mail-guard's watch	I/45,46
HAPPACHER (Philipp)	XI/14
HESSEN (André)	
French travelling clock, late 18th cent.	I/29,30,31
HOFMANN (Carolus)	XI/6
HUBER	IX/94
HUGHES (William). Coach watch c.1780	I/16
HUMPBACKED CLOCKS	II/3,10,11,14,15,16, IX/2,3,16,17,76,78,79,81,82,83, XI/13
IVORY	VII/39
JACOT (Henri)	V/14-V/17, VII/8, 24, VIII/8,9
JANVIER (Cadet). Capucine	I/35
JAPY	
Beaucourt factory mid 19th cent.	VI/4
Badeval factory as in 1971	VI/5,6,7,8,9,12
Clock, American style	VI/10,11
Jean-Charles Japy's memorial at Badeval as in 1971	VI/13
Beaucourt. Bust of Frédéric Japy	VI/14
Beaucourt. Dis-used Japy factory as in 1971	VI/15
E. Pequignot. Living link with old Badeval factory	VI/16
Other clocks	VI/17-VI/23, VII/27, VIII/24,44
JEROME	XII/49
JOHNSON (William)	IX/89,90
JOKER	XI/21, XII/51,52
JULES	II/47-II/51
JUMP	IX/76,77
KLENTSCHI (C.F.)	X/5,6,7
KULLBERG (Victor)	
Portrait	Page 302
IX/68,69,70,71,72	
LEFRANC	V/9,10, VIII/17,18
L'EPEE	
Portrait of Auguste	VI/24
Old views of factory	VI/25,26
Musical watch	VI/27
Peculiar escapement	VI/28
LEPINE	II/41,42
LE ROY/LEROY BUSINESSES	I/25, II/35, V/23,24,25, VII/32, VIII/16,40,41
LEROY (Theodore). "Giant" clock	II/37,38
LIMOGES ENAMEL	VII/C4, C5
LION TRAVELLING CLOCKS	XI/9,10
LOCKING PLATE STRIKING	II/48,54, VIII/5
LUND & BLOCKLEY	IX/60
MADELEINE	VII/18
MAIL-GUARD'S WATCH	
English, made between 1830-1837	I/45,46
MAINSPRINGS	
Manufacture 18th century	I/4
Unusual arrangements	II/8,56, XII/28-30,36-40
MAP. France and Switzerland	
Relative positions of Paris, Saint-Nicolas-d'Aliermont and the Jura	Page 82
MARBLE	VII/18,C13,C14
MARGAINE	
Price list	Page 119
Clock	VII/9, VII/19, VII/C8
MATHEY-TISSOT	X/19

481

Index of Plates

MAYER (A.W.) XI/16,17,18,19
MERCER (Thomas) IX/84,85
MICHOUDET VI/29,30,31,32
MINIATURE CARRIAGE CLOCKS
    VII/C6,C15,C16, VII/25, VIII/2, IX/51, XII/15,16,17,18
MINUTE REPEATERS V/9, VIII/8,9,14,17,18, X/19,20
MONTBELIARD
    Local scenes including castle VI/1
    Le Pays village scenes VI/2,3
MONTRES PENDULETTES DE VOYAGE X/19,20
MOON WORK
    II/1,2,10, V/7,18,24, VII/43, VIII/12,14,15,17,18, IX/2,6,32,
    35,58,76,78,79, X/18
MOULINIE X/17,18
MUDGE (Thomas)
    English travelling clock, probably pre 1769 .... I/21,22
MULTI-PIECE CASES II/29,46,49,53, VII/3,4
MUSICAL WORK (Visible)
    L'Epée watch VI/27
    J.F. Cole clock IX/5
MacDOWALL (J. Eden) IX/65,66,67
McCABE (James) .... Page 230, IX/61,62,63,64,95
NEUCHATEL TRAVELLING CLOCK, c.1820-30 .... I/37
NICOLE, NIELSEN IX/81,82,83
OBIS VII/7, VII/36
OBLONG VII/28,29
OLLIER VII/25, VII/31
ONE-PIECE CASES II/19,29,30,31,41,50, VI/17, VII/1,2
OVAL VII/11, VII/12, VII/13
PALAIS ROYAL, Paris V/22
PARACHUTE II/11,13
PARIS
    M. Pitou's workroom V/1
    The Palais Royal V/22
PAULET. English travelling clock, c.1720 .... I/20
PEDIMENT-TOPPED VII/18
PENDULES D'OFFICIER
    French, c.1780 .... I/27
    Swiss, c.1780 .... I/28
PENDULE PARIS MOVEMENT II/25,36
PENDULES PORTATIVES II/23,24,25,35,36
PEQUIGNOT (Edouard). Living link with Japy at Badevel VI/16
PIETRA DURA VII/C13
PILLARS VII/17,18,19,20,21
PINCHON II/55,56,57
PITOU
    Clocks II/17,18, V/4,5,6
    Workroom in 1971 .... V/1
    Portrait V/2
    Blancs-roulants V/3
PLATFORM ESCAPEMENTS
    I/31, II/13,42,43,47,51,52,57, V/6,10,13,14,17,21, VI/28,32,
    VIII/24,27,28,30,31,32,33,34,35, IX/14,24,25,30,31,49,50,54,
    57,62,71,72,74,88, X/7,9,12, XI/19, XII/61,65
PORCELAIN PANELS VII/C1,C7,C8,C9,C10,C11, VIII/4
PORTRAITS
    Breguet ....Page 36
    B.L. Vulliamy ....Page 301
    E.J. Dent ....Page 301
    Sir John Bennett ....Page 302
    Victor Kullberg ....Page 302
    Fritz Courvoisier ....Page 306
    M. and Mme. Delabarre IIII/19
    Frédéric Japy VI/14
    E. Péquignot VI/16
    Auguste L'Epée VI/24
QUILL (Col. H.) IX/79,80
RAINGO FRERES II/39
RAIT (D.C.) II/53,54
RANDALL VIII/32,33,34
RECLUS. Clock with Japy VI/18,19, VIII/24

R.E.D. VIII/29,30
REID & AULD. Sedan clock made post 1806 .... I/38,39
REMONTOIRE
    Train VIII/50,51
    Escapement VIII/32,33,34,35
REPEATING WORK OF SPECIAL INTEREST
    II/1,6,7,8,9,10,12,14,15,26,27,28, VIII/6,7,8,9,14, X/19,20
RETTICH XI/15
REVEIL FANTAISIE. Dessiaux IIII/2
ROBERT. Swiss travelling clock I/36
ROCOCO VII/38, VII/C1
ROUND CASE IIII/3,4, XI/24,25,26
ROWEAU (I.). Table clock c.1580, French I/8,9
RUETSCHMANN (Joseph) XI/3
SACHS (Hans). "I make the clocks for travellers", 1568 .... I/7
SAINT-NICOLAS-D'ALIERMONT
    Arms of the town IIII/1
    General IIII/1-IIII/19
    Characteristic local clocks IIII/4,10,11, VII/6,7,14,15,16,
    22,23,28,29,30,36,37
SAINTE-SUZANNE. Views VI/25,26
SAMUEL VII/11, VIII/25,26
SAULET VIII/50,51
SEDAN CLOCKS
    Reid & Auld I/38,39
    Buchanan I/40
    French hanging clocks I/42,43,44
SEIKOSHA CLOCK COMPANY XIII/1,2
SHIBYAMA PANELS VII/C12
SHIP'S BELL STRIKE VIII/10
SIDEREAL DIAL IX/73
SILVER OR SILVERED CASES
    I/20, II/10,11,14, VII/25,34,40, IX/2,76,78,79,81,85,
    X/19,20, XI/13
SINGING-BIRD CLOCKS VIII/42,43,44
SKELETON MOVEMENT XI/13, XII/33,34
SMITH & SON (S.) IX/75
SQUARE CASES VII/40, X/20
STAUFFER FAMILY X/20
STRUT CLOCK IX/7
SUNDIALS VIII/45
SUNRISE AND SUNSET IX/2
TABERNACLE CLOCKS
    16th/17th cent. upright table clocks I/10
TABLE CLOCKS
    German, drum-shaped, c.1550 .... I/5
    French, drum-shaped with detachable alarm, c.1590 .... I/6
    French, square, two-tier movement, c.1580 .... I/8,9
    German, upright, 16th/17th cent. I/10
    Flemish, square, c.1590 .... I/11
    French, round, late 16th cent. I/12
THERMOMETERS II/2, V/7, VIII/13,14,46, IX/8
THEUVILLAT VIII/32,33,34
THOMAS (Seth) XII/51,52
THORNTON (P.) IX/78
THWAITES & REED IX/58
TIEDE (Fridrich) XI/4
TOURBILLON
    Breguet II/3
    Nicole, Nielsen IX/81,83
    Auguste "tourbillon" VIII/38
    Waterbury rotating movement XII/27-40
TRAVELLING BOXES
    I/5,6,8,10,11,12,34, II/22,35, IX/9,51,64,79,81,84, XI/24,
    XII/4,5,15,45,46, Page 324
UP AND DOWN DIALS
    II/2, V/7,12, IX/46,47,48,68,69,75,81,84,85
VALLIN. Table clock c.1590, Flemish I/11
VERMONT CLOCK COMPANY
    Factory XII/62
    XII/60,61

Index of Plates

VITRAGE D'HORLOGER IIII/14
VOLANT. Table clock 16th cent., French I/12
VULLIAMY
    Workbooks IX/10
    Clocks IX/9, 11-15, 95
    Portrait . . . . Page 301
WATCH-CLOCK
    English 18th cent. verge watch movement re-used I/41
    French verge watch made into small travelling alarm V/26,27
WATERBURY
    Carriage clocks XII/1-19,21-26,31-40
    Long wind watches XII/27-30

    Alarm clocks XII/1,4,5,20
WATSON (Samuel)
    Miniature English travelling clock, c.1710 . . . . I/18,19
WEBSTER IX/51
WEISSE XI/5
WELCH (E.N.) XII/50
WESTMINSTER QUARTERS VIII/4,5, IX/39,40
WHITELAW (James) IX/52,95
WOODEN CASES II/4,5,6,7, VII/31, IX/9,11,13,95, X/1,2,3,4
YEAR-GOING CLOCKS VIII/49,50,51
ZECH (Jacob). Clock c.1525, Prague I/2,3

ALC
DEPOSÉ

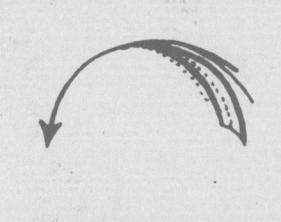

G.L   R C   A★B